Methods in Enzymology

Volume 299
OXIDANTS AND ANTIOXIDANTS
Part A

METHODS IN ENZYMOLOGY

EDITORS-IN-CHIEF

John N. Abelson   Melvin I. Simon

DIVISION OF BIOLOGY
CALIFORNIA INSTITUTE OF TECHNOLOGY
PASADENA, CALIFORNIA

FOUNDING EDITORS

Sidney P. Colowick and Nathan O. Kaplan

Methods in Enzymology

Volume 299

Oxidants and Antioxidants
Part A

EDITED BY

Lester Packer

UNIVERSITY OF CALIFORNIA
BERKELEY, CALIFORNIA

Editorial Advisory Board

Bruce N. Ames
Enrique Cadenas
Balz Frei
Matthew Grisham
Barry Halliwell
William Pryor
Catherine Rice-Evans
Helmut Sies

ACADEMIC PRESS
San Diego   London   Boston   New York   Sydney   Tokyo   Toronto

This book is printed on acid-free paper.

Copyright © 1999 by ACADEMIC PRESS

All Rights Reserved.
No part of this publication may be reproduced or transmitted in any form or by any means, electronic or mechanical, including photocopy, recording, or any information storage and retrieval system, without permission in writing from the Publisher.
The appearance of the code at the bottom of the first page of a chapter in this book indicates the Publisher's consent that copies of the chapter may be made for personal or internal use, or for the personal or internal use of specific clients. This consent is given on the condition, however, that the copier pay the stated per copy fee through the Copyright Clearance Center, Inc. (222 Rosewood Drive, Danvers, Massachusetts 01923) for copying beyond that permitted by Sections 107 or 108 of the U.S. Copyright Law. This consent does not extend to other kinds of copying, such as copying for general distribution, for advertising or promotional purposes, for creating new collective works, or for resale. Copy fees for pre-1999 chapters are as shown on the chapter title pages. If no fee code appears on the chapter title page, the copy fee is the same as for current chapters.
0076-6879/99 $30.00

Academic Press
a division of Harcourt Brace & Company
525 B Street, Suite 1900, San Diego, California 92101-4495, USA
http://www.academicpress.com

Academic Press Limited
24-28 Oval Road, London NW1 7DX, UK
http://www.hbuk.co.uk/ap/

International Standard Book Number: 0-12-182200-1

PRINTED IN THE UNITED STATES OF AMERICA
98 99 00 01 02 03 MM 9 8 7 6 5 4 3 2 1

Table of Contents

CONTRIBUTORS TO VOLUME 299 ix
PREFACE xiii
VOLUMES IN SERIES xv

Section I. Total Antioxidant Activity

1. Total Antioxidant Activity Measured by Chemiluminescence Methods — HANNU ALHO AND JANNE LEINONEN — 3

2. Ferric Reducing/Antioxidant Power Assay: Direct Measure of Total Antioxidant Activity of Biological Fluids and Modified Version for Simultaneous Measurement of Total Antioxidant Power and Ascorbic Acid Concentration — IRIS F. F. BENZIE AND J. J. STRAIN — 15

3. Automated Electron Spin Resonance Free Radical Detector Assays for Antioxidant Activity in Natural Extracts — YASUKO NODA, MASAHIRO KOHNO, AKITANE MORI, AND LESTER PACKER — 28

4. Measurement of Oxidizability of Blood Plasma — ANATOL KONTUSH AND ULRIKE BEISIEGEL — 35

5. Measurement of Oxygen Radical Absorbance Capacity in Biological Samples — GUOHUA CAO AND RONALD L. PRIOR — 50

Section II. Vitamin C

6. Analysis of Ascorbic and Dehydroascorbic Acid in Biological Samples — MARK LEVINE, YAOHUI WANG, AND STEVEN C. RUMSEY — 65

7. Analysis of Water-Soluble Antioxidants by High-Performance Liquid Chromatography with Electrochemical Detection — ANN M. BODE AND RICHARD C. ROSE — 77

8. Ascorbic Acid Recycling in Rat Hepatocytes as Measurement of Antioxidant Capacity: Decline with Age — JENS LYKKESFELDT AND BRUCE N. AMES — 83

Section III. Polyphenols and Flavonoids

9. Detecting and Measuring Bioavailability of Phenolics and Flavonoids in Humans: Pharmacokinetics of Urinary Excretion of Dietary Ferulic Acid — LOUISE C. BOURNE AND CATHERINE A. RICE-EVANS — 91

10. Simultaneous Analysis of Individual Catechins and Caffeine in Green Tea — TETSUHISA GOTO AND YUKO YOSHIDA — 107

11. Reversed-Phase High-Performance Liquid Chromatography Methods for Analysis of Wine Polyphenols — ANDREW L. WATERHOUSE, STEVEN F. PRICE, AND JEFFREY D. MCCORD — 113

12. Analysis of Antioxidant Wine Phenols by High-Performance Liquid Chromatography — DAVID M. GOLDBERG AND GEORGE J. SOLEAS — 122

13. Analysis of Antioxidant Wine Polyphenols by Gas Chromatography–Mass Spectrometry — GEORGE J. SOLEAS AND DAVID M. GOLDBERG — 137

14. Analysis of Total Phenols and Other Oxidation Substrates and Antioxidants by Means of Folin–Ciocalteu Reagent — VERNON L. SINGLETON, RUDOLF ORTHOFER, AND ROSA M. LAMUELA-RAVENTÓS — 152

15. Size Separation of Condensed Tannins by Normal-Phase High-Performance Liquid Chromatography — VÉRONIQUE CHEYNIER, JEAN-MARC SOUQUET, ERWAN LE ROUX, SYLVAIN GUYOT, AND JACQUES RIGAUD — 178

16. Resveratrol and Piceid in Wine — ROSA M. LAMUELA-RAVENTÓS AND ANDREW L. WATERHOUSE — 184

17. Antioxidant Activity by Headspace Gas Chromatography of Volatile Oxidation Products of ω-6 and ω-3 Polyunsaturated Lipids — E. N. FRANKEL — 190

18. Determination of Tea Catechins by Reversed-Phase High-Performance Liquid Chromatography — PETER C. H. HOLLMAN, DINI P. VENEMA, SANTOSH KHOKHAR, AND ILJA C. W. ARTS — 202

19. Flavonoids as Peroxynitrite Scavengers *in Vitro* — ANANTH SEKHER PANNALA, SURINDER SINGH, AND CATHERINE RICE-EVANS — 207

Section IV. Thiols

20. Determination of Oxidized and Reduced Lipoic Acid Using High-Performance Liquid Chromatography and Coulometric Detection — CHANDAN K. SEN, SASHWATI ROY, SAVITA KHANNA, AND LESTER PACKER — 239

21. Flow Cytometric Determination of Cellular Thiols	CHANDAN K. SEN, SASHWATI ROY, AND LESTER PACKER	247
22. Measurement of Glutathione, Glutathione Disulfide, and Other Thiols in Mammalian Cell and Tissue Homogenates Using High-Performance Liquid Chromatography Separation of N-(1-Pyrenyl)Maleimide Derivatives	LISA A. RIDNOUR, ROGER A. WINTERS, NURAN ERCAL, AND DOUGLAS R. SPITZ	258
23. Ratio of Reduced to Oxidized Glutathione as Indicator of Oxidative Stress Status and DNA Damage	MIGUEL ASENSI, JUAN SASTRE, FEDERICO V. PALLARDÓ, ANA LLORET, MARTIN LEHNER, JOSE GARCIA-DE-LA ASUNCION, AND JOSÉ VIÑA	267
24. Nonenzymatic Colorimetric Assay of Glutathione in Presence of Other Mercaptans	JEAN CHAUDIÈRE, NADIA AGUINI, AND JEAN-CLAUDE YADAN	276
25. Protozoological Method for Assaying Lipoate in Human Biologic Fluids and Tissue	HERMAN BAKER, BARBARA DEANGELIS, ELLIOTT R. BAKER, AND SEYMOUR H. HUTNER	287
26. Antioxidant Activity of Amidothionophosphates	OREN TIROSH, YEHOSHUA KATZHENDLER, YECHEZKEL BARENHOLZ, AND RON KOHEN	293
27. Quantitation of Anethole Dithiolthione Using High-Performance Liquid Chromatography with Electrochemical Detection	KATRINA TRABER AND LESTER PACKER	300

Section V. Vitamin E and Coenzyme Q_{10}

28. Gas Chromatography–Mass Spectrometry Analysis of Vitamin E and Its Oxidation Products	DANIEL C. LIEBLER, JEANNE A. BURR, AND AMY J. L. HAM	309
29. Determination of Vitamin E Forms in Tissues and Diets by High-Performance Liquid Chromatography Using Normal-Phase Diol Column	JOHN K. G. KRAMER, ROBERT C. FOUCHARD, AND KRISHNA M. R. KALLURY	318
30. Sensitive High-Performance Liquid Chromatography Techniques for Simultaneous Determination of Tocopherols, Tocotrienols, Ubiquinols, and Ubiquinones in Biological Samples	MAURIZIO PODDA, CHRISTINE WEBER, MARET G. TRABER, RAINER MILBRADT, AND LESTER PACKER	330

31. High-Performance Liquid Chromatography–Coulometric Electrochemical Detection of Ubiquinol 10, Ubiquinone 10, Carotenoids, and Tocopherols in Neonatal Plasma — BARBARA FINCKH, ANATOL KONTUSH, JENS COMMENTZ, CHRISTOPH HÜBNER, MARTIN BURDELSKI, AND ALFRIED KOHLSCHÜTTER — 341

32. Simultaneous Determination of Retinol, Tocopherols, Carotene, Lycopene, and Xanthophylls in Plasma by Means of Reversed-Phase High-Performance Liquid Chromatography — CLAUDE-PIERRE AEBISCHER, JOSEPH SCHIERLE, AND WILLY SCHÜEP — 348

33. Assessment of Prooxidant Activity of Vitamin E in Human Low Density Lipoprotein and Plasma — PAUL K. WITTING, DETLEF MOHR, AND ROLAND STOCKER — 362

Section VI. Carotenoids and Retinoids

34. Screening of Dietary Carotenoids and Carotenoid-Rich Fruit Extracts for Antioxidant Activities Applying 2,2′-Azinobis(3-ethylbenzothiazoline-6-sulfonic acid) Radical Cation Decolorization Assay — NICOLETTA PELLEGRINI, ROBERTA RE, MIN YANG, AND CATHERINE RICE-EVANS — 379

35. Matrix-Assisted Laser Desorption Ionization–Postsource Decay-Mass Spectrometry in Carotenoid Analysis — THOMAS WINGERATH, DIETER KIRSCH, RAIMUND KAUFMANN, WILHELM STAHL, AND HELMUT SIES — 390

36. Carotenoid Photobleaching — ALAN MORTENSEN AND LEIF H. SKIBSTED — 408

37. Interactions between Vitamin A and Vitamin E in Liposomes and in Biological Contexts — MARIA A. LIVREA AND LUISA TESORIERE — 421

38. On-Line Solid-Phase Extraction and Isocratic Separation of Retinoic Acid Isomers in Microbore Column Switching System — THOMAS E. GUNDERSEN AND RUNE BLOMHOFF — 430

39. Purification and Characterization of Cellular Carotenoid-Binding Protein from Mammalian Liver — M. R. LAKSHMAN AND MANJUNATH N. RAO — 441

40. Analysis of Zeaxanthin Stereoisomer Distribution within Individual Human Retinas — JOHN T. LANDRUM, RICHARD A. BONE, LINDA L. MOORE, AND CHRISTINA M. GOMEZ — 457

AUTHOR INDEX 469

SUBJECT INDEX 491

Contributors to Volume 299

Article numbers are in parentheses following the names of contributors.
Affiliations listed are current.

CLAUDE-PIERRE AEBISCHER (32), *Vitamins and Fine Chemicals Division, F. Hoffmann–La Roche Ltd., CH-4070 Basel, Switzerland*

NADIA AGUINI (24), *Oxis International, 94385 Bonneuil Cedex, France*

HANNU ALHO (1), *Department of Mental Health and Alcohol Research, National Public Health Institute, 00101 Helsinki, Finland*

BRUCE N. AMES (8), *Department of Biochemistry, University of California, Berkeley, California 94720-3202*

ILJA C. W. ARTS (18), *State Institute for Quality Control of Agricultural Products (Rikilt-DLO), NL-6708 PD Wageningen, The Netherlands*

MIGUEL ASENSI (23), *Department of Physiology, Faculty of Medicine, University of Valencia, 46010 Valencia, Spain*

ELLIOT R. BAKER (25), *Department of Preventive Medicine and Community Health, New Jersey Medical School, Newark, New Jersey 07107*

HERMAN BAKER (25), *Departments of Preventive Medicine and Community Health, and Medicine, New Jersey Medical School, Newark, New Jersey 07107*

YECHEZKEL BARENHOLZ (26), *Department of Biochemistry, Faculty of Medicine, The Hebrew University of Jerusalem, Jerusalem, Israel 91120*

ULRIKE BEISIEGEL (4), *Medical Clinic, University Hospital Eppendorf, D-20246 Hamburg, Germany*

IRIS F. F. BENZIE (2), *Department of Nursing and Health Sciences, Hong Kong Polytechnic University, Hung Hom, Kowloon, Hong Kong*

RUNE BLOMHOFF (38), *Institute for Nutrition Research, University of Oslo, N-0316 Oslo, Norway*

ANN M. BODE (7), *Department of Physiology, University of North Dakota School of Medicine, Grand Forks, North Dakota 58202*

RICHARD A. BONE (40), *Department of Physics, Florida International University, Miami, Florida 33199*

LOUISE C. BOURNE (9), *International Antioxidant Research Centre, UMDS–Guy's Hospital, London SE1 9RT, United Kingdom*

MARTIN BURDELSKI (31), *Kinderklinik, Universitätskrankenhaus Eppendorf, D-20246 Hamburg, Germany*

JEANNE A. BURR (28), *Department of Pharmacology and Toxicology, College of Pharmacy, University of Arizona, Tucson, Arizona 85721-0207*

GUOHUA CAO (5), *University of Connecticut, Storrs, Connecticut 06269*

JEAN CHAUDIÈRE (24), *Laboratoire de Pharmacochimie Moléculaire, Université Paris 7, 75251 Paris Cedex 05, France*

VÉRONIQUE CHEYNIER (14, 15), *Unité de Recherche Biopolymères et Arômes, Institut Supérieur de la Vigne et du Vin–IPV, INRA, 34060 Montpellier, France*

JENS COMMENTZ (31), *Kinderklinik, Universitätskrankenhaus Eppendorf, D-20246 Hamburg, Germany*

BARBARA DEANGELIS (25), *Department of Preventive Medicine and Community Health, New Jersey Medical School, Newark, New Jersey 07107*

J. GARCIA-DE-LA ASUNCION (23), *Department of Physiology, Faculty of Medicine, University of Valencia, 46010 Valencia, Spain*

NURAN ERCAL (22), *Department of Chemistry, University of Missouri–Rolla, Rolla, Missouri 65401*

BARBARA FINCKH (31), *Neurochemisches Labor/Kinderklinik, Universitätskrankenhaus Eppendorf, D-20246 Hamburg, Germany*

ROBERT C. FOUCHARD (29), *Canadian Explosives Research Laboratory, Natural Resources Canada, Nepean, Ontario, Canada K2L 4G1*

E. N. FRANKEL (17), *Department of Food Science and Technology, University of California, Davis, California 95616*

DAVID M. GOLDBERG (12, 13), *Department of Laboratory Medicine and Pathobiology, University of Toronto, Toronto, Ontario, Canada M5G 1L5*

CHRISTINA M. GOMEZ (40), *Department of Chemistry, Florida International University, Miami, Florida 33199*

TETSUHISA GOTO (10), *National Food Research Institute, MAFF, Ibaraki-ken 305-8642, Japan*

THOMAS E. GUNDERSEN (38), *Institute for Nutrition Research, University of Oslo, N-0316 Oslo, Norway*

SYLVAIN GUYOT (14, 15), *Station de Recherche Cidricole, Biotransformation des Fruits et Légumes, INRA, 35650 Le Rheu, France*

AMY J. L. HAM (28), *Department of Pharmacology and Toxicology, College of Pharmacy, University of Arizona, Tucson, Arizona 85721-0207*

PETER C. H. HOLLMAN (18), *State Institute for Quality Control of Agricultural Products (Rikilt-DLO), NL-6708 PD Wageningen, The Netherlands*

CHRISTOPH HÜBNER (31), *Kinderklinik, Virchow-Klinikum, Humboldt-Universität, D-13353 Berlin, Germany*

SEYMOUR H. HUTNER (25), *Haskins Laboratories, Pace University, New York, New York 10038*

KRISHNA M. R. KALLURY (29), *Supelco Inc., Bellefonte, Pennsylvania 16823-0048*

YEHOSHUA KATZHENDLER (26), *Department of Pharmaceutical Chemistry, The Hebrew University of Jerusalem, Jerusalem, Israel 91120*

RAIMUND KAUFMANN (35), *Institut für Physiologische Chemie I, Heinrich-Heine-Universität, D-40225 Düsseldorf, Germany*

SAVITA KHANNA (20), *University of California, Berkeley, California 94720-3200*

SANTOSH KHOKHAR (18), *State Institute for Quality Control of Agricultural Products (Rikilt-DLO), NL-6708 PD Wageningen, The Netherlands*

DIETER KIRSCH (35), *Institut für Physiologische Chemie I, Heinrich-Heine-Universität, D-40225 Düsseldorf, Germany*

RON KOHEN (26), *Department of Pharmaceutics, School of Pharmacy, The Hebrew University of Jerusalem, Jerusalem, Israel 91120*

ALFRIED KOHLSCHÜTTER (31), *Kinderklinik, Universitätskrankenhaus Eppendorf, D-20246 Hamburg, Germany*

MASAHIRO KOHNO (3), *JEOL Ltd., Tokyo, Japan*

ANATOL KONTUSH (4, 31), *Medical Clinic, University Hospital Eppendorf, D-20246 Hamburg, Germany*

JOHN K. G. KRAMER (29), *Southern Crop Protection Food Research Center, Agriculture and Agri-Food Canada, Guelph, Ontario, Canada N1G 2W1*

M. R. LAKSHMAN (39), *Lipid Research Laboratory, DVA Medical Center and George Washington University, Washington, DC 20422*

ROSA M. LAMUELA-RAVENTÓS (14, 16), *Departament de Nutrició i Bromatologia, Facultat de Farmàcia, Av. Joan XXIII, s/n 08028 Barcelona, Spain*

JOHN T. LANDRUM (40), *Department of Chemistry, Florida International University, Miami, Florida 33199*

MARTIN LEHNER (23), *Fakultät für Biologie, Universität Konstanz, Germany*

JANNE LEINONEN (1), *Laboratory of Neurobiology, University of Tampere, and Department of Mental Health and Alcohol Research, National Public Health Institute, 00101 Helsinki, Finland*

ERWAN LE ROUX (14, 15), *Unité de Recherche Biopolymères et Arômes, Institut Supérieur de la Vigne et du Vin–IPV, INRA, 34060 Montpellier, France*

MARK LEVINE (6), *Molecular and Clinical Nutrition Section, National Institutes of Health, Bethesda, Maryland 20892*

DANIEL C. LIEBLER (28), *Department of Pharmacology and Toxicology, College of Pharmacy, University of Arizona, Tucson, Arizona 85721-0207*

MARIA A. LIVREA (37), *Istituto Farmacologia e Farmacognosia, Università di Palermo, 90134 Palermo, Italy*

ANA LLORET (23), *Department of Physiology, Faculty of Medicine, University of Valencia, 46010 Valencia, Spain*

JENS LYKKESFELDT (8), *Department of Molecular and Cell Biology, University of California, Berkeley, California 94720-3202*

JEFFREY D. MCCORD (11), *E&J Winery, Modesto, California 95353*

RAINES MILBRADT (30), *Department of Biochemistry and Nutrition, The Technical University of Denmark, 2800 Lyngby, Denmark*

DETLEF MOHR (33), *Biochemistry Group, The Heart Research Institute, Camperdown NSW 2050, Australia*

LINDA L. MOORE (40), *Department of Chemistry, Florida International University, Miami, Florida 33199*

AKITANE MORI (3), *Department of Molecular and Cell Biology, University of California, Berkeley, California 94720-3200*

ALAN MORTENSEN (36), *Food Chemistry, Department of Dairy and Food Science, The Royal Veterinary and Agricultural University, DK-1958 Frederiksberg C, Denmark*

YASUKO NODA (3), *Department of Molecular and Cell Biology, University of California, Berkeley, California 94720-3200*

RUDOLF ORTHOFER (14), *Austrian Research Centre, A-2444, Seibersdorf, Austria*

LESTER PACKER (3, 20, 21, 27, 30), *Department of Molecular and Cell Biology, University of California at Berkeley, Berkeley, California 94720-3200*

FEDERICO V. PALLARDÓ (23), *Department of Physiology, Faculty of Medicine, University of Valencia, 46010 Valencia, Spain*

ANANTH SEKHER PANNALA (19), *International Antioxidant Research Centre, UMDS–Guy's Hospital, London SE1 9RT, United Kingdom*

NICOLETTA PELLEGRINI (34), *International Antioxidant Research Centre, UMDS–Guy's Hospital, London SE1 9RT, United Kingdom*

MAURIZIO PODDA (30), *Zentrum der Dermatologie, J. W. Goethe-Universität Frankfurt, D-60590 Frankfurt am Main, Germany*

STEVEN F. PRICE (11), *ETS Laboratories, St. Helena, California 94574*

RONALD L. PRIOR (5), *Agriculture Research Service and Human Nutrition Research Center on Aging, U.S. Department of Agriculture, Boston, Massachusetts 02111*

MANJUNATH N. RAO (39), *Department of Medicine, George Washington University, Washington, DC 20422*

ROBERTA RE (34), *International Antioxidant Research Centre, UMDS–Guy's Hospital, London SE1 9RT, United Kingdom*

CATHERINE A. RICE-EVANS (9, 19, 34), *International Antioxidant Research Centre, UMDS–Guy's Hospital, London SE1 9RT, United Kingdom*

LISA A. RIDNOUR (22), *Section of Cancer Biology, Radiation Oncology Center, Washington University School of Medicine, St. Louis, Missouri 63108*

JACQUES RIGAUD (14, 15), *Unité de Recherche Biopolymères et Arômes, Institut Supérieur de la Vigne et du Vin–IPV, INRA, 34060 Montpellier, France*

RICHARD C. ROSE (7), *Department of Physiology and Biophysics, Finch University/ Chicago Medical School, North Chicago, Illinois 60064*

SASHWATI ROY (20, 21), *University of California, Berkeley, California 94720-3200*

STEVEN C. RUMSEY (6), *Molecular and Clinical Nutrition Section, National Institutes of Health, Bethesda, Maryland 20892*

JUAN SASTRE (23), *Department of Physiology, Faculty of Medicine, University of Valencia, 46010 Valencia, Spain*

JOSEPH SCHIERLE (32), *F. Hoffmann–La Roche Ltd., CH-4070 Basel, Switzerland*

WILLY SCHÜEP (32), *F. Hoffmann–La Roche Ltd., CH-4070 Basel, Switzerland*

CHANDAN K. SEN (20, 21), *University of California, Berkeley, California 94720-3200*

HELMUT SIES (35), *Institut für Physiologische Chemie I, Heinrich-Heine-Universität, D-40225 Düsseldorf, Germany*

SURINDER SINGH (19), *International Antioxidant Research Centre, UMDS–Guy's Hospital, London SE1 9RT, United Kingdom*

VERNON L. SINGLETON (14), *Department of Viticulture and Enology, University of California, Davis, California 95616*

LEIF H. SKIBSTED (36), *Food Chemistry, Department of Dairy and Food Science, The Royal Veterinary and Agricultural University, DK-1958 Frederiksberg C, Denmark*

GEORGE J. SOLEAS (12, 13), *Quality Assurance, Liquor Control Board of Ontario, Toronto, Ontario, Canada M5E 1A4*

JEAN-MARC SOUQUET (14, 15), *Unité de Recherche Biopolymères et Arômes, Institut Supérieur de la Vigne et du Vin–IPV, INRA, 34060 Montpellier, France*

DOUGLAS R. SPITZ (22), *Section of Cancer Biology, Radiation Oncology Center, Washington University School of Medicine, St. Louis, Missouri 63108*

WILHELM STAHL (35), *Institut für Physiologische Chemie I, Heinrich-Heine-Universität, D-40225 Düsseldorf, Germany*

ROLAND STOCKER (33), *Biochemistry Group, The Heart Research Institute, Camperdown NSW 2050, Australia*

J. J. STRAIN (2), *Northern Ireland Centre for Diet and Health, University of Ulster, Londonderry BT52 1SA, Northern Ireland*

LUISA TESORIERE (37), *Istituto Farmacologia e Farmacognosia, Università di Palermo, 90134 Palermo, Italy*

OREN TIROSH (26), *Department of Pharmaceutics, School of Pharmacy, The Hebrew University of Jerusalem, Jerusalem, Israel 91120*

KATRINA TRABER (27), *University of California, Berkeley, California 94720-3200*

MARET G. TRABER (30), *Department of Molecular and Cell Biology, University of California, Berkeley, California 94720-3200*

DINI P. VENEMA (18), *State Institute for Quality Control of Agricultural Products (Rikilt-DLO), NL-6708 PD Wageningen, The Netherlands*

JOSÉ VIÑA (23), *Department of Physiology, Faculty of Medicine, University of Valencia, 46010 Valencia, Spain*

YAOHUI WANG (6), *Molecular and Clinical Nutrition Section, National Institutes of Health, Bethesda, Maryland 20892*

ANDREW L. WATERHOUSE (11, 16), *Department of Viticulture and Enology, University of California, Davis, California 95616*

CHRISTINE WEBER (30), *Department of Biochemistry and Nutrition, The Technical University of Denmark, 2800 Lyngby, Denmark*

THOMAS WINGERATH (35), *Institut für Physiologische Chemie I, Heinrich-Heine-Universität, D-40225 Düsseldorf, Germany*

ROGER A. WINTERS (22), *Oread Laboratories, Inc., Lawrence, Kansas 66047*

PAUL K. WITTING (33), *Biochemistry Group, The Heart Research Institute, Camperdown NSW 2050, Australia*

JEAN-CLAUDE YADAN (24), *Oxis International, 94385 Bonneuil Cedex, France*

MIN YANG (34), *International Antioxidant Research Centre, UMDS–Guy's Hospital, London SE1 9RT, United Kingdom*

YUKO YOSHIDA (10), *Tokyo Metropolitan Agricultural Experiment Station, Tachikawa-shi, Tokyo 190, Japan*

Preface

The importance of reactive oxygen and nitrogen species (ROS and RNS) and antioxidants in health and disease has now been recognized in all of the biological sciences and has assumed special importance in the biomedical sciences. Overwhelming evidence indicates that ROS play a role in most major health problems, that antioxidants play a critical role in wellness and health maintenance, and that by inhibiting oxidative damage to molecules, cells, and tissues prevent chronic and degenerative diseases.

We now know that ROS are essential for many enzyme-catalyzed reactions. Low levels of reactive oxygen and reactive nitrogen species are signaling molecules. At high concentration, these ROS are essential in the antitumor, antimicrobial, antiparasitic action, etc., of neutrophils and macrophages and contribute to oxidative damage to molecules, cells, and tissues.

In this volume all of the major natural antioxidants with respect to assays for evaluating their antioxidant activity have been included. There has been wide usage of methods to access total antioxidant activity, and some of the new methods in this area have also been included.

Many antioxidant substances have biological activities which may or may not depend on their antioxidant actions. Although this is of course relevant to understanding their actions in biological systems, we have chosen not to include such methods. Antioxidant activity can be defined as the protection against oxidative damage; however, it is becoming eminently clear that it is difficult to define an antioxidant. Antioxidants have so many different biological activities, in addition to their direct quenching of radicals or acting as redox molecules in reducing reactions, that their definition must surely be very broad.

In bringing this volume to fruition, credit must be given to experts in various specialized fields of oxidant and antioxidant research. Our appreciation is to the contributors who, with those who helped select them, have produced this state-of-the-art volume on oxidant and antioxidant methodology. The topics included were chosen on the excellent advice of Bruce N. Ames, Enrique Cadenas, Balz Frei, Matthew Grisham, Barry Halliwell, William Pryor, Catherine Rice-Evans, and Helmut Sies. To these colleagues, I extend my sincere thanks and most grateful appreciation.

Lester Packer

METHODS IN ENZYMOLOGY

VOLUME I. Preparation and Assay of Enzymes
Edited by SIDNEY P. COLOWICK AND NATHAN O. KAPLAN

VOLUME II. Preparation and Assay of Enzymes
Edited by SIDNEY P. COLOWICK AND NATHAN O. KAPLAN

VOLUME III. Preparation and Assay of Substrates
Edited by SIDNEY P. COLOWICK AND NATHAN O. KAPLAN

VOLUME IV. Special Techniques for the Enzymologist
Edited by SIDNEY P. COLOWICK AND NATHAN O. KAPLAN

VOLUME V. Preparation and Assay of Enzymes
Edited by SIDNEY P. COLOWICK AND NATHAN O. KAPLAN

VOLUME VI. Preparation and Assay of Enzymes (*Continued*)
Preparation and Assay of Substrates
Special Techniques
Edited by SIDNEY P. COLOWICK AND NATHAN O. KAPLAN

VOLUME VII. Cumulative Subject Index
Edited by SIDNEY P. COLOWICK AND NATHAN O. KAPLAN

VOLUME VIII. Complex Carbohydrates
Edited by ELIZABETH F. NEUFELD AND VICTOR GINSBURG

VOLUME IX. Carbohydrate Metabolism
Edited by WILLIS A. WOOD

VOLUME X. Oxidation and Phosphorylation
Edited by RONALD W. ESTABROOK AND MAYNARD E. PULLMAN

VOLUME XI. Enzyme Structure
Edited by C. H. W. HIRS

VOLUME XII. Nucleic Acids (Parts A and B)
Edited by LAWRENCE GROSSMAN AND KIVIE MOLDAVE

VOLUME XIII. Citric Acid Cycle
Edited by J. M. LOWENSTEIN

VOLUME XIV. Lipids
Edited by J. M. LOWENSTEIN

VOLUME XV. Steroids and Terpenoids
Edited by RAYMOND B. CLAYTON

VOLUME XVI. Fast Reactions
Edited by KENNETH KUSTIN

VOLUME XVII. Metabolism of Amino Acids and Amines (Parts A and B)
Edited by HERBERT TABOR AND CELIA WHITE TABOR

VOLUME XVIII. Vitamins and Coenzymes (Parts A, B, and C)
Edited by DONALD B. MCCORMICK AND LEMUEL D. WRIGHT

VOLUME XIX. Proteolytic Enzymes
Edited by GERTRUDE E. PERLMANN AND LASZLO LORAND

VOLUME XX. Nucleic Acids and Protein Synthesis (Part C)
Edited by KIVIE MOLDAVE AND LAWRENCE GROSSMAN

VOLUME XXI. Nucleic Acids (Part D)
Edited by LAWRENCE GROSSMAN AND KIVIE MOLDAVE

VOLUME XXII. Enzyme Purification and Related Techniques
Edited by WILLIAM B. JAKOBY

VOLUME XXIII. Photosynthesis (Part A)
Edited by ANTHONY SAN PIETRO

VOLUME XXIV. Photosynthesis and Nitrogen Fixation (Part B)
Edited by ANTHONY SAN PIETRO

VOLUME XXV. Enzyme Structure (Part B)
Edited by C. H. W. HIRS AND SERGE N. TIMASHEFF

VOLUME XXVI. Enzyme Structure (Part C)
Edited by C. H. W. HIRS AND SERGE N. TIMASHEFF

VOLUME XXVII. Enzyme Structure (Part D)
Edited by C. H. W. HIRS AND SERGE N. TIMASHEFF

VOLUME XXVIII. Complex Carbohydrates (Part B)
Edited by VICTOR GINSBURG

VOLUME XXIX. Nucleic Acids and Protein Synthesis (Part E)
Edited by LAWRENCE GROSSMAN AND KIVIE MOLDAVE

VOLUME XXX. Nucleic Acids and Protein Synthesis (Part F)
Edited by KIVIE MOLDAVE AND LAWRENCE GROSSMAN

VOLUME XXXI. Biomembranes (Part A)
Edited by SIDNEY FLEISCHER AND LESTER PACKER

VOLUME XXXII. Biomembranes (Part B)
Edited by SIDNEY FLEISCHER AND LESTER PACKER

VOLUME XXXIII. Cumulative Subject Index Volumes I–XXX
Edited by MARTHA G. DENNIS AND EDWARD A. DENNIS

VOLUME XXXIV. Affinity Techniques (Enzyme Purification: Part B)
Edited by WILLIAM B. JAKOBY AND MEIR WILCHEK

VOLUME XXXV. Lipids (Part B)
Edited by JOHN M. LOWENSTEIN

VOLUME XXXVI. Hormone Action (Part A: Steroid Hormones)
Edited by BERT W. O'MALLEY AND JOEL G. HARDMAN

VOLUME XXXVII. Hormone Action (Part B: Peptide Hormones)
Edited by BERT W. O'MALLEY AND JOEL G. HARDMAN

VOLUME XXXVIII. Hormone Action (Part C: Cyclic Nucleotides)
Edited by JOEL G. HARDMAN AND BERT W. O'MALLEY

VOLUME XXXIX. Hormone Action (Part D: Isolated Cells, Tissues, and Organ Systems)
Edited by JOEL G. HARDMAN AND BERT W. O'MALLEY

VOLUME XL. Hormone Action (Part E: Nuclear Structure and Function)
Edited by BERT W. O'MALLEY AND JOEL G. HARDMAN

VOLUME XLI. Carbohydrate Metabolism (Part B)
Edited by W. A. WOOD

VOLUME XLII. Carbohydrate Metabolism (Part C)
Edited by W. A. WOOD

VOLUME XLIII. Antibiotics
Edited by JOHN H. HASH

VOLUME XLIV. Immobilized Enzymes
Edited by KLAUS MOSBACH

VOLUME XLV. Proteolytic Enzymes (Part B)
Edited by LASZLO LORAND

VOLUME XLVI. Affinity Labeling
Edited by WILLIAM B. JAKOBY AND MEIR WILCHEK

VOLUME XLVII. Enzyme Structure (Part E)
Edited by C. H. W. HIRS AND SERGE N. TIMASHEFF

VOLUME XLVIII. Enzyme Structure (Part F)
Edited by C. H. W. HIRS AND SERGE N. TIMASHEFF

VOLUME XLIX. Enzyme Structure (Part G)
Edited by C. H. W. HIRS AND SERGE N. TIMASHEFF

VOLUME L. Complex Carbohydrates (Part C)
Edited by VICTOR GINSBURG

VOLUME LI. Purine and Pyrimidine Nucleotide Metabolism
Edited by PATRICIA A. HOFFEE AND MARY ELLEN JONES

VOLUME LII. Biomembranes (Part C: Biological Oxidations)
Edited by SIDNEY FLEISCHER AND LESTER PACKER

VOLUME LIII. Biomembranes (Part D: Biological Oxidations)
Edited by SIDNEY FLEISCHER AND LESTER PACKER

VOLUME LIV. Biomembranes (Part E: Biological Oxidations)
Edited by SIDNEY FLEISCHER AND LESTER PACKER

VOLUME LV. Biomembranes (Part F: Bioenergetics)
Edited by SIDNEY FLEISCHER AND LESTER PACKER

VOLUME LVI. Biomembranes (Part G: Bioenergetics)
Edited by SIDNEY FLEISCHER AND LESTER PACKER

VOLUME LVII. Bioluminescence and Chemiluminescence
Edited by MARLENE A. DELUCA

VOLUME LVIII. Cell Culture
Edited by WILLIAM B. JAKOBY AND IRA PASTAN

VOLUME LIX. Nucleic Acids and Protein Synthesis (Part G)
Edited by KIVIE MOLDAVE AND LAWRENCE GROSSMAN

VOLUME LX. Nucleic Acids and Protein Synthesis (Part H)
Edited by KIVIE MOLDAVE AND LAWRENCE GROSSMAN

VOLUME 61. Enzyme Structure (Part H)
Edited by C. H. W. HIRS AND SERGE N. TIMASHEFF

VOLUME 62. Vitamins and Coenzymes (Part D)
Edited by DONALD B. MCCORMICK AND LEMUEL D. WRIGHT

VOLUME 63. Enzyme Kinetics and Mechanism (Part A: Initial Rate and Inhibitor Methods)
Edited by DANIEL L. PURICH

VOLUME 64. Enzyme Kinetics and Mechanism (Part B: Isotopic Probes and Complex Enzyme Systems)
Edited by DANIEL L. PURICH

VOLUME 65. Nucleic Acids (Part I)
Edited by LAWRENCE GROSSMAN AND KIVIE MOLDAVE

VOLUME 66. Vitamins and Coenzymes (Part E)
Edited by DONALD B. MCCORMICK AND LEMUEL D. WRIGHT

VOLUME 67. Vitamins and Coenzymes (Part F)
Edited by DONALD B. MCCORMICK AND LEMUEL D. WRIGHT

VOLUME 68. Recombinant DNA
Edited by RAY WU

VOLUME 69. Photosynthesis and Nitrogen Fixation (Part C)
Edited by ANTHONY SAN PIETRO

VOLUME 70. Immunochemical Techniques (Part A)
Edited by HELEN VAN VUNAKIS AND JOHN J. LANGONE

VOLUME 71. Lipids (Part C)
Edited by JOHN M. LOWENSTEIN

VOLUME 72. Lipids (Part D)
Edited by JOHN M. LOWENSTEIN

VOLUME 73. Immunochemical Techniques (Part B)
Edited by JOHN J. LANGONE AND HELEN VAN VUNAKIS

VOLUME 74. Immunochemical Techniques (Part C)
Edited by JOHN J. LANGONE AND HELEN VAN VUNAKIS

VOLUME 75. Cumulative Subject Index Volumes XXXI, XXXII, XXXIV–LX
Edited by EDWARD A. DENNIS AND MARTHA G. DENNIS

VOLUME 76. Hemoglobins
Edited by ERALDO ANTONINI, LUIGI ROSSI-BERNARDI, AND EMILIA CHIANCONE

VOLUME 77. Detoxication and Drug Metabolism
Edited by WILLIAM B. JAKOBY

VOLUME 78. Interferons (Part A)
Edited by SIDNEY PESTKA

VOLUME 79. Interferons (Part B)
Edited by SIDNEY PESTKA

VOLUME 80. Proteolytic Enzymes (Part C)
Edited by LASZLO LORAND

VOLUME 81. Biomembranes (Part H: Visual Pigments and Purple Membranes, I)
Edited by LESTER PACKER

VOLUME 82. Structural and Contractile Proteins (Part A: Extracellular Matrix)
Edited by LEON W. CUNNINGHAM AND DIXIE W. FREDERIKSEN

VOLUME 83. Complex Carbohydrates (Part D)
Edited by VICTOR GINSBURG

VOLUME 84. Immunochemical Techniques (Part D: Selected Immunoassays)
Edited by JOHN J. LANGONE AND HELEN VAN VUNAKIS

VOLUME 85. Structural and Contractile Proteins (Part B: The Contractile Apparatus and the Cytoskeleton)
Edited by DIXIE W. FREDERIKSEN AND LEON W. CUNNINGHAM

VOLUME 86. Prostaglandins and Arachidonate Metabolites
Edited by WILLIAM E. M. LANDS AND WILLIAM L. SMITH

VOLUME 87. Enzyme Kinetics and Mechanism (Part C: Intermediates, Stereochemistry, and Rate Studies)
Edited by DANIEL L. PURICH

VOLUME 88. Biomembranes (Part I: Visual Pigments and Purple Membranes, II)
Edited by LESTER PACKER

VOLUME 89. Carbohydrate Metabolism (Part D)
Edited by WILLIS A. WOOD

VOLUME 90. Carbohydrate Metabolism (Part E)
Edited by WILLIS A. WOOD

VOLUME 91. Enzyme Structure (Part I)
Edited by C. H. W. HIRS AND SERGE N. TIMASHEFF

VOLUME 92. Immunochemical Techniques (Part E: Monoclonal Antibodies and General Immunoassay Methods)
Edited by JOHN J. LANGONE AND HELEN VAN VUNAKIS

VOLUME 93. Immunochemical Techniques (Part F: Conventional Antibodies, Fc Receptors, and Cytotoxicity)
Edited by JOHN J. LANGONE AND HELEN VAN VUNAKIS

VOLUME 94. Polyamines
Edited by HERBERT TABOR AND CELIA WHITE TABOR

VOLUME 95. Cumulative Subject Index Volumes 61–74, 76–80
Edited by EDWARD A. DENNIS AND MARTHA G. DENNIS

VOLUME 96. Biomembranes [Part J: Membrane Biogenesis: Assembly and Targeting (General Methods; Eukaryotes)]
Edited by SIDNEY FLEISCHER AND BECCA FLEISCHER

VOLUME 97. Biomembranes [Part K: Membrane Biogenesis: Assembly and Targeting (Prokaryotes, Mitochondria, and Chloroplasts)]
Edited by SIDNEY FLEISCHER AND BECCA FLEISCHER

VOLUME 98. Biomembranes (Part L: Membrane Biogenesis: Processing and Recycling)
Edited by SIDNEY FLEISCHER AND BECCA FLEISCHER

VOLUME 99. Hormone Action (Part F: Protein Kinases)
Edited by JACKIE D. CORBIN AND JOEL G. HARDMAN

VOLUME 100. Recombinant DNA (Part B)
Edited by RAY WU, LAWRENCE GROSSMAN, AND KIVIE MOLDAVE

VOLUME 101. Recombinant DNA (Part C)
Edited by RAY WU, LAWRENCE GROSSMAN, AND KIVIE MOLDAVE

VOLUME 102. Hormone Action (Part G: Calmodulin and Calcium-Binding Proteins)
Edited by ANTHONY R. MEANS AND BERT W. O'MALLEY

VOLUME 103. Hormone Action (Part H: Neuroendocrine Peptides)
Edited by P. MICHAEL CONN

VOLUME 104. Enzyme Purification and Related Techniques (Part C)
Edited by WILLIAM B. JAKOBY

VOLUME 105. Oxygen Radicals in Biological Systems
Edited by LESTER PACKER

VOLUME 106. Posttranslational Modifications (Part A)
Edited by FINN WOLD AND KIVIE MOLDAVE

VOLUME 107. Posttranslational Modifications (Part B)
Edited by FINN WOLD AND KIVIE MOLDAVE

VOLUME 108. Immunochemical Techniques (Part G: Separation and Characterization of Lymphoid Cells)
Edited by GIOVANNI DI SABATO, JOHN J. LANGONE, AND HELEN VAN VUNAKIS

VOLUME 109. Hormone Action (Part I: Peptide Hormones)
Edited by LUTZ BIRNBAUMER AND BERT W. O'MALLEY

VOLUME 110. Steroids and Isoprenoids (Part A)
Edited by JOHN H. LAW AND HANS C. RILLING

VOLUME 111. Steroids and Isoprenoids (Part B)
Edited by JOHN H. LAW AND HANS C. RILLING

VOLUME 112. Drug and Enzyme Targeting (Part A)
Edited by KENNETH J. WIDDER AND RALPH GREEN

VOLUME 113. Glutamate, Glutamine, Glutathione, and Related Compounds
Edited by ALTON MEISTER

VOLUME 114. Diffraction Methods for Biological Macromolecules (Part A)
Edited by HAROLD W. WYCKOFF, C. H. W. HIRS, AND SERGE N. TIMASHEFF

VOLUME 115. Diffraction Methods for Biological Macromolecules (Part B)
Edited by HAROLD W. WYCKOFF, C. H. W. HIRS, AND SERGE N. TIMASHEFF

VOLUME 116. Immunochemical Techniques (Part H: Effectors and Mediators of Lymphoid Cell Functions)
Edited by GIOVANNI DI SABATO, JOHN J. LANGONE, AND HELEN VAN VUNAKIS

VOLUME 117. Enzyme Structure (Part J)
Edited by C. H. W. HIRS AND SERGE N. TIMASHEFF

VOLUME 118. Plant Molecular Biology
Edited by ARTHUR WEISSBACH AND HERBERT WEISSBACH

VOLUME 119. Interferons (Part C)
Edited by SIDNEY PESTKA

VOLUME 120. Cumulative Subject Index Volumes 81–94, 96–101

VOLUME 121. Immunochemical Techniques (Part I: Hybridoma Technology and Monoclonal Antibodies)
Edited by JOHN J. LANGONE AND HELEN VAN VUNAKIS

VOLUME 122. Vitamins and Coenzymes (Part G)
Edited by FRANK CHYTIL AND DONALD B. MCCORMICK

VOLUME 123. Vitamins and Coenzymes (Part H)
Edited by FRANK CHYTIL AND DONALD B. MCCORMICK

VOLUME 124. Hormone Action (Part J: Neuroendocrine Peptides)
Edited by P. MICHAEL CONN

VOLUME 125. Biomembranes (Part M: Transport in Bacteria, Mitochondria, and Chloroplasts: General Approaches and Transport Systems)
Edited by SIDNEY FLEISCHER AND BECCA FLEISCHER

VOLUME 126. Biomembranes (Part N: Transport in Bacteria, Mitochondria, and Chloroplasts: Protonmotive Force)
Edited by SIDNEY FLEISCHER AND BECCA FLEISCHER

VOLUME 127. Biomembranes (Part O: Protons and Water: Structure and Translocation)
Edited by LESTER PACKER

VOLUME 128. Plasma Lipoproteins (Part A: Preparation, Structure, and Molecular Biology)
Edited by JERE P. SEGREST AND JOHN J. ALBERS

VOLUME 129. Plasma Lipoproteins (Part B: Characterization, Cell Biology, and Metabolism)
Edited by JOHN J. ALBERS AND JERE P. SEGREST

VOLUME 130. Enzyme Structure (Part K)
Edited by C. H. W. HIRS AND SERGE N. TIMASHEFF

VOLUME 131. Enzyme Structure (Part L)
Edited by C. H. W. HIRS AND SERGE N. TIMASHEFF

VOLUME 132. Immunochemical Techniques (Part J: Phagocytosis and Cell-Mediated Cytotoxicity)
Edited by GIOVANNI DI SABATO AND JOHANNES EVERSE

VOLUME 133. Bioluminescence and Chemiluminescence (Part B)
Edited by MARLENE DELUCA AND WILLIAM D. MCELROY

VOLUME 134. Structural and Contractile Proteins (Part C: The Contractile Apparatus and the Cytoskeleton)
Edited by RICHARD B. VALLEE

VOLUME 135. Immobilized Enzymes and Cells (Part B)
Edited by KLAUS MOSBACH

VOLUME 136. Immobilized Enzymes and Cells (Part C)
Edited by KLAUS MOSBACH

VOLUME 137. Immobilized Enzymes and Cells (Part D)
Edited by KLAUS MOSBACH

VOLUME 138. Complex Carbohydrates (Part E)
Edited by VICTOR GINSBURG

VOLUME 139. Cellular Regulators (Part A: Calcium- and Calmodulin-Binding Proteins)
Edited by ANTHONY R. MEANS AND P. MICHAEL CONN

VOLUME 140. Cumulative Subject Index Volumes 102–119, 121–134

VOLUME 141. Cellular Regulators (Part B: Calcium and Lipids)
Edited by P. MICHAEL CONN AND ANTHONY R. MEANS

VOLUME 142. Metabolism of Aromatic Amino Acids and Amines
Edited by SEYMOUR KAUFMAN

VOLUME 143. Sulfur and Sulfur Amino Acids
Edited by WILLIAM B. JAKOBY AND OWEN GRIFFITH

VOLUME 144. Structural and Contractile Proteins (Part D: Extracellular Matrix)
Edited by LEON W. CUNNINGHAM

VOLUME 145. Structural and Contractile Proteins (Part E: Extracellular Matrix)
Edited by LEON W. CUNNINGHAM

VOLUME 146. Peptide Growth Factors (Part A)
Edited by DAVID BARNES AND DAVID A. SIRBASKU

VOLUME 147. Peptide Growth Factors (Part B)
Edited by DAVID BARNES AND DAVID A. SIRBASKU

VOLUME 148. Plant Cell Membranes
Edited by LESTER PACKER AND ROLAND DOUCE

VOLUME 149. Drug and Enzyme Targeting (Part B)
Edited by RALPH GREEN AND KENNETH J. WIDDER

VOLUME 150. Immunochemical Techniques (Part K: *In Vitro* Models of B and T Cell Functions and Lymphoid Cell Receptors)
Edited by GIOVANNI DI SABATO

VOLUME 151. Molecular Genetics of Mammalian Cells
Edited by MICHAEL M. GOTTESMAN

VOLUME 152. Guide to Molecular Cloning Techniques
Edited by SHELBY L. BERGER AND ALAN R. KIMMEL

VOLUME 153. Recombinant DNA (Part D)
Edited by RAY WU AND LAWRENCE GROSSMAN

VOLUME 154. Recombinant DNA (Part E)
Edited by RAY WU AND LAWRENCE GROSSMAN

VOLUME 155. Recombinant DNA (Part F)
Edited by RAY WU

VOLUME 156. Biomembranes (Part P: ATP-Driven Pumps and Related Transport: The Na,K-Pump)
Edited by SIDNEY FLEISCHER AND BECCA FLEISCHER

VOLUME 157. Biomembranes (Part Q: ATP-Driven Pumps and Related Transport: Calcium, Proton, and Potassium Pumps)
Edited by SIDNEY FLEISCHER AND BECCA FLEISCHER

VOLUME 158. Metalloproteins (Part A)
Edited by JAMES F. RIORDAN AND BERT L. VALLEE

VOLUME 159. Initiation and Termination of Cyclic Nucleotide Action
Edited by JACKIE D. CORBIN AND ROGER A. JOHNSON

VOLUME 160. Biomass (Part A: Cellulose and Hemicellulose)
Edited by WILLIS A. WOOD AND SCOTT T. KELLOGG

VOLUME 161. Biomass (Part B: Lignin, Pectin, and Chitin)
Edited by WILLIS A. WOOD AND SCOTT T. KELLOGG

VOLUME 162. Immunochemical Techniques (Part L: Chemotaxis and Inflammation)
Edited by GIOVANNI DI SABATO

VOLUME 163. Immunochemical Techniques (Part M: Chemotaxis and Inflammation)
Edited by GIOVANNI DI SABATO

VOLUME 164. Ribosomes
Edited by HARRY F. NOLLER, JR., AND KIVIE MOLDAVE

VOLUME 165. Microbial Toxins: Tools for Enzymology
Edited by SIDNEY HARSHMAN

VOLUME 166. Branched-Chain Amino Acids
Edited by ROBERT HARRIS AND JOHN R. SOKATCH

VOLUME 167. Cyanobacteria
Edited by LESTER PACKER AND ALEXANDER N. GLAZER

VOLUME 168. Hormone Action (Part K: Neuroendocrine Peptides)
Edited by P. MICHAEL CONN

VOLUME 169. Platelets: Receptors, Adhesion, Secretion (Part A)
Edited by JACEK HAWIGER

VOLUME 170. Nucleosomes
Edited by PAUL M. WASSARMAN AND ROGER D. KORNBERG

VOLUME 171. Biomembranes (Part R: Transport Theory: Cells and Model Membranes)
Edited by SIDNEY FLEISCHER AND BECCA FLEISCHER

VOLUME 172. Biomembranes (Part S: Transport: Membrane Isolation and Characterization)
Edited by SIDNEY FLEISCHER AND BECCA FLEISCHER

VOLUME 173. Biomembranes [Part T: Cellular and Subcellular Transport: Eukaryotic (Nonepithelial) Cells]
Edited by SIDNEY FLEISCHER AND BECCA FLEISCHER

VOLUME 174. Biomembranes [Part U: Cellular and Subcellular Transport: Eukaryotic (Nonepithelial) Cells]
Edited by SIDNEY FLEISCHER AND BECCA FLEISCHER

VOLUME 175. Cumulative Subject Index Volumes 135–139, 141–167

VOLUME 176. Nuclear Magnetic Resonance (Part A: Spectral Techniques and Dynamics)
Edited by NORMAN J. OPPENHEIMER AND THOMAS L. JAMES

VOLUME 177. Nuclear Magnetic Resonance (Part B: Structure and Mechanism)
Edited by NORMAN J. OPPENHEIMER AND THOMAS L. JAMES

VOLUME 178. Antibodies, Antigens, and Molecular Mimicry
Edited by JOHN J. LANGONE

VOLUME 179. Complex Carbohydrates (Part F)
Edited by VICTOR GINSBURG

VOLUME 180. RNA Processing (Part A: General Methods)
Edited by JAMES E. DAHLBERG AND JOHN N. ABELSON

VOLUME 181. RNA Processing (Part B: Specific Methods)
Edited by JAMES E. DAHLBERG AND JOHN N. ABELSON

VOLUME 182. Guide to Protein Purification
Edited by MURRAY P. DEUTSCHER

VOLUME 183. Molecular Evolution: Computer Analysis of Protein and Nucleic Acid Sequences
Edited by RUSSELL F. DOOLITTLE

VOLUME 184. Avidin–Biotin Technology
Edited by MEIR WILCHEK AND EDWARD A. BAYER

VOLUME 185. Gene Expression Technology
Edited by DAVID V. GOEDDEL

VOLUME 186. Oxygen Radicals in Biological Systems (Part B: Oxygen Radicals and Antioxidants)
Edited by LESTER PACKER AND ALEXANDER N. GLAZER

VOLUME 187. Arachidonate Related Lipid Mediators
Edited by ROBERT C. MURPHY AND FRANK A. FITZPATRICK

VOLUME 188. Hydrocarbons and Methylotrophy
Edited by MARY E. LIDSTROM

VOLUME 189. Retinoids (Part A: Molecular and Metabolic Aspects)
Edited by LESTER PACKER

VOLUME 190. Retinoids (Part B: Cell Differentiation and Clinical Applications)
Edited by LESTER PACKER

VOLUME 191. Biomembranes (Part V: Cellular and Subcellular Transport: Epithelial Cells)
Edited by SIDNEY FLEISCHER AND BECCA FLEISCHER

VOLUME 192. Biomembranes (Part W: Cellular and Subcellular Transport: Epithelial Cells)
Edited by SIDNEY FLEISCHER AND BECCA FLEISCHER

VOLUME 193. Mass Spectrometry
Edited by JAMES A. MCCLOSKEY

VOLUME 194. Guide to Yeast Genetics and Molecular Biology
Edited by CHRISTINE GUTHRIE AND GERALD R. FINK

VOLUME 195. Adenylyl Cyclase, G Proteins, and Guanylyl Cyclase
Edited by ROGER A. JOHNSON AND JACKIE D. CORBIN

VOLUME 196. Molecular Motors and the Cytoskeleton
Edited by RICHARD B. VALLEE

VOLUME 197. Phospholipases
Edited by EDWARD A. DENNIS

VOLUME 198. Peptide Growth Factors (Part C)
Edited by DAVID BARNES, J. P. MATHER, AND GORDON H. SATO

VOLUME 199. Cumulative Subject Index Volumes 168–174, 176–194

VOLUME 200. Protein Phosphorylation (Part A: Protein Kinases: Assays, Purification, Antibodies, Functional Analysis, Cloning, and Expression)
Edited by TONY HUNTER AND BARTHOLOMEW M. SEFTON

VOLUME 201. Protein Phosphorylation (Part B: Analysis of Protein Phosphorylation, Protein Kinase Inhibitors, and Protein Phosphatases)
Edited by TONY HUNTER AND BARTHOLOMEW M. SEFTON

VOLUME 202. Molecular Design and Modeling: Concepts and Applications (Part A: Proteins, Peptides, and Enzymes)
Edited by JOHN J. LANGONE

VOLUME 203. Molecular Design and Modeling: Concepts and Applications (Part B: Antibodies and Antigens, Nucleic Acids, Polysaccharides, and Drugs)
Edited by JOHN J. LANGONE

VOLUME 204. Bacterial Genetic Systems
Edited by JEFFREY H. MILLER

VOLUME 205. Metallobiochemistry (Part B: Metallothionein and Related Molecules)
Edited by JAMES F. RIORDAN AND BERT L. VALLEE

VOLUME 206. Cytochrome P450
Edited by MICHAEL R. WATERMAN AND ERIC F. JOHNSON

VOLUME 207. Ion Channels
Edited by BERNARDO RUDY AND LINDA E. IVERSON

VOLUME 208. Protein–DNA Interactions
Edited by ROBERT T. SAUER

VOLUME 209. Phospholipid Biosynthesis
Edited by EDWARD A. DENNIS AND DENNIS E. VANCE

VOLUME 210. Numerical Computer Methods
Edited by LUDWIG BRAND AND MICHAEL L. JOHNSON

VOLUME 211. DNA Structures (Part A: Synthesis and Physical Analysis of DNA)
Edited by DAVID M. J. LILLEY AND JAMES E. DAHLBERG

VOLUME 212. DNA Structures (Part B: Chemical and Electrophoretic Analysis of DNA)
Edited by DAVID M. J. LILLEY AND JAMES E. DAHLBERG

VOLUME 213. Carotenoids (Part A: Chemistry, Separation, Quantitation, and Antioxidation)
Edited by LESTER PACKER

VOLUME 214. Carotenoids (Part B: Metabolism, Genetics, and Biosynthesis)
Edited by LESTER PACKER

VOLUME 215. Platelets: Receptors, Adhesion, Secretion (Part B)
Edited by JACEK J. HAWIGER

VOLUME 216. Recombinant DNA (Part G)
Edited by RAY WU

VOLUME 217. Recombinant DNA (Part H)
Edited by RAY WU

VOLUME 218. Recombinant DNA (Part I)
Edited by RAY WU

VOLUME 219. Reconstitution of Intracellular Transport
Edited by JAMES E. ROTHMAN

VOLUME 220. Membrane Fusion Techniques (Part A)
Edited by NEJAT DÜZGÜNEŞ

VOLUME 221. Membrane Fusion Techniques (Part B)
Edited by NEJAT DÜZGÜNEŞ

VOLUME 222. Proteolytic Enzymes in Coagulation, Fibrinolysis, and Complement Activation (Part A: Mammalian Blood Coagulation Factors and Inhibitors)
Edited by LASZLO LORAND AND KENNETH G. MANN

VOLUME 223. Proteolytic Enzymes in Coagulation, Fibrinolysis, and Complement Activation (Part B: Complement Activation, Fibrinolysis, and Nonmammalian Blood Coagulation Factors)
Edited by LASZLO LORAND AND KENNETH G. MANN

VOLUME 224. Molecular Evolution: Producing the Biochemical Data
Edited by ELIZABETH ANNE ZIMMER, THOMAS J. WHITE, REBECCA L. CANN, AND ALLAN C. WILSON

VOLUME 225. Guide to Techniques in Mouse Development
Edited by PAUL M. WASSARMAN AND MELVIN L. DEPAMPHILIS

VOLUME 226. Metallobiochemistry (Part C: Spectroscopic and Physical Methods for Probing Metal Ion Environments in Metalloenzymes and Metalloproteins)
Edited by JAMES F. RIORDAN AND BERT L. VALLEE

VOLUME 227. Metallobiochemistry (Part D: Physical and Spectroscopic Methods for Probing Metal Ion Environments in Metalloproteins)
Edited by JAMES F. RIORDAN AND BERT L. VALLEE

VOLUME 228. Aqueous Two-Phase Systems
Edited by HARRY WALTER AND GÖTE JOHANSSON

VOLUME 229. Cumulative Subject Index Volumes 195–198, 200–227

VOLUME 230. Guide to Techniques in Glycobiology
Edited by WILLIAM J. LENNARZ AND GERALD W. HART

VOLUME 231. Hemoglobins (Part B: Biochemical and Analytical Methods)
Edited by JOHANNES EVERSE, KIM D. VANDEGRIFF, AND ROBERT M. WINSLOW

VOLUME 232. Hemoglobins (Part C: Biophysical Methods)
Edited by JOHANNES EVERSE, KIM D. VANDEGRIFF, AND ROBERT M. WINSLOW

VOLUME 233. Oxygen Radicals in Biological Systems (Part C)
Edited by LESTER PACKER

VOLUME 234. Oxygen Radicals in Biological Systems (Part D)
Edited by LESTER PACKER

VOLUME 235. Bacterial Pathogenesis (Part A: Identification and Regulation of Virulence Factors)
Edited by VIRGINIA L. CLARK AND PATRIK M. BAVOIL

VOLUME 236. Bacterial Pathogenesis (Part B: Integration of Pathogenic Bacteria with Host Cells)
Edited by VIRGINIA L. CLARK AND PATRIK M. BAVOIL

VOLUME 237. Heterotrimeric G Proteins
Edited by RAVI IYENGAR

VOLUME 238. Heterotrimeric G-Protein Effectors
Edited by RAVI IYENGAR

VOLUME 239. Nuclear Magnetic Resonance (Part C)
Edited by THOMAS L. JAMES AND NORMAN J. OPPENHEIMER

VOLUME 240. Numerical Computer Methods (Part B)
Edited by MICHAEL L. JOHNSON AND LUDWIG BRAND

VOLUME 241. Retroviral Proteases
Edited by LAWRENCE C. KUO AND JULES A. SHAFER

VOLUME 242. Neoglycoconjugates (Part A)
Edited by Y. C. LEE AND REIKO T. LEE

VOLUME 243. Inorganic Microbial Sulfur Metabolism
Edited by HARRY D. PECK, JR., AND JEAN LEGALL

VOLUME 244. Proteolytic Enzymes: Serine and Cysteine Peptidases
Edited by ALAN J. BARRETT

VOLUME 245. Extracellular Matrix Components
Edited by E. RUOSLAHTI AND E. ENGVALL

VOLUME 246. Biochemical Spectroscopy
Edited by KENNETH SAUER

VOLUME 247. Neoglycoconjugates (Part B: Biomedical Applications)
Edited by Y. C. LEE AND REIKO T. LEE

VOLUME 248. Proteolytic Enzymes: Aspartic and Metallo Peptidases
Edited by ALAN J. BARRETT

VOLUME 249. Enzyme Kinetics and Mechanism (Part D: Developments in Enzyme Dynamics)
Edited by DANIEL L. PURICH

VOLUME 250. Lipid Modifications of Proteins
Edited by PATRICK J. CASEY AND JANICE E. BUSS

VOLUME 251. Biothiols (Part A: Monothiols and Dithiols, Protein Thiols, and Thiyl Radicals)
Edited by LESTER PACKER

VOLUME 252. Biothiols (Part B: Glutathione and Thioredoxin; Thiols in Signal Transduction and Gene Regulation)
Edited by LESTER PACKER

VOLUME 253. Adhesion of Microbial Pathogens
Edited by RON J. DOYLE AND ITZHAK OFEK

VOLUME 254. Oncogene Techniques
Edited by PETER K. VOGT AND INDER M. VERMA

VOLUME 255. Small GTPases and Their Regulators (Part A: Ras Family)
Edited by W. E. BALCH, CHANNING J. DER, AND ALAN HALL

VOLUME 256. Small GTPases and Their Regulators (Part B: Rho Family)
Edited by W. E. BALCH, CHANNING J. DER, AND ALAN HALL

VOLUME 257. Small GTPases and Their Regulators (Part C: Proteins Involved in Transport)
Edited by W. E. BALCH, CHANNING J. DER, AND ALAN HALL

VOLUME 258. Redox-Active Amino Acids in Biology
Edited by JUDITH P. KLINMAN

VOLUME 259. Energetics of Biological Macromolecules
Edited by MICHAEL L. JOHNSON AND GARY K. ACKERS

VOLUME 260. Mitochondrial Biogenesis and Genetics (Part A)
Edited by GIUSEPPE M. ATTARDI AND ANNE CHOMYN

VOLUME 261. Nuclear Magnetic Resonance and Nucleic Acids
Edited by THOMAS L. JAMES

VOLUME 262. DNA Replication
Edited by JUDITH L. CAMPBELL

VOLUME 263. Plasma Lipoproteins (Part C: Quantitation)
Edited by WILLIAM A. BRADLEY, SANDRA H. GIANTURCO, AND JERE P. SEGREST

VOLUME 264. Mitochondrial Biogenesis and Genetics (Part B)
Edited by GIUSEPPE M. ATTARDI AND ANNE CHOMYN

VOLUME 265. Cumulative Subject Index Volumes 228, 230–262

VOLUME 266. Computer Methods for Macromolecular Sequence Analysis
Edited by RUSSELL F. DOOLITTLE

VOLUME 267. Combinatorial Chemistry
Edited by JOHN N. ABELSON

VOLUME 268. Nitric Oxide (Part A: Sources and Detection of NO; NO Synthase)
Edited by LESTER PACKER

VOLUME 269. Nitric Oxide (Part B: Physiological and Pathological Processes)
Edited by LESTER PACKER

VOLUME 270. High Resolution Separation and Analysis of Biological Macromolecules (Part A: Fundamentals)
Edited by BARRY L. KARGER AND WILLIAM S. HANCOCK

VOLUME 271. High Resolution Separation and Analysis of Biological Macromolecules (Part B: Applications)
Edited by BARRY L. KARGER AND WILLIAM S. HANCOCK

VOLUME 272. Cytochrome P450 (Part B)
Edited by ERIC F. JOHNSON AND MICHAEL R. WATERMAN

VOLUME 273. RNA Polymerase and Associated Factors (Part A)
Edited by SANKAR ADHYA

VOLUME 274. RNA Polymerase and Associated Factors (Part B)
Edited by SANKAR ADHYA

VOLUME 275. Viral Polymerases and Related Proteins
Edited by LAWRENCE C. KUO, DAVID B. OLSEN, AND STEVEN S. CARROLL

VOLUME 276. Macromolecular Crystallography (Part A)
Edited by CHARLES W. CARTER, JR., AND ROBERT M. SWEET

VOLUME 277. Macromolecular Crystallography (Part B)
Edited by CHARLES W. CARTER, JR., AND ROBERT M. SWEET

VOLUME 278. Fluorescence Spectroscopy
Edited by LUDWIG BRAND AND MICHAEL L. JOHNSON

VOLUME 279. Vitamins and Coenzymes, Part I
Edited by DONALD B. MCCORMICK, JOHN W. SUTTIE, AND CONRAD WAGNER

VOLUME 280. Vitamins and Coenzymes, Part J
Edited by DONALD B. MCCORMICK, JOHN W. SUTTIE, AND CONRAD WAGNER

VOLUME 281. Vitamins and Coenzymes, Part K
Edited by DONALD B. MCCORMICK, JOHN W. SUTTIE, AND CONRAD WAGNER

VOLUME 282. Vitamins and Coenzymes, Part L
Edited by DONALD B. MCCORMICK, JOHN W. SUTTIE, AND CONRAD WAGNER

VOLUME 283. Cell Cycle Control
Edited by WILLIAM G. DUNPHY

VOLUME 284. Lipases (Part A: Biotechnology)
Edited by BYRON RUBIN AND EDWARD A. DENNIS

VOLUME 285. Cumulative Subject Index Volumes 263, 264, 266–284, 286–289

VOLUME 286. Lipases (Part B: Enzyme Characterization and Utilization)
Edited by BYRON RUBIN AND EDWARD A. DENNIS

VOLUME 287. Chemokines
Edited by RICHARD HORUK

VOLUME 288. Chemokine Receptors
Edited by RICHARD HORUK

VOLUME 289. Solid Phase Peptide Synthesis
Edited by GREGG B. FIELDS

VOLUME 290. Molecular Chaperones
Edited by GEORGE H. LORIMER AND THOMAS BALDWIN

VOLUME 291. Caged Compounds
Edited by GERARD MARRIOTT

VOLUME 292. ABC Transporters: Biochemical, Cellular, and Molecular Aspects
Edited by SURESH V. AMBUDKAR AND MICHAEL M. GOTTESMAN

VOLUME 293. Ion Channels (Part B)
Edited by P. MICHAEL CONN

VOLUME 294. Ion Channels (Part C)
Edited by P. MICHAEL CONN

VOLUME 295. Energetics of Biological Macromolecules (Part B)
Edited by GARY K. ACKERS AND MICHAEL L. JOHNSON

VOLUME 296. Neurotransmitter Transporters
Edited by SUSAN G. AMARA

VOLUME 297. Photosynthesis: Molecular Biology of Energy Capture
Edited by LEE MCINTOSH

VOLUME 298. Molecular Motors and the Cytoskeleton (Part B)
Edited by RICHARD B. VALLEE

VOLUME 299. Oxidants and Antioxidants (Part A)
Edited by LESTER PACKER

VOLUME 300. Oxidants and Antioxidants (Part B)
Edited by LESTER PACKER

VOLUME 301. Nitric Oxide: Biological and Antioxidant Activities (Part C)
Edited by LESTER PACKER

VOLUME 302. Green Fluorescent Protein (in preparation)
Edited by P. MICHAEL CONN

VOLUME 303. cDNA Preparation and Display (in preparation)
Edited by SHERMAN M. WEISSMAN

VOLUME 304. Chromatin (in preparation)
Edited by PAUL M. WASSERMAN AND ALAN P. WOLFFE

VOLUME 305. Bioluminescence and Chemiluminescence (Part C) (in preparation)
Edited by MIRIAM M. ZIEGLER AND THOMAS O. BALDWIN

Section I
Total Antioxidant Activity

[1] Total Antioxidant Activity Measured by Chemiluminescence Methods

By HANNU ALHO and JANNE LEINONEN

Introduction

Reactive oxygen species (ROS) have been implicated in more than 100 diseases, from malaria and hemorrhagic shock to acquired immunodeficiency syndrome.[1] This wide range of diseases implies that ROS are not something esoteric, but that their increased formation accompanies tissue injury in most, if not all, human diseases.[1] Tissue damage by disease, trauma, toxins, ischemia/reperfusion, and other causes usually leads to the formation of increased amounts of putative "injury mediators," as well as to increased ROS formation.[2,3]

Four endogenous sources appear to account for most of the oxidants produced by cells: (i) normal aerobic respiration, i.e., mitochondria, consume O_2 by reducing it in sequential steps, thus producing H_2O_2; (ii) stimulated polymorphonuclear leukocytes and macrophages release superoxide, which in turn is a source for H_2O_2, HOCl, and NO; (iii) peroxisomes, organelles responsible for degrading fatty acids and other molecules, produce H_2O_2 as a by-product; and (iii) induction of P450 enzymes can also result in oxidant by-products.[4]

Hydroxyl radical OH·, the fearsomely reactive oxygen species, has been proposed to be produced in living organisms by at least three separate mechanisms: (i) by reaction of transition metal ions with H_2O_2, the so-called superoxide-driven Fenton reaction; (ii) by peroxynitrite, a nonradical product of NO· and O_2^-, can protonate and decompose to a range of noxious products, including nitrogen dioxide and nitronium iron; and (iii) by making ·OH *in vivo* by the reaction of O_2^- with hypochlorous acid.

Exogenous sources of free radicals include tobacco smoke, ionizing radiation, certain pollutants, organic solvents, anesthetics, hyperoxic environment, and pesticides. Some of these compounds, as well as certain medications, are metabolized to free radical intermediates that have been

---

[1] B. Halliwell, *Haemostasis* **23,** 118 (1992).
[2] B. Halliwell and J. M. C. Gutteridge, eds., *in* "Free Radicals in Biology and Medicine," 2nd ed., p. 253 Clarendon Press, Oxford, 1989.
[3] S. Toyokuni, K. Okamoto, J. Yodoi, and H. Hiai, *FEBS Lett.* **358,** 1 (1995).
[4] B. N. Ames, M. K. Shigenaga, and T. M. Hagen, *Proc. Natl. Acad. Sci. U.S.A.* **90,** 7915 (1993).

shown to cause oxidative damage to the target tissues. Exposure to radiation results in the formation of free radicals within the exposed tissues.

To protect itself against the deleterious effects of free radicals, the human body has developed an antioxidant defense system that consists of enzymatic, metal-chelating, and free radical-scavenging properties. In addition to the protective effects of endogenous enzymatic antioxidant defenses, consumption of dietary antioxidants appears to be important.[4] The concentration of antioxidants in human blood plasma is important in investigating and understanding the relationship among diet, oxidative stress, and human disease. The measurement of the total antioxidant activity of biological fluids, especially plasma, serum, or serum lipoprotein fractions, is of value in estimating the capability to resist oxidative stress. Different methods applicable to this task have been reviewed previously in this series.[5] The principle of practically all of these methods is to by some means produce free radicals at a known rate and to study the capability of a sample to inhibit this radical production by a certain end point. In this study, chemiluminescence-based methods are evaluated for measuring the peroxyl radical-scavenging capacity of human plasma, low-density lipoprotein (LDL), and cerebrospinal fluid (CSF).

Methods of assessing antioxidant activity vary greatly with regard to the radical species that is generated (Table I), the reproductivity of the generation process, and the end point that is used (Table II). One of the most widely used end points is chemiluminescence. Earlier methods[6,7] were based on the inhibition of spontaneous tissue autoxidation, but Wayner[8] took advantage of their discovery that the thermal decomposition of water-soluble azo compound 2,2'-azobis([2-amidinopropane])hydrochloride (ABAP) yields peroxyl radicals at a known constant rate. The decomposition of ABAP has been shown to induce the following temporal order of consumption of plasma antioxidants: ascorbate > thiols > bilirubin > urate > α-tocopherol.

It has been shown that chemiluminescence as an end point offers a sensitive way to observe antioxidant-consuming free radical reactions against either whole plasma or LDL, and several modifications utilizing chemiluminescence have been developed. Hirayama *et al.*[9] have used a

---

[5] C. Rice-Evans and N. J. Miller, *Methods Enzymol.* **234,** 279 (1994).
[6] T. Ogasawara and M. Kan, *Tohoku J. Exp. Med.* **144,** 9 (1984).
[7] J. Stocks, J. M. C. Gutteridge, R. J. Sharp, and T. L. Dormandy, *Clin. Sci. Mol. Med.* **47,** 215 (1974).
[8] D. D. M. Wayner, G. W. Burton, K. U. Ingold, and S. Locke, *FEBS Lett.* **187,** 33 (1985).
[9] O. Hirayama, M. Tagaki, K. Hukumoto, and S. Katoh, *Anal. Biochem.* **247,** 237 (1997).

TABLE I
METHODS FOR GENERATING RADICAL SPECIES[a]

Cu^{2+}/cumene hydroperoxide (5)
Cu^{2+}/H_2O_2 (5)
HRP/H_2O_2 (11)
OPD/H_2O_2 (5)
Ferrimyoglobin radicals and ABTS (5)
Peroxyl radicals from ABAP (8, 13)
AAPH (31)
AMVN (17)
Lipoperoxides in brain homogenates (5)
Superoxide (6)
Photoinduction (12)

[a] HRP, Horseradish peroxidase; OPD, o-phenylenediamine; ABTS, 2,2'-azinobis(3-ethylbenzothiazoline 6-sulfonate); ABAP, 2,2'-azobis(2-amidinopropane hydrochloride); AAPH, 2,2'-azobis(2-amidinopropane) dihydrochloride; AMVN, 2,2'-azobis(2,4-dimethylvaleronitrile). Numbers in parentheses refer to literature cited in the text.

mixture of lipid hydroperoxides and microperoxidase to produce oxyradicals and further light emission by luminol oxidation to study the antioxidant activity of plasma and saliva. Because cumene hydroperoxide induces a rapid chemiluminescence that is followed for only 3 min, half-inhibition values of the initial chemiluminescence are determined for various anti-

TABLE II
METHODS FOR END POINT OBSERVATIONS[a]

Fluorescence inhibition (5)
Chemiluminescence (8–11, 13, 24, 32)
Oxygen uptake (32)
Absorbance change (5)
TBA-RS (5)
CO production (5)
Cell morphology (6)

[a] TBA-RS, Thiobarbituric acid-reactive substances. Numbers in parentheses refer to literature cited in the text.

oxidants and biological samples. Maxwell et al.[10] have measured the total antioxidant activity of LDL, HDL, and VLDL and Whitehead et al.[11] that of serum by measuring luminol-based chemiluminescence catalyzed by horseradish peroxidase (HRP) by the addition of phenolic enhancer compounds. One method for testing and quantification of nonenzymatic antioxidants is based on a photoinduced, chemiluminescence-accompanied, and antioxidant-inhibitable autoxidation of luminol.[12] The mean values of an integral antioxidant capacity (AC) of human blood plasma showed age-dependent patterns with maximal values with newborns. The AC of six tested animal species was lower than that of humans, with maximal values with guinea pigs and spontaneously hypertensive rats (see Ref. 12).

However, the most often used chemiluminescence-based methods are modifications of the original total peroxyl radical-trapping potential (TRAP) method of Wayner et al.[8] The problem with the original TRAP assay method lies in the oxygen electrode used to measure the end point, as it will not maintain its stability over the period of time required. However, the TRAP assay measured with a chemiluminescence modification developed by Metsä-Ketelä[13] produces an assay of considerably better precision than the original TRAP assay. This chemiluminescence-enhanced TRAP utilizes water- or lipid-soluble azo initiators as sources of a constant flux of carbon-centered peroxyl radicals. Using this approach, the ability of plasma antioxidants to inhibit the artificial propagation phase of membrane lipid peroxidation can be tested. It also lends itself to a higher degree of automation and significant numbers of samples can be processed.

We have previously reported changes of total antioxidant capacity of plasma, CSF, or LDL in control materials and various clinical situations by using this TRAP method.[14–21] In addition to the methodological com-

---

[10] S. R. J. Maxwell, O. Wiklund, and G. Bondjers, *Atherosclerosis* **11,** 79 (1994).
[11] T. P. Whitehead, G. H. G. Thorpe, and S. R. J. Maxwell, *Anal. Chim. Acta* **266,** 265 (1992).
[12] I. N. Popov and G. Lewin, *Free Radic. Biol. Med.* **17,** 267 (1994).
[13] T. Metsä-Ketelä, in "Bioluminescence and Chemiluminescence Current Status" (P. E. Stanley and L. J. Kricka, eds.), Wiley, Chirchester, 1991.
[14] R. Aejmelaeus, T. Metsä-Ketelä, P. Laippala, and H. Alho, *FEBS Lett.* **384,** 128 (1996).
[15] R. Aejmelaeus, T. Metsä-Ketelä, T. Pirttilä, A. Hervonen, and H. Alho, *Free Radic. Res.* **26,** 335 (1996).
[16] R. Aejmelaeus, P. Holm, U. Kaukinen, T. Metsä-Ketelä, P. Laippala, A. Hervonen, and H. E. R. Alho, *Free Radic. Biol. Med.* **23,** (1996).
[17] R. Aejmelaeus, T. Metsä-Ketelä, P. Laippala, T. Solakivi, and H. Alho, *Mol. Asp. Med.* **18,** 113 (1997).

ments on TRAP developed by Metsä-Ketelä,[13] this chapter presents detailed protocols for the measurement of chemiluminescence-enhanced TRAP of both plasma and LDL.

Total Peroxyl Radical-Trapping Potential

General Principle

Thermal decomposition of the water-soluble azo compound ABAP or the lipid-soluble (AMVN) generates peroxyl radicals at a known constant rate. Their reaction with the chemiluminescent substrate luminol leads to the formation of luminol radicals that emit light that can be detected by a luminometer. Antioxidants in the sample inhibit this chemiluminescence for a time that is directly proportional to the total antioxidant potential of the sample. This potential of the sample is compared to that of either water- or lipid-soluble tocopherol analogs, capable of trapping 2 moles of peroxyl radicals per 1 mole of Trolox (6-hydroxy-2,-5,7,8-tetramethylchroman-2-carboxylic acid, Aldrich, Germany).

TRAP Assay for Plasma and LDL

Prepare plasma samples by drawing venous (fasting or nonfasting, see Observations) blood into EDTA-containing Vacutainer tubes on ice, protected from light. Separate plasma by centrifugation using a temperature-controlled centrifuge at 4° after which plasma can be stored at −80° for up to 6 months without a significant change in the TRAP value. Mix 475 μl of oxygen-saturated 100 μM sodium phosphate buffer, pH 7.4, in a plastic cuvette with 50 μl of 400 mM ABAP (Polysciences, Warrington, PA) in the same buffer and with 50 μl of 10 mM luminol (5-amino-2,3-dihydro-1,4-phthalazinedione, Sigma Chemical Co., St. Louis, MO) in 20 mM boric acid–borax buffer, pH adjusted to 9.5 with 10 N HCl. After a 15-min

---

[18] M. Erhola, M. Nieminen, A. Ojala, T. Metsä-Ketelä, P. Kellokumpu-Lehtinen, and H. Alho, *J. Exp. Clin. Cancer Res.* **17**(1), 1 (1998).
[19] M. Erhola, M. Nieminen, P. Kellokumpu-Lehtinen, T. Metsä-Ketelä, T. Poussa, and H. Alho, *Free Radic. Res.* **26**, 439 (1997).
[20] M. Erhola, P. Kellokumpu-Lehtinen, T. Metsä-Ketelä, K. Alanko, and M. Nieminen, *Free Radic. Biol. Med.* **21**, 383 (1996).
[21] K. Lönnrot, T. Metsä-Ketelä, G. Molnar, J.-P. Ahonen, M. Latvala, J. Peltola, T. Pietilä, H. Alho, *Free Radic. Biol. Med.* **21**, 211 (1996).

incubation at 37° the rate of synthesis of peroxyl radicals is constant; dispense 25 μl of plasma into the cuvette. Using LKB Wallac Luminometer 1251, a PC, and software from TriStar Enterprise (Tampere, Finland), detect chemiluminescence readings at 36-sec intervals for 90 min. The linear regression line for Trolox in our laboratory is $y = 131.7x + 43.2$, where y is the inhibition time in seconds and x is the concentration of Trolox in nM.

For thorough evaluation of the antioxidant capacity, it is essential to also measure the concentrations of main chain-breaking antioxidants from the same sample (see Discussion). For the measurement of ascorbic acid, we use a final concentration of 5% of metaphosphoric acid as an additive in the plasma samples. Ascorbic acid and uric acid are then measured by HPLC according to Frei et al.[22] α-Tocopherol is measured by the modified HPLC method of Catignani et al.[23] and ubiquinol-10 from a heparin–citrate-precipitated LDL fraction according to Lang et al.[24] Protein sulfhydryl groups ($-SH$) are measured according to Ellman.[25] From the individual concentrations of measured antioxidants it is possible to derive the theoretical TRAP value (TRAP$_{Calc}$) of the sample by using the stoichiometric peroxyl radical-scavenging factors (see also Observations and Discussion) that have been established[8,13,21]: TRAP$_{Calc}$ = 2.0 [sample concentration of uric acid] + 2.0 [α-tocopherol] + 0.7 [ascorbic acid] + 0.4 [$-SH$]. It is also possible to calculate the difference between measured TRAP and TRAP$_{Calc}$, i.e., the TRAP$_{Unid}$, which is composed of actions of unmeasured and partly uncharacterized antioxidants of the sample. TRAP values are presented as micromoles of peroxyl radicals trapped by 1 liter of the sample.

For measurement of LDL TRAP (TRAP$_{LDL}$), heparin–citrate-precipitated LDL is extracted from plasma with chloroform/methanol (1, v/v). TRAP$_{LDL}$ is measured analogically to plasma TRAP by replacing water-soluble ABAP with 50 μl of 25 mM lipid-soluble AMVN (Polyscience, Warrington, PA). D-α-Tocopherol is used as an internal standard. TRAP$_{LDL}$ is expressed as picomoles of peroxyl radicals. If the concentrations of α-tocopherol and ubiquinol-10 are measured separately from the LDL extract, the TRAP$_{Calc}$ can be derived using stoichiometric factors 2.0 for

---

[22] B. Frei, L. England, and B. Ames, *Proc. Natl. Acad. Sci. U.S.A.* **86**, 6377 (1989).
[23] G. Catignani and J. Bieri, *Clin. Chem.* **29**, 708 (1993).
[24] J. K. Lang, K. Gohil, and L. Packer, *Anal. Biochem.* **157**, 106 (1986).
[25] G. Ellman, *Arch. Biochem. Biophys.* **82**, 70 (1959).

both α-tocopherol and ubiquinol-10.[8,17] Again, $TRAP_{Unid}$ is the difference between measured $TRAP_{LDL}$ and $TRAP_{Calc}$.

Observations

Methodological Aspects

The inhibition method combined with luminol chemiluminescence used in our studies demonstrates its high accuracy by the fact that the coefficient for both inter- and intraassay variability is 2% in our laboratory. For evaluation of the analytic system, artificial "plasma" was prepared in phosphate-buffered saline (100 mM, pH 7.4) by solving the major known chain-breaking antioxidants of human plasma in the following concentration ranges: urate, 125, 250, and 500 μM; SH groups (as reduced glutathione), 250, 500, and 1000 μM; Trolox C (as vitamin E), 25, 50, and 100 μM; and ascorbate, 50, 100, and 200 μM. For each combination, both experimental and theoretical TRAP values were determined. Experimental and theoretical values were almost equal, despite the combination of the tested antioxidants used in human experiments. In concentrations under 700 μM/liter, there is a tendency for $TRAP_{Calc}$ to be lower than $TRAP_{Meas}$. When artificial plasmas were reconstituted in the presence of metal chelators, Desferal (150 and 300 μM) and EDTA (artificial plasmas were made in an EDTA blood tube), no effect was found. Enzymatic antioxidant, superoxide dismutase (SOD)(5 U/ml, 20 U/ml), was also added to the artificial plasma, but no effect on the TRAP value was observed. No synergistic action of antioxidants of this artificial plasma could be observed.[16,21]

The accuracy of the TRAP method was also investigated by comparing TRAP with the commercial total antioxidant status (TAS) kit (Total Antioxidant Status, NX 2332, Randox Laboratory, Cromwell, UK). A fairly good correlation was observed between values of plasma TRAP measured by a chemiluminescence-enhanced method and values of plasma TAS in 51 healthy children ($r = 0.43$, $p = 0.001$).

The significance of fasting in the measurement of TRAP was evaluated in healthy volunteers ($n = 11$). Blood samples were taken after an overnight fast and after a standardized lunch. There was no difference in TRAP values between the time points before and 2 hr after the lunch (1390 μM vs 1400 μM, mean, $p = $ NS), although there was a clear increase in plasma ascorbic acid concentrations. We concluded that for the TRAP analysis only it is not necessary to have a fasting blood sample, but because it is essential to also perform the analysis of individual components of TRAP, which may change during the day, a fasting blood sample for all TRAP

TABLE III
MEAN CONCENTRATIONS OF KNOWN MAJOR PLASMA AND CEREBROSPINAL
FLUID CHAIN-BREAKING ANTIOXIDANTS AND THEIR STOICHIOMETRIC
PEROXYL RADICAL-SCAVENGING FACTORS[a]

Factor	In plasma		In cerebrospinal fluid	
	SF	CO	SF	CO
SH groups	0.4	450–700	0.4	90–100
Ascorbic acid	0.7	20–80	0.4	160–230
Bilirubin	2	2–20	?	>0.001
Uric acid	2	120–350	2	25–35
Ubiquinol-10	2	0.5–1	2	>0.009
α-Tocopherol	2	9–30	2	>0.03

[a] CO, Concentration (μM); SF, stoichiometric factor, combined from Refs. 8, 13, 21, 27, 32.

analyses is recommended. It has also been demonstrated that dietary antioxidant supplementation does not have significant effects on the plasma TRAP value.[21,26] Supplementation of ascorbic acid (500–1000 mg/day) or ubiquinone (150–300 mg/day) for 4 weeks increased plasma concentrations significantly, but surprisingly did not increase plasma TRAP significantly. Smoking also does not seem to have any significant effect on plasma TRAP, as demonstrated in miscellaneous attendants of a health care center [(1241 μM vs 1220 μM; smokers ($n = 16$) vs nonsmokers ($n = 83$), mean, $p = $ NS)].

The stoichiometric peroxyl radical-scavenging factor (SF) for individual antioxidants has proposed to be concentration dependent. Stoichiometric factors of ascorbic acid, uric acid, SH groups, and α-tocopherol were tested in concentrations appearing in plasma and CSF. Indeed, the SF for ascorbic acid was concentration dependent, in the concentration appearing in liquor (approximately 10 times higher), was only 0.4, which was 0.7 in a concentration appearing in plasma. The scavenging factors of the other TRAP components in CSF were identical to plasma (Table III).

Observations of TRAP in Normal Human Population

The percentage contributions of TRAP components in a normal Finnish healthy population are given in Table IV. The largest contribution is given by uric acid and the smallest by α-tocopherol. When the components of $TRAP_{LDL}$ were taken into consideration, the contribution of α-tocopherol was 73 ± 1.5%, ubiquinol 2.5 ± 0.9% and of unidentified antioxidants 24.5 ±

[26] C. W. Mullholland and J. J. Strain, *Int. J. Vit. Nutr. Res.* **63**, 27 (1993).

TABLE IV
CONTRIBUTIONS OF TRAP COMPONENTS[a]

Component	Contribution (%) of components from total	
	TRAP	$TRAP_{LDL}$
Uric acid	43–52	NA
SH groups	13–22	NA
α-Tocopherol	1–4	65–70
Ascorbic acid	2–3	NA
Ubiquinol-10	<1	1.5–3.5
$TRAP_{Unid}$	25–35	30–35

[a] In a random Finnish healthy population. Combined data from Refs. 14–16.

0.5%. CSF TRAP is approximately five times lower than in plasma (240 μM vs 1150 μM). The concentration of uric acid is approximately 8 times, α-tocopherol 1000 times, ubiquinol-10 100 times, and of SH groups 5 times lower than in plasma. Only ascorbic acid is considerably higher in CSF than in plasma.[21] Age and gender have major effects on TRAP and its components (Table V). In females, TRAP increased consistently with age in all age groups ($r = 0.668$, $r^2 = 0.447$, $p = 0.0001$). In males, however, an increase in TRAP from age 34 to 74 was followed by a significant ($p = 0.02$) decline in the oldest age group from 75 to 96 years. Plasma TRAP of healthy children (age 14–16, $n = 33$) is relatively low (1110 ± 128 μM),

TABLE V
CHANGES IN TRAP WITH AGE[a]

Factor	Sex	Age			
		18–34	35–50	51–74	75–96
TRAP	M	1051 ± 84	1231 ± 56*	1300 ± 53*	1126 ± 43*
	F	988 ± 36	1076 ± 32	1216 ± 35**	1288 ± 42***
$TRAP_{Calc}$	M	739 ± 57	804 ± 37	799 ± 43	809 ± 44
	F	672 ± 39	694 ± 31	762 ± 30	839 ± 30
$TRAP_{Unid}$	M	312 ± 31	426 ± 38	501 ± 27	317 ± 30
	F	316 ± 47	392 ± 21	454 ± 26**	449 ± 42**

[a] A random healthy Finnish population. TRAP expressed in μM, mean ± SE. M, Male; F, female. *$p \leq 0.05$, **$p \leq 0.02$, ***$p \leq 0.001$ compared with the youngest age group, data combined from Refs. 14–16.

mainly due to the low plasma uric acid (122 ± 29 μM). Men have higher plasma TRAP than women in all age groups under 75 years (Table V).

Observations of TRAP in Clinical Conditions

TRAP, $TRAP_{LDL}$, and their main components have been used to study the effects of acute infection, diabetes, immobilization, Alzheimer's disease, coronary heart disease, and cancer. It was observed that plasma antioxidant defenses seem to respond to the basic metabolic rate and the challenges caused by physiological or pathological stress: during acute infection and immobilization, levels of ascorbic acid and α-tocopherol remained unchanged, whereas the amount of $TRAP_{Unid}$ in TRAP declined sharply,[14–16] in those diabetic patients with coronary heart disease, TRAP was increased, but diabetes and Alzheimer's disease did not affect plasma TRAP, and in those lung cancer patients with a poor response to treatment, TRAP was reduced.[18,19]

Discussion

The chemiluminescence-enhanced TRAP method has proven to be a reliable tool for evaluating the total activity of chain-breaking antioxidants in biological fluids. As observed by many laboratories, human plasma possesses a significant ability to scavenge peroxyl radicals. Chemiluminescence-enhanced TRAP possesses advantages over the original method. The measurement is not laborious, and up to 24 samples can be processed using one LKB Wallac luminometer in 8 hours.

In the TRAP assay, the strongest contribution is given by uric acid because of its high plasma concentration (120–350 μM in healthy adults). However, the concentration of uric acid may be raised markedly in a gouterous state and noninsulindependent diabetes mellitus. High concentrations of uric acid are not beneficial, and uric acid, although capable of scavenging superoxide, peroxyl radicals, and hydroxyl radicals, can give rise to urate-derived radicals that are potentially harmful. Changes in the levels of other antioxidants may not be reflected in the TRAP value due to the high contribution of uric acid. Lissi *et al.*[27] have introduced a new parameter derived from TRAP, the total antioxidant reactivity (TAR), which describes the ability of an antioxidant to instantaneously reduce

---

[27] E. Lissi, M. Salim-Hanna, C. Pascual, and M. D. del Castillo, *Free Radic. Biol. Med.* **18,** 153 (1995).

chemiluminescence in the TRAP assay. Although the molar stoichiometric factor for both uric acid and Trolox is 2.0, the TAR of uric acid has been reported to be nearly 18 times less than that of Trolox. Albumin has been reported to be the most important antioxidant in the plasma TAS assay.[28] Albumin does not, however, cause significant suppression of ABAP-derived peroxyl radicals in the TRAP assay.[27] This may be due to the fact that chemiluminescence is not interfered by the light-absorbing properties of albumin or other macromolecules.

As a measure of the amount of systemic oxidative stress plasma TRAP may not be the tool of choice. Free radical production would probably have to be very extensive to disturb the system's steady-state level of antioxidants. It must also be emphasized that antioxidant systems are highly compartmentalized to act on the cell membrane (e.g., α-tocopherol and β-carotene), in the cytoplasm (e.g., glutathione, superoxide dismutase, catalase), or in extracellular fluids (e.g., ascorbic acid, uric acid, albumin). The most important antioxidant function of extracellular fluids may be the binding of free transition metals such as Fe^{2+} and Cu^{2+} ions by actions of transferrin, ferritin, and ceruloplasmin.[2] However, lipid peroxidation and thus peroxyl radicals are likely to exist in plasma, and in this respect the measurement of TRAP gives a useful estimate of the ability of the chain-breaking antioxidants to prevent the chain reaction on membrane lipids.

The rate of lipid peroxidation induced by azo initiators has been shown to be dependent on the pH of the used buffer.[29] In the case of ABAP, an increase in pH results in an enhanced rate of peroxidation in a pH range of 5–8.5 due to the changes in the reaction kinetics of ABAP-derived radical species. It is thus essential to control the pH of the TRAP reaction buffer in order to control the amount of peroxyl radicals produced. The choice of an anticoagulant for isolation of plasma may also be of importance in some settings. Heparin and EDTA possess an antioxidant activity,[30] but this may be unimportant in TRAP assays, considering the low concentrations of EDTA or heparin used.

Of measured plasma TRAP, 25–35% seems to be composed of actions of unmeasured or even totally uncharacterized antioxidants. This suggestion is supported by the finding of the lack of any significant synergistic action

---

[28] N. J. Miller, C. Rice-Evans, M. J. Davies, V. Gopinathan, and A. Milner, *Clin. Sci.* **84**, 407 (1993).
[29] M. C. Hanlon and D. W. Seybert, *Free Radic. Biol. Med.* **23**, 712 (1997).
[30] H. F. Goode, N. Richardson, D. S. Myers, P. D. Howdle, B. E. Walker, and N. R. Webster, *Ann. Clin. Biochem.* **32**, 413 (1995).

of antioxidants added to a saline at physiologic concentrations.[16] Studies have shown that $TRAP_{Unid}$ seems to react sharply in cases of oxidative stress. In acute infection the amount of classical antioxidants remained unchanged, whereas the amount of unidentified antioxidants declined significantly.[14] Immobilized patients with a decreased basic metabolism and free radical production showed a significant decrease in TRAP primarily because of a decline in the concentration of unknown antioxidants, which may be in connection with the basic metabolic rate.[15] Other results from our laboratory also support the idea that $TRAP_{Unid}$ is composed of actions of endogenous products and may be related to basic metabolism; in addition in heavy exercise, a significant increase in $TRAP_{Unid}$ can be detected.[17–20]

When looking for the possible candidate molecules responsible for $TRAP_{Unid}$, several conditions must be fulfilled. Such a compound(s) must be present in plasma in sufficient concentrations and must be capable of scavenging peroxyl radicals. $TRAP_{Unid}$ has been proposed to be composed of actions of endogenous compounds, such as albumin, glucose, cholesterol, steroids, and melatonin, but antioxidants from nutritional origin such as flavonoids or benzylisoquinoline alkaloids cannot yet be totally ruled out.[31] Finally, the nature and role of $TRAP_{Unid}$ are the focal concerns of further investigations, and the wide variety of candidates ensure that it will be an interesting challenge.

In conclusion, chemiluminescence-enhanced TRAP has revealed important information in evaluating the antioxidant status of human plasma. Plasma antioxidant defenses seem to respond to needs related to the basic metabolic rate and to the challenges caused by physiological or pathological stress. According to our studies, an important, possibly endogenous, antioxidant remains to be identified. However, the total antioxidative potential cannot be evaluated reliably by measuring only TRAP because of its dependence on uric acid. TRAP combined with its main components is likely to give more valid information in determining the total antioxidant status.

---

[31] R. J. Reiter, D. Melchiorri, E. Sewerynec, B. Poeggler, L. Barlow-Walden, J. Chuang *et al., J. Pineal Res.* **18,** 1 (1995).

[32] D. D. M. Wayner, G. W. Burton, K. U. Ingold, L. R. C. Barclay, and S. J. Locke, *Biochim. Biophys. Acta* **924,** 408 (1987).

Acknowledgments

This work was supported by Medical Research Fund of Tampere University Hospital, Y. Jahnsson Foundation, International Graduate School in Neuroscience at University of Tampere, Finnish Diabetes Research Foundation, and TEKES. We thank Dr. David Sinclair for revision of the English and invaluable comments.

[2] Ferric Reducing/Antioxidant Power Assay: Direct Measure of Total Antioxidant Activity of Biological Fluids and Modified Version for Simultaneous Measurement of Total Antioxidant Power and Ascorbic Acid Concentration

By IRIS F. F. BENZIE and J. J. STRAIN

Introduction

The ferric reducing/antioxidant power (FRAP) assay[1,2] is a recently developed, direct test of "total antioxidant power." Other tests of total antioxidant power used to date are indirect methods[3-8] that measure the ability of antioxidants in the sample to inhibit the oxidative effects of reactive species purposefully generated in the reaction mixture. In inhibition assays, antioxidant action induces a lag phase; exhaustion of antioxidant power is denoted by a change in signal, such as rate of oxygen utilization, fluorescence, or chemiluminescence. Measurement of these signals requires specialized equipment, and such tests can be time-consuming, technically demanding, and may lack sensitivity.[9]

In contrast to other tests of total antioxidant power, the FRAP assay is simple, speedy, inexpensive, and robust. The FRAP assay uses antioxidants as reductants in a redox-linked colorimetric method, employing an

---

[1] I. F. F. Benzie and J. J. Strain, U.S. Patent Pending (1997).
[2] I. F. F. Benzie and J. J. Strain, *Anal. Biochem.* **239,** 70 (1996).
[3] D. D. M. Wayner, G. W. Burton, K. U. Ingold, L. R. C. Barclay, and S. J. Locke, *Biochim. Biophys. Acta* **924,** 408 (1987).
[4] T. P. Whitehead, G. H. G. Thorpe, and S. R. J. Maxwell, *Anal. Chim. Acta* **266,** 265 (1992).
[5] N. J. Miller and C. A. Rice-Evans, *Redox Rep.* **2,** 161 (1996).
[6] G. Cao, H. M. Alessio, and R. G. Cutler, *Free Radic. Biol. Med.* **14,** 303 (1993).
[7] E. Lissi, M. Salim-Hanna, C. Pascual, and M. D. Del Castillo, *Free Radic. Biol. Med.* **18,** 153 (1995).
[8] A. Ghiselli, M. Serafini, G. Maiani, E. Azzini, and A. Ferro-Luzzi, *Free Radic. Biol. Med.* **18,** 29 (1995).
[9] D. Schofield and J. M. Braganza, *Clin. Chem.* **42,** 1712 (1996).

easily reduced oxidant present in stoichiometric excess. Unlike the many indirect radical scavenging tests designed to measure total antioxidant power, the FRAP assay does not use a lag phase type of measurement. In the FRAP assay, sample pretreatment is not required, stoichiometric factors are constant, linearity is maintained over a wide range, reproducibility is excellent, and sensitivity is high.[2] The FRAP assay does not need highly specialized equipment or skills, or critical control of timing and reaction conditions. The FRAP assay can be performed using automated, semiautomated, and manual versions, and, in a modified version known as the ferric reducing/antioxidant power and ascorbic acid concentration (FRASC[10]) assay, supplies three indices of antioxidant status—the total reducing (antioxidant) power, the absolute concentration of ascorbic acid, and the relative contribution of ascorbic acid to the total antioxidant power of the sample—virtually simultaneously.

Concept of FRAP Assay

A biological antioxidant has been defined as "any substance that, when present at low concentrations compared to those of an oxidisable substrate, significantly delays or prevents oxidation of that substrate."[11] However, unless an antioxidant prevents the generation of an oxidizing species, for example, by metal chelation or enzyme-catalyzed removal of a potential oxidant, a redox reaction still generally occurs, even in the presence of an antioxidant. The difference is that the oxidizing species reacts with the antioxidant instead of the "substrate," i.e., the antioxidant reduces the oxidant. In simple terms then, electron-donating antioxidants can be described as reductants, and inactivation of oxidants by reductants can be described as redox reactions in which one reactive species is reduced while another is oxidized. In this context, therefore, "total antioxidant power" may be referred to analogously as total reducing power.

Principle of FRAP Assay

At low pH, reduction of a ferric tripyridyltriazine (Fe^{III}-TPTZ) complex to the ferrous form, which has an intense blue color, can be monitored by measuring the change in absorption at 593 nm. The reaction is nonspecific, in that any half-reaction that has a lower redox potential, under reaction conditions, than that of the ferric/ferrous half-reaction will drive the ferric (Fe^{III}) to ferrous (Fe^{II}) reaction. The change in absorbance, therefore, is

[10] I. F. F. Benzie and J. J. Strain *Redox Rep.* **3,** 233 (1997).
[11] B. Halliwell and J. M. C. Gutteridge, *Free Radic. Biol. Med.* **18,** 125 (1995).

directly related to the combined or "total" reducing power of the electron-donating antioxidants present in the reaction mixture.

Materials and Methods

Reagent Preparation

Mix 300 mM acetate buffer, pH 3.6 [3.1 g sodium acetate trihydrate (Riedel-de Haen, Germany), plus 16 ml glacial acetic acid (BDH Laboratory Supplies, England) made up to 1 liter with distilled water]; 10 mM TPTZ (2,4,6-tripyridyl-s-triazine, Fluka Chemicals, Switzerland) in 40 mM HCl (BDH); and 20 mM $FeCl_3 \cdot 6H_2O$ (BDH) in the ratio of 10:1:1 to give the working FRAP reagent. Prepare working reagent fresh as required.

The following antioxidants are used to evaluate the FRAP assay. Solid L-(+)-ascorbic acid extra pure crystals (Merck, Germany); uric acid, solid (BDH); albumin, solid (bovine serum albumin, fraction V, Sigma Chemical Co., St. Louis, MO); bilirubin calibrator solution (Sigma); and Trolox, the water-soluble analog of α-tocopherol (Aldrich Chemical Co., Milwaukee, WI), are used in aqueous solutions of known concentrations. DL-α-Tocopherol (Merck) is diluted in ethanol (Merck) to give required concentrations.

Standards and Controls

Aqueous solutions of known Fe(II) concentration ($FeSO_4 \cdot 7H_2O$; Riedel-de Haen) and/or freshly prepared aqueous solutions of a pure antioxidant, such as ascorbic acid (extra pure crystals, Sigma), are used for calibration of the FRAP assay. Reaction of Fe(II) represents a one electron exchange reaction and is taken as unity, i.e., the blank corrected signal given by 100 μM solution of Fe(II) is equivalent to a FRAP value of 100 μM. Typical Fe(II) standard concentrations used in our laboratories are in the range of 100–1000 μM. Ascorbic acid has a constant stoichiometric factor of 2.0 in the FRAP assay, i.e., direct reaction of Fe(II) gives a change in absorbance half that of an equivalent molar concentration of ascorbic acid (see later). An ascorbic acid standard of 1000 μM, therefore, is equivalent to 2000 μM of antioxidant power as FRAP.

Pooled, aged plasma stored at $-70°$ and thawed overnight at $4°$, or frozen, aliquoted commercially available QC serum can be used to monitor precision. In addition, freshly prepared solutions of pure antioxidants, such as uric acid, Trolox, and ascorbic acid, in known concentration in aged plasma, commercially available QC serum, or in aqueous solution can be used to monitor accuracy and precision. For ease of use and reliability, aqueous ascorbic acid solutions at 100, 250, 500, and 1000 μM (equivalent to 200, 500, 1000, and 2000 μM FRAP) prepared fresh daily and aged QC serum freshly spiked with ascorbic acid are recommended as quality control samples. These should be run in parallel with test samples to actively monitor the performance of the test and to ensure comparability with previous results.

Samples

The FRAP assay can be performed on a wide range of complex biological fluids, including plasma, serum, saliva, tears, urine, cerebrospinal fluid, exudates, transudates, and aqueous and ethanolic extracts of drugs, foods, and plants, as well as on simple and heterogeneous solutions of pure antioxidants. If, using the reaction conditions described, the FRAP value of a sample is >3000 μM, it is recommended that the sample be diluted in water or ethanol, as appropriate, and the test repeated, with the additional dilution factor allowed for during the final calculation of the FRAP value.

Procedure for Automated FRAP Assay

The FRAP assay can be performed using any type of automated analyzer that permits blank corrected readings at 593 nm to be taken at predetermined intervals after sample–reagent mixing. In our laboratories the Cobas Fara centrifugal analyzer is used, and the user-defined test program is presented in Table I. The 0- to 4-min reaction time window is used for data capture for plasma total antioxidant power. Absorbance change is translated into a FRAP value (in μM) by relating the $\Delta A_{593\ nm}$ of test sample to that of a standard solution of known FRAP value [e.g., 1000 μM Fe(II)] shown in Eq. (1):

$$\frac{0\text{- to 4-min }\Delta A_{593\ nm}\text{ test sample}}{0\text{- to 4-min }\Delta A_{593\ nm}\text{ standard}} \times \text{FRAP value of standard }(\mu M) \quad (1)$$

This additional step can be added easily to the analyzer test program if desired in order to give a direct printout of FRAP values in μM.

Procedure for Manual FRAP Assay

To perform the FRAP assay manually, the same working reagent, standards, controls, and test samples are used; reagent and sample volumes are simply increased pro rata to give a volume large enough for manual handling/transfer of reaction mixtures. For example, 3.0 ml of working FRAP reagent is mixed with 100 μl test sample or standard in a test tube; this is vortex mixed, and the absorbance at 593 nm is read against a reagent

TABLE I
Cobas Fara Test Program for Automated FRAP Assay[a]

Measurement mode	Abs
Reaction mode	R1-I-S-A
Reagent blank	reag/dil
Wavelength	593 nm
Temperature	37°
R1	300 μl
M1	1.0 sec
Sample volume	10 μl
Diluent name	H$_2$O
Volume	30 μl
Readings	
First	0.5 sec
Number	17
Interval	15 sec
Reaction direction	Increase
Number of steps	1
Calculation	End point
First	M1
Last	17 (i.e., 4 min) for FRAP
	5 (i.e., 1 min) for ascorbic acid

[a] Reproduced with permission from I. F. F. Benzie and J. J. Strain, *Redox Rep.* **3**, 233 (1997).

blank at a predetermined time after sample–reagent mixing. If the test is performed at 37°, the 0- to 4-min reaction time window is used for plasma; if performed at room temperature, a 0- to 6-min reaction time window is preferable as the reaction of uric acid is slightly slower at lower temperatures. The calculation of results is the same as for the automated method.

Results

Results are presented for pure solutions of ascorbic acid, α-tocopherol, and uric acid; for more complex aqueous mixtures of pure antioxidants; and for fresh, fasting plasma from apparently healthy adults. Preliminary data on FRAP values of selected beverages are also given.

FRAP Reaction Characteristics of Pure Antioxidants

Figure 1 shows the post sample–reagent mixing change in absorbance at 593 nm for equimolar solutions of different antioxidants compared to the monitored absorbance of working FRAP reagent only using the automated FRAP assay. Ascorbic acid and α-tocopherol react very quickly, with a

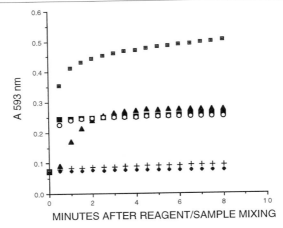

FIG. 1. FRAP reaction kinetics with individual antioxidants; rate of increase in absorbance at 593 nm for 100 μM solutions of bilirubin (⊞), ascorbic acid (■), uric acid (▲), α-tocopherol (○), albumin (+), and reagent alone (♦). Reproduced with permission from I. F. F. Benzie and J. J. Strain, *Anal. Biochem.* **239**, 70 (1996).

plateau reached within a few seconds. The reaction of uric acid reaches an end point after 3 min, but the reaction with bilirubin and with albumin does not appear to have reached an end point at 6 min. The reaction of bilirubin is fast in the first few seconds, however, and the continued slow increase in absorbance at 593 nm is due to the albumin content of the bilirubin preparation used.[2]

Figure 2 shows the dose–response characteristics of individual antioxidants in the FRAP assay. Whereas different antioxidant "efficiencies" can be seen, the dose–response line of each individual antioxidant tested is linear, showing that antioxidant efficiency is not concentration dependent. The relative activity is 2.0, i.e., direct reaction of Fe(II) gave a change in absorbance half that of an equivalent molar concentration, for Trolox, α-tocopherol, ascorbic acid, and uric acid. After correcting for albumin content, the stoichiometric factor of bilirubin in the FRAP assay is estimated to be 4.0. The activity of albumin in the FRAP assay is very low.[2] This is an advantage of FRAP, as the contribution of and changes in nonprotein antioxidants can be measured without the muffling and possibly misleading effect of the dominating contribution of protein seen in other methods of total antioxidant power.[3–6]

When the FRAP assay is performed with plasma and with aqueous solutions of antioxidants but with no Fe(III) added to the working reagent, no color develops. This indicates that there is no detectable free Fe(II) in

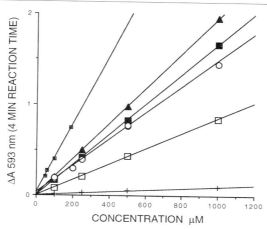

FIG. 2. Linearity of FRAP: dose–response lines for solutions of bilirubin preparation (⊞), uric acid (▲), ascorbic acid (■), α-tocopherol and Trolox (○), Fe(II) (□), and albumin (+). Reproduced with permission from I. F. F. Benzie and J. J. Strain, *Anal. Biochem.* **239,** 70 (1996).

plasma and that there is no detectable agent in normal plasma that reacts directly with TPTZ to form the blue ferrous-TPTZ chromogen. Monitoring the FRAP working reagent with no sample addition shows that no color develops (Fig. 1), indicating negligible spontaneous Fe(III) reduction in the absence of added reductants (antioxidants).

There is no apparent interaction between antioxidants in the FRAP assay. When known amounts of individual antioxidants were mixed and the FRAP value measured, there was good recovery of antioxidant power (91–112%), and good agreement was seen between anticipated and measured FRAP values ($r = 0.990$; $P < 0.001$) when known amounts of individual antioxidants were added to plasma and to water (Fig. 3). In addition, the FRAP dose–response relationship was the same, i.e., parallel lines were obtained, when uric acid solutions of different concentrations were tested with and without the presence of 100 μM ascorbic acid and when different concentrations of ascorbic acid were tested with and without 200 μM uric acid (Fig. 4).

Precision and Sensitivity of FRAP Assay

Precision is excellent: within-run coefficients of variation (CVs) are <1.0% at FRAP values of 100, 200, and 900 μM and the between-run CV is <3.0% at 960 μM. The limit of detection of the FRAP assay is <2 μM reducing/antioxidant power.

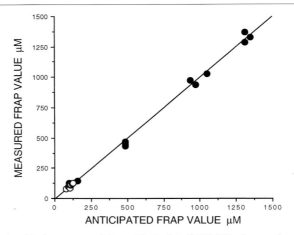

FIG. 3. Relationship between anticipated (calculated) FRAP values and measured FRAP values when known amounts of pure antioxidants were added to plasma (●) and water (○); $r = 0.99$, $P < 0.001$. Reproduced with permission from I. F. F. Benzie and J. J. Strain, *Anal. Biochem.* **239,** 70 (1996).

FRAP Values Obtained on Biological Fluids

The mean (SD) FRAP value of fresh, fasting plasma from 68 apparently healthy, consenting adults was 1035 (226) μM (range 638–1634 μM). There was a significant correlation between FRAP values and plasma uric acid

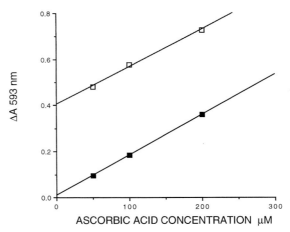

FIG. 4. Study of interaction between uric acid and ascorbic acid in the FRAP assay: the FRAP dose–response relationship of ascorbic acid in water (■) and in an aqueous 200 μM uric acid solution (□).

TABLE II
RELATIVE ANTIOXIDANT ACTIVITY OF INDIVIDUAL ANTIOXIDANTS AND THEIR ESTIMATED CONTRIBUTION TO FRAP VALUE OF FRESH FASTING PLASMA[a]

Plasma antioxidant	Relative activity "antioxidant efficiency" (measured range)	Expected fasting plasma concentration (μM)	Estimated contribution to total FRAP (%)
Ascorbic acid	2.0 (1.9–2.1)	30–100	15
α-Tocopherol	2.0 (1.7–2.1)	15–40	5
Uric acid	2.0 (2.0–2.4)	150–450	60
Bilirubin	4.0 (4.2–4.6)	<20	5
Protein	0.10 (0.1–0.15)	800–1,100	10
Others	—	—	5

[a] Modified and reproduced with permission from I. F. F. Benzie and J. J. Strain, *Anal. Biochem.* **239**, 70 (1996).

concentrations ($r = 0.913$, $P < 0.001$; $n = 68$), with uric acid contributing around 60% of the total antioxidant power of plasma. However, as stoichiometric factors are constant in the FRAP assay, it is a simple matter to subtract the contribution of uric acid to "total" antioxidant power; the measured concentration (in μM) of uric acid multiplied by its stoichiometric factor gives the μM antioxidant power of the sample due to uric acid. Subtracting this from the FRAP value of the sample gives the nonuric acid FRAP value,[10,12] which may offer a more sensitive index of antioxidant status in uric acid-rich fluids such as plasma. The relative contribution to total reducing/antioxidant power of other major antioxidants in plasma has also been calculated (Table II). Table III shows the representative range of FRAP values found in teas, wines, and orange juices, plant-based dietary agents that are rich in polyphenolic antioxidant compounds.

FRASC–FRAP and Ascorbic Acid Measurements in One Combined Assay

The measurement of ascorbic acid can be problematical due to its instability in plasma, and the specific measurement of ascorbic acid has generally required pretreatment of plasma to stabilize its ascorbic acid content, followed by high-performance liquid chromatography (HPLC) analysis.[13,14] The FRASC assay is a modification of the FRAP assay that

---

[12] I. F. F. Benzie and J. J. Strain, *Redox Rep.* **2**, 231 (1996).
[13] C. J. Bates, A. Bailey, H. van der Berg, F. van Schaik, C. Coudray, A. Faviet, R. Farré, A. Frigola, H. Heseker, G. Maiani, A. Ferro-Luzzi, K. Pietrzik, and D. I. Thurnham, *Intl. J. Vit. Nutr. Res.* **64**, 283 (1994).
[14] L. A. Pachla, D. L. Reynolds, and P. T. Kissinger, *J. Assoc. Anal. Chem.* **68**, 1 (1985).

TABLE III
ANTIOXIDANT POWER OF TEAS AND WINES[a]

Sample	Amount	Antioxidant power (FRAP value) (μmol)
Fermented (black) teas, 1% infusion	200 ml	500–900
Semifermented teas, 1% infusion	200 ml	1000–1400
Nonfermented (green) teas, 1% infusion	200 ml	1600–2200
Red wines	150 ml	2900–3700
White wines	150 ml	380–520
Fresh orange juices (prepacked)	200 ml	500–600
Pure ascorbic acid (vitamin C)	1 g	11,364

[a] Range of values found in typical servings of teas and wines compared with orange juice and pure ascorbic acid.

permits the virtually simultaneous, specific measurement of ascorbic acid concentration and total reducing/antioxidant power (as FRAP) in one simple, rapid, automated test.[1,10] In FRASC, ascorbic acid in one of a pair of sample aliquots is selectively destroyed[15] by ascorbate oxidase [EC 1.10.3.3 (Sigma)]. Reagents are otherwise the same as in the FRAP assay. Reduction of the FRAP (FRASC) working reagent by ascorbic acid is complete within a few seconds of sample–reagent mixing. The 0- to 1-min post sample–reagent mixing absorbance change (at 593 nm) of a sample to which ascorbate oxidase (40 μl of a 4 U/ml solution added to a 100-μl sample) was added is subtracted from the absorbance change of a matching aliquot of sample to which water (40 μl added to a 100-μl sample), rather than ascorbate oxidase, was added; the difference is due specifically to ascorbic acid.[10,15] In FRASC, the 0- to 4-min absorbance change of the aliquot diluted in water is due to the combined reductive activity of all the reacting antioxidants present in the sample, i.e., the "total antioxidant capacity," or ferric reducing/antioxidant power (FRAP) value.[2] The 0- to 4-min and paired 0- to 1-min absorbance changes are translated into μM of FRAP and ascorbic acid, respectively, by comparison with those of standard solutions of Fe(II) or ascorbic acid of the appropriate molar concentration. It must be remembered that if Fe(II) standards are used for the calculation of ascorbic acid concentration, ascorbic acid has a stoichiometric factor of 2.0 in the FRAP assay; i.e. 1000 μM Fe(II) is equivalent to 1000 μM of FRAP but to only 500 μM of ascorbic acid. Similarly, if ascorbic acid standards are used for the calculation of FRAP values, these

---

[15] I. F. F. Benzie, *Clin. Biochem.* **29,** 111 (1996).

values are double the ascorbic acid concentration; i.e., 1000 μM ascorbic acid is equivalent to 2000 μM FRAP.

By monitoring the 0- to 4-min absorbance change of paired aliquots of water- and ascorbate oxidase-treated samples run in parallel on a Cobas Fara centrifugal analyzer and using the automated FRAP assay program detailed in Table I, all necessary data are gathered to obtain FRAP values and ascorbic acid concentrations of up to 13 pairs of test samples in one run. The concept of FRASC is represented in Fig. 5.

Calculation of Results

Using water-diluted samples,

$$\text{FRAP} (\mu M) \text{ value} = \frac{\text{0- to 4-min } \Delta A_{593 \text{ nm}} \text{ test sample}}{\text{0- to 4-min } \Delta A_{593 \text{ nm}} \text{ standard}} \times [\text{FRAP}]_{\text{std}} (\mu M) \quad (2)$$

Using paired water ($-$ao)- and ascorbate oxidase-diluted ($+$ao) samples, the ascorbic acid concentration is calculated as follows:

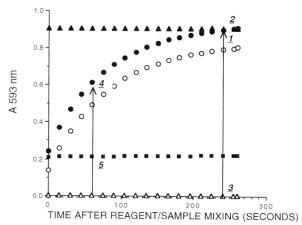

FIG. 5. Measuring concept of FRASC showing the absorbance change due to Fe(III)-TPTZ reduction by antioxidants in the sample. Calculation of the FRAP value is done by taking the 0- to 4-min ΔA for test sample (●, 1) and relating it to the 0- to 4-min ΔA for the Fe(II) standard (▲, 2), with a reagent blank correction (△, 3) for both. Calculation of ascorbic acid results is by subtracting the 0- to 1-min ΔA reading of the ascorbate oxidase-treated test sample (○) from the matching water-treated sample (●, 4); this signal is then related to that given by a standard solution of Fe(II) (▲) (or ascorbic acid, ■, 5) of appropriate concentration. Reproduced with permission from I. F. F. Benzie and J. J. Strain, *Redox Rep.* **3**, 233 (1997).

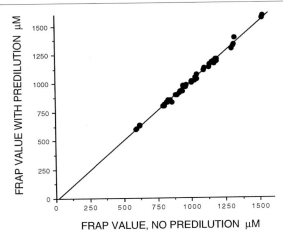

FIG. 6. Relationship between measured FRAP values on EDTA plasma with no predilution step and after predilution of 100 μl plasma with 40 μl water. Excellent agreement was seen ($r = 0.998$; $y = 1.03x - 23$), showing no net loss or gain of reductive activity with dilution. Reproduced with permission from I. F. F. Benzie and J. J. Strain, *Redox Rep.* **3**, 233 (1997).

$$\text{0- to 1-min ascorbic acid related } \Delta A_{593 \text{ nm}} = (\text{0- to 1-min } \Delta A_{593 \text{ nm}} \text{ sample } -\text{ao}) \\ - (\text{0- to 1-min } \Delta A_{593 \text{ nm}} \text{ sample } +\text{ao}) \quad (3)$$

$$\text{Ascorbic acid concentration } (\mu M) = \\ \frac{\text{0- to 1-min ascorbic acid related } \Delta A_{593 \text{ nm}} \text{ of test sample}}{\text{0- to 1-min ascorbic acid related } \Delta A_{593 \text{ nm}} \text{ of standard}} \times [\text{ascorbic acid}]_{\text{std}} \quad (4)$$

Data presented in Fig. 6 show FRAP values of 25 EDTA plasma samples measured with and without predilution in water (100-μl sample plus 40 μl of water). All samples were measured twice, in separate runs. FRAP values (μM) were obtained with reference to a Fe(II) standard solution run in

TABLE IV
FRAP VALUES AND ASCORBIC ACID CONCENTRATIONS (MEAN; MEDIAN; SD), USING FRASC, OF FRESH FASTING EDTA PLASMA FROM HEALTHY SUBJECTS[a]

Parameter	All ($n = 130$)	Men ($n = 66$)	Women ($n = 64$)
Age (years)	43; 43; 16.4	42; 42; 16.3	43; 44; 16.6
FRAP (μM)	1018; 1004; 198	1086; 1077; 189	948; 927; 183
Ascorbic acid (μM)	51; 48; 17.9	49; 48; 13.8	52; 50; 21.3

[a] Reproduced with permission from I. F. F. Benzie and J. J. Strain, *Redox Rep.* **3**, 233 (1997).

parallel and treated identically to the samples in each case, i.e., prediluted or nonprediluted in water as appropriate. Results obtained by the original FRAP assay (no predilution) and by FRASC show very good agreement ($r = 0.998$; $p < 0.0001$); mean (SD) FRAP values of the neat and prediluted samples were 998 (217) and 1010 (225) μM.

Precision of FRASC is good: within- and between-run CVs are, respectively, <1 and <3% at 900 and 1800 μM for FRAP values; for ascorbic acid, within- and between-run CVs are <5% at 25, 50, 100, and 440 μM.

Ferric reducing/antioxidant power values and ascorbic acid concentrations were measured,[10] using FRASC, in fresh EDTA plasma samples from 130 apparently healthy, fasting adults (66 men, 64 women) aged 21–74 years from whom informed consent had been obtained. Results are shown in Table IV.

Conclusions and Clinical Applications

The FRAP assay, in both its original[2] and modified, as FRASC,[10] versions, is robust, sensitive, simple, and speedy and will facilitate experimental and clinical studies investigating the relationship among antioxidant status, dietary habits, and risk of disease.[16–18] Measurement of the total antioxidant power of fresh biological fluids, such as blood plasma, can be measured directly, the antioxidant content of various dietary agents can be measured objectively and reproducibly, and their potential for improving the antioxidant status of the body investigated and compared. The FRAP assay is also sensitive and analytically precise enough to be used in assessing the bioavailability of antioxidants in dietary agents, to help monitor longitudinal changes in antioxidant status associated with an increased intake of dietary antioxidants, and to investigate the effects of disease on antioxidant status.

Acknowledgment

The authors thank the Hong Kong Polytechnic University for funding this work.

---

[16] B. Frei B. *Crit. Rev. Food Sci. Nutr.* **35,** 83 (1995).
[17] K. F. Gey, *J. Nutr. Biochem.* **6,** 206 (1995).
[18] B. Halliwell B. *Free Radic. Res.* **25,** 57 (1996).

[3] Automated Electron Spin Resonance Free Radical Detector Assays for Antioxidant Activity in Natural Extracts

By Yasuko Noda, Masahiro Kohno, Akitane Mori, and Lester Packer

Introduction

There is now increasing interest in the antioxidant activity of phytochemicals present in the diet, in health food supplements (neutraceuticals), and in topical preparations for protection of the skin (cosmaceuticals) from environmental exposure.[1,2]

i. A simple and rapid estimation of hydroxyl and superoxide anion radical scavenging activities by aqueous extract from natural sources can be made using a new computerized JEOL ESR system.

ii. The relative free radical scavenging activities of various samples are able to be evaluated based on normalizing electron spin resonance (ESR) signals relative to the standard activity of L-ascorbic acid 2-[3,4-dihydro-2,5,7,8-tetramethyl-2-(4,8,12-trimethyl-2H-1-benzopyran-6-yl hydrogen phosphate] potassium salt (EPC-K_1) as a scavenger of hydroxyl radical and copper-zinc superoxide dismutase (SOD) as a superoxide anion radical scavenger.

iii. Treatment of extracts with ascorbate oxidase reveals that in some cases the presence of vitamin C partially accounts for hydroxyl and superoxide anion radical scavenging activities. Treatments with centrifuge-type filters [Ultrafree-MC filters: 10,000 nominal molecular weight limit (NMWL) regenerated cellulose membrane or 100,000 NMWL polysulfone membrane] divide antioxidant activities in a sample to low molecular weight materials and high molecular materials (such as enzymes), and heat treatment also shows the contribution of heat-inactivated components to antioxidant activities in a sample.

Superoxide Anion Radical Scavenging Activity

A standard assay is performed using 130 μl of measuring sample in a quartz flat ESR cell. In the sample, 50 μl of 2 mM hypoxanthine (HPX),

---

[1] L. Packer, "Proceedings of the International Symposium of Natural Antioxidant, Molecular Mechanisms and Health Effects," AOCS Press, Champaign, IL, 1996.

[2] Y. Noda, K. Anzai, A. Mori, M. Kohno, M. Shimmei, and L. Packer, *Biochem. Mol. Biol. Int.* **42,** 35 (1997).

50 μl of 0.4 units/ml xanthine oxidase (XOD), 30 μl of dimethyl sulfoxide (DMSO), 20 μl of 4.5 M 5,5-dimethyl-1-pyrroline N-oxide (DMPO), and 50 μl of sample are mixed. When 1 unit of copper-zinc SOD, i.e., 50 μl of SOD solution containing 20 units/ml, is added to the reaction system, the ESR signal of DMPO-OOH is scavenged about 50%. If the concentration of DMPO is adjusted to 0.35 mM, the superoxide anion scavenging activity estimated by the ESR spin trapping method[3,4] coincides with the values of superoxide anion radical scavenging activity, measured by the cytochrome c method.[5]

Hydroxyl Radical Scavenging Activity

EPC-K_1 is used for standard material of hydroxyl radical scavenging activity, as it reacts selectively with hydroxyl radical.[6] Therefore, the calibration curve is obtained, and activity is given as the concentration of the standard EPC-K_1.

Sample Preparation

Samples from natural extracts are prepared by dissolving in 0.1 M potassium phosphate buffer (pH 7.4), and water-soluble components are examined for scavenging activities. *Ginkgo biloba* extract EGb 761 (I.P.S.E.N., Paris, France), pine bark extract (Pycnogenol, M. W. International, Inc., Hillside, NJ), and green tea extract (the highest grade, Matsucha, Fukujuen Co., Kyoto, Japan) are chosen as examples.

Chemicals

HPX is obtained from Sigma Chemical Co. (St. Louis, MO), XOD (1 U/mg, from cow milk) is from Boehringer Mannheim Corp. (Indianapolis, IN), and DMPO is from Labotec Co. (Tokyo, Japan). EPC-K_1 is from Labotec Co. All other chemicals are of the highest grade.

Electron Spin Resonance Measurements

Electron spin resonance spectra are recorded with a JES-FR30 ESR spectrometer (JEOL Co. Ltd., Tokyo, Japan), which is controlled by a

---

[3] K. Mitsuta, Y. Mizuta, M. Kohno, M. Hiramatsu, and A. Mori, *Bull. Chem. Soc. Jpn.* **63**, 187 (1990).
[4] M. Kohno, Y. Mizuta, M. Kusai, T. Masumizu, and K. Makino, *Bull. Chem. Soc. Jpn.* **67**, 1085 (1994).
[5] J. M. McCord and I. Fridovich, *J. Biol. Chem.* **244**, 6049 (1969).
[6] A. Mori, R. Edamatsu, M. Kohno, and S. Ohmori, *Neurosciences* **15**, 371 (1989).

computer system. Instrument settings are as follows: magnetic field, 335.5 ± 5 mT; power, 4.0 mW; modulation frequency, 9.41 GHz; modulation width, 0.1 × 0.63 or 1 × 0.1 mT; sweep width, 5.0 mT; sweep time, 2 min; response time, 0.1 sec; and amplitude, 1 × 200. Electron spin resonance spectra are measured at 23°. The intensity of DMPO-OH and DMPO-OOH spin adducts that are generated from the reaction between DMPO and hydroxyl or superoxide anion radical is expressed as a ratio of the signal intensity at the lowest magnetic field to that of Mn^{2+} in MnO used as an internal standard.

Procedures

Hydroxyl Radical Scavenging Activity

The hydroxyl radicals are generated from the Fenton reaction.[7] A sample solution and other reagents are mixed in a glass tube, and ESR spectra are measured as follows.

1. Place the sample solution in 0.1 M potassium phosphate buffer (pH 7.4) (50 μl) in a glass tube (capacity: 10 ml, 16 × 75 mm).
2. Add 0.18 M DMPO to the buffer (pH 7.4) (50 μl).
3. Add 2 mM H_2O_2 to the buffer (pH 7.4) (50 μl).
4. Add 0.2 mM $FeSO_4$ in distilled water (50 μl).
5. Mix for exactly 10 sec with a vortex mixer.
6. Transfer the solution into a quartz flat cell (capacity: 200 μl).
7. Measure ESR spectra exactly 30 sec after the addition of $FeSO_4$.

EPC-K_1 Standard. There is a strong correlation between EPC-K_1 concentration and ($I_0/I - 1$) (correlation coefficient $\gamma = 0.997$), where I_0 = RP for the DMPO-OH ESR signal in the $FeSO_4/H_2O_2$ system without EPC-K_1, and I = RP for the DMPO-OH ESR signal in the $FeSO_4/H_2O_2$ system with various concentrations of EPC-K_1 added (Fig. 1). In a comparison of the initial concentration of the sample solution and the calibration curve, the hydroxyl radical scavenging activity can be expressed as EPC-K_1 μmol/mg of sample.

Superoxide Anion Radical Scavenging Activity

The superoxide anion radicals are generated by a hypoxanthine–xanthine oxidase system.[7] The reaction is initiated by the addition of XOD. All solutions except DMSO are dissolved in 0.1 M potassium phosphate

[7] J. Lui, R. Edamatsu, H. Kabuto, and A. Mori, *Free Radic. Biol. Med.* **9,** 451 (1990).

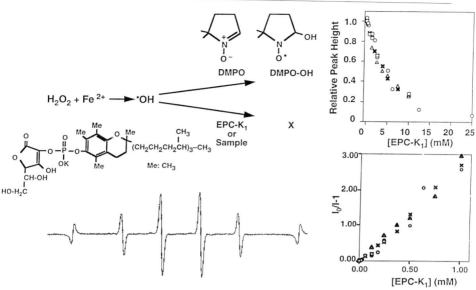

Fig. 1. Principle of measurement of scavenging activity of hydroxyl radicals. The hydroxyl radicals are generated by the Fenton reaction. ESR spectra of the DMPO-OH spin adduct are recorded. EPC-K$_1$ is used as a standard. The hydroxyl radical scavenging activity is calculated from the calibration curve of EPC-K$_1$, and results are expressed in units of EPC-K$_1$ μmol per milligram. Data points represent three separate experiments. From Y. Noda, K. Anzai, M. Kohno, M. Shimmei, and L. Packer, *Biochem. Mol. Biol. Int.* **42**, 35 (1997).

buffer (pH 7.4). A sample solution and other reagents are mixed in a glass tube, and ESR spectra are measured as follows.

1. Put 4 mM HPX dissolved in 0.1 M potassium phosphate buffer (pH 7.4) (50 μl) in a glass tube (capacity: 10 ml, 16 × 75 mm).
2. Add DMSO (30 μl).
3. Add sample solution to the buffer (pH 7.4) (50 μl).
4. Add 4.5 M DMPO to the buffer (pH 7.4) (20 μl).
5. Add XOD (0.4 units/ml) to the buffer (pH 7.4) (50 μl).
6. Mix for exactly 10 sec with a vortex mixer.
7. Transfer the mixture into a quartz flat cell (capacity: 200 μl).
8. Measure ESR spectra exactly 30 sec after the addition of XOD.

Superoxide anion radical scavenging activity has the same meaning of that of SOD-like activity or SOD mimic activity. Using the ESR spin trapping method, the intensity (I) of the spin adduct of DMPO-OOH is

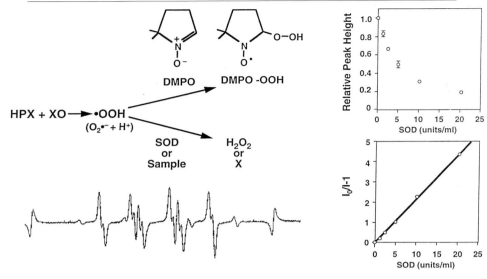

FIG. 2. Principle of measurement of scavenging activity of superoxide anion radicals. The generation of superoxide anion radicals are performed by reaction of the hypoxanthine (HPX)–xanthine oxidase (XOD) system. ESR spectra of the DMPO-OOH spin adduct are analyzed. SOD-like activity is expressed as SOD equivalent units per milligram. From Y. Noda, K. Anzai, A. Mori, M. Kohno, M. Shimmei, and L. Packer, *Biochem. Mol. Biol. Int.* **42**, 35 (1997).

plotted for the SOD concentration. The standard curve is shown in Fig. 2. With this calibration curve, the SOD-like activity of the scavenger can be measured, which is shown as units per milliliter or units per milligram converted to SOD concentration.

Participation of Ascorbate, Low and High Molecular Weight Molecules in Antioxidant Activities in Sample

To define the possible participation of ascorbate in antioxidant activity, molecular weights of antioxidants, and denaturation by heat treatment of samples, the following treatments are performed.

Ascorbate Oxidase. Ascorbate oxidase (Sigma Chemical Co. from *Cucurbita* species lyophilized powder) dissolved in 0.1 M potassium phosphate (pH 7.4) is dialyzed in a dialysis tube (Spectra/Por 3, molecular weight cutoff 3500, Spectrum Medical Industries, Inc., Gardena, CA) at 4° to eliminate artificially added sucrose as a stabilizer. Twenty microliters of ascorbate oxidase (200 units/ml of 0.1 M phosphate buffer, pH 7.4) is added to an aliquot (200 μl) of sample solution in 0.1 M phosphate buffer (pH 7.4) and incubated at 23° for 60 min.

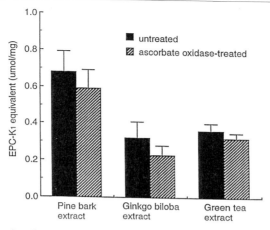

Fig. 3. Hydroxyl radical scavenging activity of pine bark extract (Pycnogenol), *Ginkgo biloba* extract, and green tea extract. Samples are untreated or ascorbate oxidase treated. Data are expressed as means ± SEM ($n = 6$ for untreated samples and $n = 5$ for ascorbate oxidase-treated samples). *$p < 0.05$, compared to untreated samples.

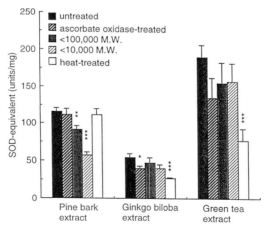

Fig. 4. Superoxide anion radical scavenging activity of pine bark extract (Pycnogenol), *Ginkgo biloba* extract and green tea extract. Samples are untreated, ascorbate oxidase treated, filtered (<100,000 or <10,000 MW), or heat treated (at 100° for 10 min). Data are expressed as means ± SEM [$n = 6$ for all untreated samples, except green tea extract ($n = 4$), $n = 5$ for all treated samples, except green tea extract, $n = 2$ for ascorbate oxidase-treated, filtered (<100,000 and <10,000 MW) and $n = 3$ for heat-treated sample]. *, different from untreated, $p < 0.05$; **, different from untreated, $p < 0.01$; and ***, different from untreated, $p < 0.001$. From Y. Noda, K. Anzai, A. Mori, M. Kohno, M. Shimmei, and L. Packer, *Biochem. Mol. Biol. Int.* **42,** 35 (1997).

Molecular Weight Filter. An aliquot (400 µl) of sample solution is passed through a centrifuge-type filter (capacity: 400 µl) (Ultrafree-MC filters: 10,000 NMWL regenerated cellulose membrane or 100,000 NMWL polysulfone membrane, Millipore, Bedford, MA) by centrifugation at 5000 g, at 4° for 20–40 min.

Heat Treatment. An aliquot (600 µl) of sample solution is put into a microcentrifuge tube (capacity: 1.5 ml), and the sample is kept at 100° in boiling water for 10 min to clarify the possible participation of protein in antioxidant activity in the sample.

Statistics

Data are expressed as means ± SEM. Statistical analysis is performed using the Student's *t* test.

Comments

With this method, the relative free radical scavenging activities of various samples are normalized from ESR signals relative to the standard activity of EPC-K_1 as a scavenger of hydroxyl radical or copper-zinc SOD as a superoxide anion radical scavenger. Ascorbate oxidase treatment provides a simple and convenient way to discriminate the contribution of ascorbate activity to antioxidant activity.

To make the estimation of radical scavenging activity more reliable and quantitative, various other strategies are considered: with heat treatment (100° for 10 min), for example, much of the activity is heat labile for superoxide anion radical scavenging activity.

Solubilization of the components of natural extracts is another factor that must be taken into consideration when evaluating results. Radical scavenging activities reported here only refer to the solubilized components or of those substances that have passed through Millipore filters, i.e., molecular weights lower than 100,000 or 10,000. Hence, most of the contribution toward radical scavenging assays from extracts is by low molecular weight materials. In some cases, preparations are treated to remove insoluble components such as green tea extract.

Examples of analysis of natural source antioxidants are shown in Figs. 3 and 4.

[4] Measurement of Oxidizability of Blood Plasma

By ANATOL KONTUSH and ULRIKE BEISIEGEL

Introduction

The hypothesis that the oxidation of low density lipoproteins (LDL) plays a central role in atherogenesis has received much attention over the past decade.[1,2] More recently, it has also been found that other plasma lipoproteins, such as high density lipoprotein (HDL), very low density lipoprotein (VLDL), and chylomicrons, can be oxidized *in vitro*[3-6] and are able to carry substantial amounts of lipid peroxidation products in human plasma *in vivo*.[5,7] The oxidation hypothesis of atherosclerosis implies that LDL oxidation occurs in the intima of the arterial wall.[1] It has been currently established that the arterial intima contains, in addition to LDL, other plasma lipoproteins, such as HDL and VLDL, as well as postprandial chylomicron remnants.[8,9] These findings suggest that the oxidation of all major plasma lipoproteins, not just LDL oxidation alone, should be considered when discussing the oxidative hypothesis of atherosclerosis.

The role of the lipoprotein oxidation in atherogenesis implies that lipoprotein oxidizability by physiological oxidants is of a great importance for this process. The oxidizability of isolated lipoproteins is determined by their content of components directly involved in the oxidation process, namely oxidizable substrates and antioxidant compounds.[10,11] However, the interstitial fluid of the arterial wall, where lipoprotein oxidation is thought to take place, also contains high amounts of water-soluble antioxidants, such

---

[1] D. Steinberg, S. Parthasarathy, T. E. Carew, J. C. Khoo, and J. L. Witztum, *N. Engl. J. Med.* **320,** 915 (1989).
[2] B. Halliwell, *Am. J. Clin. Nutr.* **61,** 670S (1995).
[3] S. Parthasarathy, J. Barnett, and L. G. Fong, *Biochim. Biophys. Acta* **1044,** 275 (1990).
[4] D. Mohr and R. Stocker, *Arterioscler. Thromb.* **14,** 1186 (1994).
[5] I. Staprans, J. H. Rapp, X.-M. Pan, K. Y. Kim, and K. R. Feingold, *Arterioscler. Thromb.* **14,** 1900 (1994).
[6] A. Kontush, A. Kohlschütter, and U. Beisiegel, *Biochem. Mol. Biol. Int.* **37,** 707 (1995).
[7] V. W. Bowry, K. K. Stanley, and R. Stocker, *Proc. Natl. Acad. Sci. U.S.A.* **89,** 10316 (1992).
[8] B. G. Nordestgaard and L. B. Nielsen, *Curr. Opin. Lipidol.* **5,** 252 (1994).
[9] B. G. Nordestgaard, R. Wootton, and B. Lewis, *Arterioscler. Thromb.* **15,** 534 (1995).
[10] B. Frei and J. M. Gaziano, *J. Lipid Res.* **34,** 2135 (1993).
[11] A. Kontush, C. Hübner, B. Finckh, A. Kohlschütter, and U. Beisiegel, *Free Radic. Res.* **24,** 135 (1996).

as ascorbate, urate, and albumin,[12,13] all of which are capable of inhibiting lipoprotein oxidation *in vitro*.[14–16] It is therefore conceivable that the lipoprotein susceptibility to oxidation *in vivo* is related not only to their intrinsic properties, but also to the antioxidant properties of the surrounding medium, i.e., those of the interstitial fluid. It has been shown that the antioxidant composition of human interstitial fluid is similar to that of human plasma.[12] These data indicate that the lipoprotein oxidation induced *in vitro* in blood plasma can be expected to represent a relevant model of the lipoprotein oxidation in the arterial wall. The *in vitro* oxidation of isolated lipoproteins, e.g., LDL,[2,17,18] is, in contrast, a rather unphysiological method.

Today, no standard method exists that allows characterization of the lipoprotein oxidizability directly in blood plasma. It is also unclear whether the *in vitro* oxidizability of lipoproteins measured in plasma reflects the oxidizability of isolated lipoproteins and whether it is of physiological and clinical importance. This chapter describes a method that allows us to measure the oxidizability of lipoproteins in diluted blood plasma and presents data supporting its physiological and clinical relevance. Because lipoprotein oxidation occurring in plasma represents a measure of the oxidation of plasma itself, the lipoprotein oxidizability measured in such a system will be further referred to as the oxidizability of blood plasma.

Methods to Characterize Plasma Oxidizability

Generally, the oxidizability of blood plasma is a measure of its susceptibility to oxidation that can be described by changes in various chemical or biological parameters during oxidation. Plasma oxidation *in vitro* is known to be seen as an accumulation of water-soluble and lipid-associated oxidation products and the consumption of hydrophilic and lipophilic antioxidants and fatty acids, as well as modification of proteins and enzymes.[19] All of these processes can be used theoretically to characterize plasma oxidizability; however, of all of them, the oxidation of plasma lipoproteins is thought to be of the most critical importance for atherogenesis. Therefore,

---

[12] A. J. Dabbagh and B. Frei, *J. Clin. Invest.* **96**, 1958 (1995).
[13] C. Suarna, R. T. Dean, J. May, and R. Stocker, *Arterioscler. Thromb. Vasc. Biol.* **15**, 1616 (1995).
[14] B. Halliwell, *Biochem. Pharmacol.* **37**, 569 (1988).
[15] B. Frei, L. England, and B. N. Ames, *Proc. Natl. Acad. Sci. U.S.A.* **86**, 6377 (1989).
[16] Y.-S. Ma, W. L. Stone, and I. O. LeClair, *Proc. Soc. Exp. Biol. Med.* **206**, 53 (1994).
[17] H. Puhl, G. Waeg, and H. Esterbauer, *Methods Enzymol.* **233**, 425 (1994).
[18] R. Stocker, *Curr. Opin. Lipidol.* **5**, 422 (1994).
[19] H. Esterbauer, J. Gebicki, H. Puhl, and G. Jürgens, *Free Radic. Biol. Med.* **13**, 341 (1992).

a clinically relevant method to measure plasma oxidizability should describe the oxidizability of lipoproteins.

This assumption implies that the kind of detection used must be adequate to measure lipoprotein oxidation. Time-dependent measurements of lipoprotein-derived lipophilic (lipid hydroperoxides, oxycholesterols) and hydrophilic (short-chain aldehydes) oxidation products, lipophilic (i.e., lipoprotein-associated) antioxidants, fatty acids, and oxidized apolipoproteins can all be utilized for this purpose. However, all of these methods are very time-consuming due to quite sophisticated sample processing. In addition, they include time-dependent collection of a number of aliquots out of an oxidizing sample. A clinical method should allow continuous registration of oxidation and also be simple enough to be used routinely in the laboratory.

Continuous photometrical detection of lipid hydroperoxides having a conjugated diene structure represents an approach that, being simple enough for routine use, allows us to adequately characterize lipoprotein oxidation. This method has been used successfully for isolated LDL[20] and HDL.[21] Other possibilities to continuously registrate lipoprotein oxidation include measurements of oxygen uptake and fluorescence at 360/430 nm,[17] both of which are not as suitable for clinical applications as the conjugated diene method. Simplicity and accuracy of this method, which utilizes an absorbance increase at 234 nm as a measure of accumulation of lipid hydroperoxides in isolated lipoproteins, were the main reasons why it became probably the most widely used to measure lipoprotein oxidizability.

The choice of conditions used to oxidize plasma samples is another important feature of the assay. The most common LDL oxidation assay employs high amount of Cu(II) to catalyze oxidation.[20] The conjugated diene method has also been reported to be applicable to plasma samples under strong oxidative conditions of Cu(II)-induced oxidation.[21,22] However, the physiological importance of Cu(II) as an oxidant for plasma lipoproteins still remains controversial.[18] In addition, strong oxidative conditions are often considered to be too unphysiological to adequately model lipoprotein oxidation *in vivo*. For these reasons, other oxidants and milder oxidative conditions were tested for their ability to oxidize LDL and plasma.[18,23] As it is still not clear which oxidant(s) induces lipoprotein

---

[20] H. Esterbauer, G. Striegl, H. Puhl, and M. Rotheneder, *Free Radic. Res. Commun.* **6**, 67 (1989).

[21] E. Schnitzer, I. Pinchuk, M. Fainaru, Z. Schafer, and D. Lichtenberg, *Biochem. Biophys. Res. Commun.* **216**, 854 (1995).

[22] J. Regnström, K. Ström, P. Moldeus, and J. Nilsson, *Free Radic. Res. Commun.* **19**, 267 (1993).

[23] A. Kontush, B. Finckh, B. Karten, A. Kohlschütter, and U. Beisiegel, *J. Lipid Res.* **37**, 1436 (1996).

oxidation *in vivo* and what the oxidative conditions are in the arterial wall, the choice of oxidants and their concentrations for an *in vitro* assay of lipoprotein oxidizability should be broad enough to cover the physiologically relevant range.

The method described here is a modification of the common LDL oxidation assay[20] that characterizes the oxidizability of lipoproteins in diluted blood plasma, i.e., the plasma oxidizability, as an absorbance increase of plasma samples at 234 nm measured under various oxidative conditions in the presence of different oxidants.

Measurement of Plasma Oxidizability by Conjugated Diene Method

Standard Procedure

To obtain plasma, blood is placed into tubes containing heparin, ethylenediaminetetraacetic acid (EDTA), citrate (15 IU heparin, 1.6 mg EDTA, or 10.6 μmol citrate/ml blood), or any anticoagulant (Sarstedt, Nümbrecht, Germany) after an overnight fast and immediately centrifuged at 4° for 10 min.

Plasma or serum samples (20 μl) are diluted with phosphate-buffered saline (PBS), pH 7.4, containing 0.16 M NaCl (2950 μl) preliminarily equilibrated at 37° for at least 15 min. Phosphate-buffered saline is made up in double-distilled deionized water treated with Chelex 100 ion-exchange resin (Bio-Rad, München, Germany) to remove transition metal ions. A high dilution (150-fold) of plasma is necessary to provide absorbance low enough to be measured reliably and is similar to that used by Regnström *et al.*[22] and Schnitzer *et al.*[21] to characterize the time course of oxidation in human plasma and serum.

To oxidize plasma or serum, an oxidant is added as a 30-μl aqueous solution. Cu(II) (as a sulfate), 2,2'-azobis(2-amidinopropane) hydrochloride (AAPH), or soybean lipoxygenase is used as an oxidant. Blank samples are supplemented with 30 μl PBS. To correct for the absorbance change at 234 nm produced by AAPH or enzymes, corresponding blanks containing only these oxidants in PBS are measured and, together with blanks containing plasma or serum but no oxidant (see earlier), are subtracted from the samples under investigation. Samples are transferred into 3-ml quartz spectrophotometrical cuvettes and incubated at 37° in a UV2 dual-beam spectrophotometer equipped with a thermostated auto cell holder for eight cuvettes (ATI Unicam, Cambridge, Great Britain). Sample absorbance at 234 nm is measured every 5 min over a period of up to 20 hr. Kinetic data are analyzed using a Vision Software supplied together with the spectrophotometer.

Typical Oxidation Kinetics

When plasma or serum samples are incubated in the presence of oxidants under the assay conditions described, their absorbance in the UV region between 220 and 280 nm is found to increase consistently. The absorption increase has a maximum at about 234 nm, in accordance with previously reported data.[22]

The absorbance increase at 234 nm is measured with all the oxidants and all kinds of plasma and serum used in our study. However, its exact kinetics depend critically on the kind and concentration of the oxidant employed as well as on the anticoagulant used (Figs. 1 and 2). The absorbance increase at 234 nm of oxidizing heparin plasma (Fig. 1), citrate plasma, and serum (Fig. 2) in the presence of high concentration of Cu(II) (50 μM) is characterized by three consecutive phases comparable to the lag, propagation, and decomposition phases of LDL oxidation.[17] This is in accordance with data reported previously.[21,22] The oxidizability of the samples under these conditions is therefore characterized using the same parameters that are typically used to characterize the oxidizability of plasma LDL, namely lag phase and propagation phase duration and maximal oxidation rate (maximal oxidation rate measured within the propagation phase)[17] (Fig. 1).

When heparin plasma, citrate plasma, or serum is incubated in the presence of AAPH (330–1650 μM), lipoxygenase (25–250 U/ml), or low concentrations of Cu(II) (10 μM), no lag phase of the absorbance increase at 234 nm is found in most incubations (Figs. 1 and 2). Samples oxidized under such conditions normally exhibit a monotonic increase in their absorbance at 234 nm. Oxidation curves obtained under these conditions are therefore described using a single parameter, i.e., the mean oxidation rate within the initial monotonic phase of the oxidation (Fig. 2). The mean oxidation rate measured within the lag phase in the presence of 50 μM Cu(II) and the mean oxidation rate within the initial monotonic phase of the oxidation induced by AAPH, lipoxygenase, and lower concentrations of Cu(II) are interpreted as initial oxidation rates because these phases began immediately after the addition of the oxidants.

Plasma samples prepared with different anticoagulants exhibit differences in their oxidation kinetics. Comparison of initial oxidation rates does not reveal any consistent difference among heparin plasma, citrate plasma, and serum obtained from the same donor, whereas EDTA plasma is oxi-

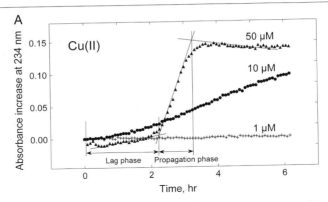

FIG. 1. Typical oxidation kinetics of plasma from a healthy donor measured in the presence of different oxidants. Oxidation was evaluated as an increase in sample absorbance at 234 nm after addition of the oxidant. Heparin plasma (20 μl) was diluted 150-fold with PBS containing 0.16 M NaCl and oxidized at 37° by Cu(II) (A), AAPH (B), or lipoxygenase (C). Numbers on the curves show final oxidant concentrations in the reaction mixture. Straight lines show the method of calculation of the lag phase and propagation phase duration measured in the presence of 50 μM Cu(II).

dized at a lower rate (Fig. 2). In contrast, the lag phase of Cu(II)-induced oxidation is similar in serum and heparin plasma and is considerably shorter in citrate plasma. The absorbance increase seen in EDTA plasma incubated with 50 μM Cu(II) is slow and does not show any discernible phases.

The shape of oxidation curves does not depend on the plasma dilution in the range between 50- and 300-fold. Lower dilutions lead to absorbance values outside the range of the spectrophotometer. In accordance with data reported previously, the duration of the lag phase measured in the presence of 50 μM Cu(II) is independent of plasma dilution in the range between 100- and 300-fold, whereas initial plasma oxidation rates by AAPH (330 μM) and Cu(II) (10 μM) are higher at lower dilutions (data not shown).

Because freezing of plasma samples is often necessary in large-scale population studies, we also investigated the influence of freezing at −80° and thawing on the oxidation kinetics. Storage at −80° for 1–3 days has no influence on the oxidation parameters measured in heparin and EDTA plasma in the presence of any of the oxidants used in our study, whereas a similar treatment of citrate plasma and serum prolongs the lag phase measured with 50 μM Cu(II) and, in the case of citrate plasma, also lowers oxidation rates (data not shown).

The reproducibility of the oxidation kinetics was investigated using heparin plasma samples. The oxidation kinetics are reproducible with all the oxidants used. The interassay variability of the measurement of the

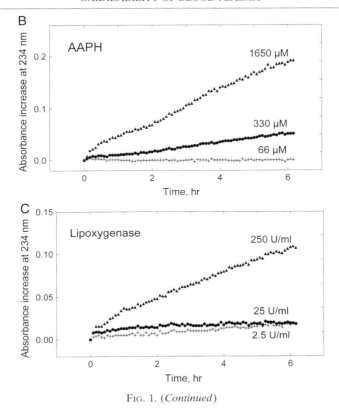

FIG. 1. (*Continued*)

initial oxidation rate calculated for a selected frozen heparin plasma sample is in the range of 4.5 (330 μM AAPH, $n = 8$) to 8.3% (25 U/ml lipoxygenase, $n = 7$). The interassay variability of the measurement of the lag phase duration with 50 μM Cu(II) is 1.7% ($n = 8$). The intrassay variability studied in five different analyses of the same sample on 5 different days within a week gave coefficients of variance of 9.2 and 7.8% for initial oxidation rates measured in the presence of 330 μM AAPH and 25 U/ml lipoxygenase, respectively.

Comparison with Other Methods

Relationship between Absorbance Increase at 234 nm and Other Indices of Lipid Peroxidation. An absorbance increase at 234 nm is known to reflect the accumulation of conjugated dienes in oxidizing LDL[20] and HDL.[21] In plasma, an absorbance increase at 234 nm has also been shown to be

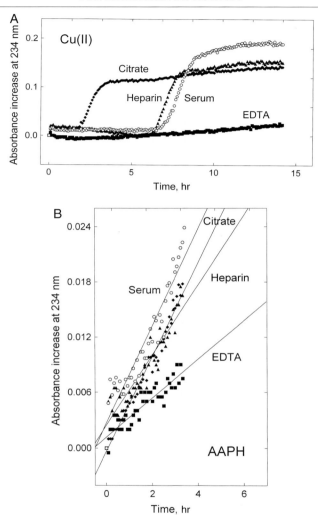

FIG. 2. Typical oxidation kinetics of serum and different types of plasma from a healthy donor. Oxidation was evaluated as an increase in sample absorbance at 234 nm after addition of the oxidant. Heparin plasma (▲), EDTA plasma (■), citrate plasma (◆), or serum (○), 20 µl of each, was diluted 150-fold with PBS containing 0.16 M NaCl and oxidized at 37° by 50 μM Cu(II) (A), 330 μM AAPH (B), or 25 U/ml lipoxygenase (C). Straight lines show the method of calculation of the initial oxidation rates in the presence of AAPH and lipoxygenase as slopes of linear regressions performed on each data set.

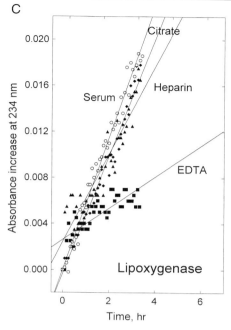

FIG. 2. (*Continued*)

indicative of the oxidation of plasma lipoproteins and to correlate with other indices of lipid peroxidation.[21,22] To prove these findings, we characterized the oxidation time course in a selected heparin plasma sample diluted 150-fold and incubated with 50 μM Cu(II) using three different indices of lipid peroxidation, namely absorbance increase at 234 nm, accumulation of oxycholesterols, and consumption of polyunsaturated fatty acids (PUFAs).[24] Oxycholesterols and PUFAs were measured directly in the samples using gas chromatography with mass selective and flame ionization detectors, respectively.[23,25] The time course of absorbance at 234 nm paralleled those of oxycholesterol formation and PUFA consumption (Fig. 3). When plasma samples were diluted 2-fold and incubated with 3.75 mM Cu(II) to achieve the same Cu(II)/plasma ratio as in 150-fold diluted plasma incubated with 50 μM Cu(II), the time courses of oxycholesterol formation and PUFA and antioxidant consumption were also characterized by a

---

[24] A. Kontush, T. Spranger, A. Reich, S. Djahansouzi, B. Karten, J. H. Braesen, B. Finckh, A. Kohlschütter, and U. Beisiegel, *BioFactors* **6,** 99 (1997).

[25] A. Kontush, S. Meyer, B. Finckh, A. Kohlschütter, and U. Beisiegel, *J. Biol. Chem.* **271,** 11106 (1996).

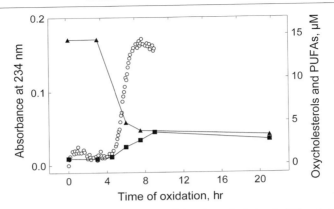

FIG. 3. Accumulation of conjugated dienes (○) and oxycholesterols (■) and consumption of PUFAs (▲) in plasma of a healthy donor. Heparin plasma was diluted 150-fold with PBS containing 0.16 M NaCl and oxidized by 50 μM Cu(II) at 37° in a spectrophotometrical cuvette to measure conjugated dienes or in a water bath to measure oxycholesterols and PUFAs. From A. Kontush, T. Spranger, A. Reich, S. Djahansouzi, B. Karten, J. H. Braesen, B. Finckh, A. Kohlschütter, and U. Beisiegel, *BioFactors* **6**, 99 (1997), with kind permission from IOL Press, Amsterdam, The Netherlands.

distinct lag phase.[26] In addition, gel electrophoresis showed that plasma lipoproteins, such as LDL and HDL, were modified oxidatively during the incubation. In a separate experiment, we measured conjugated diene accumulation in another plasma sample using an absorbance increase at 234 nm and HPLC with UV detection at the same wavelength and found that both methods gave the same oxidation time course (data not shown). Taken together, these data suggest that the measurement of absorbance at 234 nm may be used as an index of conjugated diene accumulation in human plasma.

Relationship between Oxidizabilities of Blood Plasma and Isolated LDL. To clarify whether the plasma oxidizability was related to the oxidizability of isolated LDL, both were measured in a group of 23 normolipidemic donors.[24] Indices of oxidizability of heparin plasma were measured in the presence of 50 μM Cu(II), whereas indices of LDL oxidizability were measured under conditions of the typical Cu(II)-based oxidation assay.[20] It was found that the maximal plasma oxidation rate measured within the propagation phase of the oxidation correlated positively with the mean LDL oxidation rate measured within the lag phase of LDL oxidation. The maximal plasma oxidation rate also correlated negatively with the LDL

[26] B. Karten, U. Beisiegel, G. Gercken, and A. Kontush, *Chem. Phys. Lipids* **88**, 83 (1997).

lag phase duration. The existence of a relationship between the propagation phase of plasma oxidation and the lag phase of LDL oxidation was confirmed by a positive correlation between the plasma oxidation rate measured after 180–270 min of oxidation (i.e., at the end of the lag phase and the beginning of the propagation phase) and the LDL oxidation rate measured after 30–90 min of oxidation (i.e., within the lag phase).

Relationship between Oxidizability and Chemical Composition of Plasma. To elucidate whether the plasma oxidizability was related to its antioxidant and fatty acid composition, the latter was characterized in the same group of 23 normolipidemic donors.[27] Plasma lipophilic antioxidants were measured using reversed-phase high-performance liquid chromatography with electrochemical detection, hydrophilic antioxidants were determined photometrically, and fatty acids were measured using gas chromatography.[23,25,27] Indices of plasma oxidizability by Cu(II) revealed series of significant correlations with the initial plasma antioxidant and fatty acid content. Plasma lag phase duration correlated positively with its SH group and albumin content. Plasma propagation phase duration correlated positively with its albumin content. The plasma maximal oxidation rate correlated negatively with its albumin, ascorbate, bilirubin, α-tocopherol, and ubiquinol-10 content (if the latter was expressed as a percentage of total ubiquinol-10 + ubiquinone-10) and positively with its ubiquinone-10 content. When major plasma fatty acids were expressed as a percentage of total plasma fatty acids, positive correlations between propagation phase duration and plasma monounsaturated and saturated fatty acids were found. A similar correlation calculated for plasma PUFAs was found to be negative. Of individual fatty acids, the plasma content of oleic (18:1) and docosahexaenoic (22:6) acids revealed correlations of comparably high significance.

To prove the relationship between plasma oxidizability and its antioxidant content, we supplemented heparin plasma obtained from two healthy donors with physiological amounts of major plasma antioxidants *in vitro* and measured the oxidizability of the samples in the presence of Cu(II).[26] The plasma oxidizability was found to decrease following *in vitro* supplementation with albumin, urate, ascorbate, bilirubin, α-tocopherol, and ubiquinol-10. The lag phase and propagation phase were longer in samples supplemented with albumin, ascorbate, urate, bilirubin, and α-tocopherol. The oxidation rate in the lag phase was lower in samples supplemented with urate, bilirubin, α-tocopherol, and ubiquinol-10.

---

[27] T. Spranger, B. Finckh, R. Fingerhut, A. Kohlschütter, U. Beisiegel, and A. Kontush, *Chem. Phys. Lipids* **91,** 39 (1998).

Physiological Relevance

Plasma Oxidizability in Hyperlipidemia and Coronary Heart Disease. To evaluate whether the plasma oxidizability is indicative of the presence of pathological conditions linked to increased oxidative stress, this parameter was measured in hyperlipidemic patients whose oxidative status is *a priori* expected to be impaired. Forty-four hyperlipidemic patients and 14 healthy normolipidemic age-matched subjects were recruited for this study.[24] Fifteen hyperlipidemic patients had angiographically proved CHD and were analyzed as a separate group. When the initial oxidation rates of heparin plasma samples were compared between the subject groups studied, they were found to be higher in hyperlipidemic patients than in healthy controls. The plasma initial oxidation rate increased in the order of healthy controls < hyperlipidemic patients without CDH < hyperlipidemic patients with CHD when lipoxygenase or myeloperoxidase was used to oxidize plasma. The plasma initial oxidation rate was also higher in patients with CHD than in control subjects when plasma was oxidized by Cu(II) or AAPH. Differences in the initial oxidation rate between patients and controls were seen as a trend when Cu(II) or myeloperoxidase was used as an oxidant and reached significance with AAPH and lipoxygenase. In contrast, no difference in the duration of the lag phase or propagation phase of the oxidation between the patient and control groups in the samples oxidized by Cu(II) was found. The maximal oxidation rate was found to increase in the order of healthy controls < hyperlipidemic patients without CHD < hyperlipidemic patients with CHD in the samples oxidized by Cu(II). An increase in the initial oxidation rate in plasma samples obtained from hyperlipidemic patients compared to healthy controls was accompanied by a decrease in their content of lipophilic antioxidants, such as α-tocopherol, ubiquinol-10, α-carotene, and β-carotene (all normalized to plasma lipids).

Because oxidative stress has been implicated in the development of not only atherosclerosis but also several other pathologies, such as Alzheimer's and Parkinson's disease, we also applied our method to measure plasma oxidizability to patients with Alzheimer's disease and found that EDTA plasma obtained from these patients was significantly more susceptible to *in vitro* oxidation induced by AAPH or lipoxygenase than plasma from age-matched neurological controls.[28]

Plasma Oxidizability in Animal Model of Atherosclerosis. To prove whether plasma oxidizability reflects atherogenesis *in vivo,* we measured this parameter in Watanabe heritable hyperlipidemic (WHHL) rabbits fed

---

[28] S. Schippling, A. Kontush, and U. Beisiegel, unpublished data (1997).

with different antioxidants at dosages comparable to those used in humans (4.3 mg ubiquinone-10, 4.3 mg vitamin E, 15 mg probucol, or 75 mg carvedilol/kg body weight daily) for 12 months. Each feeding group consisted of 10 animals. At the beginning of the study and after 3 and 6 months of feeding the oxidizability of rabbit heparin plasma was measured in the presence of Cu(II). At the end of the study the development of atherosclerosis was assessed as the arterial wall thickness in each rabbit. It was found that supplementation with all the antioxidants prolonged the lag phase of plasma oxidation by Cu(II) and lowered the maximal oxidation rate.[24] The lag phase duration measured after 6 (but not after 0 or 3) months of supplementation revealed a significant negative correlation with the intima-to-media ratio calculated for rabbit aorta.[29]

Conclusions

The plasma oxidation assay was employed to characterize the oxidizability of plasma lipoproteins in our study. This assay uses an increase in absorbance of plasma at 234 nm after the addition of oxidants as a measure of plasma lipoprotein oxidation. In accordance with data reported by others,[21,22] we found that the time course of absorbance at 234 nm paralleled those of other indices of lipoprotein oxidation in blood plasma. The oxidizability of human plasma measured in the plasma oxidation assay in the presence of Cu(II) was found to correlate with the oxidizability of LDL isolated from the same plasma and measured in the commonly used LDL oxidation assay. These data indicate that the measurement of absorbance at 234 nm reflects the lipoprotein oxidation induced *in vitro* in blood plasma.

We also found that the plasma oxidizability correlated negatively with plasma antioxidant and positively with plasma oxidizable fatty acid content. Supplementation of human plasma with different antioxidants (albumin, urate, ascorbate, bilirubin, α-tocopherol, and ubiquinol-10) *in vitro* decreased its oxidizability. Supplementation of WHHL rabbits with different antioxidants (vitamin E, ubiquinone-10, probucol, and carvedilol) *in vivo* lowered the oxidizability of rabbit plasma in comparison to rabbits fed a standard diet. These data demonstrate that the plasma oxidizability is determined critically by its chemical composition and that the plasma antioxidants and fatty acids seem to contribute most importantly to this parameter. It should be kept in mind that, as a result of plasma dilution, concentrations of hydrophilic antioxidants in the assay are highly reduced in comparison with their initial plasma concentrations. However, an inhibitory effect of the supplementation with hydrophilic antioxidants on plasma oxi-

---

[29] J. H. Braesen, S. Djahansouzi, U. Beisiegel, and A. Kontush, unpublished data (1997).

dation, as well as negative correlations between the plasma oxidation rate and its hydrophilic antioxidants, indicates that they can influence plasma oxidizability even at such low concentrations.

When plasma from hyperlipidemic patients with or without coronary heart disease and from age-matched healthy controls was studied, the plasma oxidizability was found to be highest in patients with coronary heart disease and lowest in controls. Increased oxidizability of plasma from hyperlipidemic and atherosclerotic patients is in good agreement with the increased oxidizability of plasma LDL from such patients reported previously by others.[30] Lipophilic antioxidants α-tocopherol, ubiquinol-10, α-carotene, and β-carotene were found to be lower in plasma of hyperlipidemic patients than control subjects. The increased initial oxidation rate of plasma from patients with hyperlipidemia and CHD can be explained correspondingly by its decreased content of major lipophilic antioxidants.

Using blood plasma to characterize the oxidizability of plasma lipoproteins offers several important advantages over assays that use isolated lipoproteins, including the most common Cu(II)-based LDL oxidation assay.[17,20] Most importantly, lipoproteins in plasma are oxidized in the presence of hydrophilic antioxidants, which are also present in the interstitial fluid of the arterial wall, where lipoprotein oxidation is thought to take place.[1,12] A correspondence was found between the lag phase of LDL oxidation and the propagation phase of plasma oxidation, i.e., between early stages of LDL oxidation and later stages of plasma oxidation. This may reflect an additional inhibition of lipoprotein oxidation by hydrophilic antioxidants in plasma compared to isolated LDL. In addition to providing additional protection against oxidation, hydrophilic antioxidants are important modulators of the antioxidant activity of α-tocopherol, the major antioxidant in human lipoproteins. α-Tocopherol functions as an antioxidant in the presence of such hydrophilic antioxidants as ascorbate, which are able to recycle the potentially harmful α-tocopheroxyl radical back to α-tocopherol, thereby eliminating its prooxidant activity.[31] However, this antioxidant activity may evolve into a prooxidant under conditions of mild oxidation, when ascorbate is consumed.[18,23,31] The presence of such coantioxidants for α-tocopherol in the assay mixture makes the assay more physiological compared to the oxidation of isolated lipoproteins and allows the evaluation of a concerted action of all plasma antioxidants.

Another important feature of the plasma oxidation assay is that it characterizes the oxidizability of all plasma lipoproteins rather than that of an isolated lipoprotein class. The assay also employs much lower and

[30] D. Tribble, *Curr. Opin. Lipidol.* **6**, 196 (1995).
[31] V. W. Bowry and R. Stocker, *J. Am. Chem. Soc.* **115**, 6029 (1993).

therefore more physiological amounts of oxidants (lipoxygenase, myeloperoxidase, and AAPH) than the LDL oxidation assay. The assay requires only small plasma volumes (10–20 μl) and can therefore be used as a micromethod. Other advantageous features of the plasma oxidation assay include fast, simple, and efficient sample processing; avoidance of artifactual oxidation during lipoprotein isolation; and simple photometric registration of the oxidation course.

A major drawback of the assay includes a high dilution of plasma, which results in unphysiologically low concentrations of its constituents in the reaction mixture. Such a dilution is necessary for the photometrical detection used to measure lipid peroxidation. The relative unspecificity of the detection is also a disadvantage. A future improvement of methods aimed at detecting oxidation products in undiluted plasma should overcome this problem. However, it is important to emphasize that even using diluted plasma, we were able to detect clinically meaningful differences between samples with the help of simple photometrical registration. Another potential problem is the use of anticoagulants. Heparin provides a comparability of results obtained in fresh and frozen plasma (see also Ref. 22), but is known to possess antioxidant properties. The use of EDTA makes it impossible to evaluate plasma susceptibility to oxidation induced by transition metal ions. Additional studies are necessary to determine conditions of blood sampling that are optimal to measure plasma oxidizability.

Taken together, our data indicate that the plasma oxidation assay (i) provides information similar to that obtained using the common LDL oxidation assay, (ii) upgrades the latter, taking into account the effect of hydrophilic antioxidants on lipoprotein oxidation and characterizing the oxidizability of all plasma lipoproteins, and (iii) offers important practical advantages, such as fast and simple sample processing, the low amount of plasma required, and avoidance of artifactual oxidation during lipoprotein isolation. We propose that the measurement of plasma oxidizability at 234 nm is an adequate screening method in characterizing the oxidizability of plasma lipoproteins.

[5] Measurement of Oxygen Radical Absorbance Capacity in Biological Samples

By GUOHUA CAO and RONALD L. PRIOR

Introduction*

Several methods[1-8] have been developed to assess the total antioxidant capacities of various biological samples, particularly complex matrices such as plasma, serum, wine, fruits, vegetables, and animal tissues. These methods have been needed due to (i) the difficulty in measuring each antioxidant component separately and (ii) the potential interactions among different antioxidant components in complex biological samples. The total peroxyl radical trapping parameter (TRAP) assay of Wayner et al.[1] was the most widely used assay of antioxidant capacity during the last decade. The major problem with the original TRAP assay lies in the oxygen electrode end point; an oxygen electrode will not maintain its stability over the period of time required.[9] The method of Glazer measures the decrease in fluorescence of B- or R-phycoerythrin (PE) in the presence of reactive species (RS) and relates a lag phase or a rate constant of PE fluorescence decay to antioxidant capacity of an added antioxidant sample.[2] The procedure reported by Ghiselli et al.[3] is basically a duplicate of that part of Glazer's method[2] that uses a peroxyl radical generator, 2,2'-azobis(2-amidinopropane) dihydrochloride (AAPH). The Trolox equivalent antioxidant capacity (TEAC) assay of Miller et al.[4,9] is based on the inhibition by antioxidants of the absorbance of the radical cation of 2,2'-azinobis(3-ethylbenzothiazoline 6-sulfonate) (ABTS) and has been commercialized by Randox Labora-

---

* Mention of a trade name, proprietary product, or specific equipment does not constitute a guarantee by the U.S. Department of Agriculture and does not imply its approval to the exclusion of other products that may be suitable.

[1] D. D. M. Wayner, G. W. Burton, K. U. Ingold, and S. Locke, *FEBS Lett.* **187**, 33 (1985).
[2] A. N. Glazer, *Methods Enzymol.* **186**, 161 (1990).
[3] A. Ghiselli, M. Serafini, G. Maiani, E. Assini, and A. Ferro-Luzzi, *Free Radic. Biol. Med.* **18**, 29 (1994).
[4] N. J. Miller, C. Rice-Evans, M. J. Davies, V. Gopinathan, and A. Milner, *Clin. Sci.* **84**, 407 (1993).
[5] T. P. Whitehead, G. H. G. Thorpe, and S. R. J. Maxwell, *Anal. Chim. Acta* **266**, 265 (1992).
[6] G. Cao, H. M. Alessio, and R. G. Cutler. *Free Radic. Biol. Med.* **14**, 303 (1993).
[7] G. Cao, C. P. Verdon, A. H. B. Wu, H. Wang, and R. L. Prior, *Clin. Chem.* **41**, 1738 (1995).
[8] I. F. F. Benzie and J. J. Strain, *Anal. Chem.* **239**, 70 (1996).
[9] C. Rice-Evans, N. J. Miller, *Methods Enzymol.* **234**, 279 (1994).

tories Ltd. (Crumlin, UK). The ferric-reducing ability of plasma (FRAP) assay of Benzie and Strain[8] measures the ferric to ferrous iron reduction in the presence of antioxidants.

We have developed a method called oxygen radical absorbance capacity (ORAC) assay[6,7] based largely on the work reported by Glazer's laboratory, which depends on the unique properties of PE.[2] The ORAC assay is, to date, the only method that takes RS reaction to completion and uses an "area under the curve" (AUC) technique for quantitation, thus combining both inhibition time and inhibition percentage of the RS action by antioxidants into a single quantity.[7]

General Principles of ORAC Assay

The ORAC assay depends on the detection of chemical damage to R- or B-PE through the decrease in its fluorescence emission. The fluorescence of PE is highly sensitive to the conformation and chemical integrity of the protein. Under appropriate conditions, the loss of PE fluorescence in the presence of RS is an index of oxidative damage of the protein. The inhibition of the RS action by an antioxidant, which is reflected in the protection against the loss of PE fluorescence in the ORAC assay, is a measure of its antioxidant capacity against the RS.

Two elements need to be considered in measuring the inhibition of RS action by an added antioxidant sample: the time that the inhibition lasts and the percentage that the inhibition displays at different times. A specific antioxidant may display 100% inhibition of a specific RS action over a period of time and thus produce a lag phase. However, this lag phase is not a general condition with all antioxidants. Some antioxidants, such as reduced glutathione (GSH) (Fig. 1) and melatonin,[10] do not even show a lag phase in inhibiting the oxidation of PE by peroxyl radicals. The lack of a lag phase was also demonstrated for albumin in the TEAC assay.[11] In the presence of a hydroxyl radical generator, such as ascorbate plus Cu^{2+} or H_2O_2 plus Cu^{2+}, none of the antioxidants tested ever give a lag phase in inhibiting the PE oxidation. In addition, the loss of PE fluorescence in the presence of RS, which include peroxyl radicals generated from AAPH, does not follow zero-order kinetics (i.e., linear with time), as shown in Fig. 1. Therefore, any methods based on the assumption of zero-order kinetics[2,3] will inevitably have technical difficulties in measuring the lag phase in the loss of PE fluorescence. This becomes even more important when we

---

[10] C. Pieri, M. Maurizio, M. Fausto, R. Recchioni, and F. Marcheselli, *Life Sci.* **55**, PL271 (1994).
[11] D. Schofield and J. M. Braganza, *Clin. Chem.* **42**, 1712 (1996).

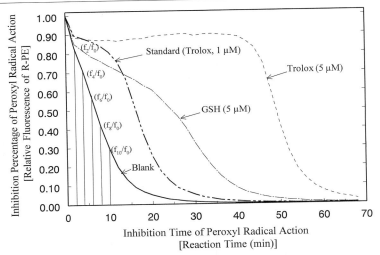

FIG. 1. Principle of the ORAC assay with R-PE as a target for free radical action and AAPH as a peroxyl radical generator.

consider that the kinetics of PE fluorescence loss may be increasingly complex in the presence of an antioxidant mixture.

Also, relating the percentage inhibition observed at a specific time point to the antioxidant capacity of an antioxidant, as used in the TEAC assay, is not ideal. Two compounds having the same inhibition percentage at one time point may exhibit different inhibition percentages at another time point. This can be clearly seen for ascorbate and urate in inhibiting the production of ABTS radical cation in the TEAC assay[11] and for GSH and Trolox in inhibiting peroxyl radical-induced PE oxidation in the ORAC assay.[12] Therefore, AUC technique was developed for quantitation of the results in the ORAC assay. By integrating inhibition percentages over the whole inhibition time period, the ORAC assay successfully overcomes all related problems in quantitation of the antioxidant capacity of a biological sample.

ORAC Assay for Assessing Antioxidant Capacity against Peroxyl Radicals

Sample Preparation

Blood Plasma or Serum and Other Biological Fluids. Blood plasma is prepared by using heparin. Heparin at the concentration used for preparing

[12] G. Cao, E. Sofic, and R. L. Prior, *Free Radic. Biol. Med.* **22**, 749 (1997).

blood plasma do not have significant effects on the ORAC assay that uses peroxyl radicals. However, it is recommended that the blank and standard always include the reagents contained in the samples. Blood plasma or serum needs to be diluted 100- to 200-fold with phosphate buffer before it is used in the ORAC assay. To measure the ORAC in the nonprotein fraction of serum, dilute serum with 0.5 M perchloric acid (PCA) (1:1, v/v) or acetone (1:4, v/v), centrifuge at 4° for 10 min, and recover the supernatant for the ORAC assay after suitable dilution with buffer (e.g., 1:4, v/v). Treating serum with PCA preserves ascorbate; fresh serum should be used to prepare the PCA extract. Blood plasma or serum can be stored at $-80°$ with the understanding that ascorbate may be oxidized. Other biological fluids, including urine, can be used in the ORAC assay either directly after suitable dilution or after removing their protein components.

Animal Tissues. Crude tissue extracts are prepared by homogenizing the tissues in phosphate buffer (1:4 for liver, kidney and lung, 1:2 for brain tissues; w/v) and separating the soluble fractions by a two-step centrifugation process (12,000 g for 10 min followed by 100,000 g for 15 min at 4°). Tissue extracts prepared this way generally contain at least 0.5% (w/v) protein. Samples are then diluted with buffer to contain 0.02% protein for the ORAC assay. For determination of antioxidant capacity in the nonprotein fraction, crude tissue extracts are adjusted to contain 0.5% protein and then treated with PCA or acetone for the ORAC assay, as described earlier for serum samples.

Tea, Fruits, Vegetables, and Other Natural Products. Green tea is brewed for 30 min in deionized water (1:60, w/v, 95–100°). The edible portion of a fruit or vegetable is weighed and then homogenized by using a blender after adding deionized water (e.g., 1:2, w/v). The brewed tea or the fruit or vegetable homogenate is then centrifuged at 34,000 g for 30 min (4°). The supernatant (water-soluble fraction) is recovered. The water-insoluble fraction (pulp) is washed with deionized water, and the recovered supernatant is pooled with the supernatant obtained from the first centrifugation step. The pooled supernatant is measured for its volume and is used directly for the ORAC assay after suitable dilution with phosphate buffer. The pulp is then further extracted by using pure acetone [1:4, w (pulp)/v] with shaking at room temperature for 30 min. The acetone extract is recovered after centrifugation (34,000 g, 10 min, 4°) and is used for the ORAC assay after suitable dilution with phosphate buffer. The ORAC activity of a tea, a fruit, or a vegetable is calculated by adding the ORAC activity from its water-soluble fraction and its acetone-extracted pulp fraction. Red wine and fruit or vegetable juices can be used in the ORAC assay directly after suitable dilution if there are no obvious precipitates; otherwise, these samples need to be centrifuged before use.

Automated Procedure on Cobas Fara II

Reagents

1. Phosphate buffer (75 mM, pH 7.0). To make the phosphate buffer, prepare 0.75 M K_2HPO_4 and 0.75 M NaH_2PO_4 first and then mix them (K_2HPO_4/NaH_2PO_4, 61.6:38.9, v/v) and dilute with deionized water (1:9, v/v). The pH of the buffer made in this manner should be around 7.0. However, the pH should be checked using a pH meter. The 0.75 M K_2HPO_4 and 0.75 M NaH_2PO_4 can be stored at 4° for several months.

2. Phycoerythrin (3.73 mg/liter). Phycoerythrin can be purchased from Sigma (St. Louis, MO) or Prozyme (San Leandro, CA). A stock solution (0.17 g/liter) is prepared by dissolving 1 mg PE in 5.90 ml phosphate buffer. The stock solution can be stored at 4° for several months. To make a working PE solution (3.73 mg/liter), dilute the stock solution with phosphate buffer. For example, to run 12 samples in duplicate on the Cobas, dilute 300 μl of the stock PE solution with 13.4 ml of phosphate buffer. The working PE solution should be preincubated in a water bath at 37° for 15 min before loading into the Cobas reagent rack.

3. 2,2'-Azobis(2-amidinopropane) dihydrochloride (320 mM). AAPH can be purchased from Wako Chemicals USA, Inc. (Richmond, VA). To make this solution, weigh 868 mg AAPH and add phosphate buffer to 10 ml. We usually weigh a given amount of AAPH in a plastic weighing boat and add buffer accordingly (86.8 mg/1 ml). This reagent needs to be prepared fresh and kept in ice water until it is loaded into the Cobas reagent rack.

4. 6-Hydroxy-2,5,7,8-tetramethylchroman-2-carboxylic acid (Trolox) (20 μM). Trolox can be obtained from Aldrich (Milwaukee, WI). To make a stock solution (100 μM), dissolve 5.0 mg Trolox in 200 ml of phosphate buffer. Aliquot the stock solution and store at $-80°$. Trolox at this concentration is stable at $-80°$ for at least 4 months. Prepare a working solution (20 μM) by diluting the stock solution with buffer (1:4, v/v).

Cobas Fara II Protocol

A Cobas Fara *II* centrifugal analyzer (Roche Diagnostic System Inc., Branchburg, NJ) is used for the automated procedure. The Cobas Fara *II* is programmed to maintain a temperature of 37°, and a two-reagent system (Reaction Mode 3, P-I-SRI-A) is used. This reaction mode pipettes and transfers sample (20 μl), phosphate buffer (5 μl), and main reagent (PE) (365 μl, 3.73 mg/liter) in parallel (P) into the main reagent wells of their respective cuvette rotor positions, spins, mixes, and incubates (I) for the programmed time of 1 min and records the initial fluorescence (F_0) (excitation 540 nm, emission 565 nm; the emission filter was purchased from

Andover Corporation, Salem, NH). When the rotor stops spinning, a start reagent (SRI), 5 μl of AAPH (320 mM) plus 5 μl of buffer, is pipetted into the appropriate start reagent wells in the cuvette rotor. At this point, the sample makes up 5% of the reaction volume, and the final concentrations of PE and AAPH are 3.40 mg/liter and 4 mM, respectively. Between transfers, sample and reagent transfer pipettes are washed with buffer to eliminate sample cross-contamination. When the analyzer starts spinning, it causes mixing of sample/PE with AAPH and the reaction starts. Fluorescence readings are taken every 2 min (F_2, F_4, F_6, ...) (excitation 540 nm, emission 565 nm) for up to 70 min (see Fig. 1). If the fluorescence of the last reading does not decline to less than 5% of the first reading, the dilution of the sample analyzed is adjusted accordingly and the sample is reanalyzed. The reaction direction is selected as a "decrease" and the conversion factor is set to "1". To determine the maximum voltage for the photomultiplier tube (PM adjust), AAPH is omitted and replaced with buffer and the analyzer is run for a period of 10 min using the ORAC assay program. The maximum voltage is adjusted on the Cobas automatically. The run time needs to be changed back to 70 min after the PM adjustment in the ORAC assay program. The PM adjustment procedure should be performed when a different PE or the same PE with a different lot number is used.

Phosphate buffer is used as a blank and Trolox (20 μM) is used as a standard, which is added in a manner similar to the samples to give a final concentration of 1 μM. The final results (ORAC value) are expressed using Trolox equivalents.

$$\text{ORAC value } (\mu M) = 20k(S_{\text{Sample}} - S_{\text{Blank}})/(S_{\text{Trolox}} - S_{\text{blank}}) \quad (1)$$

where k is the sample dilution factor and S is the area under the fluorescence decay curve of the sample, Trolox, or blank that is calculated as follows (see Fig. 1):

$$S = (0.5 + f_2/f_0 + f_4/f_0 + f_6/f_0 + \cdots + f_{68}/f_0 + f_{70}/f_0) \times 2 \quad (2)$$

where f_0 is the initial fluorescence at 0 min and f_i is the fluorescence measurement at time i.

Samples and standards are always analyzed in duplicate using a "forward-then-reverse" ordering as follows: blank, standard, sample 1, sample 2, ..., sample 2, sample 1, standard, and blank. This arrangement is adopted to correct for the signal "drift" that correlates with the position of each sample in the Cobas Fara II cuvettes.

Data Collection

Data are sent electronically from the Cobas Fara II through the RS-232 serial interface to a PC computer system running Crosstalk software

(Digital Communications Associates, Alpharetta, GA). Data are analyzed using Microsoft Excel (Microsoft Corporation, Roselle, IL) to apply Eq. (2) to the area under the fluorescence decay curve (AUC) and Eq. (1) to the final ORAC value. The basis for the ORAC value calculation is presented schematically in Fig. 1.

Manual Procedure

For a manual procedure, the following reagents are used: phosphate buffer (75 mM, pH 7.0), 1750 μl; PE (68 mg/liter), 100 μl; AAPH (160 mM), 50 μl; and sample, 100 μl. Phycoerythrin and AAPH solutons are prepared with phosphate buffer. Phosphate buffer and the PE solutions are preincubated at 37° for 15 min. The assay is carried out at 37° in fluorimeter cuvettes. A blank and a standard are assayed during each run. For the blank, 100 μl buffer instead of sample is used. For the standard, 100 μl of 20 μM Trolox solution instead of sample is used. The reaction is started by the addition of 50 μl of 160 mM AAPH. Once AAPH is added, the cuvette is vortexed briefly and the fluorescence is measured immediately (emission 565 nm, excitation 540 nm) using a fluorescence spectrophotometer (e.g., Perkin-Elmer, Norwalk, CT). The fluorescence is recorded every 5 min until the fluorescence of the last reading has declined to less than 5% of the first reading. One blank, one standard, and a maximum of eight samples can be analyzed at the same time when a cuvette rack is used. The cuvette rack should be kept at 37° in a water bath. For calculation of the results, Eq. (2) is modified as follows:

$$S = (0.5 + f_5/f_0 + f_{10}/f_0 + f_{15}/f_0 + \cdots + f_{65}/f_0 + f_{70}/f_0) \times 5 \qquad (3)$$

where f_0 is the initial fluorescence at 0 minutes and f_i is the fluorescence measurement at time i.

Comments

Reproducibility of $ORAC_{ROO \cdot}$ Assay and ORAC of Some Common Biological Samples. The intra- and interassay coefficients of variation (CV) of the automated procedure and the manual procedure are all less than 7.0%.[6,7] The ORAC activity against peroxyl radicals ($ORAC_{ROO \cdot}$, Trolox equivalents) measured in serum of healthy human subjects (age: 66.9 ± 0.6 years) was 3.10 ± 0.20 mM (mean ± SD, $n = 8$). The $ORAC_{ROO \cdot}$ of the PCA nonprotein fraction of the plasma of healthy human subjects (age: 20–80 years) was 0.58 ± 0.11 mM ($n = 34$). The $ORAC_{ROO \cdot}$ of rat (adult, Fischer 344, $n = 5$–6) serum, serum PCA fraction, and serum acetone fraction were 4.39 ± 0.33, 0.65 ± 0.06, and 0.44 ± 0.11 mM, respectively. The liver $ORAC_{ROO \cdot}$ of these rats were 142 ± 10.6 and 62.4 ± 11.8 nmol/

mg protein for cytosolic samples, nonprotein (PCA) cytosolic samples, respectively. The $ORAC_{ROO\cdot}$ of common fruits and vegetables varies considerably; it can be as high as 19.4 μmol/g for garlic and as low as 0.5 μmol/g for cucumber.[13,14] The high $ORAC_{ROO\cdot}$ found in some fruits and vegetables is thought to be mainly from phenolic compounds, including flavonoids[12,15] and phenolic acids.[16] When an HPLC procedure coupled with a coulometric array detection system was developed to analyze these phenolic compounds in fruits and vegetables, a significant positive linear correlation was observed between $ORAC_{ROO\cdot}$ and electrochemical data.[16]

Sensitivity of PE to Peroxyl Radical Damage and Effects of Changing PE or AAPH Concentration on Final ORAC Activity. Commercially available PE includes B-PE and R-PE. Both of these are mixtures of PE molecules with different molecular weights and can be used in the ORAC assay. B-PE and R-PE are distinct spectroscopically, reflecting the content and ratio of different bilins in their subunits $[(\alpha\beta)_6\gamma]$. The same excitation wavelength (540 nm) and emission filter (565 nm) can be used to monitor the fluorescence of both B-PE and R-PE, although they may not be optimal for a specific PE. However, the fluorescence intensity and the sensitivity to peroxyl radical damage can be different even for the same PE with different lot numbers. It appears that the difference in the sensitivity of PE to peroxyl radical damage is not due to contamination by compounds of low molecular weight, as dilution of the product and dialysis of the product against deionized water does not affect the sensitivity. The ORAC assay procedures described earlier are based on using B- or R-PE that loses more than 90% of its fluorescence within 30 min in the presence of 4 mM AAPH. When PE is relatively resistant to peroxyl radical damage, the concentration of AAPH and Trolox standard can be increased accordingly in the ORAC assay. As shown in Fig. 2, the kinetics of PE fluorescence loss in the presence of AAPH is mainly dependent on AAPH concentration and is less affected by the concentration of PE itself. The ORAC values (Trolox equivalents) of Trolox, ascorbic acid, uric acid, GSH, and quercetin are basically independent of the concentrations of AAPH (Table I). However, the ORAC value of albumin and blood serum is affected by the selection of PE and the concentration of AAPH (Table I). Therefore, a control sample should always be analyzed when a PE product of a different lot number is used. We recommend that PE of a single lot number be used for a planned project.

---

[13] G. Cao, E. Sofic, and R. L. Prior, *J. Agric. Food Chem.* **44**, 3426 (1996).
[14] H. Wang, G. Cao, and R. L. Prior, *J. Agric. Food Chem.* **44**, 701 (1996).
[15] H. Wang, G. Cao, and R. L. Prior, *J. Agric. Food Chem.* **45**, 304 (1997).
[16] C. Guo, G. Cao, E. Sofic, and R. L. Prior, *J. Agric. Food Chem.* **45**, 1787 (1997).

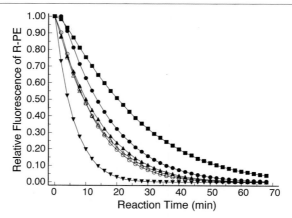

FIG. 2. Effects of AAPH and R-PE concentration on the kinetics of R-PE fluorescence loss. When the AAPH concentration is fixed (4 mM), different R-PE concentrations (●, 3.40; ○, 1.70; ▲, 1.13; and △, 0.85 mg/liter) produce similar areas (17.4, 12.4, 13.6, and 12.7) under the R-PE decay curves. However, when the R-PE concentration is fixed (1.70 mg/liter), different AAPH concentrations (■, 2; 0.4; and ▼, 8 mM) produce very different areas (24.9, 12.4, and 6.1) under the R-PE decay curves; actually, an excellent correlation exists between the area (y) and the AAPH concentration (x) ($y = 50.14\ x^{-1}$, $r = 1.0$).

Comparison of $ORAC_{ROO}$. Assay with TEAC and FRAP Assays. The ORAC, TEAC, and FRAP assays were developed most recently. A significant linear, but low, correlation was found between human serum $ORAC_{ROO}$. and FRAP values. There was no agreement between these assays.[17] This is not surprising as they use different technologies. Both ORAC and TEAC assays are inhibition methods: A sample is added to a free radical-generating system, and the inhibition of the free radical action is measured. The $ORAC_{ROO}$. assay uses AAPH as a free radical-generating system. AAPH undergoes spontaneous decomposition and produces peroxyl radicals with a rate determined primarily by temperature. The analyzed antioxidant samples are not likely to affect this rate, particularly when the chemical structure of AAPH and the very high molar ratio (more than 2000) of AAPH to an antioxidant sample are considered[7]; therefore, the $ORAC_{ROO}$. assay measures the capacity of an antioxidant to directly quench peroxyl radicals. The high molar ratio between the free radical generator and an antioxidant sample in the $ORAC_{ROO}$. assay also indicates a high specificity. The $ORAC_{ROO}$. assay is more sensitive than TEAC and FRAP assays, as the final standard concentration required in the ORAC assay is much lower than those required in TEAC and FRAP assays. The TEAC

---

[17] G. Cao and R. L. Prior, *Clin. Chem.* **44** (in press) 1998.

TABLE I
EFFECT OF ASSAY CONDITION ON ORAC RESPONSE[a,b]

Source	RA1[c]	BA2[d]	BA3[e]
Trolox	1.00 ± 0.02[f]	1.00 ± 0.02	1.01 ± 0.02
Vitamin C	0.66 ± 0.01	0.69 ± 0.04	0.67 ± 0.04
GSH	0.44 ± 0.03	0.51 ± 0.02	0.55 ± 0.04
Uric acid	1.15 ± 0.12	0.91 ± 0.03	0.93 ± 0.03
Quercetin	3.39 ± 0.08	3.35 ± 0.04	—
Albumin	5.48 ± 0.27	8.61 ± 0.59	12.0 ± 0.33
Serum[g]	—	2174 ± 193	2247 ± 97

[a] Linear regression coefficients were obtained from plots of ORAC value (μM Trolox equivalents) vs antioxidant concentration (μM).

[b] All y intercepts are not significantly different from zero ($p > 0.05$). All regression coefficients (slopes) are significantly different from zero ($p < 0.01$). All r coefficients are greater than 0.99.

[c] R-PE (Sigma P8912, Lot 10H40582), 3.40 mg/liter; AAPH, 4 mM; standard, 1 μM; samples 0.25–2.0 μM (for bovine albumin, sample concentrations are 0.094–0.376 μM).

[d] B-PE (Sigma P1286, lot 27H4111), 3.40 mg/liter; AAPH, 16 mM; standard, 2.5 or 5 μM; samples, 1.25–10 μM (for quercetin and bovine albumin, sample concentrations are 0.5–2.0 and 0.188–0.753 μM, respectively. For serum, sample concentrations are 1/800, 1/1600, 1/3200, and 1/6400 (v/v)).

[e] B-PE (Sigma P1286, Lot 27H4111), 3.40 mg/liter; AAPH, 32 mM; standard, 5 μM; samples, 1.25–10 μM (for bovine albumin, sample concentrations are 0.188–0.753 μM. For serum, sample concentrations are 1/800, 1/1600, 1/3200, and 1/6400 (v/v)).

[f] Standard error of estimate.

[g] Data for serum are based on undiluted sample, not μM.

assay uses ABTS and H_2O_2 to generate ABTS radical cations in the presence of metmyoglobin as a peroxidase.[4,9] Only a lag phase or inhibition percentage at a fixed time point can be quantified as the result in the TEAC assay.[9] Added antioxidants quench ABTS radical cations formed by the interaction of H_2O_2 with metmyoglobin. However, the direct interactions of an added antioxidant sample with the reagents cannot be totally excluded, as the molar ratio of H_2O_2 : metmyoglobin : ABTS : Trolox standard, for example, in the Randox TEAC assay is only 10.2 : 0.25 : 25 : 1 (Lot 21440). These interactions may reduce the production of radical species and not quenching them. These interactions may also even promote the production of reactive species. A number of problems have been found with the Randox TEAC assay methodology, which include a significant increase in the final plasma TEAC value as a result of sample dilution, the marked increase of plasma TEAC value with reaction time, the low recovery of a known quantity of Trolox added to plasma samples, and the different kinetics of chromophore generation in the presence of ascorbate, albumin, and urate.[11] The FRAP assay measures the ferric reducing ability of a sample. It is very simple and convenient in terms of its operation. However, the FRAP assay is totally different from ORAC and TEAC assays because there are no free radicals or oxidants applied in the assay. The antioxidant capacity of an antioxidant against a free radical does not necessarily match its ability to reduce Fe^{3+} to Fe^{2+}. This is why the FRAP value for GSH was almost zero and the FRAP value for quercetin was much lower than its ORAC and TEAC values.[17] Using Fe^{2+} as a final indicator in the FRAP assay may cause problems when an analyzed antioxidant, such as ascorbic acid, cannot only reduce Fe^{3+} to Fe^{2+}, but can also react with Fe^{2+} to further generate free radicals.

ORAC Assay for Assessing Antioxidant Capacity against Reactive Species Produced in H_2O_2–Cu^{2+} System

The reaction between H_2O_2 and transition metals is an important source of RS, particularly hydroxyl radicals in biological systems. The H_2O_2–Cu^{2+} system has been shown to cause oxidative damage *in vitro,* such as oxidation of protein amino acid residues, hemolysis, lipid peroxidation, and DNA base modification and cleavage. This H_2O_2–Cu^{2+} system-induced oxidative damage is generally attributed to the formation of hydroxyl radicals, the most harmful free radicals produced in biological systems.

Sample Preparation

A sample preparation for the ORAC assay using this hydroxyl radical generator is basically the same as that for the ORAC assay using the peroxyl radical generator, AAPH. However, blood plasma should not be prepared by using metal chelators, such as EDTA.

Automated Procedure on Cobas Fara II

  Reagents

  Phosphate buffer, 75 mM, pH 7.0
  Phycoerythrin, 3.78 mg/liter
  6-Hydroxy-2,5,7,8-tetramethylchroman-2-carboxylic acid (Trolox), 40 μM
  Hydrogen peroxide (H_2O_2), 24%
  Cupric sulfate, 0.72 mM

  Cobas Fara II Protocol

  The Cobas protocol is basically the same as that described for the ORAC assay using AAPH, except for the following: the Cobas pipettes and transfers 20 μl sample, 5 μl phosphate buffer, and 360 μl main reagent (PE, 3.78 mg/liter) into the main reagent wells and 10 μl H_2O_2–Cu^{2+} mixture and 5 μl buffer into the start reagent wells. H_2O_2 (24%) and cupric sulfate (0.72 mM) are mixed (1:1, v/v) before loading into the Cobas reagent rack. The sample makes up 5% of the reaction volume, and the final concentrations of PE, H_2O_2, and Cu^{2+} are 3.40 mg/liter, 0.3%, and 9 μM, respectively. The calculation of results in Eq. (1) is modified as follows:

$$\text{ORAC value } (\mu M) = 40k(S_{\text{Sample}} - S_{\text{Blank}})/(S_{\text{Trolox}} - S_{\text{blank}}) \quad (4)$$

Manual Procedure

The manual procedure is basically the same as that described for the ORAC assay using AAPH, but the following reagents and volumes are used: phosphate buffer (75 mM, pH 7.0), 1750 μl; PE (68 mg/liter), 100 μl; H_2O_2–Cu^{2+} mixture, 50 μl; and sample, 100 μl. The H_2O_2–Cu^{2+} mixture is made by mixing 1 vol of H_2O_2 (24%) and 1 vol of cupric sulfate (0.72 mM) before adding to the cuvettes. For the standard, 100 μl of 40 μM Trolox solution is used. The reaction is started by the addition of 50 μl of H_2O_2–Cu^{2+} mixture. Equations (4) and (3) are used in the calculation of results.

Comments

Either B-PE or R-PE can be used in the ORAC assay. The sensitivity of B- or R-PE to hydroxyl radical damage may be different even for the same PE with different lot numbers. The concentrations of Cu^{2+} and standard (Trolox) can be adjusted, when it is necessary. The aforementioned procedures are based on using B- or R-PE that loses more than 90% of its fluorescence within 30 min under the described conditions.

ORAC Assay for Assessing Antioxidant Capacity against Transition Metals

Cu^{2+} is used as an oxidant in the ORAC assay for assessing antioxidant capacity against transition metals. However, Trolox cannot be used as a standard, as it has no antioxidant activity against Cu^{2+}. Also, when the result is negative, it actually measures the Cu^{2+}-initiated prooxidant capacity of a sample. Both the automated procedure and the manual procedure are similar to those for ORAC assay using H_2O_2–Cu^{2+}. The start reagent is cupric sulfate instead of the H_2O_2–Cu^{2+} mixture. The final concentration of cupric sulfate is 18 μM, which can be adjusted according to the sensitivity of B- or R-PE to Cu^{2+}-induced oxidative damage. For the calculation of results, the Eq. (5) is used:

$$\text{ORAC value (units)} = [k(S_{\text{Sample}} - S_{\text{blank}})/(S_{\text{Blank}})] \times 100 \qquad (5)$$

One unit equals the antioxidant (when the result is positive) or prooxidant (when the result is negative) activity that increases or reduces the area under the PE decay curve by 1% in the ORAC assay.

Section II

Vitamin C

[6] Analysis of Ascorbic Acid and Dehydroascorbic Acid in Biological Samples

By MARK LEVINE, YAOHU WANG, and STEVEN C. RUMSEY

Introduction

Accurate measurements of ascorbic acid (ascorbate, vitamin C) and its reducible metabolite dehydroascorbic acid are essential for the proper interpretation of function. Unfortunately, inaccurate measurements are sometimes responsible for confusing or misleading biological conclusions regarding vitamin C. Ascorbate and dehydroascorbic acid measurement techniques and their difficulties were comprehensively reviewed elsewhere.[1,2]

Ascorbic acid (ascorbate, vitamin C) and dehydroascorbic acid analyses in biological samples should account for sensitivity, specificity, substance interference, and stability.[1,2] For plasma and especially cells, an assay should be sensitive enough to measure 10 pmol as absolute mass, or an assay concentration less than or equal to 1 μM. The assay should be specific for ascorbate and not inadvertently include other reducing substances. Other nonreducing but interfering compounds should be minimized or eliminated. Finally, any assay for ascorbate and dehydroascorbic acid must account for stability. Ascorbate will oxidize in solution as a function of light, ionic strength, temperature > 4°, low concentrations, or some divalent cations (i.e., iron, copper).[3] Similarly, dehydroascorbic acid will hydrolyze in solution as a function of increasing pH, low concentration, light, temperature above 4°, and iron/copper. Stability must be accounted for when samples are prepared, stored, and assayed.

Ascorbate Measurements

High-Performance Liquid Chromatography with Electrochemical Detection: Assay Principles

This chapter focuses on ascorbate assays because these techniques are substantially more advanced than those for dehydroascorbic acid and be-

---

[1] P. W. Washko, R. W. Welch, K. R. Dhariwal, Y. Wang, and M. Levine, *Anal. Biochem.* **204,** 1 (1992).

[2] S. C. Rumsey and M. Levine, *in* "Modern Analytical Methodologies on Fat and Water Soluble Vitamins" (W. O. Song and G. R. Beecher, eds.). Wiley, New York, 1998.

[3] J. R. Cooke and R. E. D. Moxon, *in* "Vitamin C" (J. N. Counsell and D. H. Hornig, eds.), p. 167. Applied Science Publishers, London, 1981.

cause most techniques for the latter are based on those of the former. High-performance liquid chromatography (HPLC) with electrochemical (EC) detection is the ascorbate assay technique that provides the highest sensitivity, specificity, and accounts for substance interference. In HPLC assays, the general separation principles of HPLC are utilized to separate ascorbate from other substances. A mobile phase, or the solution for chromatography, is selected that optimizes both separation and detection. Once separation is accomplished, ascorbate is detected by one of two distinct types of EC detectors. The principle of both detectors is that they pass a voltage across an area. Current generated by the voltage and the components of the mobile phase are measured.

Amperometric EC detectors are "flow by" detectors, in which the solution containing ascorbate and the components of the mobile phase for HPLC separation flow around the detector.[4] As little as 1% of the mobile phase and analyte is oxidized by the selected voltage. Advantages of amperometric detectors are (1) that electrodes can be repolished and reused and (2) that alterations of the mobile-phase separation conditions may be well tolerated with respect to background current. A disadvantage of amperometric detectors is that the sensitivity can be decreased by contaminants on the electrode surface, which has a relatively small surface area.

Coulometric detectors are "flow through" detectors.[1,5] The mobile phase and ascorbate flow through the detector, which is porous. Although coulometric detectors in theory should oxidize 100% of the mobile phase, in practice they oxidize approximately 70% of the mobile phase and analyte. An advantage of coulometric detectors is that they are exquisitely sensitive, providing that the background current from the mobile phase is minimized. A disadvantage is that slight impurities in the mobile phase can increase background current markedly, thereby decreasing sensitivity. Coulometric detectors often require ultrapure water to minimize background.

HPLC with Coulometric Electrochemical Detection

We prefer coulometric to amperometric detection for several reasons.[6–9] With optimized conditions, coulometric detectors may be 10-fold more

---

[4] I. N. Mefford, *Methods Biochem. Anal.* **31,** 221 (1985).
[5] K. Stalik and V. Pacakova, "Electroanalytical Measurements in Flowing Liquids." Halsted Press, New York, 1987.
[6] P. W. Washko, W. O. Hartzell, and M. Levine, *Anal. Biochem.* **181,** 276 (1989).
[7] K. R. Dhariwal, P. W. Washko, and M. Levine, *Anal. Biochem.* **189,** 18 (1990).
[8] K. R. Dhariwal, W. O. Hartzell, and M. Levine, *Am. J. Clin. Nutr.* **54,** 712 (1991).
[9] M. Levine, C. Conry-Cantilena, Y. Wang, R. W. Welch, P. W. Washko, K. R. Dhariwal, J. B. Park, A. Lazarev, J. Graumlich, J. King, and L. R. Cantilena, *Proc. Natl. Acad. Sci. U.S.A.* **93,** 3704 (1996).

sensitive than amperometric detectors, which is advantageous for samples containing low concentrations or amounts of ascorbate. Stability conditions have been characterized clearly for coulometric detectors, including stability during initial sample preparation, during storage, and during actual assay. Published data concerning these aspects of stability are incomplete for amperometric detectors. Because HPLC with either amperometric or coulometric detection requires sophisticated equipment and trained operators, we feel that coulometric detection provides a better return on invested time and money.

Coulometric detection facilitates routine detection of as little as 0.5 pmol of ascorbate per sample. Samples are injected into the chromatography apparatus using an autosampler. Because injection volumes range from as little as 5 μl to as much as 100 μl, sample concentrations as low as 5 nM can be detected routinely. Lower concentrations or amounts can be detected if necessary, with a practical limit of approximately 1 nM or 0.1 pmol/injection volume.

For coulometric detection, samples are treated with stabilizing solutions of methanol/water/1 mM EDTA to provide stability at all phases of analysis. Once the stabilizing solution is added, samples should be processed without delay. Samples should be protected from direct light, and copper and iron concentrations must be minimized. In addition to providing ascorbate stability, stabilizing solutions also conveniently precipitate protein, which is necessary prior to HPLC. Precipitated protein should be removed by centrifugation without delay. With appropriate sample handling, ascorbate is fully stabilized during sample preparation. At the end of the sample preparation, samples should be placed on dry ice for rapid cooling prior to freezing at $-70°$. Frozen human plasma, cell, and urine samples processed in this manner are stable for at least 1 year. Once samples are thawed for analysis, stability is dependent on the sample type and the autosampler type. In a 4° autosampler that accepts amber vials, stability is at least 12 hr for plasma, urine, and cell samples. Because sample stability is a function of tissue source, autosampler stability should be tested if unknown samples are kept in a 4° autosampler for more than a few hours. EDTA is essential for stability: without it, oxidation occurs.

Ascorbate Assay Method

Instrumentation

Chromatography systems (Alliance 2690 Separations Module, includes dual piston pump, temperature-controlled autosampler, and data analysis system interface; Waters, Milford, MA; or Series 1050 or later, includes pump and temperature-controlled autosampler and data analysis system interface, Hewlett-Packard, Wilmington, DE)

Coulometric electrochemical detector (5200 Series, flow cell 5011, ESA, Chelmsford, MA)

Data analysis system (Millenium; Waters)

Water system (Milli-RO/Milli-Q water purification system; Millipore, Bedford, MA)

Materials

Sodium phosphate monobasic and sodium acetate anhydrous (Mallinckrodt, Paris, KY)

Dodecyltrimethylammonium chloride (Eastman-Kodak, Rochester, NY)

Tetraoctylammonium bromide and orthophosphoric acid (Fluka, Ronkonkoma, NY)

L-Ascorbic acid, D-isoascorbic acid, and EDTA (Sigma, St. Louis, MO)

L-[1-^{14}C]Ascorbic acid (NEN Life Science Products, Boston, MA)

HPLC grade methanol (Baker, Phillipsburg, NJ)

HPLC grade water from water purification system as described earlier

Chromatography tubing: internal tubing diameter is 0.010 mm except in the following locations: between the guard column and column and between the column and detector the internal tubing diameter is 0.007 mm. Tubing can either be stainless steel or PEEK (Upchurch, Oak Harbor, WA)

Columns: Ultrasphere ODS-DABS, C18, 5 μm 250 × 4.6 mm (Beckman, Fullerton, CA) or Columbus or Luna C_{18}, 5 μm, 250 × 4.6 mm (Phenomenex, Torrance, CA)

Guard column: C_{18} 7 μm 3.2 × 1.5 mm (Applied Biosystems, San Jose, CA)

Amber vials (for Waters Chromatography Systems, Waters, Milford, MA; for Hewlett-Packard Chromatography Systems, Sunbrokers, Wilmington, NC)

Graphite filters (ESA)

Mobile phase filters (Durapore Hydrophilic 0.22 μm, Millipore)

All other reagents/materials are HPLC grade

Chromatography Conditions. The mobile phase has 24–60% methanol, dependent on sample source and the compound analyzed (Fig. 1). For ascorbate, the mobile phase is 25% for plasma and urine samples and 30% for other samples. The higher the percentage methanol, the faster ascorbate elutes. The advantage of 25% mobile phase is distinct separation from uric

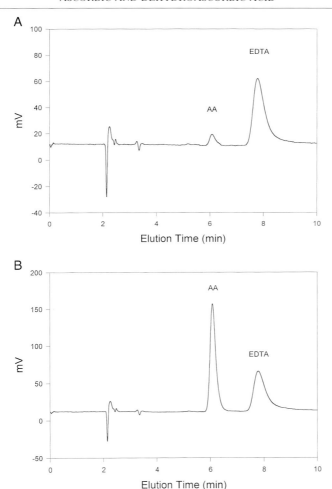

FIG. 1. Chromatography and electrochemical detection of ascorbic acid in standards (A–C), human neutrophils (D), and human plasma (E). The ascorbate peak is indicated by AA in all figures. The peak following ascorbate is EDTA. Ascorbate standards were prepared in 30% methanol/water/1 mM EDTA, injection volume 10 μl; A: 1 pmol/10 μl, B: 10 pmol/10 μl, C: 100 pmol/10 μl. Human neutrophils (D) (3×10^6) were extracted in 500 μl 60% methanol/water/1 mM EDTA. After centrifugation the supernatant was processed as described in the text and 10 μl was injected. The ascorbate peak corresponds to approximately 20 pmol ascorbic acid. Human plasma (E) (200 μl) was extracted with 90% methanol/water/1 mM EDTA (800 μl) and processed as described in the text. The ascorbate peak represents 44 pmol/10 μl injection, corresponding to an original plasma concentration of 22 μM. The peak preceding ascorbate is uric acid.

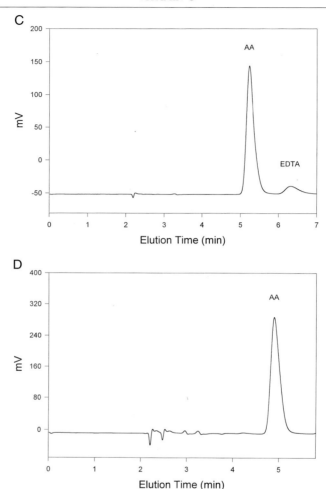

FIG. 1. (*Continued*)

acid, which elutes approximately 1–2 min before ascorbate. For isoascorbic acid, use 24% methanol and one of the Phenomenex columns (Fig. 2). For 6-chloroascorbate, use 60% methanol.

The mobile phase contains 0.05 M sodium phosphate monobasic, 0.05 M sodium acetate anhydrous, 189 μM dodecyltrimethylammonium bromide, and 36.6 μM tetraoctylammonium bromide. All concentrations are final concentrations. The methanol concentration is 24–60%, and pH is adjusted to 4.8. The flow rate is 1.0 ml/min.

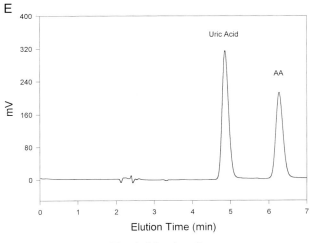

FIG. 1. (*Continued*)

To make 1 liter 25% mobile phase, add 6.899 g sodium phosphate monobasic, 6.8 g sodium acetate anhydrous, and 0.05 g dodecyltrimethylammonium chloride in 750 ml water. Separately, add 0.02 g tetraoctylammonium bromide to 250 ml methanol. Add both solutions together and adjust

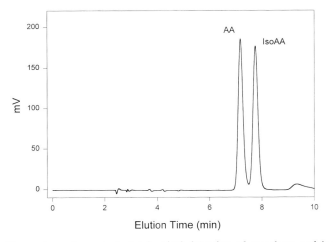

FIG. 2. Chromatography and electrochemical detection of ascorbate and isoascorbate, which were present in solution together each at a concentration of 10 μM in 60% methanol/water/1 mM EDTA. Ten microliters of solution (100 pmol of ascorbate and isoascorbate) was injected. Ascorbate eluted at 7 min (AA) and isoascorbate at 8 min (IsoAA). The mobile phase was 24% methanol.

to pH 4.8 with orthophosphoric acid. Before use, filter the final solution with 0.22-μm filters. For higher percentage methanol, adjust methanol and water accordingly.

The column should be equilibrated at least 17 hr (i.e., overnight) in mobile phase at 1 ml/min that is not recycled. For isoascorbic acid separation, equilibration should be for at least 48 hr. The column should be washed once every 2–3 weeks using 2 liters of 30% methanol/water at 1 ml/min without recycling.

The retention time of ascorbate is dependent on column batch, column type, percentage methanol, and days in mobile phase. After column equilibration in 30% methanol, the ascorbate retention time is usually between 4 and 5 min and in 25% methanol is 6–7 min.

The guard column is placed after the autosampler and before the column. Graphite filters are placed immediately before the guard column, the column, and the detector cell.

Passivation. Acid cleaning (passivation) is necessary on initial use of the chromatography system and once every 6–12 months thereafter. Prior to passivation, 1 liter of 30% methanol/water is run through the system and the column and guard column are removed. Passivation is accomplished with all flow rates at 1 ml/min as follows: H_2O for 1 hr, 6 N nitric acid for 45 min, H_2O flush for 1 hr, 6 N nitric acid for 45 min, and H_2O for at least 30 hr. After passivation the system is placed in 30% methanol/water for 1 hr and the columns are placed on line. For Waters instrumentation, the autosampler purge button must be activated every 15 min until the second hour of the 30-hr H_2O flush and then every hour until completion, except overnight.

Detector Use. Detector settings are 0.00 V, positive polarity, for detector 1 and 0.25 V, positive polarity, for detector 2. Range settings are dependent on pmol ascorbate/10-μl injection. As examples, the range is 500 nA for 1 pmol/10 μl and 2–5 μA for 100 pmol/10 μl.

Standard and Sample Preparation. Ascorbate standards are prepared in 30 or 60% methanol/water with 1 mM EDTA. Standards are prepared for 0.5–200 pmol/10 μl to be injected, depending on the desired sensitivity. Standards are run before and after samples to verify assay stability conditions. Amber vials are always used for samples and standards.

Plasma, serum, or urine samples are diluted 1 volume of sample: 4 volumes 90% methanol/water/1 mM EDTA (i.e., 200 μl plasma, 800 μl methanol/water/EDTA).[9] Solutions are placed on light-protected ice for 5 min to precipitate protein and are spun at 16,000 g for 10 min. The supernatant is either placed on dry ice for 5 min and then stored at $-70°C$ or a 100-μl aliquot is pipetted to an amber vial and placed in the 4° autosampler for analysis within 12 hr. The pellet can be saved for protein analysis if desired.

For cell samples, cells are pelleted and a convenient volume (i.e., 500 μl) of 60% methanol/water/1 mM EDTA is added. The sample is vortexed and processed as described earlier. For plated cells, 60% methanol/water/1 mM EDTA is added directly to tissue culture wells.[10,11]

For urine samples containing small amounts of ascorbate it may be necessary to decrease uric acid with uricase. Urine is diluted as described earlier with 90% methanol/water/1 mM EDTA. The sample is mixed with an equal volume (i.e., 50 μl) of 50 mM HEPES, pH 8.5, and then 10 μl uricase (0.5 U/100 μl H_2O, Sigma) is added. After incubating the light-protected mixture at room temperature for 10 min, it is placed in the autosampler for analysis.

Dehydroascorbic Acid Measurements

Assay Principles

Dehydroascorbic acid assays are not as advanced as those for ascorbate. As for ascorbate, HPLC is the most sensitive and specific means to measure dehydroascorbic acid.[1,2] The difficulty with dehydroascorbic acid assays is detection. Although ascorbate is detected directly with electrochemistry, there is no direct sensitive method to detect dehydroascorbic acid. An ideal dehydroascorbic acid detection method does not exist, and investigators must understand current limitations and choose detection accordingly.

One detection principle for dehydroascorbic acid is to derivatize it and to detect the derivatized product.[1,12,13] Although derivatization is effective for detecting myriad compounds, we feel the technique should not be used with dehydroascorbic acid. Because dehydroascorbic acid and ascorbate can be interconverted nonenzymatically by oxidation or reduction, derivatization may inadvertently shift equilibrium between the two species. In particular, artificially high dehydroascorbic acid concentrations may result. Other compounds in samples can affect derivatization or generate interfering substances. On-line HPLC derivatization has been reported.[13,14] However, this procedure is difficult, is usually relatively insensitive for biological samples, is subject to similar problems as off-line derivatization, and may be affected by dehydroascorbic acid formation during the procedure.

A second principle for dehydroascorbic acid detection is to measure ascorbate in reduced and unreduced samples. Subtraction of the latter from

---

[10] P. W. Washko, D. Rotrosen, and M. Levine, *J. Biol. Chem.* **264**, 18996 (1989).
[11] R. W. Welch, P. Bergsten, J. D. Butler, and M. Levine, *Biochem. J.* **294**, 505 (1993).
[12] F. Tessier, I. Birlouez-Aragon, C. Tjani, and J. Guilland, *Int. J. Vit. Nutr. Res.* **66**, 166 (1996).
[13] I. Koshiishi and T. Imanari, *Anal. Chem.* **69**, 216 (1997).
[14] J. T. Vanderslice, D. J. Higgs, G. R. Beecher, H. E. Higgs, and J. Bouma, *Intl. J. Vitam. Nutr. Res.* **62**, 101 (1992).

the former yields dehydroascorbic acid. Advantages of this method are that it uses known ascorbate HPLC assay and detection principles. Disadvantages are that dehydroascorbic acid detection is indirect and subtraction is necessary. Because ascorbate is present in great excess of dehydroascorbic acid in most biological samples, a large number is subtracted from a large number for dehydroascorbic acid determination, introducing the possibility of error.

Dehydroascorbic Acid Measurement: HPLC with Coulometric Electrochemical Detection

The disadvantages of current derivatization assays currently outweigh the disadvantages of reduction assays, and reduction assays are those of choice despite their imperfections. Reduction methods are compatible with amperometric and coulometric detection systems for ascorbate. We prefer the HPLC reduction assay because it utilizes coulometric electrochemical detection.[7] Specific advantages of this technique are that the coulometric assay most clearly addresses issues of stability, sensitivity, interference, and specificity. A disadvantage of the reduction technique with coulometric electrochemical detection is that the reducing agent currently used must be extracted before assay. To avoid artifact, extraction must be practiced and performed carefully, preferably with an ascorbate control as described next.

Dehydroascorbic Acid Assay Method

Because dehydroascorbic acid is reduced and detected as ascorbate, the techniques are those described earlier for ascorbate with respect to instrumentation, materials, chromatography conditions, and detector use.

Additional Materials

2,3-Dimercapto-1-propanol and ethyl ether (Aldrich, Milwaukee, WI)
Bromine (Fluka)

Dehydroascorbic Acid Preparation. The stock dehydroascorbic acid solution is always prepared fresh from ascorbate by bromine oxidation.[15] An ascorbate solution final concentration (20 mM) is prepared in 1 ml H_2O, pH 7. Bromine (5 μl) is added, and the solution is vortexed for 15 sec and purged gently with nitrogen gas for 10 min. Within 5 min the solution color should change from yellow to clear. Bromine addition and purging are performed on ice. The stock solution is not stored for more than 1 hr on light-protected ice. The stock solution is diluted as needed for standards

---

[15] P. W. Washko, Y. Wang, and M. Levine, *J. Biol. Chem.* **268**, 15531 (1993).

using 30% methanol/water/1 mM EDTA. Diluted standards should be reduced immediately as described later and then processed as ascorbate samples in the autosampler. Because commercial dehydroascorbic acid varies in purity, we do not recommend its use.

Dehydroascorbic Acid Reduction. Concentrated 2,3-dimercapto-1-propanol is diluted to give a final concentration of 10 mM in 30% methanol/water/1 mM EDTA. An equal volume of dehydroascorbic standard and 10 mM 2,3-dimercapto-1-propanol is mixed and incubated in the dark for 10 min at room temperature. The solution is extracted with 3 volumes of water-saturated ethyl ether to remove 2,3-dimercapto-1-propanol. Three extractions are performed. The solution is purged with nitrogen gas by gently bubbling for 2 min, transferred to an amber vial, and placed in the autosampler. Samples are reduced in the same way as standards using convenient volumes (i.e., 0.2 ml).

Practice is necessary to control for inadvertent evaporation of water/methanol. We suggest that ascorbate standards are prepared as described earlier and are treated as dehydroascorbic acid samples with reduction and extraction. Recovery of ascorbate should be 100%.

Detection of Radiolabeled Ascorbate and Dehydroascorbic Acid

[^{14}C]Ascorbate and [^{14}C]dehydroascorbic acid can be detected using an in-line radioactivity detector alone (Packard, Downers, Grove, IL) (Fig. 3)

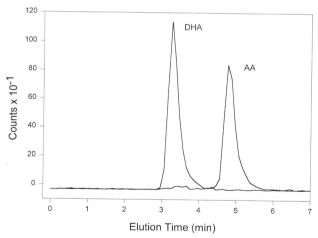

FIG. 3. Ascorbate (AA) and dehydroascorbic acid (DHA) chromatography with radiomatic detection. Pure standards were prepared of [^{14}C]dehydroascorbic acid (28 μM) or [^{14}C]ascorbate (23 μM) in 60% methanol/water/1 mM EDTA, and 50 μl was injected using mobile phase 30% methanol. DHA elutes at approximately 3.3 min and AA at 5 min.

or an electrochemical detector followed by an in-line radioactivity detector. Assay conditions are the same as those described earlier. Sensitivity is dependent on specific activity, but is usually not less than 20 pmol/injection. Although of limited use, this method can be quite useful when it is necessary to determine purity or intracellular processing of radiolabeled materials.

Future Goals

There are drawbacks to ascorbate and dehydroascorbic acid assays, and those of the latter are certainly more substantial. There is no direct method to detect dehydroascorbic acid independent of ascorbate detection, and such detection is needed. Without it, assays must continue to rely on reduction and subtraction. The dehydroascorbic acid reduction method could be improved by finding a reducing agent that does not require extraction and also does not interfere with chromatography and electrochemical detection. Disadvantages of ascorbate assays are that a dedicated operator is needed for electrochemical detection and that initial start-up costs are high. There is no ascorbate assay using a simpler detection method that also accounts for stability, sensitivity, specificity, and substance interference. Until a direct dehydroascorbic acid detection method is developed, one possible solution to these problems is to search for a dehydroascorbic acid reducing agent that does not interfere with chromatography and detection. Ideally, such a reducing agent should be compatible not only with electrochemical detection, but with a simpler ascorbate detection method, such as ultraviolet (UV) detection.[16] Although ultraviolet detection sacrifices high sensitivity, it remains possible that an HPLC UV assay could still account for stability, specificity, and substance interference. Ultraviolet detection should still be sensitive enough to measure ascorbate in most plasma and urine samples and perhaps in some cell samples if the cell number is not limiting. We look forward to solutions to these problems.

---

[16] S. A. Margolis and T. P. Davis, *Clin. Chem.* **34**, 2217 (1988).

[7] Analysis of Water-Soluble Antioxidants by High-Performance Liquid Chromatography with Electrochemical Detection

By ANN M. BODE and RICHARD C. ROSE

Introduction

Water-soluble antioxidants such as ascorbic acid and glutathione are receiving substantial attention because of their potential role in maintaining optimal health. The endogenous level of one or more antioxidants may reflect the health of a particular tissue. Therefore, reliable, quantitative measurements that combine sample stability, detection sensitivity, specificity, and little or no interference by other substances in the sample would have considerable application. High-performance liquid chromatography coupled with electrochemical detection (HPLC/EC) for analysis of water-soluble antioxidants has been available for several years. These methods have become quite popular due to advances in the technology of electrochemical detectors, ease of use, and the ability to automate the analysis. Compared to colorimetric or spectrophotometric methods, HPLC/EC provides a relatively fast, quantitative method that is reliable, highly sensitive, specific, and allows no interference from nonrelated substances within the sample and is currently the method of choice.

The numerous methods and techniques used over the years to quantitate ascorbic acid and dehydroascorbic acid (DHAA) have been reviewed.[1] Levine and co-workers[2] were the first group to use HPLC coupled with coulometric electrical detection for measuring ascorbate and DHAA in a variety of cells[3–5] and in serum and plasma.[6] The measurement of ascorbate and DHAA has been described in homogenates from whole tissue samples.[7] The simultaneous measurement of ascorbate, reduced glutathione, and uric

---

[1] P. W. Washko, R. W. Welch, K. R. Dhariwal, Y. Wang, and M. Levine, *Anal. Biochem.* **204**, 1 (1992).
[2] P. W. Washko, W. O. Hartzell, and M. Levine, *Anal. Biochem.* **181**, 276 (1989).
[3] R. W. Welch *et al.*, *J. Biol. Chem.* **270**, 12584 (1995).
[4] J. D. Butler, P. Bergsten, R. W. Welch, and M. Levine, *Am. J. Clin. Sci.* **54**, 1144S (1991).
[5] P. Bergsten, R. Yu, J. Kehrl, and M. Levine, *Arch. Biochem. Biophys.* **317**, 208 (1995).
[6] K. R. Dhariwal, W. O. Hartzell, and M. Levine, *Am. J. Clin. Nutr.* **54**, 712 (1991).
[7] D. A. Schell and A. M. Bode, *Biomed. Chromatogr.* **7**, 267 (1993).

acid has been successful in liver tissue[8] and in vitreous humor[9]; these three plus cysteine and tyrosine were measured concurrently.

Materials and Equipment

Chemicals should be HPLC grade or of the highest grade possible. Coulometric electrochemical detection requires highly purified water completely free of electrochemically active contaminants that can increase background current and decrease sensitivity. Water should therefore be as pure as possible, preferably deionized, subjected to carbon adsorption and ultrafiltration, and doubly glass distilled. In addition, before use, all solutions (e.g., mobile phase, metaphosphoric acid) are filtered through a 0.22-μm filter (Millipore Corp., Bedford, MA).

List of Materials

  HPLC grade monobasic potassium phosphate
  Potassium hydroxide
  Bromine
  Metaphosphoric acid
  HPLC grade orthophosphoric acid
  Ethylenediaminetetraacetic acid (EDTA)
  Thiourea
  Monobasic and dibasic sodium phosphate
  2-Mercaptoethanol

HPLC/EC System

The HPLC system consists of minimally, one high-pressure pump; a hand injector or, preferably, an automatic injector with a 20-μl injection loop; a 2 cm × 2 mm ID guard column containing 40-μm particles; and a Waters radial compression separation system (Waters Associates, Milford, MA) consisting of a radial compression module model RCM-100 that contains a C_{18} (10 μm) reversed-phase Waters cartridge type column or equivalent. Data are collected with an analog interface module and a personal computer. The computer is used to integrate and interpret the resulting chromatogram, with the aid of software specific to the HPLC system. The samples are injected onto the column in a mobile phase consisting of 0.2 M KH_2PO_4, pH 3.0, at a flow rate of 1 ml/min. The ascorbic acid content is detected using an electrochemical detector such as the ESA Model 5100A

---

[8] R. C. Rose and A. M. Bode, *Biochem. J.* **306,** 101 (1995).
[9] R. C. Rose, R. Gogia, and S. P. Richer, *Exp. Eye Res.* **64,** 807 (1997).

Coulochem multielectrode electrochemical detector (ESA, Inc., Bedford, MA) and an ESA Model 5010 dual electrode analytical cell (ESA, Inc.). The ESA analytical cell contains two porous graphite in-line working electrodes encapsulated with associated reference electrodes, counter electrodes, and heat sink. In-line high-pressure carbon filters (ESA, Inc.) are placed between the analytical column and the electrode.

Hydrodynamic Voltammetry Plot

The appropriate detector settings for optimal antioxidant (e.g., ascorbic acid) sensitivity and selectivity are determined by generating a graph of response or current versus potential or voltage. By reviewing the hydrodynamic voltammetry (HDV) plot, the optimal voltage is selected for both electrodes so a maximum response is obtained while minimizing background noise from interfering compounds. Each antioxidant of interest for assay must have its own HDV curve generated in order to obtain the proper detector settings. The procedure for making an HDV plot begins with the preparation of a standard containing the antioxidant of interest. Although the method for the generation of the HDV curve is specific to the electrochemical detector used and is available from the manufacturer, some generalities apply. For example, the Model 5010 analytical cell contains two coulometrically efficient, porous graphite working electrodes placed in series. In preparing the HDV plot, detector 1 is set at a very low potential (-0.5 V) because the optimal potential for detector 2 needs to be determined. The potential at detector 2 is varied with each injection. To begin the determinations, detector 2 is set at -0.4 V, the antioxidant solution is injected, and the response is recorded. For the next injection, the potential at detector 2 is increased by 0.05 to -0.35 V, the sample is injected, and the response is recorded. This procedure is repeated until the response reaches its maximum, as evidenced by no change in response (peak height) for increases in potential. Figure 1 illustrates a sample HDV plot generated for the determination of optimal electrochemical detector settings for ascorbic acid. Results indicate that detector 1 should be set at -0.3 V and detector 2 be set at 0.0 V.

Sample Preparation

Probably the most important component of an accurate analysis is proper sample preparation. Factors that accelerate degradation of the component of interest should be minimized during the entire procedure. Like many redox compounds, ascorbic acid and DHAA are unstable and un-

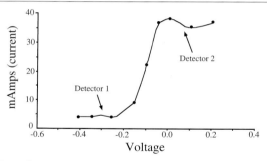

FIG. 1. Hydrodynamic voltammetry plot for ascorbic acid. Optimal detector settings were determined by injecting a standard solution of ascorbic acid (1 mM). In preparing the HDV plot, detector 1 was set at −0.5 V and the potential at detector 2 was varied with each injection. To begin the determinations, detector 2 was set at −0.4 V, ascorbate injected, and the response recorded. For the next injection, the potential at detector 2 was increased by 0.05 to −0.35 V, the sample injected, and the response recorded. The procedure was repeated until the response reached its maximum, as evidenced by no change in response (peak height) for increases in potential.

dergo spontaneous oxidation or hydrolysis, respectively.[10] The degradation of samples may be influenced by temperature, concentration within the sample, light, pH, oxygen (air), and the presence of oxidizing or hydrolyzing substances within the sample itself. For optimal stabilization, the sample is extracted on ice with a small amount of 0.3–2% metaphosphoric acid containing EDTA (0.1 mM) and thiourea (1 mM). For measurement of both ascorbate and DHAA, we have found that using 0.3% metaphosphoric acid stabilizes the sample but also allows complete reduction of the sample for the subsequent measurement of total ascorbic acid. Small samples are minced immediately and homogenized on ice in *preweighed* vials containing 0.1–0.5 ml (approximately 3:1 acid to sample) 0.3% metaphosphoric acid, EDTA, and thiourea. Sample and tube are then reweighed to determine sample weight as the difference between the preweight of the vial and the vial containing the sample. Larger samples may be frozen in liquid nitrogen, weighed, and then homogenized in the acid solution at a ratio of 3 ml acid/ 1 g tissue. Following homogenization, a small amount of an inert gas (e.g., nitrogen or argon) may be blown over the sample to displace oxygen in the tube and then tubes are capped. Samples are centrifuged at 19,000g for 20 min in a refrigerated centrifuge and the pellet is discarded and the supernatant fraction saved for analysis. Samples may be analyzed immediately or frozen for 2–3 days at −70°; once samples are thawed, they should be analyzed immediately. This method may also be used for the preparation

[10] A. M. Bode, L. Cunningham, and R. C. Rose, *Clin. Chem.* **36,** 1807 (1990).

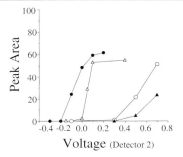

FIG. 2. Hydrodynamic voltammetry plot for simultaneous measurement of ascorbic acid (●), uric acid (△), cysteine (peak area/10) (○), and glutathione (▲). Detector 1 was set at −0.1 V and detector 2 as indicated.

of samples for analysis of other water-soluble antioxidants, including glutathione, cysteine, tyrosine, and uric acid.

Analysis of Ascorbic Acid/DHAA

For analysis of ascorbic acid, the prepared sample is diluted 1:10 with 10% metaphosphoric acid, placed into vials, and analyzed by HPLC/EC.

Whereas ascorbic acid can be detected directly by coulometric detection, DHAA must be measured indirectly. One aliquot of the sample is first assayed for ascorbic acid. A second aliquot is removed and added to a vial containing 10 mM 2-mercaptoethanol, which reduces DHAA to ascorbic acid for detection. This step is performed in a sodium phosphate buffer (pH 7.4) with 1 mM thiourea and 0.1 mM EDTA. The mixture is allowed to react for 10 min at room temperature and then is stopped with the addition of 20% MPA (one-half original reaction volume). The difference in ascorbic acid content between the two samples is the DHAA content of the sample. This method has the advantage of complete reduction with 100% recovery of DHAA; no loss of ascorbic acid or DHAA occurs during the reduction; and no extraction of 2-mercaptoethanol is required prior to analysis by HPLC/EC due to the different retention time of 2-mercaptoethanol (~8 min) and ascorbate (~5 min).[7,11] Reduction of samples with 2,3-dimercapto-1-propanol has also shown to result in 100% recovery of DHAA and no loss of sensitivity; however, the dimercapto-1-propanol must be extracted prior to analysis.[12] In addition, both glutathione and glutathione disulfide can be measured in homogenates or extracts of whole tissues using

[11] A. M. Bode and R. C. Rose, *J. Micronutr. Anal.* **8,** 55 (1990).
[12] K. R. Dhariwal, P. W. Washko, and M. Levine, *Anal. Biochem.* **189,** 18 (1990).

this technique. Reduction of the sample with 2-mercaptoethanol results in the reduction of GSSG to GSH, and recovery of GSSG is approximately 100%. This method was not as successful in determining both the reduced form of uric acid and its oxidized by-product.

Simultaneous Measurement of Water-Soluble Antioxidants

An emerging concept in medicine is that as long as antioxidant defenses are sufficient to prevent damage by reactive oxygen species, an organism remains healthy. Tissue damage and disease occur when the prooxidant activity is greater than the antioxidant protection. The redox state within a cell is influenced by the ratio of reduced substances to reduced plus oxidized substances. The relative amounts of reduced and oxidized forms of antioxidants therefore give an indication of tissue or cellular health. The simultaneous measurement of certain water-soluble antioxidants could be very useful in determining the redox state of a tissue.

Prior to measurement, the response of the electrochemical detector to each antioxidant of interest is determined by making repeated injections of standards to generate HDV plots. Detector 1 is usually maintained at -0.1 V and, depending on the antioxidants measured, detector 2 is set at various potentials from 0.2 to 0.8 V. When measuring glutathione, uric acid, and ascorbate simultaneously, glutathione requires the most positive potential (0.7–0.8 V) whereas the others are detectable at lower potentials. For complete and accurate measurement of antioxidants simultaneously, detector 2 should be set at the highest positive potential as determined from the HDV curve. Figure 2 illustrates an HDV curve generated for ascorbate, uric acid, cysteine, and glutathione.

This technique has been applied to the analysis of the antioxidant state in ocular fluids.[9] Identification and quantification of electrochemically active components of ocular fluids are particularly challenging because of the small size of samples that can be obtained from many animal sources. For example, the vitreous humor sample from rat eyes is usually less than 20 μl total volume. The sensitivity and selectivity of HPLC/EC are appropriate for evaluating small tissue samples. Vitreous or aqueous humor samples are obtained by a needle connected to a syringe. The size of the needle depends on the animal species and whether vitreous humor or aqueous humor is being obtained. Equal volumes of humor fluid and HPLC buffer (pH 3.0) are mixed and centrifuged in a refrigerated microfuge for 2 min at 14,000g to give a nominally protein-free sample. The sample is stable under these conditions. Following centrifugation, samples are analyzed by HPLC/ECD.

Acknowledgment

This work was supported by grants from the National Institutes of Health, Illinois Society for Prevention of Blindness, and American Optometric Foundation and NIH Grant DK-47953.

[8] Ascorbic Acid Recycling in Rat Hepatocytes as Measurement of Antioxidant Capacity: Decline with Age

By JENS LYKKESFELDT and BRUCE N. AMES

Introduction

Oxidative damage is considered to be partly responsible for the development of degenerative diseases such as cancer and atherosclerosis and even aging in general.[1,2] Concentrations of ascorbic acid and its oxidized metabolite dehydroascorbic acid, the most important water-soluble antioxidant couple,[3] have for some time been considered biomarkers of oxidative stress.[4-7] Ascorbic acid analyses have been performed with plasma and tissue in a wide variety of studies. However, a more detailed assessment of the antioxidant capacity on the cellular level is sometimes necessary for the interpretation of the often subtle differences observed. The measurement of ascorbic acid recycling, i.e., the reduction of dehydroascorbic acid to ascorbic acid, provides valuable information about the redox capacity of cells under basal, activated, and stressed conditions.

Ascorbic acid recycling has previously been measured in cyanobacteria,[8] human neutrophils[9] and erythrocytes,[10,11] liver homogenate,[12] and intact hepatocytes.[13] In these reports, dehydroascorbic acid concentrations rang-

---

[1] M. K. Shigenaga, T. M. Hagen, and B. N. Ames, *Proc. Natl. Acad. Sci. U.S.A.* **91,** 10771 (1994).
[2] K. B. Beckman and B. N. Ames, *Physiol. Rev.* **78,** 547 (1998).
[3] B. Frei, L. England, and B. N. Ames, *Proc. Natl. Acad. Sci. U.S.A.* **86,** 6377 (1989).
[4] R. C. Rose, *Biochem. Biophys. Res. Commun.* **169,** 430 (1990).
[5] K. Koyama, K. Takatsuki, and M. Inoue, *Arch. Biochem. Biophys.* **309,** 323 (1994).
[6] J. Lykkesfeldt, H. Priemé, S. Loft, and H. E. Poulsen, *Br. Med. J.* **313,** 91 (1996).
[7] J. Lykkesfeldt, S. Loft, J. B. Nielsen, and H. E. Poulsen, *Am. J. Clin. Nutr.* **65,** 959 (1997).
[8] E. Tel-Or, M. Huflejt, and L. Packer, *Biochem. Biophys. Res. Commun.* **132,** 533 (1985).
[9] P. W. Washko, Y. Wang, and M. Levine, *J. Biol. Chem.* **268,** 15531 (1993).
[10] J. M. May, Z. C. Qu, and R. R. Whitesell, *Biochemistry* **34,** 12721 (1995).
[11] J. M. May, Z. C. Qu, R. R. Whitesell, and C. E. Cobb, *Free Radic. Biol. Med.* **20,** 543 (1996).
[12] L. E. Rikans, T. R. Lopez, and K. R. Hornbrook, *Mech. Aging Dev.* **91,** 165 (1996).
[13] L. Braun, F. Puskas, M. Csala, E. Gyorffy, T. Garzo, J. Mandl, and G. Banhegyi, *FEBS Lett.* **390,** 183 (1996).

ing from micromolar to millimolar have been used in recycling experiments of a few minutes to 1-hr duration. Because of the unstable nature of dehydroascorbic acid, it may be most reliable to use a short incubation period when no lag time is observed in a system. In our experiments, we were interested in the response under physiologically relevant conditions and therefore minimized the dehydroascorbic acid concentration, although still maintaining a linear response throughout the incubation period.

This chapter reports a method for measuring ascorbic acid recycling in hepatocytes that is fast and easy and the method is applied to study the effect of aging on ascorbic acid metabolism.

Materials and Methods

All materials are of analytical or better quality. Dehydroascorbic acid varies in quality in commercially available products. Groups that routinely work with dehydroascorbic acid prepare the compound fresh from ascorbic acid, e.g., by bromination[9] or by using ascorbic acid oxidase.[14,15] We tested the commercially available dehydroascorbic acid by GC-MS (unpublished observations) and found that products from Aldrich (Milwaukee, WI) and ICN Biochemicals (Costa Mesa, CA) (methanol complex) are impure and unsuitable for our experiments whereas the dimer sold by Fluka (Ronkonkoma, NY) is >97% pure. The dimer associates readily in water to the bicyclic hemiketal form, which is the preferred form in aqueous solution[16] (Fig. 1) and has been used successfully throughout our studies.

Ascorbic acid itself is also a relatively unstable compound and meticulous and consistent sample preparation is important in order to ensure reproducible data. Ascorbic acid oxidizes at room temperature, even in acidified samples, so a minimum of time should be spent on sample handling.

Hepatocyte Isolation

Cells are isolated from livers from male rats (Simonsen, Gilroy, CA) by collagenase perfusion as described.[17,18] This isolation method typically results in a cell suspension that maintains the proper representation of the cell types found in the intact organ. Three consecutive washes with 150 ml Krebs–Henseleit buffer, pH 7.4, are employed followed by aspiration to

[14] J. C. Deutsch and J. F. Kolhouse, *Anal. Chem.* **65,** 321 (1993).
[15] D. Every, *Anal. Biochem.* **242,** 234 (1996).
[16] K. Pfeilsticker, F. Marx, and M. Bockkisch, *Carbohydr. Res.* **45,** 269 (1975).
[17] T. M. Hagen, D. L. Yowe, J. C. Bartholomew, C. M. Wehr, K. L. Do, J.-Y. Park, and B. N. Ames, *Proc. Natl. Acad. Sci. U.S.A.* **94,** 3064 (1997).
[18] P. Moldéus, J. Hogberg, and S. Orrenius, *Methods Enzymol.* **52,** 60 (1978).

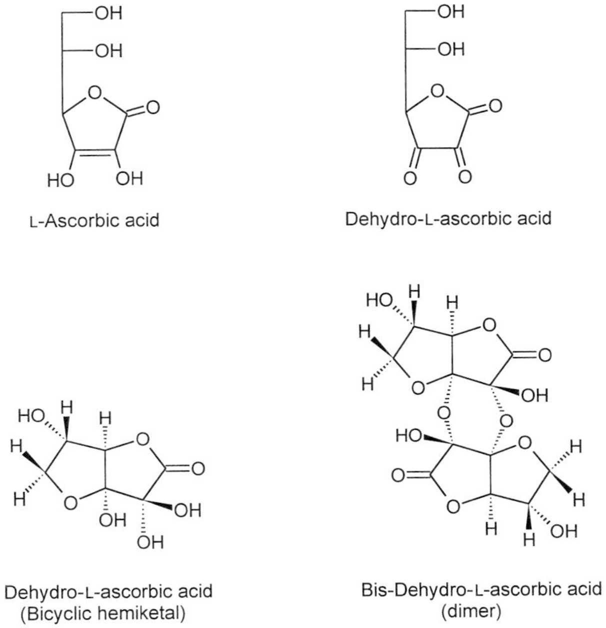

FIG. 1. Ascorbic acid and dehydroascorbic acid. The hydrated bicyclic hemiketal form is the preferred structure of dehydroascorbic acid in aqueous solution (>99%).

remove nonparenchymal cells, extracellular ascorbic acid, and excessive volume. Cells are kept in suspension in a pear-shaped flask rotating at 50 rpm at room temperature until used for the experiments. The cell number is assessed by using a hemocytometer, and viability is determined by the ability of cells to exclude trypan blue. All data are corrected for viability. In general, about 6×10^8 parenchymal cells are obtained from the isolation procedure.

Ascorbic Acid Recycling

Ascorbic acid is oxidized by quenching radicals and is rapidly reduced back to ascorbic acid intracellularly. In normal cells, no measurable concentration of dehydroascorbic acid is found. In experiments with dehydroascorbic acid, the instability of the compound must be considered. The half-life of dehydroascorbic acid at pH 7.0 is 6 min.[19] However, lowering pH improves stability considerably and by keeping the solution on ice at pH 4.0, stability

[19] A. M. Bode, L. Cunningham, and R. C. Rose, *Clin. Chem.* **36**, 1807 (1990).

is ensured for at least 1 hr. Because the assay pH is 7.0, the short half-life of dehydroascorbic acid was taken into account in planning the assay time to ensure that saturating conditions were being used.

Recycling Assay

The assay is based on the difference in ascorbic acid concentration in aliquots taken from an incubation mixture at zero and 5 min after the addition of dehydroascorbic acid. The dehydroascorbic acid solution is made immediately prior to the experiments as a 2-mg/ml solution in 5 mM sodium acetate buffer, pH 4.0. To 975 μl of cell suspension (2 × 10^6 cells/ml final concentration) is added 25 μl of dehydroascorbic acid solution (~280 μM final concentration; Fluka) and immediately after gentle mixing, a 200-μl aliquot of incubation mixture is added to an equal volume of freshly prepared 10% (w/v) metaphosphoric acid. This sample is then vortex mixed, centrifuged for 1 min at high speed, and stored at −20° until analysis. Meanwhile, the remaining cell suspension is incubated with dehydroascorbic acid for 5 min in an incubator set at 37° with gentle rocking, after which a second 200-μl aliquot is quenched as described earlier.

The stability of ascorbic acid under different storage conditions has been studied extensively.[20-23] No significant change was found in ascorbic acid concentration in acidified samples stored at −20° for at least 6 months.

Note that the cells are not homogenized, lysed, and otherwise disrupted. Experiments have shown (unpublished observations) that cells become permeable to ascorbic acid on treatment with metaphosphoric acid and that additional measures to disrupt cells only decrease the recovery of this labile compound.

Analysis is best performed by high-performance liquid chromatography (HPLC) and several methods are available.[24] The system we use involves reversed-phase HPLC with coulometric detection as described previously.[22] Briefly, the samples are thawed gently and 50 μl of acidified sample is mixed with 50 μl of 200 mM Trizma buffer (Sigma, St. Louis, MO), pH 9.0, giving a sample pH of about 2.5. The samples are placed in a chilled (2°) auto sampler for analysis. The area corresponding to ascorbic acid is integrated and compared to an authentic standard. The amount of reduced ascorbic acid is assessed by subtraction of the zero sample from the 5-min

---

[20] S. H. Rubin, E. DeRitter, and J. B. Johnson, *J. Pharm. Sci.* **65,** 963 (1976).
[21] M. A. Petersen and H. Berends, *Z. Lebensm. Forsch.* **197,** 546 (1993).
[22] J. Lykkesfeldt, S. Loft, and H. E. Poulsen, *Anal. Biochem.* **229,** 329 (1995).
[23] G. W. Comstock, E. P. Norkus, S. C. Hoffman, M. W. Xu, and K. J. Helzlsouer, *Cancer Epidemiol. Biomark. Prev.* **4,** 505 (1995).
[24] P. A. Motchnik, B. Frei, and B. N. Ames, *Methods Enzymol.* **234,** 269 (1994).

sample and can be expressed as nanomoles per minute per million cells. Control experiments showed no ascorbic acid production during the incubation period without the addition of dehydroascorbic acid. A reference range from 2.1 to 3.0 (nmol/min/10^6 cells) has been calculated based on samples from 10 rats of varying age.

Application with Young and Old Rats

The method just described was applied to a cell model in which parenchymal cells (hepatocytes) isolated from young (3–5 months) and old (25–29 months) male rats (Fischer 344, outbred albino, National Institute of Aging animal colonies) were subjected to oxidative stress. Five animals were used in each group.

To investigate the ability to respond to elevated oxidative stress, cells were preincubated for 1 hr at 37° in the presence 0, 300, and 500 μM *tert*-butyl hydroperoxide (Sigma) prior to assessment of ascorbic acid recycling. All data were corrected for viability.

Figure 2 shows the ascorbic acid recycling in hepatocytes isolated from young and old rats. There was no difference in recycling capacity between unstressed cells isolated from young and old animals. This is in agreement with previously published results based on experiments with liver homogenates.[12] Linear regression analysis showed no significant influence of the

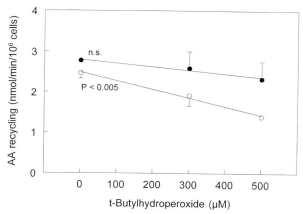

FIG. 2. Ascorbic acid (AA) recycling in hepatocytes isolated from young (●) or old (○) rats. Cells were preincubated at 37° in the presence of 0, 300, or 500 μM *tert*-butyl hydroperoxide, and ascorbic acid recycling was measured by incubating for 5 min with dehydroascorbic acid. Cells isolated from old rats showed a significant inverse correlation with *tert*-butyl hydroperoxide concentrations whereas cells isolated from young rats were not affected significantly by the treatment. Error bars are SEM.

tert-butyl hydroperoxide concentration on the recycling capacity in cells isolated from young animals (intercept = 2.80; slope = -0.00087; $r = 0.29$; $p = 0.37$), indicating an ability to respond to the increased oxidant load. However, in hepatocytes isolated from old animals, a significant inverse correlation between ascorbic acid recycling and *tert*-butyl hydroperoxide concentration was observed (intercept = 2.48; slope = -0.0021; $r = 0.75$, $p < 0.005$). Results indicate that old animals are capable of maintaining the same basal level of ascorbic acid metabolism as young animals but are unable to respond to a massive oxidative insult. Thus, based on these results it seems that susceptibility to oxidative stress increases with age. This is consistent with previous work from our laboratory that demonstrated that increased oxidative damage, as well as the decline in vital cellular functions, is consistently observed with age.[1,17,25]

Conclusions

The assessment of ascorbic acid recycling is a useful extension to the evaluation of ascorbic acid metabolism. The experiment is easy to carry out, but attention should be paid to sample instability. When carried out with appropriate caution, the assay provides new information about response and flexibility of the recycling of this important antioxidant and can be used as a measurement of antioxidant capacity. The susceptibility to oxidative stress, as measured by the ability of the hepatocytes to respond to an oxidative insult, is shown to decline with age.

Acknowledgments

The assistance of Dr. Tory M. Hagen and Vladimir Vinarsky in isolation of the rat hepatocytes is gratefully appreciated. The work was supported by Danish Natural Science Research Council Grant SNF9502434 to J.L. and National Cancer Institute Outstanding Investigator Grant CA39910 and NIEHS Center Grant ESO1896 to B.N.A.

[25] B. N. Ames, M. K. Shigenaga, and T. M. Hagen, *Biochim. Biophys. Acta* **1271**, 165 (1995).

Section III

Polyphenols and Flavonoids

[9] Detecting and Measuring Bioavailability of Phenolics and Flavonoids in Humans: Pharmacokinetics of Urinary Excretion of Dietary Ferulic Acid

By LOUISE C. BOURNE and CATHERINE A. RICE-EVANS

Introduction

Three major classes of antioxidants contribute to the protection of biomolecules from damage induced by oxidative and nitrative stress *in vivo*: preventive antioxidants that suppress free radical formation; antioxidants that scavenge free radicals, inhibiting their reactivity; and those that are involved in the repair process. The health benefits of fresh fruit and vegetables have been attributed largely to their constituent antioxidant vitamins and carotenoids, although they may derive from contributions from a large number of other diverse components and their individual or combined activities. Definitive evidence for the antioxidant properties of dietary agents[1] *in vivo* is derived from (i) evidence for their reactivity as scavengers of reactive oxygen and nitrogen species *in vitro* and/or for their ability to prevent the formation of reactive oxygen species by, for example, metal chelation; (ii) evidence for their uptake, absorption, and bioavailability in humans; and (iii) evidence for their bioactivity as antioxidants *in vivo*.

Over the last decade evidence has been accumulating that polyphenolic components of higher plants are highly efficacious antioxidants *in vitro*, functioning effectively as chain-breaking antioxidants[2-9] and scavengers of

---

[1] C. Rice-Evans, *in* "Antioxidant Food Supplements and Human Health." Academic Press, in press.

[2] H. Mangiapane, J. Thomson, A. Slater, S. Brown, G. D. Bell, and D. A. White, *Biochem. Pharmacol.* **43,** 445 (1992).

[3] S. Miura, J. Watanabe, M. Sano, T. Tomita, T. Osawa, Y. Hara, and I. Tomita, *Biol. Pharmacol. Bull.* **18,** 1 (1995).

[4] J. A. Vinson, Y. A. Dabbagh, M. M. Serry, and J. H. Jang, *J. Agric. Food Chem.* **43,** 2800 (1995).

[5] D. G. Yan, M. Zhou, and Y. Chen, *Acta Biochem. Biophys. Sinica* **28,** 106 (1996).

[6] L. Mathieson, K. E. Malterud, M. S. Nenseter, and R. B. Sund, *Pharmacol. Toxicol.* **78,** 143 (1996).

[7] N. Salah, N. J. Miller, G. Paganga, L. Tijburg, G. P. Bolwell, and C. Rice-Evans, *Arch. Biochem. Biophys.* **322,** 339 (1995).

[8] C. V. de Whalley, S. M. Rankin, J. R. Hoult, W. Jessup, and D. S. Leake, *Biochem. Pharmacol.* **39,** 1743 (1990).

[9] W. Jessup, S. M. Rankin, C. V. De Whalley, J. R. Hoult, and D. S. Leake, *Biochem. J.* **265,** 399 (1990).

peroxynitrite,[10,11] as well as direct scavengers of superoxide radical, hydroxyl radical, and singlet oxygen.[12-17] The catechol o-dihydroxy structure in the B ring of flavonoids has been emphasized as a key functional group for metal ion chelation[18-21]; however, the ability to redox cycle and promote prooxidant properties may confound these effects in flavonoids with specific structural features.[21] Until recently there has been a paucity of information on the absorption of flavonoids in humans (Table I).

Absorption of Flavonoids

The antioxidant functions of flavonoids and phenolic compounds *in vivo* will depend on the ways in which they are metabolized in the gastrointestinal tract and the structures of the resulting metabolites. Conjugation with glucuronide or with sulfate is generally considered to be the most common final step in the metabolism of intact flavonoids. A number of early studies hypothesized that flavonoids would not enter the circulation, either as natural glycosides or as aglycone hydrolysis products,[22] but would be cleaved by enzymes of the intestinal bacteria at the central heterocyclic ring, generating products with no antioxidant activity.[23-25] Furthermore, it has been proposed that glycosidic linkages may not be able to withstand

[10] A. Pannala, C. Rice-Evans, B. Halliwell, and S. Singh, *Biochem. Biophys. Res. Commun.* **232**, 164 (1997).
[11] A. Pannala, R. Razaq, B. Halliwell, S. Singh, and C. Rice-Evans, *Free Radic. Biol. Med.* **24**, 594 (1998).
[12] B. Havsteen, *Biochem. Pharmacol.* **32**, 1141 (1983).
[13] N. Cotelle, J. L. Bernier, J. P. Hénichart, J. P. Catteau, E. Gaydou, and J. C. Wallet, *Free Radic. Biol. Med.* **13**, 211 (1992).
[14] S. R. Husain, J. Cillard, and P. Cillard, *Phytochemistry* **26**, 2489 (1987).
[15] G. Sichel, C. Corsaro, M. Scalia, A. J. Di Bilio, and R. P. Bonomo, *Free Radic. Biol. Med.* **11**, 1 (1991).
[16] C. Tournaire, S. Croux, M. T. Maurette, I. Beck, M. Hocquaux, A. M. Braun, and E. Oliveros, *J. Photochem. Photobiol. B. Biol.* **19**, 205 (1993).
[17] J. Robak and R. J. Gryglewski, *Biochem. Pharmacol.* **36**, 317 (1987).
[18] M. Thompson, C. R. Williams, and G. E. P. Elliot, *Anal. Chim. Acta* **85**, 375 (1976).
[19] I. B. Afanas'ev, I. Dorozhko, V. Brodskii, V. A. Kostyuk, and A. I. Potapovitch, *Biochem. Pharmacol.* **38**, 1763 (1989).
[20] G. Paganga, H. Al-Hashim, H. Khodr, B. C. Scott, O. I. Aruoma, R. C. Hider, B. Halliwell, and C. Rice-Evans, *Redox Rep.* **2**, 359 (1996).
[21] J. Brown, H. Khodr, R. C. Hider, and C. Rice-Evans, *Biochem. J.* **330**, 1173 (1998).
[22] W. G. Clark and E. M. Mackay, *J. Am. Med. Assoc.* **143**, 1411 (1950).
[23] A. M. Hackett, *in* "Plant Flavonoids in Biology and Medicine" (V. Cody, E. Middleton, and J. Harborne, eds.), p. 177. A. R. Liss, New York, 1996.
[24] A. R. Ibrahim and Y. J. Abul-Hajj, *Xenobiotoca* **20**, 363 (1990).
[25] S. Baba, T. Furtuta, M. Fujioka, and T. Goramaru, *J. Pharm. Sci.* **72**, 1115 (1983).

TABLE I
EVIDENCE FOR FLAVONOIDS AND PHENOLICS IN HUMANS

Supplement	Observation	Ref.
Broccoli: 500 g daily for 12 days; rich in kaempferol and quercetin glycosides	Urine: kaempferol conjugates	27
Gingko biloba	Urine: detection of substituted benzoic acids at <30% of flavonoid administration	29
Grapefruit juice 20 ml ≡ 621 μM naringen-7-glucoside	Urine: naringenin conjugates	30
Citrus juice, grapefruit, pure compounds naringin, hesperidin	Urine: naringenin and hesperetin conjugates	31
Catechin 92.3 mg/kg body weight	Urine: 26% of administered dose excreted mainly as *m*-hydroxyphenylpropionic acid	32
3-*O*-Methylcatechin 1.72 g	Urine: 46% excreted as glucuronides and sulfates; 11–18 μg/ml in plasma within 2 hr	33
Decaffeinated green tea 88 mg epigallocatechin gallate 82 mg epigallocatechin 33 mg epicatechin gallate 32 mg epicatechin	Plasma 46–286 ng/ml epigallocatechin gallate 82–206 ng/ml epigallocatechin Epicatechin gallate not detected 40–80 ng/ml epicatechin Urine: only epicatechin and epigallocatechin detected (sulfated)	34
25 mg of flavonol glycoside equivalents consumed hourly for 5 hr Apricot Tomato Raspberries Apple	Urine Caffeic acid Caffeic acid, ferulic acid, *p*-coumaric acid } anthocyanidin conjugates/derivatives	47
Subjects on regular diet rich in fruit and vegetables	Plasma Quercetin glycosides Phloretin conjugate Anthocyanidin conjugate	48
Quercetin 64-mg aglycone equivalents from onion	Plasma concentration ~1 μM at 2 hr	50
Diosmin	Plasma: diosmetin 400 ng/ml peak level at 1 hr Urine: mainly glucuronide of *m*-hydroxyphenylpropionic acid	51

TABLE II
CHARACTERISTICS OF VOLUNTEERS

Subject	BMI (kg/m^2)	Age (years)	Gender	Tomatoes ingested (g)	Tomatoes ingested (g)/ body weight (kg)	Ferulic acid ingested (mg)
T101	18.26	25	F	360	8	21.71
T102	18.82	28	F	400	8	24.12
T103	19.15	22	F	368	8	22.19
T104	28.4	22	F	728	8	43.89
Mean	21.2 ± 4.8	24.3 ± 2.9				28.0 ± 10.6

the acidic environment of the stomach.[26] However, attention has been increasingly concentrated on the detection of excreted dietary polyphenols in humans *in vivo*. Ingestion of 500 g of broccoli, rich in kaempferol and quercetin, daily for 12 days by two volunteers revealed the presence of kaempferol conjugates in the urine, detected as free kaempferol on high-performance liquid chromatography (HPLC) analysis of the hydrolyzed urine and confirmed by liquid chromatography-mass spectrometry (LC-MS) analysis.[27] This procedure failed to detect quercetin, suggesting that the uptake of dietary quercetin from this source is limited in some way compared to that of kaempferol or that quercetin is metabolized more extensively during or after absorption.

Single bolus consumption of *Gingko biloba* extract (rich in flavonol glycosides) demonstrated extensive metabolism with excretion of substituted benzoic acids in the urine, accounting for less than 30% of flavonoids administered.[28,29] Other workers have shown that after administration of grapefruit juice (20 ml ≡ 621 μM naringin–naringen-7-glucoside/kg body weight), naringenin glucuronides were detected in the urine,[30] suggesting that cleavage of the sugar moiety by glycosidase from intestinal bacteria[12] is an early step in the metabolism of glycosides of naringenin. Such findings are supported by other studies on the absorption of flavonoids derived from naringin and hesperidin (hesperetin glucoside) from the gastrointestinal tract after the oral ingestion of pure compound, citrus juice, or whole

[26] K. R. Markham, *in* "Methods in Plant Biochemistry" (P. M. Dey and J. B. Harborne, eds.), p. 197. Academic Press, New York, 1989.
[27] S. E. Nielsen, M. Kall, U. Justesen, A. Schou, and L. O. Dragsted, *Cancer Lett.* **114,** 173 (1997).
[28] P. G. Pietta, P. L. Mauri, A. Bruno, and A. Rava, *J. Chromatogr.* **553,** 233 (1991).
[29] P. G. Pietta, C. Gardana, and P. L. Mauri, *J. Chromatogr. B* **693,** 249 (1997).
[30] U. Fuhr and A. L. Kummert, *Clin. Pharmacol. Ther.* **58,** 365 (1995).

TABLE III
CHROMATOGRAPHIC CONDITIONS FOR DETERMINATION
OF HYDROXYCINNAMATES

Component	Reversed-phase HPLC		
Column	Stainless steel (250 × 4.6 mm)		
Stationary phase	NovaPak C_{18} 4 µm (Waters, Watford, UK)		
Mobile phase	Solvent A: 20% methanol, 0.1% HCl		
	Solvent B: Acetonitrile		
Gradient (linear)	Time	%A	%B
	0	95	5
	10	95	5
	45	50	50
	55	95	5
Column temperature	30°		
Flow	0.8 ml/min		
Injection	30 µl		
Detection	320 nm for hydroxycinnamates		
Calculation	External standard method, peak area		
Standard	Ferulic acid (Extrasynthese, Paris)		
Retention time	Ferulic acid 16.2 min		
	Internal standard 26.6 min		
Run time	60 min		

grapefruit,[31] although it is not clear whether they are absorbed as glycosides or cleaved prior to absorption.

The uptake of catechin, a major constituent of red wine, and its gallate forms, major components of green tea, have been well studied, as have their urinary excretion. In 1971, Das[32] reported m-hydroxyphenylpropionic acid as the major urinary metabolite of catechin. Administration of radiolabeled 3-O-methylcatechin to three human volunteers revealed 46% in the urine in the form of glucuronides and sulfates within 2 hr of administration.[33] Studies investigating the excretion of constituents of decaffeinated green tea, mainly epicatechin, epigallocatechin, epicatechin gallate, and epigallocatechin gallate, have identified the two former flavonols in the sulfated form in the urine of the four supplemented volunteers but not those of their gallate esters.[34]

---

[31] B. Ameer, R. A. Weintraub, J. V. Johnson, R. A. Yost, and R. L. Rouseff, *Clin. Pharmacol. Ther.* **60,** 34 (1996).
[32] N. P. Das, *Biochem. Pharmacol.* **20,** 3435 (1971).
[33] A. M. Hackett, L. A. Griffiths, and M. Wermeille, *Xenobiotica* **15,** 907 (1985).
[34] M.-J. Lee, Z.-Y. Yang, H. Li, L. Chen, Y. Sun, S. Gobbo, D. A. Balentine, and C. S. Yang, *Cancer Epidemiol. Biomark. Prevent.* **4,** 33393 (1995).

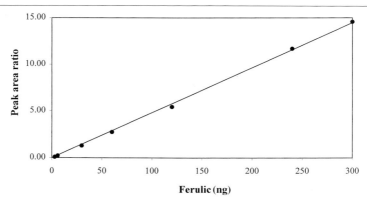

FIG. 1. Ferulic acid calibration plot.

Hydroxycinnamates

Hydroxycinnamic acids are among the most widely distributed phenylpropanoids in plant tissues, namely coumaric acid (*p*-hydroxycinnamic acid), caffeic acid (3,4-dihydroxycinnamic acid) and its quinic acid ester chlorogenic acid, and ferulic acid (4-hydroxy-3-methoxycinnamic acid), synthesized in plants from the shikimate pathway from phenylalanine or L-tyrosine. The antioxidant properties of the hydroxycinnamates have been demonstrated against peroxidizing polyunsaturated fatty acids in lipid systems and in low-density lipoproteins (LDL).[35–37] In addition to scavenging alkoxyl and peroxyl radicals, they have also been demonstrated to scavenge initiating free radical species such as ferrylmyoglobin.[35,38] Although much attention has focused on the detection of dietary flavonoids in humans, very few studies have focused on hydroxycinnamates.

The ability of ferulic acid to increase the resistance of LDL to oxidation mediated by metmyoglobin has been demonstrated to be operative through the scavenging of peroxyl radicals and to be effective from the aqueous phase.[39] The influence of this hydroxycinnamate to inhibit LDL oxidation induced by copper ions is dependent on concentration, at lower concentrations exerting a prooxidant effect.[37] Further investigations on the partitioning of ferulic acid in plasma have elucidated its association with the

---

[35] C. Castelluccio, G. Paganga, N. Melikian, G. P. Bolwell, J. Pridham, J. Sampson, and C. Rice-Evans, *FEBS Lett.* **368,** 188 (1995).
[36] J. Laranjinha, O. Vierira, L. Almeida, and V. Maderia, *Biochem. Pharmacol.* **51,** 395 (1996).
[37] L. Bourne and C. Rice-Evans, *Free Radic. Res.* **27,** 337 (1997).
[38] J. Laranjinha, L. Almeida, and V. Maderia, *Free Radic. Biol. Med.* **19,** 329 (1995).
[39] C. Castelluccio, G. P. Bolwell, C. Gerrish, and C. A. Rice-Evans, *Biochem. J.* **316,** 691 (1996).

aqueous phase and its greater efficacy as an antioxidant against LDL oxidation than the hydrophilic ascorbic acid.[39] Simple phenolics are also effective scavengers of reactive nitrogen species. They have been shown to protect against the nitration of tyrosine induced by peroxynitrite through mechanisms of competitive nitration or electron donation, depending on the specific structural characteristics of the hydroxyl groups.[10,11]

Ferulic acid and its dimers are ubiquitous components of the primary cell walls of plants.[40] The monomer is found conjugated covalently with mono- and disaccharides, plant cell wall polysaccharides, glycoproteins, polyamines, lignin, and hydroxy fatty acids in suberin and cutin. Several roles for ferulic acid in plants have been proposed, especially following dimerization. Thus it cross-links vicinal pentosan chains of arabinoxylans and hemicellulose in cell walls[41–43] associated with the cessation of cell wall expansion. Such cross-linking is also essential in the formation of barriers to invading pathogens.[44] Ferulate has also been used as a model compound in studying the formation of dehydrogenation polymers *in vitro* to understand the nature of these reactions in the wall.[45,46]

Analytical Approaches to Detection of Dietary Phenolics in Humans

HPLC analysis of urinary phenolics is performed according to Bourne and Rice-Evans.[47]

Volunteers

Four healthy volunteers (female), mean age 24.3 ± 2.9 years, BMI 21.2 ± 4.8 kg/m^2, consumed a single bolus of 360–728 g tomatoes (equivalent to 8 g/kg body weight), providing approximately 21–44 mg ferulic acid (Table II). Volunteers fast for 12 hr prior to the study and refrain from taking antioxidant supplements for 1 week prior to the study and from specific phenolic-rich dietary agents for 48 hr. These include bran cereals; whole grain products; vegetables such as broad beans, broccoli, brussel sprouts, cabbage, celery, chives, endive, eggplant, French beans, garlic, kale, leeks, lettuce, onions, parsley, radish, spinach, sweet peppers, and tomatoes;

---

[40] E. Graf, *Free Radic. Biol. Med.* **13**, 435 (1992).
[41] S. C. Fry, *Annu. Rev. Plant Physiol.* **37**, 165 (1986).
[42] K. Iiyama, T. B. T. Lam, and B. A. Stone, *Plant Physiol.* **104**, 315 (1994).
[43] S. C. Fry, *Annu. Rev. Plant Physiol. Mol. Biol.* **46**, 497 (1995).
[44] G. P. Bolwell, *Int. Rev. Cytol.* **146**, 261 (1993).
[45] J. Ralph, S. Quideau, J. H. Grabber, and D. Hatfield, *J. Chem. Soc. Perkin Trans. I* 3485 (1994).
[46] A. Zimmerlin, P. Wojtaszk, and G. P. Bolwell, *Biochem. J.* **299**, 747 (1994).
[47] L. Bourne and C. Rice-Evans, *Free Radic. Res.*, in press.

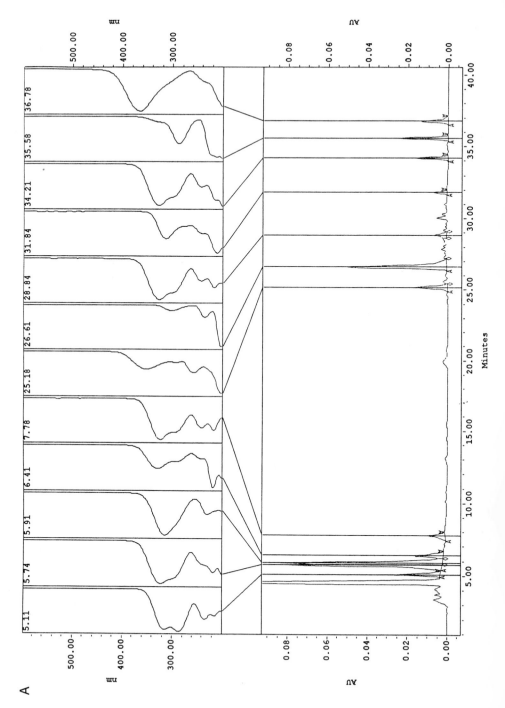

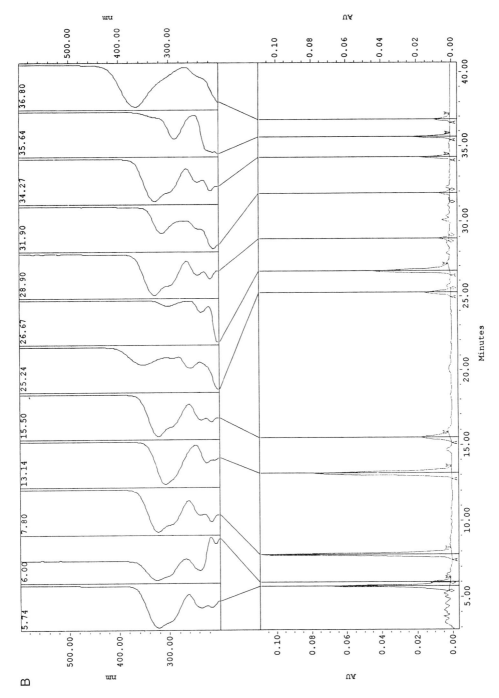

fruits such as apples, apricot, blackberries, cranberries, grapes, grapefruit, pears, plums, and strawberries; and beverages such as beer, coffee, fruit juices, tea, and wine. Urine samples are collected in sterile tubes immediately prior to and for 24 hr post tomato consumption and stored at $-70°$ until analysis.

Chemicals

Methanol and acetonitrile, all HPLC grade, are obtained from Rathburn Chemicals (Walkerburn, Scotland). All ferulic acid and all other hydroxycinnamates are obtained from Extrasynthese (ZI Lyon Nord, BP 62, 69730 Genay, France). Tomatoes (Flavia variety) are from a major local United Kingdom supermarket (Sainsbury). Elgastat UHP double-distilled water (18 MΩ grade) is used in all experiments. β-Glucosidase is from ICN Biomedicals Inc. (Ohio 44202). Salicyclic acid is purchased from Sigma (UK).

Standards

Stock solutions of the standards are prepared by dissolving 1–2 mg of sample into either methanol or mobile phase (20% methanol, 0.1% HCl). Stock solutions are stored at 0–4° and used within 4 weeks from the date of preparation, after HPLC analysis to check stability. Urine standards are prepared by the addition of ferulic acid stock solution to blank pooled human urine. The standards range from 0 to 300 ng of ferulic acid. Salicylic acid is used as an internal standard for the HPLC analysis.

Sample Preparation

The fruit is chopped into small pieces, placed in a round-bottom flask, the weight recorded, and freeze dried. The sample is lyophilized under liquid nitrogen, and the weight of the dried sample is recorded. For aqueous extraction, 0.5–1 g of freeze dried material is transferred to a 100-ml round-bottom flask, and water (15 ml), methanol (15 ml), salicylic acid (300 μl of stock solution, concentration 2 mg/ml) are added. The contents are refluxed for 30 min on a heating mantle (40°). After cooling the mixture is filtered using a Buchner flask and Whatman No. 4 filter paper. The extract is then transferred to a 100-ml round-bottom flask, and the methanol is removed from the filtrate by rotary evaporation under vacuum at 40–50°. After transferring to an amber HPLC vial the sample is ready for analysis.

FIG. 2 (see pp. 98–99). HPLC analysis of phenolics in tomatoes before (A) and after (B) β-glucosidase treatment.

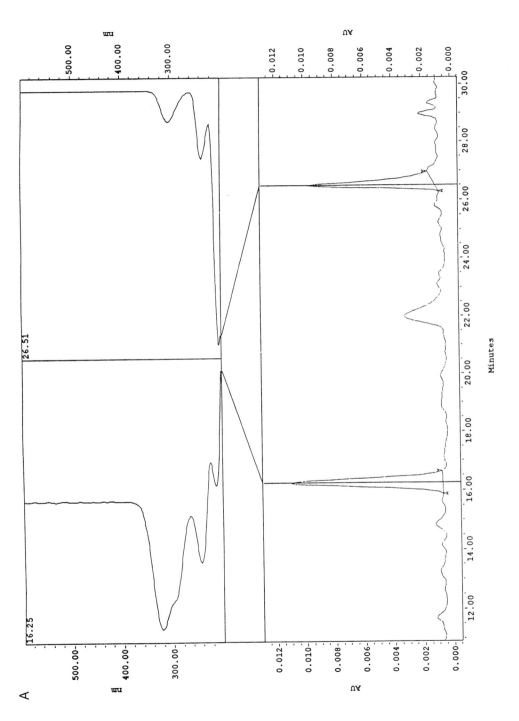

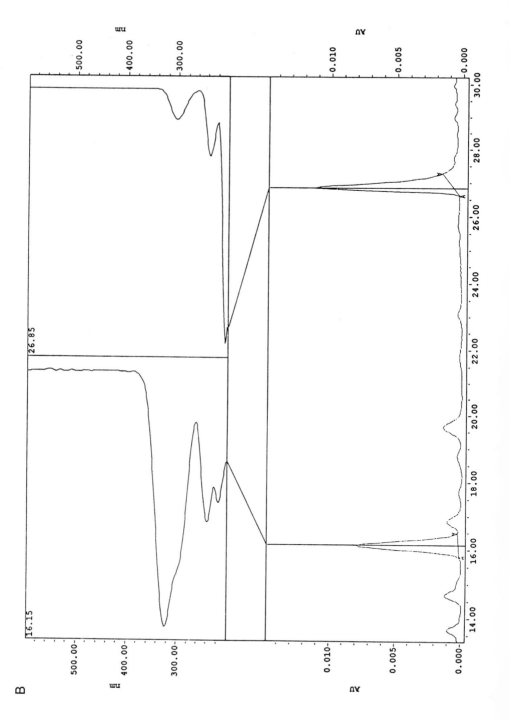

For enzymic hydrolysis, 1 ml of the aqueous extract is incubated with 4084 units of β-glucosidase in a stoppered culture tube for 1 hr at 37°. The extract is then diluted 1 : 1 (v/v) with mobile phase into an amber HPLC vial, ready for injection.

Urine Sample Preparation

Urine samples are thawed and mixed well. A 1-ml sample is diluted into a 5-ml disposable culture tube containing 100 μl of HCl (5 M) and 5 μl salicylic acid as an internal standard (stock solution 2 mg/ml). To this, 2.4 ml of methanol is added, and the sample is then stoppered and mixed for 30 sec. Samples are centrifuged at 800g for 10 min at 4°, the supernatant is collected, and the methanol is removed by rotary evaporation under vacuum at 40°. The resultant aqueous fraction is filtered using a Flowpore 0.22-μm sterile nonpyrogenic membrane filter (Whatman, UK) directly into a HPLC vial.

Analysis of Phenolics by Gradient HPLC

HPLC analysis (Table III) is conducted according to the method of Paganga and Rice-Evans.[48] The HPLC system consists of an autosampler with a Peltier temperature controller, a 626 pump with a 600S controller, a photodiode array detector, and a software system that controls all the equipment and processes the data. A Nova-Pak C_{18} column (4.6 × 250 mm) with a 4-μm particle size is used, and the temperature is maintained by the column oven set at 30°. The injection is by means of an autosampler, with a fixed loop, and the volume injected is 30 μl. Elution (0.5 ml/min) is performed using a solvent system composed of solvent A (20% methanol in 0.1% hydrochloric acid) and acetonitrile (solvent B) mixed using a linear gradient held at 95% solvent A for 10 min and then decreasing to 50% solvent A at 50 min, back to 95% solvent A at 55 min, and held at these conditions for a further 5 min. There is a 10-min delay before the next injection to ensure reequilibration of the column. The chromatograms are obtained with detection at 320 nm. All injections are performed in duplicate. Sample identification is through retention time relative to standards, spec-

---

[48] G. Paganga and C. Rice-Evans, *FEBS Lett.* **401,** 78 (1997).

FIG. 3. HPLC analysis of (A) urinary excretion of ferulic acid after tomato consumption and of (B) ferulic acid standard in basal urine.

TABLE IV
TOTAL URINARY EXCRETION OF FERULIC ACID IN RELATION TO INTAKE FROM DIETARY ENHANCEMENT FROM TOMATO

Subject	Amount of ferulic acid excreted (mg)	Total urinary volume (ml)	Total ferulic acid (24 hr) (μM)
T101	1.23	2310	2.7
T102	0.82	2260	1.9
T103	1.15	1760	3.4
T104	0.87	1715	2.6
Mean	1.02 ± 0.2		2.65 ± 0.6

tral matching relative to standard, and spiking of samples with suspected compounds for confirmation.

Quantification is carried using calibration of the ferulic acid standard. Four calibration runs with ferulic acid are executed routinely (Fig. 1). For the urine assay, calibration is performed by following the procedures for the standard solutions as described for urine samples. A linear regression calculation is performed on the resulting plot of peak area versus amount of ferulic acid, and the regression line established is used to calculate the amount of ferulic acid present. The HPLC method described here has been

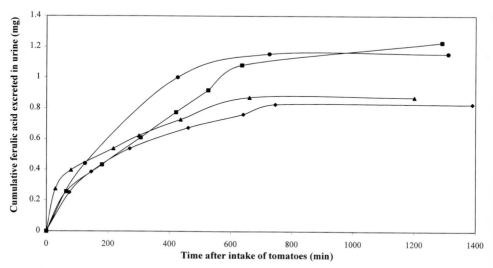

FIG. 4. Cumulative amounts of ferulic acid excreted in urine as a function of time postdietary enhancement (not adjusted for volume). ■, T101; ◆, T102; ●, T103; ◆, T104.

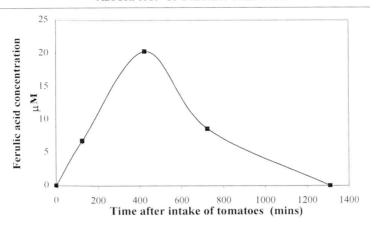

FIG. 5. Individual concentrations of ferulic acid excreted as a function of time (volunteer T103).

applied successfully to LC-MS-MS for the detection and identification of ferulic acid dimers.[49]

Results

HPLC analysis of tomato extract, before and after glucosidase treatment, is illustrated in Figs. 2A and 2B, respectively. Identifications of the peaks from retention times and spectra and confirmation by spiking with suspected compounds are, for the glucosidase-treated tomato extract, 5.74, 7.80, 13.14, and 15.50 min for the hydroxycinnamates chlorogenic, caffeic, p-coumaric, and ferulic acid, respectively (Fig. 2B). These correspond in the nonenzyme-treated extract (Fig. 2A) to 5.74 min chlorogenic acid and 5.91 min p-coumaroylglycoside, with peaks at 5.11, 6.41, and 7.78 min representing conjugates of hydroxycinnamates. In addition, peaks around 28.8 and 34.2 min are indicative of other hydroxycinnamic acid conjugates, but clearly not glucosides as they are resistant to glucosidase hydrolysis over a range of enzyme concentrations. Rutin (quercetin 3-rutinoside) and naringenin are identified with retention times at around 25 and 35.6 min,

[49] L. Bourne, N. Rendall, G. Taylor, G. P. Bolwell, and C. Rice-Evans, submitted for publication.
[50] P. C. H. Hollman, M. V. D. Gaag, M. J. B. Mengelers, J. M. P. van Trijp, J. H. M. De Vries, and M. B. Katan, *Free Radic. Biol. Med.* **21,** 703 (1996).
[51] D. Cova, L. De Angelis, F. Giavarini, G. Palladini, and R. Perego, *Int. J. Clin. Pharmacol. Ther. Toxicol.* **30,** 29 (1992).

respectively, in both enzyme- and nonenzyme-treated extracts, with the internal standard, salicylic acid, at 26.6 min.

Figure 3A illustrates the HPLC analysis (from 12 to 30 min) of the urinary excretion of ferulic acid (R_t 16.15 min) from subjects on the tomato supplementation regime compared with (Fig. 3B) the standard ferulic acid added to basal urine relative to the internal standard. The cumulative amount of ferulic acid excreted in the urine from the subjects over 24 hr is shown in Table IV and, as a function of time after ingestion of the tomatoes, in Fig. 4. Results show that the cumulative excretion of ferulic acid increases with time postsupplementation and continues progressively up to 8–10 hr, after which it reaches a plateau, showing no further excretion. The total concentration of free ferulic acid excreted in 8–10 hr was of the order of 1.9–3.4 μM in all subjects (mean 2.7 ± 0.6 μM), which is equivalent to a mean value of 1.02 ± 0.2 mg ferulic acid excreted over 24 hr, approximately 4–5% of the ferulic acid ingested. The individual concentrations excreted at specific time points for volunteer T103 are also demonstrated (Fig. 5). On the concentration versus time plot, the time at which maximal excretion occurs is 425 min.

The identification of flavonoids and hydroxycinnamates in human urine is a fundamental feature in investigating the relationship between dietary phenolics and disease. The approach taken here provides information on the bioavailability of these compounds. To function as antioxidants they must not only be absorbed intact, but also have an appropriate lifetime in the general circulation. By studying the pharmacokinetics of the excretion of these compounds, we can begin to approach the elucidation of the likely bioactivity *in vivo*.

The study described here adds to the growing weight of evidence for the absorption and bioavailability of dietary phenolics, but especially, in this case, for the hydroxycinnamate ferulic acid. Few studies have investigated the uptake of hydroxycinnamates from the diet. Evidence for the bioactivity of these phenolics *in vivo* should also be sought to support the notion that their antioxidant properties contribute to their proposed role in health protection and disease prevention.

Acknowledgments

We acknowledge Zeneca Plant Sciences and the Ministry of Agriculture, Fisheries, and Food (UK) for funding this research and all the volunteers for their willing participation in the study.

[10] Simultaneous Analysis of Individual Catechins and Caffeine in Green Tea

By TETSUHISA GOTO and YUKO YOSHIDA

Introduction

Teas produced from the leaf of *Camellia sinensis* can be divided into three groups: fermented, late fermented, and nonfermented (green) tea.[1] Most of the tea produced in the world is fermented, particularly black teas. This fermentation process involves the enzymatic oxidation of polyphenols and other components found in tea leaves. Late-fermented teas have an additional step in processing that involves fermentation via microorganisms. In contrast, the initial step of processing in nonfermented teas, i.e., green teas, involves enzymatic inactivation by heat through either dry pan frying or steam heating of the leaves.

Catechins are a group of polyphenol compounds found in the leaves of the green tea *C. sinensis*. These compounds may be contained in up to 30% of the leaf dry weight and are an important factor in the taste of the tea. In black teas, polymerized catechins such as theaflavins and thearubigins that result from the fermentation process are important factors in determining the overall quality of the tea. Caffeine is a plant alkaloid contained in some popular beverages such as tea, coffee, and cocoa and is known for its stimulatory effect.

Of the many polyphenol compounds found in food commodities, tea catechins, particularly those found in green tea, have been found to possess antioxigenic[2] and antimutagenic[3] properties and may exert prophylactic effects against hypertension.[4] Previously, the structural similarity of the various tea catechins (Fig. 1) made the quantitation and analysis of individual catechins difficult.[5-7] As each catechin possesses distinct properties, a simple and rapid method that could be used for analysis of individual catechins in a complex mixture would be advantageous. A relatively simple

---

[1] T. Goto, Y. Yoshida, I. Amano, and H. Horie, *Foods Food Ingred. J. Japan* **170**, 46 (1996).
[2] T. Matsuzaki and Y. Hara, *Nippon Nougei Kagaku Kaishi* **59**, 129 (1985).
[3] K. Shimoi, Y. Nakamura, I. Tomita, Y. Hara, and T. Kada, *Mutat. Res.* **173**, 239 (1986).
[4] Y. Hara, T. Matsuzaki, and T. Suzuki, *Nippon Nogei Kagaku Kaishi* **61**, 803 (1987).
[5] A. C. Hoefler and P. Coggon, *J. Chromatogr.* **129**, 460 (1976).
[6] B. Risch, R. Galensa, and K. Herrmann, *J. Chromatogr.* **448**, 291 (1988).
[7] S. Terada, Y. Maeda, T. Masui, Y. Suzuki, and K. Ina, *Nippon Shokuhin Kogyo Gakkaishi* **34**, 20 (1987).

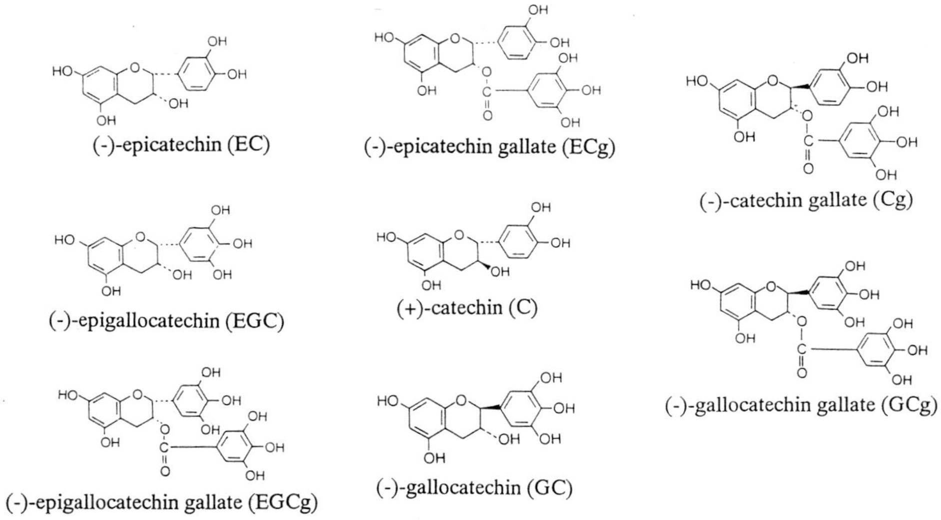

FIG. 1. Structure of catechins.

and quick high-performance liquid chromatography (HPLC) analysis method is presented in which caffeine and several catechins may be analyzed simultaneously.[8] Caffeine and the following eight catechins may be quantitated by this method: epicatechin (EC), epicatechin gallate (ECg), epigallocatechin (EGC), epigallocatechin gallate (EGCg), catechin (C), gallocatechin (GC), gallocatechin gallate (GCg), and catechin gallate (Cg).

Instruments

The HPLC system consists of Shimadzu LC-6A pumps with a two pump high-pressure gradient system (Shimadzu Co. LTD, Japan), a Shimadzu SPD-M10A diodearray detector (200–300 nm), an SSC3500 column oven (Senshu Kagaku, Japan), and a Rheodyne 7125 sample injector with a 20-μl sample loop (Rheodyne Inc.). The column is Develosil ODS-HG-5 (4.6 × 150 mm, Nomura Chemical Co., Seto, Japan) equipped with a guard column (4 × 10 mm, Nomura). The octadecyl-bonded silica gel (ODS) for this column is synthesized from a trifunctional silane agent, and materials are well end capped with trimethylsilylating agents. The flow rate of the mobile phase is 1 ml/min. Class M10A software (Shimadzu) is used for

[8] T. Goto, Y. Yoshida, M. Kiso, and H. Nagashima, *J. Chromatogr. A* **749**, 295 (1996).

Chemicals and Samples

Eight catechin standards (EC, ECg, EGC, EGCg, C, GC, GCg, and Cg), a partially purified green tea extract (ca. Polyphenon 60, 60% catechin content), and caffeine are available from various commercial sources. The acetonitrile used for the mobile phase is HPLC grade or better, and all the other chemicals are GR grade and are used without further purification. Usual green tea samples are milled by a Cyclon Sample Mill (Udy) with a 1-mm mesh screen. The Matcha tea, which is a powdered green tea, is extracted without further manipulation.

Sample Extraction

Green tea samples are extracted by the method of Suematsu *et al.*[9] with slight modifications. The samples (500 mg) are extracted with 100 ml of acetonitrile–water (1 + 1, v/v) at 30° for 40 min with constant shaking. The extract is diluted 2- to 10-fold with water. Before analysis, the extracts are filtered through a cartridge-type sample filtration unit with a polytetrafluoroethylene (PTFE) membrane with a hydrophilic surface. Several types of filtration membranes are utilized and, as shown in Table I, the composition of the filter affected the efficiency of recovery. For this purpose, the PTFE membrane is the most efficient.

Separation Conditions of Standard Catechins

A water–acetonitrile–phosphoric acid solvent system with two-step linear gradients of acetonitrile concentration successfully separates all nine chemicals within 20 min (Fig. 2B). The mobile phase composition used is (A) water–acetonitrile–85% phosphoric acid (95.45 + 4.5 + 0.05, v/v/v) and (B) water–acetonitrile–85% phosphoric acid (49.95 + 50.0 + 0.05, v/v/v). The solvent composition starts at 90% solvent A and 10% solvent B and is maintained for 5 min, then increases linearly to 30% solvent B in 3 min. This condition is maintained for 2 min followed by a linear increase of solvent B to 80% in 5 min. Final conditions are held for an additional 5 min before returning to the original condition for the next injection 10 min after reaching this condition.

---

[9] S. Suematsu, Y. Hisanobu, H. Saigo, R. Matsuda, and Y. Komatsu, *Nippon Shokuhin Kagaku Kogaku Kaishi* **42**, 419 (1995).

TABLE I
RECOVERY OF MAJOR CATECHINS AND CAFFEINE AFTER FILTRATION[a] (%)

Group[b]	Material of membrane	EC	EGC	ECg	EGCg	Caffeine
A	PTFE[c]	100.5	101	100.5	100	101
A	PTFE	100	99.5	99.5	100	100
A	PVDF[d]	98	97.5	21	43.5	99.5
A	RC[e]	95.5	93.5	93	91.5	100
B	PTFE	101	101	101	101	101
B	PVDF	98	98.5	95	95.5	98
B	RC	99	99	99.5	99	99

[a] Two milliliters of Polyphenon 60 solution (15 mg/100 ml) was filtered through each cartridge and the recovery of each chemicals was measured by HPLC.
[b] Group A: Sample was dissolved in 5% acetonitrile solution. Group B: Sample was dissolved in 50% acetonitrile solution.
[c] Polytetrafluoroethylene with hydrophilic coating.
[d] Polyvinyl difluoride.
[e] Regenerated cellulose.

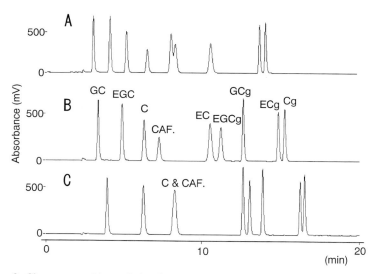

FIG. 2. Chromatographic analysis of standard catechins and caffeine and the effect of column temperature on their separation: (A) at 50°; (B) at 40°, standard analysis condition; and (C) at 30°.

The temperature of the column oven strongly affects separation. As the column temperature increases, the retention times of all nine chemicals decrease, but the magnitude of the change varies for each compound (Fig. 2). For convenience, 40° is selected as the standard oven temperature for subsequent analysis as satisfactory separation of all nine compounds is achieved at this temperature.

Quantitative Analysis

For all eight catechins and caffeine, the coefficient of variation value of retention time was less than 0.5% and for quantitation was less than 2.5% for five injections using the Polyphenon 60 solution as the sample. Calibration curves for the four major catechins and caffeine are shown in Fig. 3. At 231 nm, calibration curves of the catechins are linear from 2 to 2000 ng (Figs. 3A and 3B). Similar results are obtained for the other four catechins. For caffeine, linearity over this range is obtained at 274 nm (Fig. 3C). The minimum detection limit is approximately 0.2 ng for all nine compounds. The sensitivity of the analysis for catechins can be enhanced by using 207 nm for detection; however, as shown in Fig. 3D, the linearity

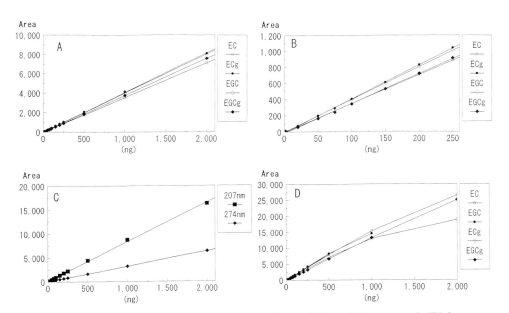

FIG. 3. Calibration curves: (A) four major catechins at 231 nm (250 ng to 2 μg), (B) four major catechins at 231 nm (2 to 250 ng), (C) caffeine, and (D) four major catechins at 207 nm.

of the calibration curves is not comparable to the curves generated at 231 nm. In addition, the baseline of chromatograms shifted with the concentration of acetonitrile at the shorter wavelength, and the potential for interference by other chemicals in complex mixtures cannot be ignored. Thus 231 nm is the appropriate wavelength for the detection and quantitation of catechins in tea leaves.

Analysis of Green Tea Sample

Different types of green tea, both high and low grades for each tea, were analyzed for individual catechins and caffeine. Matcha, the tea traditionally used in the Japanese tea ceremony and Gyokuro, is produced from new shoots by growing the plants under a sunshade. Sencha, the most popular type of Japanese green tea, is produced from new shoots or young leaves by growing the plant in an outdoor field. Hojicha, the light roast green tea, is produced from relatively mature and coarse leaves.

In general, the catechin content was higher in Sencha than in Matcha tea, and the lower grade teas contained more catechins and less caffeine than the higher grade teas.[1] Detectable amounts of C were found in most

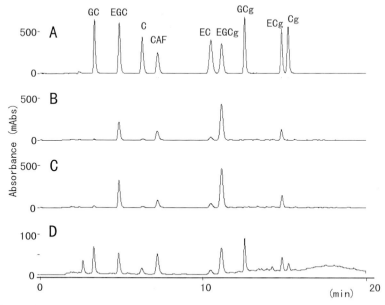

FIG. 4. Chromatograms of eight catechins and caffeine: (A) standards (20 μg/ml each catechins and caffeine), (B) Gyokuro, (C) Sencha, and (D) Hojicha.

of these tea samples. In addition, Sencha samples contained detectable amounts of Cg, GC, and GCg, whereas Matcha tea samples did not. Chromatograms of some samples are shown in Fig. 4. No extraneous peaks interfered with the chromatographic analysis of the extracts. The purity of each catechin peak in the tea samples was more than 98% as calculated from the peak spectrum. Minimum quantitation limits for each catechin and for caffeine were less than 0.02% of the tea dry weight, which is more than adequate for the analysis of catechins and caffeine in green teas.

[11] Reversed-Phase High-Performance Liquid Chromatography Methods for Analysis of Wine Polyphenols

By ANDREW L. WATERHOUSE, STEVEN F. PRICE, and JEFFREY D. MCCORD

Introduction

The analysis of phenolic compounds in any sample is a complex issue for a number of reasons. First, the extraction of these compounds is complicated by the fact that they bind strongly to protein, so foods that contain protein are difficult to extract cleanly. Wine, however, contains only low levels of protein and is a homogeneous solution, so extraction is unnecessary. The second complexity issue is based on the large number of diverse compounds that comprise "phenolics." They span a wide range of polarities as well as molecular weight ranges. To date there is no method that can separate and quantify all phenolic compounds in wine as individual components.[1]

There are a number of well-established procedures designed to analyze all phenolic compounds as a single value (total phenol) or specific classes (i.e., total flavan-3-ol), but in every case there are differential responses (based on mass or molarity) related to different functional groups. The variability of the response can be small or large depending on the sample. One such procedure is described in the chapter by Singleton et al.[1a]

---

[1] K. R. Markham and S. J. Bloor, in "Flavonoids in Health and Disease" (C. A. Rice-Evans and L. Packer, eds.), p. 1. Dekker, New York, 1998.
[1a] V. L. Singleton, R. Orthofer, and R. M. Lamuela-Raventós, Methods Enzymol. **299** [14] (1998).

Chromatographic separations of phenolic compounds provide data on the concentration of specific components. The methods detailed in this chapter are based on reversed-phase chromatography. This type of chromatographic system is quite capable of separating small phenolic compounds, typically monomers and some oligomers. The separation fails progressively as the molecular size increases beyond dimeric flavonoids. In wine samples, the failure is due to the large number of species present of approximately the same polarity and size. These compounds typically elute as a group in a large broad peak at the end of the chromatogram. This peak can be used to quantify all large polyphenolics in the sample, but to discriminate among these large molecules, other methods should be used to properly separate and analyze the higher molecular weight compounds, such as size exclusion or normal-phased chromatography methods.[2]

Separating small, largely monomeric phenolic compounds in wine is also challenging because of the wide range of polarities, from the very polar gallic acid to the nonpolar quercetin. One solution to deal with this complexity is to fractionate the sample prior to analysis into subfractions to simplify the chromatograms. The methods described in this chapter do not utilize this preliminary step, but attempt to separate all components in a single gradient chromatographic run. Because the samples contain anthocyanins, a very low pH (<2) must be used to ensure ideal behavior. At higher pH values, anthocyanins give very broad, poorly resolved peaks. This chapter presents two solutions. One is to carry out the analysis at a low pH and the other is to shift the pH during the analysis.

There are several variations on the basic method of an acidic solvent with a solvent gradient of increasing hydrophobicity. Because each laboratory tends to adopt these procedures to their specific requirements, it is not possible to describe a single procedure that is an accepted standard. Cheynier et al.[3] use formic acid with an acetonitrile gradient (pH not specified). Roggero and colleagues[4] describe a very long analysis that separates many different components. It is also possible to modify the separation conditions, particularly pH and the lipophilicity gradient, to facilitate the separation of minor components, such as resveratrol.[5] The three procedures

---

[2] J. Rigaud, M. T. Escribano-Bailon, C. Prieur, J. M. Souquet, and V. Cheynier, *J. Chromatogr. A* **654**, 255 (1993).

[3] V. Cheynier, I. Hidalgo Arellano, J. M. Souquet, and M. Moutounet, *Am. J. Enol. Vitic.* **48**, 225 (1997).

[4] J. P. Roggero, P. Archier, and S. Coen, *ACS Symp. Ser.* **661**, 6 (1997).

[5] R. M. Lamuela-Raventós, A. I. Romero-Pérez, A. L. Waterhouse, and M. C. de la Torre-Boronat, *J. Agric. Food Chem.* **42**, 281 (1995).

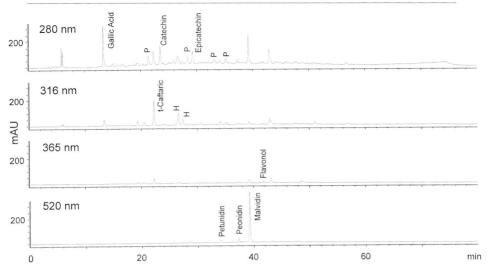

FIG. 1. pH shift. Reversed-phase HPLC separation of a 1995 California Pinot noir monitored at four wavelengths, using procedure A. P, a presumed proanthocyanidin; H, hydroxycinnamate.

described in this chapter include ones based on a previous report by Lamuela and Waterhouse[6] and one by Price et al.[7]

Methods

Procedure A: pH Shift

This procedure (Fig. 1) uses a three solvent gradient, dropping the pH during the run to ensure proper chromatographic behavior of the anthocyanins, and is the one employed at UC Davis.

Standards. (+)-Catechin, (−)-epicatechin, and *p*-coumaric acid are from Aldrich Chemical Co. (Milwaukee, WI) Rutin and caffeic acid are from Sigma Chemical Co. (St. Louis, MO) Chlorogenic acid (5′-caffeoylquinic acid) is from Fluka Chemical Corp. (Ronkonkoma, NY) Gallic acid is from MCB Manufacturing Chemists Inc. (Cincinnati, OH), and malvin is from Pfaltz & Bauer (Waterbury, CT).

---

[6] R. M. Lamuela-Raventós and A. L. Waterhouse, *Am. J. Enol. Vitic.* **45,** 1 (1994).
[7] S. F. Price, P. J. Breen, M. Valladao, and B. T. Watson, *Am. J. Enol. Vitic.* **46,** 187 (1995).

Prior to analysis, samples are filtered through 0.45-μm polytetrafluoroethylene (PTFE) syringe tip filters (Gelman Sciences, Ann Arbor, MI) into flint glass high-performance liquid chromatography (HPLC) vials equipped with PTFE-lined crimp caps. A Hewlett-Packard (Palo Alto, CA) Model 1090 high-performance liquid chromatograph with three low pressure solvent pumps and a photo diode array ultraviolet (UV)-visible detector coupled to HPChemstation software are used. A Hewlett-Packard LiChrosphere C_{18} column (4 × 250 mm, 5 μm particle size) kept at 40° is used as the stationary phase. The flow rate of the mobile phase is 0.5 ml/min. The mobile phase consists of three solvents: solvent A, 50 mM dihydrogen ammonium phosphate adjusted to pH 2.60 with orthophosphoric acid; solvent B, 20% solvent A and 80% acetonitrile; and solvent C, 0.20 M orthophosphoric acid adjusted with NaOH to pH 1.50. A list of the HPLC mobile phase multilinear gradient is shown in Table I. Four wavelengths are monitored: 280 nm for catechins and benzoic acids, 316 nm for hydroxycinnamates, 365 nm for flavonols, and 520 nm for anthocyanins. The bandwidth of each wavelength is 2 nm.

Procedure B: Low pH

This procedure (Fig. 2) uses a low pH solvent with a C_{18} reversed-phase column and is the one employed at E&J Gallo Winery.

The sample is filtered through a 0.45-μm PTFE filter, and 25 μl is injected with the pump running at 0.8 ml/min with the temperature controlled at 20° using a HP 1100 with DAD detector. The mobile phase consists of three solvents: solvent A, water adjusted to pH 1.8 with phosphoric acid (approximately 16 ml/liter); solvent B, solvent A with 40% (v/v) methanol;

TABLE I
MULTILINEAR HPLC GRADIENT

Time	Solvent composition		
	%A	%B	%C
0	100	0	0
5	100	0	0
8	92	8	0
20	0	14	86
25	0	16.5	83.5
35	0	21.5	78.5
70	0	50	50
75	100	0	0
80	100	0	0

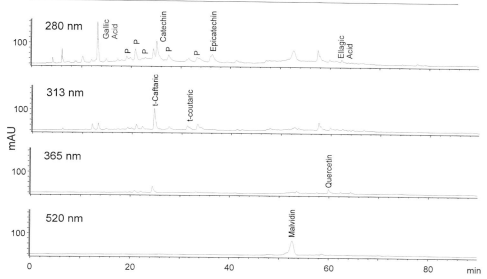

FIG. 2. Low pH. Reversed-phase HPLC separation of a 1995 California Pinot noir monitored at four wavelengths, using procedure B. P, a presumed proanthocyanidin.

and solvent C, 90% (v/v) methanol. Gradient points are shown in Table II. Four wavelengths are monitored: 280, 313, 365, and 520 nm. The stationary phase is a Zorbax C_{18} (4.6 × 250 mm) column.

Procedure C: Polystyrene Column

This procedure (Fig. 3) uses a polystyrene reversed-phase column. This column has the advantage of being extremely stable under all pH conditions and is the procedure employed at ETS Laboratories.

Prior to analysis, samples are centrifuged at 10,000g for 3 min. A Hewlett-Packard (Palo Alto, CA) Model 1090 high-performance liquid chromatograph and a photo diode array UV–visible detector coupled to HPChemstation software are used. A Polymer Labs PLRP-S 100-Å, 5-μm reversed-phase polystyrenedivinyl benzene column (4.6 × 250 mm, 5 μm particle size) kept at 25° is used as the stationary phase. The flow rate of the mobile phase is 1 ml/min. The mobile phase consists of two solvents: solvent A, 1.5% (v/v) phosphoric acid and solvent B, 1.5% (v/v) in acetonitrile. A list of the HPLC mobile phase multilinear gradient is shown in Table III, and the column is reequilibrated with solvent A for 55 min at the beginning of each run. Four wavelengths are monitored: 280 nm for catechins and benzoic acids, 316 nm for hydroxycinnamates, 365 nm for flavonols, and 520 nm for anthocyanins. Full spectral scans are acquired every 1.6 sec.

TABLE II
MULTILINEAR GRADIENT

Time	Solvent composition		
	%A	%B	%C
0	100	0	0
10	100	0	0
15	50	50	0
20	40	60	0
40	30	70	0
70	0	50	50
75	0	20	80
80	0	20	80
82	0	0	100
85	0	0	100
90	100	0	0

Discussion

Although some compounds are quantified by calibration with standards, there are numerous instances where standards are not available, so most components are quantified using the peak areas of a standard with similar

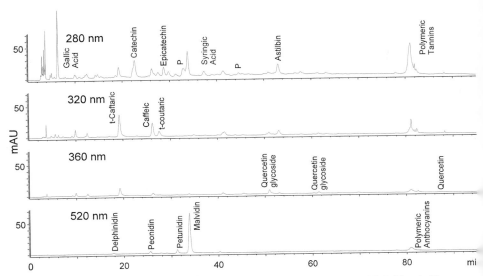

FIG. 3. Polystyrene column. Reversed-phase HPLC separation of a 1995 California Pinot noir monitored at four wavelengths, using procedure C. P, a presumed proanthocyanidin.

TABLE III
PROCEDURE C MULTILINEAR GRADIENT

Time	Solvent composition	
	%A	%B
0	95	5
85	78	22
88	50	50
95	50	50
100	95	5

spectral characteristics, and quantities are reported as equivalent amounts of that standard. For the best observation of spectra, a UV-clear solvent and acid system should be used, and phosphoric acid in aqueous acetonitrile appears to have the best properties for this purpose.

For instance, catechin is one of the monomeric flavan-3-ols and is typically used as the standard for all members of this family. This family of compounds is notorious for the large number of individual components, and few standards, if any, exist for oligomers (procyanidins) and polymers (condensed tannins). These RP-HPLC procedures reveal several of the oligomeric proanthocyanidins on the chromatograms, and peaks having this spectral identity have been marked with the letter P in the figures. Similarly, compounds identified only as anthocyanins are identified with the letter A and hydroxycinnamates with the letter H.

In data presented here, each chromatogram has a slightly different set of peaks with known identity. This is because each laboratory had a different selection of standards available. In most cases, the standards most important to a specific laboratory will be obtained from colleagues who have them, as many are not readily available from commercial sources. Specialty suppliers include Extrasynthese (France), Roth (Germany), Atomergic, Roth's United States distributor, Apin (United Kingdom), and Indofine (United States).

Spectra taken of eluting peaks can also be used to assist in the identification of individual members of specific classes. The best application is for the analysis of wine anthocyanins, where different B-ring substitutions have a measurable effect on the absorption maximum[6] and where the presence of cinnamate esters on glucosides are detected easily. However, because a differing glycoside substitution can have a greater effect on the chromophore than changes in B-ring substitution, spectra of the flavonol family, with a diversity of glycosides, cannot be used to ascertain specific identity

(B-ring substitution). The specific anthocyanin composition of wines has been described as a method to authenticate the grape variety used to produce the wine.[8] Aside from wine, the authenticity of other processed fruit products, especially juices, is often carried out by analyzing phenolic composition, and anthocyanins are important in colored juices due to the wide diversity of specific anthocyanins in different fruits.[9]

Because there are a significant number of compounds in different phenolic classes, each of which has a different absorption maximum, it is common to quantify each class at the wavelength of its absorption maximum. In these examples, four wavelengths were used. Hydroxycinnamates, such as caftaric acid, have a minimum at about 320 nm; flavonols, such as quercetin, have a maximum at about 365 nm; and red anthocyanins have a maximum at about 520 nm. Most all phenolics absorb light at 280 nm, and flavan-3-ols have a maximum at the same wavelength, so that is used to quantify flavan-3-ols, but also as the single wavelength where detectors cannot monitor more than one. Traces taken at 280 nm are also the best for evaluating overall sample complexity.

Additional information can be obtained by the use of different detectors. Procedure B has been used with absorbance detection at 254 nm in addition to the other wavelengths. This assists in selectivity with compounds that have a strong 254-nm absorbance, such as ellagic acid. In addition, a four-channel ESA coulometric detector set at 40, 120, 300, and 650 mV can be used for the detection of tyrosol and can easily distinguish catechin and epicatechin.

Copigmentation, an effect caused by pi-complexation of electron-poor and electron-rich aromatic compounds, is important in wine for several reasons. Principally, this effect increases the solubility of specific wine flavonoids, particularly condensed tannin and flavonols by interaction with anthocyanins; the converse is important as well. For the reversed-phase analysis, its effect is most notable for quercetin, where the concentration in wine often far exceeds the amount that can be dissolved into a model-wine solvent for a standard. Because of its insolubility, quercetin standards must be dissolved in a high alcohol solvent to attain levels observed during the analysis, although this generally leads to a nonideal peak shape.

A single wine was analyzed using each of the three systems. It was a 1997 Pinot noir wine from California. The major components observed were typical of wine. As noted in the figures, hydroxycinnamates, measured at 320 nm, included the fruit-derived caftaric acid as well as its hydrolysis

---

[8] R. Eder, S. Wendelin, and J. Barna, *Mitt. Klosterneuburg*, **44**, 201 (1994).

[9] S. Nagy, J. A. Attaway, and M. E. Rhodes, "Adulteration of Fruit Juice Beverages." Dekker, New York, 1988.

product caffeic acid. Some minor peaks were identified in some cases as coumaric derivatives, but several peaks were clearly hydroxycinnamates by spectral characteristics. Hydroxycinnamates were also prominent at 280 nm.

Flavonols, observed in the 365-nm trace, were at very low levels in this wine, with a trace of quercetin present, and low levels of a few glycosides were labeled simply as flavonol. Malvidin-3-glucoside was by far the predominant anthocyanin, and the major peak was at 520 nm, with only small amounts of peonidin, petunidin, and delphinidin. Abundant anthocyanins are also prominent at 280 nm.

The flavan-3-ol family, observed only on the 280-nm trace, included the monomeric catechin and epicatechin; however, the majority of the major components not specifically identified were proanthocyanidins from this family, which are labeled with "P" on the figures. In Fig. 3, astilbin is identified at 53 min, which may be the unidentified compound at 43 min in Fig. 1 and at 58 min in Fig. 2.

Comparing the three methods, the two silica gel reversed-phase systems have very similar elution orders, whereas the polystyrene column has a quite different pattern. In the latter case, polar anthocyanins and hydroxycinnamates come out earlier, whereas nonpolar flavonols come out later. This may be due to the greater hydrophobicity of the stationary phase, which lacks any polar silanol sites. In addition, because the polystyrene column tends to retain "polymeric" fractions until the solvent polarity is reduced quickly in the column flush cycle, the resulting peak can be used to assess the levels of polymeric tannin, visible at 280 nm, as well as any tannin that reacted with anthocyanins to form "polymeric" anthocyanin.

In conclusion, current HPLC methods can separate major small phenolic constituents of wine in hour-long separations. Detection at multiple wavelengths assists in distinguishing specific classes and in reducing the interferences for those compounds. Spectra of peaks can be used to assign observed components to particular classes of compounds when standards are not available.

[12] Analysis of Antioxidant Wine Polyphenols by High-Performance Liquid Chromatography

By DAVID M. GOLDBERG and GEORGE J. SOLEAS

Introduction

Although gas chromatography–mass spectrometry (GC-MS) techniques have many advantages in the assay of polyphenols, including unequivocal identification by mass spectral characteristics, their utilization by the wine industry is virtually restricted to those that can be performed by benchtop instruments that are limited in the molecular mass of the compounds (usually trimethylsilyl [TMS] derivatives) that can enter the analytical column. High-performance liquid chromatography (HPLC) procedures have been used since the 1970s and the required instrumentation is much more widely available in enological laboratories. Moreover, all classes of polyphenols present in wine can, in principle, be separated to allow qualitative recognition of their presence, and frequently quantitation, provided that authentic standards are available.

Many publications have described the simultaneous assay of a number of polyphenols by HPLC using ultraviolet (UV) absorption,[1,2] fluorescence intensity,[3] and electrochemical detection.[4] As the number of compounds increases, it becomes almost mandatory to use diode array detection for the spectral identification of peaks eluting in close proximity and to quantitatively separate the absorbance counts attributable to overlapping peaks.

Until recently, it was deemed necessary to carry out multiple organic extractions of around 100 ml of wine to concentrate the compounds of interest and to free them from others that might adversely affect chromatographic resolution.[5,6] It is now well established that with a guard column preceding the analytical column, as little as 20 μl of wine can be injected

---

[1] J. Oszmaianski, T. Ramos, and M. Bourzeix, *Am. J. Enol. Vitic.* **39,** 259 (1988).
[2] J. M. Ricardo da Silva, J.-P. Rosec, M. Bourzeix, and N. Heredia, *J. Sci. Food Agric.* **53,** 85 (1990).
[3] R. Pezet, V. Pont, and C. Cuenat, *J. Chromatogr. A.* **663,** 191 (1994).
[4] D. McMurtrey, J. Minn, K. Probanz, and T. P. Schultz, *J. Agric. Food Chem.* **42,** 2077 (1994).
[5] C. W. Nagel and L. W. Wulf, *Am. J. Enol. Vitic.* **30,** 111 (1979).
[6] M.-H. Salagoity-Auguste and A. Bertrand, *J. Sci. Food Agric.* **35,** 1241 (1984).

directly,[7,8] although a single solid-phase separation of a small volume of wine (typically 1 ml) may precede injection.

Several papers have reported HPLC methods that can simultaneously measure a number of polyphenols from different classes.[6,9,10] These do not usually include a comprehensive profile of the procyanidins or anthocyanins that are unlikely to be absorbed in the human intestinal tract, and whose biological implications for human health have not been satisfactorily established. We have described a HPLC method[11] that accurately and simultaneously measures the concentration in wine of eight polyphenols (four trihydroxystilbenes and four flavonols) that rank as the most potent from the perspective of their antioxidant and biological activities among those yet evaluated.[12–14] Subsequently, a ninth analyte, p-coumaric acid, was added.[15] This method has operated well in our laboratory for over a year, but in the past several months we have developed a method (unpublished) that, in addition to the original nine polyphenols, can measure the concentrations of other phenolic acids and some of their esters. It has been applied to >650 wines, giving satisfactory results in all cases, and also works well with distilled spirits. These two methods will now be presented.

Method 1

Wines

Commercial wines in 750-ml bottles were opened and 10 ml was withdrawn for storage at 4° in a glass vial filled to completion and protected by foil against sunlight. Analyses were completed within a 3- to 5-day period.

---

[7] J.-P. Roggero, P. Archier, and S. Coen, *J. Liq. Chromatogr.* **14,** 533 (1991).
[8] R.-M. Lamuela-Raventos, A. I. Romero-Perez, A. L. Waterhouse, and C. de la Torre, *J. Agric. Food Chem.* **43,** 281 (1995).
[9] J.-P. Roggero, S. Coen, and P. Archier, *J. Liq. Chromatogr.* **13,** 2593 (1990).
[10] J.-P. Roggero, P. Archier, and S. Coen, *in* "Wine: Nutritional and Therapeutic Benefits" (T. R. Watkins, ed.), p. 6. American Chemical Society, Washington, D.C., 1997.
[11] D. M. Goldberg, E. Tsang, A. Karumanchiri, E. P. Diamandis, G. Soleas, and E. Ng, *Anal. Chem.* **68,** 1688 (1996).
[12] J. Kanner, E. Frankel, R. Grant, B. German, and J. E. Kinsella, *J. Agric. Food Chem.* **42,** 64 (1994).
[13] E. N. Frankel, A. L. Waterhouse, and P. L. Teissedre, *J. Agric. Food Chem.* **43,** 890 (1995).
[14] G. J. Soleas, G. Tomlinson, E. P. Diamandis, and D. M. Goldberg, *J. Agric. Food Chem.* **45,** 3995 (1997).
[15] D. M. Goldberg, E. Tsang, A. Karumanchiri, and G. J. Soleas, *Am. J. Enol. Vitic.* **49,** 142 (1998).

Standards

The following were purchased from Sigma (St. Louis, MO) and used for calibration: catechin, epicatechin, *trans*-resveratrol, rutin, *p*-coumaric acid, and quercetin. *cis*-Resveratrol is prepared from the *trans* isomer by UV irradiation.[16] *trans*-Polydatin is isolated from the dried roots of *Polygonum cuspidatum* and a portion is converted to the *cis* isomer by UV irradiation.[16] All standards are dissolved in white wine at a range of final concentrations described in the next section. The absorbance of the native wine was deducted from the values of each standard peak in the white wine matrix.

Instrumentation

In our work, an ODS Hypersil 5-μm column, 250 × 4-mm ID comprised the stationary phase and was preceded by a guard column of LiChrospher 100 RP-18, 5 μm, 4 × 4 mm. Both were purchased from Hewlett Packard (Mississauga, Ontario, Canada). The chromatography equipment, all from Hewlett Packard, comprised the Series 1050 automatic sample injector; solvent degasser; quaternary pump; and diode array detector coupled to the HP Chem-Station utilizing the manufacturer's 2.05 software package.

Procedure

Twenty microliters of wine or calibration standard is injected directly onto the column and eluted with a gradient comprising acetic acid 33% (v/v) in water (pump A), methanol (pump B), and water (pump C). Zero-time conditions are 5% A, 15% B, and 80% C at a flow rate of 0.4 ml/min. After 5 min the pumps are adjusted to 5% A, 20% B, and 75% C at a flow rate of 0.5 ml/min and at 30 min to 5% A, 45% B, and 50% C at 0.5 ml/min until termination of the run at 40 min. This is followed by a 10-min equilibrium period with the zero-time solvent mixture prior to injection of the next sample. Detection is accomplished routinely by monitoring the absorbance signals at 265, 280, 306, 317, and 369 nm with a band width of 5 nm. Match and purity checks are performed for all peaks of interest, as described in the next section. A composite standard dissolved in white wine is injected after every 5 samples, with the background absorbance of the native wine deducted from each standard peak. After every 16 samples the column is washed for 2 hr with water at 1 ml/min followed by 2 hr with methanol at the same rate. Concentrations are generated by comparing the absorbance counts of unknowns with those of standards, provided that the results fall within the established range of linearity for that particular

---

[16] D. M. Goldberg, E. Ng, A. Karumanchiri, J. Yan, E. P. Diamandis, and G. J. Soleas, *J. Chromatogr. A* **708,** 89 (1995).

compound (*vide infra*). Where that range is exceeded, the wine sample has to be reanalyzed on an appropriate dilution in 12% (v/v) ethanol. The column can be used for at least 200 assays; the earliest manifestation of deterioration is inadequate resolution of the *cis*-polydatin and *trans*-resveratrol peaks. Best results are obtained at a constant temperature of 25°.

Chromatographic Resolution

Figure 1 demonstrates the satisfactory resolution accomplished for the major peaks of interest when the eluate is monitored at wavelengths of 280 and 306 nm. It should be noted that the composite dry standard in white wine (Fig. 1A) shows higher peaks than those in the authentic red wine sample (Fig. 1B), but the scale of the former was expanded 3.85-fold. In principle, sensitivity can be adjusted over a very wide range; although expansion of the smaller peaks may push several of the larger peaks off-scale, the absorbance counts are retained in the computer and employed to calculate concentrations of all nine polyphenols without reference to the graphic display. Although some peaks show minor shoulders, the software can apply adequate corrections during integration, which can also be done manually for sharper definition of the true peaks.

Match-factor spectral analysis of each peak assigned a value between 0 and 1000 for concordance between the spectrum of the peak and that of the pure standard of the compound which it was assumed to be. According to the manufacturer, values above 990 indicate near identity and those below 900 suggest that spectra are different. We use a value of 950 as our criterion for acceptability on the basis that in analyses of 100 wines, 95% of all peaks gave a higher match factor. Examples of peaks with acceptable and unacceptable match factors are illustrated in the original publication.[11] Purity checks are also performed at the inflexion points and apex of each peak, and the peak purity plot comprising the three spectra is drawn in a normalized and overlayed mode. By the same criterion as employed to define the acceptable limit for match factor, purity factors >950 led us to exclude hidden impurities and to consider the peak to be consistent with the presence of a single component. Examples of satisfactory (100%) and unsatisfactory (47%) purity factors are provided in Figs. 2A and 2B, respectively.

Linearity

Data for each polyphenol were pooled from at least three experiments in which the constituent was analyzed over a range of 6–10 concentrations individually, in a mixture of all eight dissolved in methanol, and added to a white wine matrix. The absorbance of the native white wine was deducted from that of all standard peaks. Linear regression analyses were performed

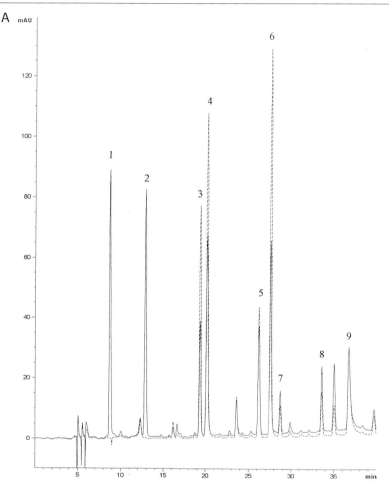

FIG. 1. (A) Chromatogram of composite polyphenol standards in white wine monitored at 280 (—) and 305 (---) nm. Peaks are as follows: 1, catechin; 2, epicatechin; 3, *trans*-polydatin; 4, *p*-coumaric acid; 5, rutin; 6, *trans*-resveratrol; 7, *cis*-polydatin; 8, *cis*-resveratrol; and 9, quercetin. (B) Chromatogram of authentic red wine sample monitored at 280 (—) and 305 (---) nm. Peaks are numbered in the same order as (A). Note that the scale of this chromatogram has a greater amplitude (0–500 mAU) than A (0–130 mAU).

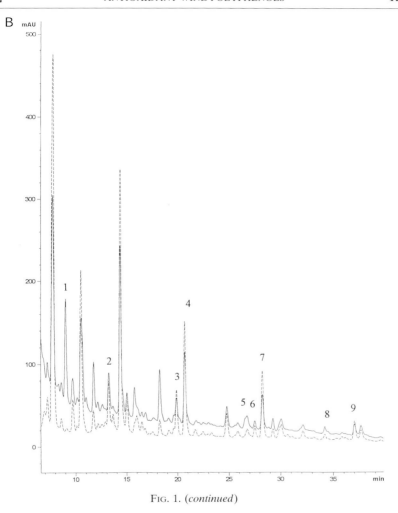

Fig. 1. (*continued*)

using the formula: $y = mx + b$. Table I shows that the slope of the calibration curve was almost perfectly linear in all instances, and that for all but two, the correlation coefficient differed from unity (if at all) only in the third decimal place. The range of concentrations covered those seen in >95% of all red wines, although many white wines have lower concentrations than the lowest used for several constituents. The ranges of the isomer and glucoside of *cis*-resveratrol could not be strictly predetermined because they were obtained by UV irradiation of pure solutions of the *trans* isomers, but all values were subsequently assigned on the basis of analysis by our

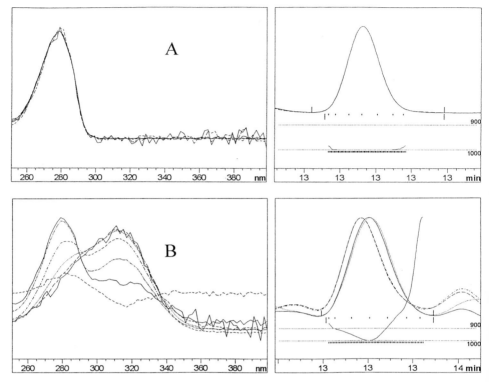

Fig. 2. Purity check spectral analysis of peak eluting at retention time of epicatechin. (A) Purity confirmed (100%). (B) Purity not confirmed (47%).

previously published GC-MS[17-19] and normal-phase HPLC[16] methods and were confirmed by the fact that the sum of the isomers and glucosides after irradiation was equal to the initial concentration of the *trans* isomer.

Only with *cis*- and *trans*-resveratrol did the intercept (positive) differ significantly from zero ($p < 0.04$ for each). This is consistent with the notion of background or baseline interference or lack of complete resolution in some matrices. Numerically, this was <2% of the average concentration of these compounds in red wine and was insufficient to adversely affect other

[17] D. M. Goldberg, J. Yan, E. Ng, E. P. Diamandis, A. Karumanchiri, G. Soleas, and A. L. Waterhouse, *Anal. Chem.* **66,** 3959 (1994).
[18] D. M. Goldberg, A. Karumanchiri, E. Ng, J. Yan, E. P. Diamandis, and G. J. Soleas, *J. Agric. Food Chem.* **43,** 1245 (1995).
[19] G. J. Soleas, D. M. Goldberg, E. P. Diamandis, A. Karumanchiri, J. Yan, and E. Ng, *Am. J. Enol. Vitic.* **46,** 346 (1995).

TABLE I
LINEARITY OF ASSAYS FOR NINE PHENOLIC CONSTITUENTS OF WINE ASSESSED BY
LINEAR REGRESSION ANALYSIS[a]

Constituent	Slope	Intercept	SE	p^b	Range[c]
Catechin	0.9968	33.7	42.3	0.48	10–200
Epicatechin	0.9994	−33.1	20.5	0.18	10–100
Rutin	0.9995	−43.4	25.6	0.16	5–50
trans-Resveratrol	0.9997	42.7	14.7	0.04	0.4–8.5
cis-Resveratrol	0.9998	23.1	7.6	0.04	0.6–13.2
Quercetin	0.9985	−241.7	92.4	0.06	5–50
trans-Polydatin	0.9689	−65.9	127.3	0.64	0.9–6.8
cis-Polydatin	0.9624	59.4	132.6	0.68	1.4–10.2
p-Coumaric acid	0.9945	10.2	20.8	0.62	0.5–8.4

[a] Y = area count; X = concentration as mg/liter.
[b] Probability that intercept is significantly different from zero.
[c] Lowest and highest concentration as milligrams per liter.

analytical variables such as sensitivity, recovery, and precision. Moreover, the match factor and purity checks eliminated any data that did not meet stringent criteria and presumably protected against errors due to such interference, especially at low concentrations.

Recovery

This was evaluated for each constituent by adding two concentrations to red (wine A), rosé (wine B), and white (wine C) commercial wines and performing quadruplicate assays before and after each addition. Excellent recovery (mean ± SD) was obtained, which on average ranged from 95.2 ± 5.5% for catechin to 105.5 ± 4.3% for *trans*-resveratrol. There was no consistent difference in recovery from any of the three wine matrices.

Precision

Ten replicate analyses were performed on four wines for each of the nine phenolics, selected from a number previously assayed so as to cover a reasonable range for each constituent. Rutin was not measurable in two and quercetin in one because of match or purity factors that did not meet the defined criteria. The mean coefficients of variation (CV) based on all four experiments indicated in Table II ranged from 1.17 to 3.38%, and the highest single value was for *cis*-reveratrol (5.7%) at a concentration of 0.79 mg/liter. This represents excellent precision for all constituents throughout the range tested.

TABLE II
PRECISION OF HPLC ASSAYS FOR NINE WINE PHENOLICS[a]

Constituent	Range of means (mg/liter)	Overall[b] mean CV (%)
Catechin	34–296	1.25
Epicatechin	20–108	1.17
cis-Polydatin	0.95–1.7	2.35
trans-Polydatin	1.32–4.4	2.10
cis-Resveratrol	0.79–4.7	3.38
trans-Resveratrol	1.66–4.3	1.95
Quercetin	9.5–20.6	2.23
Rutin	9.2–11.6	2.45
p-Coumaric acid	1.1–4.7	2.59

[a] Each value is derived from 10 replicate analyses on up to four wines of varying concentration for each constituent.
[b] Obtained by averaging the CV for all wine samples, i.e., $n = 4$ for all except quercetin ($n = 3$) and rutin ($n = 2$).

Detection Limit

Basing this on the lowest value that could be distinguished from zero (baseline) at $p < 0.001$ with a CV of $\leq 10\%$ and acceptable spectral criteria (match and purity factors), the lowest measurable values and best wavelengths were as follows: catechin, 1.5 mg/liter (280 nm); epicatechin, 1.2 mg/liter (280 nm); cis-polydatin, 75 μg/liter (285 nm); trans-polydatin, 48 μg/liter (306 nm); cis-resveratrol, 135 μg/liter (285 nm); trans-resveratrol, 30 μg/liter (306 nm); rutin, 0.8 mg/liter (265 nm); quercetin, 0.4 mg/liter (369 nm); and p-coumaric acid, 0.2 mg/liter (306 nm).

Method 2

Standards

In addition to those described for Method 1, the following may be used with satisfactory results: gallic acid, 5-(hydroxymethyl)furfural, 3,4-dihydroxybenzoic acid, 2-furoic acid, p-hydroxyphenethyl alcohol, p-hydroxyphenylacetic acid, vanillic acid, caffeic acid, vanillin, syringic acid, syringaldehyde, ellagic acid, all from Aldrich (Milwaukee, WI); quercitrin and myricetin are both from Sigma. The ethyl esters of caffeic and p-coumaric acids were prepared by reacting the pure compounds with absolute alcohol in the presence of minute amounts of sulfuric acid over a period of 18–24 hr. Isoquercitrin was a gift of Andres Wines Ltd; it was isolated from white grape juice by preparative HPLC and its purity was confirmed by analytical HPLC with spectral analysis using the diode array detector and by GC-MS. Glycosides of caffeic and *p*-coumaric acids were identified by their complete disappearance on treatment with β-glucosidase.[16] Simultaneously, there was an increment in the absorbance peaks of the parent phenolic acid that exactly matched the number of counts attributable to the intact glucoside peak and their absorbance spectra were virtually identical. Free phenolic acids are therefore used as calibrators for the glucosides, but when values are expressed as milligrams per liter as opposed to micromoles per liter, an adjustment is made to take into account the higher molecular weights of the glucosides.

Instrumentation

The HPLC equipment used was all from Waters Associates, comprising the 600 controller, 717 autosampler, 616 pumps, inline degasser, and the 996 photodiode array detector. The system is interfaced with a Millennium 2010 Chromatography Manager. An ET 250/4 Nucleosil 100-5 C_{18} column (25 cm × 4 mm) from Macherey-Nagel, Duren, Germany, with a 5-μm particle size was used as the stationary phase and was maintained at 25° during operation. It was preceded by a guard column comprising KS 11/4 Nucleosil 120-5 C_{18} (2 cm × 4 mm) of 5-μm particle size from the same source.

Procedure

A 0.5-ml aliquot of a wine sample is passed through a Sep-Pak C_{18} column (500 mg/6 ml) by gravity. The cartridge is dried in one of the following ways: under a water aspirator for 1 hr or under a stream of nitrogen at 10 psi for 10 min. Two milliliters of 20% (v/v) ethanol in ethyl acetate is added and the eluate is reduced to dryness under nitrogen at 10 ml/min. The residue is dissolved in 0–5 ml of 20% (v/v) ethanol in water. Twelve microliters is injected into the HPLC and the polyphenols are eluted in reverse phase using a gradient mobile phase of 0.2 ml phosphoric acid and 2 ml acetic acid in 1 liter of deionized water, pH 2.1 (A), together with 0.2 ml phosphoric acid and 2 ml acetic acid in 1 liter of acetonitrile, pH 2.4 (B), according to the program defined in Table III. The solvents must be filtered through a Millipore 0.45-μm PFTBA membrane filter (Millipore, Canada) and degassed inline. The polyphenols are monitored over the range of wavelengths from 250 to 390 nm.

TABLE III
GRADIENT ELUTION PROGRAM FOR METHOD 2[a]

Time (min)	Flow rate (ml/min)	A (%)	B (%)	Curve
0	0.30	90	10	—
20	0.30	87	13	8
50	0.30	85	15	11
70	0.30	84	16	11
80	0.33	80	20	11
90	0.30	78	22	11
110	0.30	75	25	11
120	0.30	73	27	10
155	0.60	50	50	6
157	0.65	90	10	4
165	0.65	90	10	—

[a] Compositions of eluants A and B are described in the text.

Chromatographic Resolution

This is demonstrated for a composite standard in white wine in Fig. 3A and in an authentic red wine sample in Fig. 3B.

Linearity

Linearity was 0.24–500 mg/liter, except for ellagic acid, which was 0.03–0.6 mg/liter.

Recovery

Recovery ranged from 87 (syringaldehyde) to 106% (furoic acid), except for gallic acid (70%).

Precision

This ranged from 1.6 (gallic acid) to 9.6% (catechin), except for furoic acid (16.0%) and 5-(hydroxymethyl)furfural (16.2%).

Detection Limit

Detection was 0.01 mg/liter, except for quercitrin, myricetin, and quercetin (0.05 mg/liter).

Precautions

The column was primed prior to usage for 1-hr with 50% A (H_2O):50% B (acetonitrile) followed by 90% A:10% (B) for 30 min. Because of the duration of the elution, some drift (up to 10 min), especially for late eluting peaks, must be expected. This makes online spectral identity of all peaks of interest mandatory, and the eluate must be monitored by a diode array detector. The software program incorporated with the instrumental system allows calculation of purity and match factors by spectral analyses described and illustrated for Method 1. Manual integration is advisable for peaks that are not fully resolved, in which case much greater accuracy is obtained.

This method has been applied without significant problems to 650 individual wines. It is also suitable for analyzing the polyphenol content of distilled spirits with the modification that 20 μl of the sample is injected directly, omitting the Sep-Pak extraction.

Comparison of Methods 1 and 2

Method 2 clearly allows a greater range of compounds to be separated but at the price of extending the run time four-fold. Both require diode array detection for peak identification and to achieve accurate quantitation of overlapping peaks, especially if this is performed manually. At present, not all of the compounds separated and quantitated by Method 2 are of enological interest; some have been included in Fig. 3 because they are important quantitatively in distilled spirits and may be useful in the chemical fingerprinting of such beverages. Gallic acid, caffeic acid, vanillic acid and vanillin, syringic acid, isoquercitrin, quercitrin, and myricetin are potentially important wine antioxidants measurable by Method 2 but not by Method 1.

The sensitivity of Method 2 is superior to that of Method 1 for the nine polyphenols measurable by the latter by one to two orders of magnitude apart from *trans*-resveratrol (3-fold) and *trans*-polydatin (5-fold). The range of linearity was greater with Method 2 for all polyphenols, notably trihydroxystilbenes and *p*-coumaric acid, where the differences were as much as 10- to 50-fold. However, the precision and recovery of most constituents were marginally superior for Method 1, although these characteristics for Method 2 were better than those of most published HPLC procedures for polyphenols.

Other Methods

For those wishing to focus upon the analysis of more specific classes of polyphenols, a number of HPLC methods are available for wine, grapes or grape juices. The best of these are well described and seem to allow good resolution as well as satisfactory quantitation of the major constituents in that particular class. A major problem with these assays is that few of the compounds are available in a sufficiently pure form from commercial

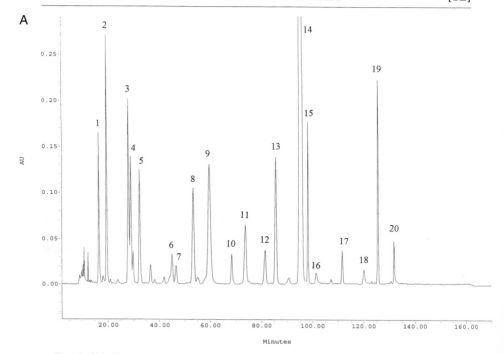

FIG. 3. (A) Chromatogram of composite dry standard in white wine: **1,** gallic acid; **2,** 5-(hydroxymethyl)furfural[a]; **3,** 3,4-dihydroxybenzoic acid; **4,** furoic acid[a]; **5,** *p*-hydroxyphenethyl alcohol; **6,** *p*-hydroxyphenyl acetic acid; **7,** (+)catechin; **8,** vanillic acid; **9,** caffeic acid; **10,** (—) epicatechin; **11,** vanillin; **12,** syringaldehyde; **13,** *p*-coumaric acid; **14,** sorbic acid[a] (this constituent is naturally present in white wines at high concentration); **15,** isoquercitrin; **16,** ellagic acid; **17,** quercitrin; **18,** myricetin; **19,** *trans*-resveratrol; and **20,** quercetin. All compounds are phenolic except for those labeled with a superscript a. The absorbance (AU) is integrated over the range 250- to 390-nm range for each peak and the scale of this printout extends from 0 to 0.30 AU. (B) Chromatogram of typical red wine demonstrating the following peaks: **1,** gallic acid; **2,** 5-(hydroxymethyl)furfural[a]; **3,** 3,4-dihydroxybenzoic acid; **4,** furoic acid[a]; **5,** caffeic acid glucoside; **6,** *p*-hydroxyphenylacetic acid; **7,** *p*-hydroxyphenethyl alcohol; **8,** (+)catechin; **9,** *p*-coumaric acid glucoside; **10,** vanillic acid; **11,** caffeic acid; **12,** (—)epicatechin; **13,** syringic acid; **14,** vanillin; **15,** syringaldehyde; **16,** *p*-coumaric acid; **17,** *trans*-polydatin; **18,** isoquercitrin; **19,** *cis*-polydatin; **20,** quercitrin; **21,** myricetin; **22,** *trans*-resveratrol; **23,** caffeic acid ethyl ester; **24,** *cis*-resveratrol; **25,** quercetin; and **26,** *p*-coumaric acid ethyl ester. All compounds are phenolic except for those labeled with superscript a. The absorbance (AU) is integrated over the 250- to 390-nm range for each peak and the scale of this printout extends from 0 to 0.18 AU.

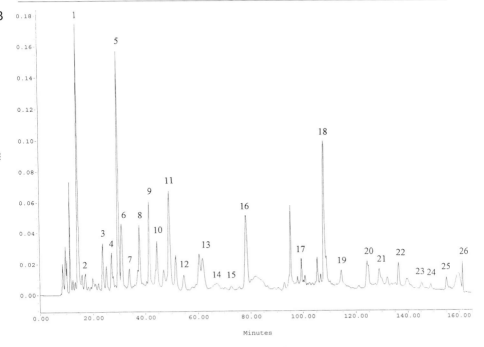

FIG. 3. (*continued*)

suppliers to serve as standards. In most cases, pure samples can only be obtained from scientific colleagues in the form of synthesized or naturally extracted compounds that can be provided in very limited quantities if at all.

Anthocyanins

In a run-time of approximately 90 min,[20] 8 were satisfactorily resolved and quantitated using ε_{520} for malvidin-3-glucoside or malvidin-3,5-diglucoside for all of the mono and diglucosides, respectively. Simultaneously, another group using a different solvent system for gradient elution were able to quantitate 10 anthocyanins as well as their aglycones after enzymatic hydrolysis.[21] Earlier, Wulf and Nagel[22] had developed a method requiring an extensive extraction procedure prior to the chromatographic step. In a

[20] K. Yokotsuka and V. L. Singleton, *Am. J. Enol. Vitic.* **48**, 13 (1997).
[21] J. D. Wightman, S. F. Price, B. T. Watson, and R. E. Wrolstad, *Am. J. Enol. Vitic.* **48**, 39 (1997).
[22] L. W. Wulf and C. W. Nagel, *Am. J. Enol. Vitic.* **29**, 42 (1978).

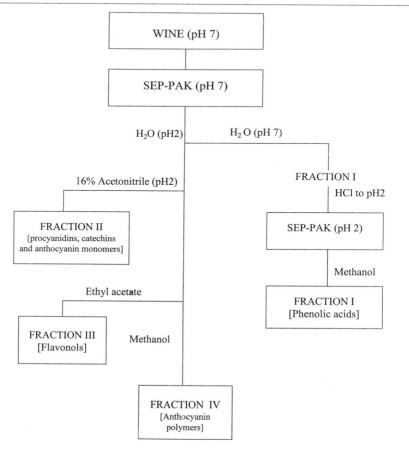

FIG. 4. Flow diagram of the four-stage HPLC separation of phenolic compounds of red wine. Redrawn from J. Oszmianski, T. Ramos, and M. Bourzeix, *Am. J. Enol. Vitic.* **39**, 259 (1988).

run time of 45 min they could separate up to 17 anthocyanins, but the resolution and peak heights of several were rather poor, and no analytical evaluation of the method was performed. Lamuela-Raventos and Waterhouse[23] described a qualitative method whereby 10 anthocyanins were resolved with good sensitivity, although quantitation was not attempted.

[23] R. M. Lamuela-Raventos and A. L. Waterhouse, *Am. J. Enol. Vitic.* **45**, 1 (1994).

Procyanidins

One of the earliest and seemingly best methods allowed good resolution of four procyanidins as well as a useful range of flavonoids and phenolic acids during a 60-min run time. The analytical characteristics were described fully and appear to be very satisfactory apart from somewhat low extraction efficiency for the procyanidins.[6]

An ambitious and somewhat lengthy procedure based on multiple elutions at different pH values and with three different mobile phases separated four fractions (Fig. 4) containing A (phenolic acids), B (anthocyanin polymers), C (flavonols), and D (procyanidins, catechins, and anthocyanin monomers). Three procyanidins gave discrete peaks, but the anthocyanin monomers and polymers were resolved poorly, the wavelength chosen being 280 nm.[1] In a method requiring a run time of 150 min, three procyanidins were identified qualitatively but were not quantitated.[10] Of the two most comprehensive methods described to date, one requires three consecutive elutions of the same column, the third of which yields 6 procyanidins that are well resolved and readily quantitated.[2] The other calls for successive chromatographic separations on two discrete columns, but 10 procyanidins can be assayed quantitatively in this procedure.[24] Good extraction efficiency was reported for the first[2] but details were not provided for the second.[24] At present, these seem to be the most comprehensive and recommendable methods for the specific measurement of procyanidins in grapes and their products, but they call for greater care and expertise than most HPLC procedures.

[24] T. Fuleki and J. M. Ricardo da Silva, *J. Agric. Food Chem.* **45**, 1156 (1997).

[13] Analysis of Antioxidant Wine Polyphenols by Gas Chromatography–Mass Spectrometry

By GEORGE J. SOLEAS and DAVID M. GOLDBERG

Introduction

Polyphenolic components of wine comprise on average only 1200 mg/liter in red wines and one-sixth of that concentration in white wines.[1] They

[1] V. L. Singleton, *in* "University of California Davis Grape and Wine Centenial Symposium Proceedings" (A. D. Webb, ed.), p. 215. University of California Press, Davis, CA, 1982.

fall into one of two major classes. Nonflavonoids comprise hydroxybenzoates and hydroxycinnamates, whose concentrations do not differ dramatically between the two classes of wines. Flavonoids include flavonols (e.g., quercetin, myricetin), flavan-3-ols (e.g., catechin and epicatechin), as well as polymers of the latter defined as procyanidins, and anthocyanins that are the pigments responsible for the color of red wines; collectively they are 20-fold higher in red than in white wines. More recently, a family of trihydroxystilbenes (the isomers of resveratrol and their glucosides) have been identified in significant concentrations in red but not white wine.[2]

These compounds impart to wine its color, astringence, aging potential, and some of its flavors. They are important in judging wine quality and its improvement during fermentation and the subsequent stages of production. Larger wineries with appropriate analytical facilities usually measure the concentrations of some of these polyphenols as part of their quality control programs. Moreover, interest has grown rapidly in the antioxidant properties of wine polyphenols, together with their potential to decrease the risk of atherosclerosis (coronary heart disease and ischemic stroke), cancer, and inflammatory disorders.[3,4] Their antioxidant activity as well as their other favorable biological properties show wide variation, and a relatively small number of the more than 150 eligible compounds that can be assayed are especially noteworthy in these respects.[5,6] We have developed a gas chromatography–mass spectrometry (GC-MS) procedure that specifically measures the concentrations in wine of low molecular weight polyphenols whose trimethylsilyl (TMS) derivatives are <800 amu.[7] These include compounds with the highest antioxidant and biological activities among nonflavonoids (hydroxybenzoates, hydroxycinnamates, trihydroxystilbenes) and flavonoids (flavonols, flavan-3-ols). By virtue of their molecular masses, procyanidins and anthocyanins are beyond the scope of the bench-top instrumentation employed by the wine industry, but for that very reason their absorption in the human intestine is improbable; additionally, evidence favoring their beneficial biological effects is nowhere so compelling as that for the 15 polyphenols that we have selected for this proposed method, which has been in routine use in our laboratory for approximately 15 months.

[2] G. J. Soleas, E. P. Diamandis, and D. M. Goldberg, *Clin. Biochem.* **30,** 91 (1997).
[3] D. M. Goldberg, S. E. Hahn, and J. G. Parks, *Clin. Chim. Acta* **237,** 155 (1995).
[4] G. J. Soleas, E. P. Diamandis, and D. M. Goldberg, *J. Clin. Lab. Anal.* **11,** 287 (1997).
[5] P. L. Teissedre, E. N. Frankel, A. L. Waterhouse, H. Peleg, and J. B. German, *J. Sci. Food Agric.* **70,** 55 (1996).
[6] G. J. Soleas, G. Tomlinson, E. P. Diamandis, and D. M. Goldberg, *J. Agric. Food Chem.* **45,** 3995 (1997).
[7] G. J. Soleas, E. P. Diamandis, A. Karumanchiri, and D. M. Goldberg, *Anal. Chem.* **69,** 4405 (1997).

Materials and Methods

Wines

Ideally, wines should be opened and analyzed within 24 hr. In the event that samples require storage, they may be kept for up to 5 days at 4° in a glass vial filled to completion and protected by foil against sunlight. Under no circumstances should wine be frozen, as this causes destability and precipitation of polyphenols when thawed.

Standards, Chemicals, and Solvents

Purity and dilution solvents, as well as recommended suppliers for all phenolic standards, are listed in Table I. cis-Resveratrol may be prepared from the *trans* isomer by ultraviolet (UV) irradiation.[8] trans-Polydatin has to be isolated from the dried roots of *Polygonum cuspidatum* and a portion can be converted to the *cis* isomer by UV irradiation.[9] A stock standard solution of 1 g/liter is convenient apart from myricetin (400 mg/liter) and *trans*-resveratrol (5 g/liter).

We use N,O-bis(trimethylsilyl)trifluoroacetamide (BSTFA) from Regis Technologies, Inc. (Morton Grove, IL), which must be stored under nitrogen at all times. Trimethylchlorosilane 1% (TMCS) from Pierce (Rockford, IL) was used in some preliminary experiments.

Instrumentation

A DB-5HT capillary column (J & W Scientific, Folsom, CA) with 5% phenyl-substituted methylpolysiloxane nonpolar stationary phase, crosslinked and double bonded to the capillary wall with excellent thermal stability and low bleed levels of dimensions 30 m × 0.25 mm I.D. × 0.10-μm film thickness was used. This should be preceded by a 1-m hollow silica capillary guard column of the same inside diameter connected to the analytical column via a chromfit glass connector (Chromatographic Specialties Inc., Brockville, Ontario, Canada). A regular DB-5 column was unsatisfactory because of excessive bleed and inadequate resolution of late-eluting peaks.

A Hewlett-Packard (HP) GC-5890 was used for the analysis, equipped with a split/splitless injection port, interfaced to a MSD-5970. The GC/MSD is controlled by an HP Vectra 486/50N PC utilizing the MS ChemStation software-G1034C (DOS series) and reporting to a HP laserjet IV printer.

---

[8] D. M. Goldberg, E. Ng, A. Karumanchiri, J. Yan, E. P. Diamandis, and G. J. Soleas, *Chromatogr. A.* **708,** 89 (1995).

[9] D. M. Goldberg, E. Tsang, A. Karumanchiri, E. P. Diamandis, G. Soleas, and E. Ng, *Anal. Chem.* **68,** 1688 (1996).

TABLE I
Source of Phenolic Standards and Dissolving Solvent(s)[a]

Compound (Chemical name)	Molecular weight (g)	Supplier	Purity (%)	Dissolving solvent (v/v)
Caffeic acid (3,4-dihydroxycinnamic acid)	180.2	Sigma[b]	98	80% ethyl acetate 20% acetone
(+)-Catechin ([+]-cyanidol-3)	290.3	Sigma	98	Ethyl acetate
m-Coumaric acid (3-hydroxycinnamic acid)	164.2	Sigma	N/A	95% ethyl acetate 5% acetone
p-Coumaric acid (4-hydroxycinnamic acid)	164.2	Sigma	N/A	95% ethyl acetate 5% acetone
(−)-Epicatechin (cis-2-[3,4-dihydroxyphenyl]-3,4-dihydro-2H-1-benzopyran-3,5-triol)	290 3	Sigma	>95	20% ethanol 80% acetone
Ferulic acid (4-hydroxy-3-methoxycinnamic acid)	194.2	Sigma	>96	95% ethyl acetate 5% acetone
Fisetin (3,3',4',7-tetrahydroxyflavone)	286.24	Aldrich[c]	98	40% acetone 60% ethyl acetate
Gallic acid (3,4,5-trihydroxybenzoic acid)	170.1	Sigma	98	95% ethyl acetate 5% acetone
Gentisic acid (2,5-dihydroxybenzoic acid)	154.12	Lancaster[d]	99	Ethyl acetate
Morin (2',3,4',5,7-pentahydroxyflavone)	302.2	Sigma	99	95% ethyl acetate 5% acetone
Myricetin (3,3',4',5,5',7-hexahydroxyflavone)	318.2	Sigma	85	Absolute ethanol
Quercetin (3,3',4',5,7-pentahydroxyflavone) dihydrate	338.3	Sigma	>98	Methanol
(+)-Rutin trihydrate (quercetin-3-rutinoside)	610.52	Lancaster	97	15% methanol 85% ethyl acetate
trans-Resveratrol (trans-3,4',5-trihydroxystilbene)	228.2	Sigma	99	Absolute ethanol
Syringic acid (4-hydroxy-3,5-dimethoxybenzoic acid)	198.2	Sigma	98	Absolute ethanol
Vanillic acid (4-hydroxy-3-methoxybenzoic acid)	168.2	Sigma	>96	95% ethyl acetate 5% acetone

[a] All stocks standards were prepared around 1000 mg/liter.
[b] Sigma-Aldrich Canada, Ltd, Mississauga, Ontario, Canada.
[c] Aldrich Chemical Company, Inc., Milwaukee, WI.
[d] Lancaster Synthesis Inc., Windham, NH.

Ultra-high purity helium with an inline Supelpure moisture trap and hydrocarbon trap should be used as carrier gas. The carrier gas-line pressure is set at 60 psi, column head pressure at 8 psi, and the septum purge is at 2.4 ml/min.

Gas Chromatography Temperature Information

Injector: 280°
Detector (transfer line): 320°
Oven equilibration time: 1.0 min
Initial temperature: 80°
Initial time: 1.0 min
Oven temperature program:

Level	Rate (degrees/min)	Final temperature (degrees)	Final time (min)
1	20.0	250	1.0
2	6.0	300	2.0
3	20.0	320	4.0

Total run time: 25.8 min.

Injector and detector temperatures are set based on previous experience with the analysis of resveratrol in wine and juices.[10] The GC oven temperature program is designed to elute all phenolic compounds at a fast rate without jeopardizing their resolution from interferences, giving sharp peaks, flat baseline, and good sensitivity. It is necessary to introduce a 4-min baking period at the end of each run to ensure the elimination of ghost peaks and a low signal-to-noise ratio.

Gas Chromatography Injector Information

Injection source: autoinjector
Sample washes: three
Sample pumps: three
Sample volume injected: 1 μl
Solvent A washes: 4; solvent A: acetone
Solvent B washes: 4; solvent B: pyridine
Injection port: Splitless with a double gooseneck glass insert and a gold-plated injector seal and a Viton O-ring for high temperatures (all purchased from Hewlett-Packard) are beneficial in improving overall sensitivity of analyses

---

[10] G. J. Soleas, D. M. Goldberg, E. P. Diamandis, A. Karumanchiri, J. Yan, and E. Ng, *Am. J. Enol. Vitic.* **46,** 346 (1995).

Mass Spectrometry Information

    Acquisition mode: Selective ion monitoring (SIM)
    Solvent delay: 7.80 min
    Electron multiplier voltage (EMV) = 1400
    EMV offset = 200
    Resulting EMV = 1600

Instrument Maintenance

Cleaning of the MSD ion source components, including the lens stack assembly, is necessary approximately every 400 injections.

The injector double gooseneck glass insert and gold-plated injector seal should be dedicated for this analysis and cleaned by sonication in dichloromethane followed by methanol every 80–100 injections. The viton O-ring also needs to be replaced at the time of cleanup. Septa were replaced on a daily basis. The MSD was tuned, using a customized tune for this analysis with perfluorotributylamine, utilizing three ions within the range of the TMS derivative ions (219, 414, and 502 amu) prior to analyzing each batch of samples.

The moisture content of the carrier gas as well as from other sources must be monitored by a customized tune immediately after the replacement of gas cylinders, columns, and ion-source cleanups. It is advisable to monitor carrier gas line pressure, column head pressure, and septum purge on a weekly basis.

Extraction and Derivatization Procedure

Sep-Pak C_8 cartridges (Waters) are preconditioned with 3 ml ethyl acetate, 3 ml 60% absolute ethanol, and 3 ml deionized water followed by 2 ml of the latter. Wine samples are diluted with an equal volume of deionized water to bring the alcohol level to approximately 6%, and exactly 1 ml of diluted sample is injected onto the preconditioned Sep-Pak and allowed to drain by gravity flow (3–5 min). A gentle flow of nitrogen is then introduced over the sample with simultaneous gradual suction on a vacuum manifold (Millipore) at 100 kPa for 45 min.

Phenolic compounds are extracted by eluting the dry Sep-Pak with 3 ml of ethyl acetate. The eluate is collected in a centrifuge tube spiked previously with fisetin as an internal standard at 1.0 mg/liter. The extract is then evaporated to dryness on a nitrogen evaporator (Meyer Organomation Associates Inc., S. Berlin, MA). This model was selected because it allows the evaporation of 12 extracts simultaneously in a temperature-controlled bath and allows for the fine adjustment of the nitrogen flow. To ensure complete removal of water, 0.5 ml of dichloromethane is added, vortexed, and evaporated to dryness (azeotropic removal of water). Extracts are dried further in an oven at 70° for 15 min and derivatized by incubating with 1.0 ml of 1:1 BSTFA/pyridine using vigorous vortexing and incubating at 70° for 30 min.

Columns of Sep-Pak C_{18} and C_1 as well as Extrelut (diatomaceous earth) are unsatisfactory for this extraction process, even when the wine samples are dealcoholized or adjusted to various pH values ranging from 2 to 7 and elution is attempted with a variety of organic solvents. Preliminary experiments demonstrated that with stacked C_8 columns, no polyphenols could be detected beyond the first column, suggesting their complete absorption under the conditions employed. This is supported by the high values for overall recovery given by the method (*vide infra*).

Dealcoholization of the wine sample in a rotary evaporator required a larger volume and a time of 30 min, but improved the recovery of several polyphenols greatly. Diluting the sample with an equal volume of purified water achieves the same increase in recovery without adding to the duration of the assay and also reduces the required sample volume to 0.5 ml. At the same time, matrix interference and the relative standard deviation (RSD) for most analytes are decreased, as are problems due to MSD overranging of certain polyphenols.

The derivatization reagent used in the recommended procedure is less prone to matrix interference and provides improvement in recovery and RSD compared with two alternative reagents: BSTFA alone and BSTFA with 1% trimethylchlorosilane.

Because some of these phenolics tend to polymerize once exposed to light, all necessary precautions must be taken during the analysis to avoid light exposure. All samples should be kept in the dark, and the extraction apparatus should preferably be kept in a dark box customized for this analysis.

Moisture is a major competitor of phenolic hydroxyl groups during derivatization with BSTFA:pyridine and can produce low recoveries. To avoid this problem, all glassware must be washed in acetone during extraction, and nitrogen passed through a moisture trap is introduced from the top of the Sep-Pak for the duration of the extraction. Finally, dichloromethane is added to the dry extract and evaporated to dryness before derivatization is performed.

Identification of Phenolic Compound Characteristic Ions

Individual phenol stock standards are diluted to individual working standards of approximately 10 mg/liter. Each working standard is dried

TABLE II
Selective Ion Monitoring of Target and Two Qualifier Ions for Each Phenolic Compound

Compound	Retention time (min)	Target ion[a] (m/z)	Qualifier ions (m/z)
Fisetin	17.90	471.00	399.0, 559.8 (55) (150)[b]
Vanillic acid	8.29	297.35	253.0, 312.4 (58) (67)
Gentisic acid	8.39	355.4	356.5, 357.4 (87) (40)
m-Coumaric acid	8.87	249.0	293.0, 308.0 (184) (178)
p-Coumaric acid	9.27	249.0	293.0, 308.0 (184) (178)
Gallic acid	9.40	282.0	443.6, 460.0 (36) (55)
Ferulic acid	10.07	338.4	323.4, 293.3 (57) (34)
Caffeic acid	10.33	396.5	381.5, 307.4 (25) (12)
cis-Resveratrol	11.83	444.7	445.6, 446.7 (41) (18)
trans-Resveratrol	14.24	444.7	445.6, 446.7 (41) (18)
(−)-Epicatechin	15.66	369.5	355.5, 368.5 (105) (233)
(+)-Catechin	15.89	369.5	355.5, 368.5 (87) (300)
Morin	16.47	648.0	649.0, 560.0 (57) (10)
Quercetin	18.70	648.0	649.0, 559.8 (61) (14)
cis-Polydatin	20.40	361.0	444.0, 372.0 (107) (59)
trans-Polydatin	23.93	361.0	444.0, 372.0 (66) (43)

[a] Target ion was taken to be 100%.

[b] Numbers in parentheses represent the target ion : qualifier ion ratio expressed as a percentage.

and derivatized following the procedure as described earlier. One microliter of each derivatized extract is injected separately onto the GC-MSD with the instrument on full scan mode, from 50 to 800 amu. This allows the establishment of the retention time and the characteristic TMS derivative spectrum of each phenolic compound. A target and two qualifying ions per compound have been chosen on the basis of their abundance, reproducibility, freedom from interference, and specificity to the compound. The molecular ion (M$^+$) was preferred when found in appreciable abundance (Table II). Phenolic compounds were divided into seven groups of ions (Table III), with each group containing the ions of one, two, or three compounds. The dwell time was set at 100 msec/ion.

Chromatographic Resolution

A composite dry standard of all substances tested, after derivatization and analysis, demonstrated excellent resolution between all compounds of interest (Fig. 1a). Similarly, satisfactory resolution was obtained with authentic wine samples (Fig. 1b). Myricetin, rutin, and isoquercitrin displayed poor sensitivity even at concentrations as high as 20 mg/liter and, therefore, are not measurable by this method as their TMS derivatives exceed the limit of 800 amu for this instrument. A method blank showed very low background noise (5000 abundance units for phenolic acids and 2500 abundance units for the remaining chromatogram).

Although some wine extracts show minor interferences, the software can apply adequate corrections during intergration, which can also be done manually for sharper definition of true peaks. The chromatogram baseline is very stable and column bleed is usually not noticeable, even after 500 injections. The earliest manifestation of column deterioration is a decrease in the sensitivity of late eluters (catechin, epicatechin, morin, fisetin, quercetin, and polydatins), usually after 200–300 injections, and cropping of the

TABLE III
GC-MSD SELECTIVE ION-MONITORING PARAMETERS[a]

Group	Group start time (min)	Ions in group (amu)
1	8.00	253.0, 297.4, 312.4, 355.4, 356.5, 357.4
2	8.70	293.0, 249.0, 308.0, 282.0, 443.6, 460.0
3	9.80	338.4, 323.4, 293.3, 396.5, 381.5, 307.4, 268.0
4	11.20	444.7, 445.6, 446.7
5	14.80	368.5, 355.5, 369.5
6	16.20	648.0, 649.0, 560.0, 471.0, 399.0
7	19.10	361.0, 444.0, 372.0

[a] Dwell time: 100 msec/ion.

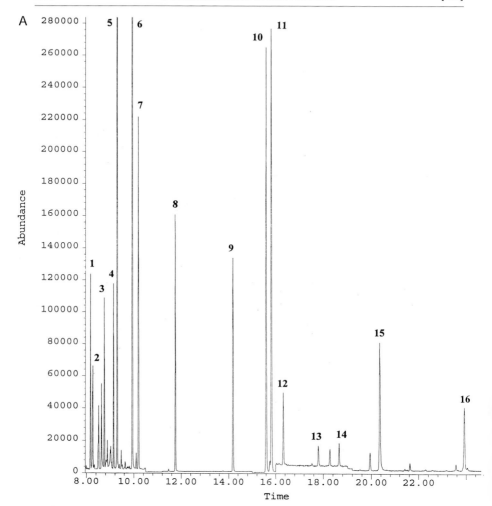

FIG. 1. (A) TIC of a composite dry standard in SIM. 1, vanillic acid; 2, gentisic acid; 3, *m*-coumaric acid; 4, *p*-coumaric acid; 5, gallic acid; 6, ferulic acid; 7, caffeic acid; 8, *cis*-resveratrol; 9, *trans*-resveratrol; 10, epicatechin; 11, catechin; 12, morin; 13, fisetin; 14, quercetin; 15, *cis*-polydatin; and 16, *trans*-polydatin. (B) TIC of a 1994 Cabernet Sauvignon wine extract in SIM. 1, vanillic acid; 2, gentisic acid; 3, *m*-coumaric acid; 4, *p*-coumaric acid; 5, gallic acid; 6, ferulic acid; 7, caffeic acid; 8, *cis*-resveratrol; 9, *trans*-resveratrol; 10, epicatechin; 11, catechin; 12, morin; 13, fisetin; 14, quercetin; 15, *cis*-polydatin; and 16, *trans*-polydatin.

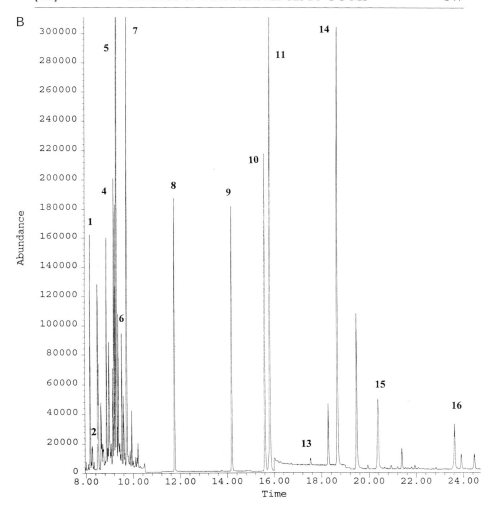

FIG. 1. (*continued*)

injector end of the column then becomes necessary. The 1-m guard column at the injector end improves the lifetime of the column and reduces the need for column cropping.

Detection of the compounds of interest is based on their retention time (RT), the presence of both qualifier ions, and the predetermined ratio between the target ion and each qualifier ($\pm 25\%$ tolerance limit) (Table II). A standard comprising a composite spiked extract is injected after every five samples. Fisetin is used as an internal standard (ISTD) in all extracts and standards at a concentration of 1.00 mg/liter. The response of fisetin is not used to correct results but rather to monitor unusual instrument fluctuation, probably due to matrix, and most often in compounds in the epicatechin-to-quercetin window. Compounds overranging the instrument or whose concentrations fall outside the linearity range are diluted and reanalyzed against a standard with an appropriate dilution factor.

Method Validation

Linearity. Data for each constituent were pooled from three experiments in which the constituent was analyzed over a range of six to nine concentrations individually, in a mixture of all 15 constituents dissolved in absolute alcohol, and added to a red wine matrix. Linear regression analyses were performed using the formula $y = mx + b$. The slope of the calibration curve was almost perfectly linear for all compounds except for catechin and epicatechin, and the square of the regression coefficient differed from unity by more than 0.020 only in the case of the former (0.029).

Recovery. This was evaluated for each constituent by adding three concentrations to simulated wine and analyzing each wine independently six

TABLE IV
OVERALL RECOVERY (%) OF 15 PHENOLIC CONSTITUENTS ADDED TO SIMULATED WINE AT THREE CONCENTRATION LEVELS

Compound	Overall mean recovery[b] (%)	RSD (%)
Vanillic acid	94.2	4.6
Gentisic acid	96.1	4.7
m-Coumaric acid	98.5	3.6
p-Coumaric acid	98.3	7.7
Gallic acid	99.3	3.9
Ferulic acid	102.3	4.8
Caffeic acid	104.6	5.8
cis-Resveratrol	90.7	5.4
trans-Resveratrol	99.9	6.4
(−)-Epicatechin	97.1	7.5
(+)-Catechin	96.5	7.7
Morin	72.2	9.7
Quercetin	92.6	13.5
cis-Polydatin	96.3	3.8
trans-Polydatin	102.7	5.3

[a] Each wine was assayed independently six times and the average result is presented.
[b] Obtained by pooling all recovery data at three levels, i.e., $n = 18$.

times. The overall recovery was obtained by pooling all recovery data, i.e., $n = 18$ (Table IV). Excellent recovery was obtained, which on average ranged from $90.7 \pm 5.4\%$ for cis-resveratrol to $104.6 \pm 5.8\%$ for caffeic acid. The exception was morin at $72.2 \pm 9.7\%$.

Precision. Six replicate analyses of four red wines (different cultivars) of varying concentrations of each constituent were performed. cis-Resveratrol was detected only in three and morin and cis- and trans-polydatin in only two of the four samples. The overall mean CV ranged from 4.0 (gentistic acid) to 10.3% (trans-resveratrol). Morin and quercetin were the exceptions at 16.1 and 16.0%, respectively (Table V).

Combined Variance for Detector and Derivatization. Extracts of 12 red wine samples were pooled together. Ten 1-ml aliquots of the combined extracts were derivatized independently and analyzed for all 15 phenolic constituents (Table VI). The percentage relative standard deviation ranged from 2.0 to 10.2%, with the highest single value for quercetin at 15.9%.

Detection Limit. This was based on three standard deviations of the mean assay value of each phenolic compound analyzed at the lowest of

TABLE V
PRECISION OF GC-MSD ASSAY FOR 15 WINE PHENOLICS[a]

Compound	Overall mean[c] (mg/liter)	Overall mean[b] CV(%)
Vanillic acid	2.58	4.5
Gentisic acid	0.42	4.0
m-Coumaric acid	0.52	6.3
p-Coumaric acid	0.97	8.6
Gallic acid	22.50	5.6
Ferulic acid	0.28	7.3
Caffeic acid	2.73	5.0
cis-Resveratrol	0.38	7.7
trans-Resveratrol	0.73	10.3
(−)-Epicatechin	33.20	5.8
(+)-Catechin	48.30	4.8
Morin	0.48	16.1
Quercetin	0.37	16.0
cis-Polydatin	0.73	7.6
trans-Polydatin	1.27	6.5

[a] Each value is derived from six replicate analyses on four wines of varying concentration for each constituent.
[b] Obtained by averaging the CV for all wine samples, i.e., $n = 4$ for all except morin, cis- and trans-polydatin ($n = 2$), and cis-resveratrol ($n = 3$).
[c] Obtained by averaging the concentrations for all wine samples, i.e., $n = 4$ for all except morin, cis- and trans-polydatin ($n = 2$), and cis-resveratrol ($n = 3$).

TABLE VI
COMBINED VARIANCE FOR DETECTOR AND DERIVATIZATION[a]

Compound	Amount (mg/liter)	RSD (%)
Vanillic acid	1.24	5.1
Gentisic acid	0.47	5.5
m-Coumaric acid	1.29	5.2
p-Coumaric acid	1.26	9.2
Gallic acid	1.21	2.9
Ferulic acid	1.29	2.0
Caffeic acid	1.62	3.0
cis-Resveratrol	1.02	6.5
trans-Resveratrol	0.80	6.6
(−)-Epicatechin	4.72	7.7
(+)-Catechin	4.82	7.4
Morin	4.84	10.2
Quercetin	4.87	15.9
cis-Polydatin	0.68	6.9
trans-Polydatin	1.92	8.2

[a] The extracts of 12 samples were pooled together. Ten 1-ml aliquots of the combined extracts were derivatized independently and analyzed by GC-MSD.

the three concentration levels (Table V), satisfying both qualifier and target ions and the correct abundance ratio. These limits were as follows (in mg/liter): vanillic acid, 0.063; gentisic acid, 0.024; m-coumaric acid, 0.051; p-coumaric acid, 0.117; gallic acid, 0.048; ferulic acid, 0.063; caffeic acid, 0.111; cis-resveratrol, 0.111; trans-resveratrol, 0.084; epicatechin, 0.324; catechin, 0.336; morin, 0.309; quercetin, 0.843; cis-polydatin, 0.015; and trans-polydatin, 0.132.

Day-to-Day Variation. Seven bottles of red wine picked from the same case were stored in the dark and analyzed on seven separate occasions. The CV for all polyphenols ranged from 4.7 [(+)-catechin] to 12.5% (trans-resveratrol). Values were not significantly different from CV data for simultaneously analyzed replicates (Table V).

Comments

Although GC-MS analysis has been used to measure certain trihydroxystilbenes[2] and some other polyphenols in wine,[11] the present method is the first fully developed to permit simultaneous quantitative determination of

---

[11] J. L. Wolfender and K. Hostettmann, *J. Chromatogr.* **647**, 191 (1993).

a wide array of compounds, including most of those that have been shown to possess significant biological properties. All analytical characteristics required for a thorough evaluation of the method have been provided for each constituent analyzed.

Putting aside capital and service costs, reagent costs per 100 samples, for 15 analytes (including recycling of the Sep-Pak cartridge twice), approximate $300. The current cost of the columns per 1000 samples, taking account of their longevity under the operating conditions specified, is $800. Instrument run time is 26 min. Because the system can function unattended with automatic sample injection, 50 samples can be analyzed in ~22 hr. The time spent in solid-phase extraction and derivatization, both of which are batch processes, is modest. Finally, the employment of mass spectra characteristics in identification and quantitation leads to greater confidence in the accuracy of the assay and reduces the requirements for calibration and standardization compared to high-performance liquid chromatography (HPLC) methods.

Few investigators have reported the use of MS methods to analyze the polyphenol content of other beverages and foodstuffs, a task for which HPLC has been employed more often. Exceptions include thermospray LS-MS analysis of polyphenols from tea,[12] a similar approach to screen for polyphenols in plant extracts,[11] and a pyrolysis GC-MS technique that has been proposed as applicable for the analysis of wine polyphenolics but not yet validated.[13] The present procedure has been used to analyze the same polyphenols in extracts of solid vitaceous materials such as stems, leaves, skins, and pips after exhaustive pulverization and homogenization in ethanol and adjustment of the final concentration of the latter to 6% (v/v) prior to solid-phase separation. It should be equally suitable for analyzing these polyphenols, and potentially many others, in any plant or food material provided that extraction is complete and possible matrix interference by the solvents employed on the solid-phase separation and derivatization steps are excluded or circumvented. Furthermore, the excellent sensitivity and selectivity coupled with the small sample volume required (0.5 ml) for this assay render it potentially useful for the analysis of biological fluids, although this application has not yet been validated.

---

[12] A. Keine and U. H. Engelhardt, *Z. Levensm Unters Forsch.* **202**, 48 (1996).
[13] G. C. Galletti and A. Antonelli, *Rapid Commun. Mass Spectrom.* **7**, 656 (1993).

[14] Analysis of Total Phenols and Other Oxidation Substrates and Antioxidants by Means of Folin–Ciocalteu Reagent

By VERNON L. SINGLETON, RUDOLF ORTHOFER, and ROSA M. LAMUELA-RAVENTÓS

Introduction

Phenols occurring in nature and the environment are of interest from many viewpoints (antioxidants, astringency, bitterness, browning reactions, color, oxidation substrates, protein constituents, etc.). In addition to simple benzene derivatives, the group includes hydroxycinnamates, tocopherols, and flavonoids in plants and foods from them, tyrosine and DOPA derivatives in animal products, and additives such as propyl gallate in foods. Estimation of these compounds as a group can be very informative, but obviously not simple to accomplish. Isolative methods such as high-performance liquid chromatography (HPLC) are difficult to apply to such a diverse group having, furthermore, many individual compounds within each subgroup. Interpretation of such results is even more difficult.

Phenols are responsible for the majority of the oxygen capacity in most plant-derived products, such as wine. With a few exceptions such as carotene, the antioxidants in foods are phenols. Among those added to prevent oxidative rancidity in fats are the monophenols (benzene derivatives with a single free hydroxyl group) 2,6-di-*tert*-butyl-4-hydroxytoluene (BHT) and its monobutylated anisole analog (BHA). *tert*-Butyl substituents function mainly to increase the lipid solubility. In aqueous solution the parent monophenols and others can also function as antioxidants. Therefore, it is important that total phenol assays include monophenols as well as more easily oxidized polyphenols.

An antioxidant effect can be from competitive consumption of the oxidant, thus sparing the target molecules being protected, and from quenching the chain reaction propagating free radical oxidation. Antioxidants become oxidized as they interfere with the oxidation of lipids and other species. Paradoxically, because of coproduction of hydrogen peroxide as an antioxidant phenol or ascobic acid reacts with oxygen, coupled oxidation can occur of substrates (ethanol, for example) that would not react readily with oxygen alone.[1]

---

[1] H. L. Wildenradt and V. L. Singleton, *Am. J. Enol. Vitic.* **25,** 119 (1974).

If one electron is removed (oxidized) from a phenolate anion, the product is a semiquinone free radical. Removal of a second electron from *ortho*- or *para*-diphenols produces a quinone. A mixture of phenol and quinone equilibrates to produce semiquinone intermediates. The molecule accepting a removed electron is, of course, reduced. Free radicals are very reactive molecules with an unpaired electron. Encountering another free radical from any source (its own type, lipoidal, etc.), the two combine to form a new covalent bond, terminating any chain reaction caused by extraction by the free radical of an electron from an intact molecule to generate another free radical. The unpaired electron in a semiquinone can resonate among the former hydroxyl and the positions *ortho* and *para* to it (two, four, or six of the ring). A mixture of dimerized products results as the new bonds form. If the new bond is to one of the ring carbons, the phenolate is regenerated. Oxidation may then not only be repeated, but the regenerated phenol is often oxidized more easily than the original one. If the important property of oxidizability is to be the basis for the quantitation of phenols, the reaction must be brought quickly to a conclusion to minimize such regenerative polymerization.

That the phenolate ion is important is shown by the fact that the uptake of oxygen by phenols can be rapidly complete near or above the pK of the phenol (usually about pH 10).[2-4] Because of the relative ease of removing an electron from its phenolate, the less acidic the particular phenol the easier its oxidation. At lower pH the reaction appears proportionate to the pH, but as low as pH 3 equilibria supply enough phenolate among natural phenols of low acidity to allow slow reaction (with, of course, additional total oxygen uptake from regenerative polymerization and any other slow, competing reaction). Reaction at alkaline pH is indicated for assay purposes.

A method based on these considerations can be very useful provided it is reproducible, its basis understood, and its applicability verified. The proper use of the reagent proposed by Otto Folin and Vintila Ciocalteu[5] is such a method. The resultant total value is often directly comparable and informative among different samples, e.g., the total phenol content of commercial wines can range from about 50 to 5000 mg/liter.[6] This 100-fold range not only distinguishes white, pink, and red wines as groups, but enables the evaluation of high versus low astringency, browning tendency,

---

[2] J. A. Rossi, Jr. and V. L. Singleton, *Am. J. Enol. Vitic.* **17,** 231 (1966).
[3] V. L. Singleton, *Am. J. Enol. Vitic.* **38,** 69 (1987).
[4] J. J. L. Cillers and V. L. Singleton, *J. Agric. Food Chem.* **37,** 890 (1989).
[5] O. Folin and V. Ciocalteu, *J. Biol. Chem.* **73,** 627 (1927).
[6] V. L. Singleton and P. Esau, *Adv. Food Res., Suppl.* **1,** 1 (1969).

and other characteristics within a group. Determination of the total content before and after a treatment to remove or inactivate specific subgroups of reactants can give more specific information. Determination of single substances by HPLC and calculation of the contribution of that content to the total can lead to a balance sheet of contributors to the total and indicate the magnitude of any remaining unknown fraction.[2] This latter adaptation appears capable of more exploitation than has been made so far.

Analyses made with reagents of the Folin and Ciocalteu (FC) type are often numerically appreciably different than those obtained with other methods purported to determine total phenols. Nevertheless, relative values usually correlate well among these methods, as long as samples of similar type are being compared. This correlation may be somewhat illusory and not found among samples of widely different types because qualitatively and relatively the particular mixture of different positive reactants may be rather constant in samples of a given product.

Considering the heterogeneity of natural phenols and the possibility of interference from other readily oxidized substances, it is not surprising that several methods have been used for total phenol determination and none are perfect. Among such methods competing with FC are permanganate titration, colorimetry with iron salts, and ultraviolet absorbance. Oxidation with potassium permanganate is more difficult to standardize among different analysts and is subject to greater interferences, particularly from sugars. Several direct comparisons of FC methods with those based on $KMnO_4$ have shown the preferability of the FC. Colorimetry with iron salts has the problem, from the viewpoint of total phenol determination, that monophenols generally do not react and under some conditions vicinal diphenols and vicinal triphenols give different colors. Because of apparently less interference from dextrins, melanoidins, and proteins, ferrous colorimetry is often used for beer analysis, particularly heavy, dark beers. In most other direct comparisons FC has been found preferable.

Ultraviolet absorbance is difficult to apply to total phenol analysis not only because of potential interference from other compounds, which absorb at similar maxima, but because individual natural phenols differ greatly in both wavelength of maximum absorbance and their molar absorbance.[7]

Reagent

Folin and Denis[8] first proposed a heteropoly phosphotungstate–molybdate reagent to react with tyrosine to give a blue color proportionate

---

[7] A. Scalbert, in "Plant Polyphenols" (R. W. Hemingway and P. E. Laks, eds.), p. 259. Plenum Press, New York, 1992.
[8] O. Folin and W. Denis, *J. Biol. Chem.* **12**, 239 (1912).

to the protein content. Application of this reagent is troubled by the occasional production of a white precipitate that interferes with direct colorimetry. It can be removed, but an extra step is required and some blue may be adsorbed on a filter. Improvements by Folin and Ciocalteu increased the proportion of molybdate and prevented this precipitation by adding lithium sulfate to the reagent. Comparison of the Folin–Denis (FD) procedure with that of Folin–Ciocalteu gives somewhat greater sensitivity and reproducibility for the FC.[9] Nevertheless, in consideration of the desirability of constant procedures for official regulatory purposes, the FD reagent is sometimes still used.[10] With proper technique and standards, results are very similar as the chemical basis is the same.

To prepare the FC reagent (FCR),[5,9] dissolve 100 g of sodium tungstate ($Na_2WO_4 \cdot 2H_2O$) and 25 g sodium molybdate ($Na_2MoO_4 \cdot 2H_2O$) in about 700 ml of distilled water. Add 100 ml of concentrated HCl and 50 ml of 85% phosphoric acid. Boil under reflux for 10 hr (this time should not be shortened appreciably, but need not be continuous). Stop the heating and rinse down the condenser with a small amount of water and dissolve 150 g of $Li_2SO_4 \cdot 4H_2O$ in the slightly cooled solution. The resultant solution should be clear and intense yellow without a trace of green (blue). Any blue results from traces of reduced reagent and will cause elevated blanks. Refluxing for a short time after adding a couple of drops of bromine followed by removal of the excess bromine by open boiling (in a hood, of course) will correct this problem. If excess is avoided, a small amount of 30% hydrogen peroxide can be substituted for the bromine. Make the final solution to 1 liter. Filtration through a sintered glass filter removes insolubles, if necessary. Commercially prepared FCR is often employed. If protected from reducants, the reagent is ordinarily stable indefinitely, even if diluted.

Procedure

The manual method[9] calls for 1.00 ml of sample, blank, or standard in water (or dilute aqueous solution) added to at least 60 ml of distilled water in a 100-ml volumetric flask. Add FCR (5.0 ml) and mix. After 1 min and before 8 min, add 15 ml of 20% sodium carbonate solution, adjust the volume to 100.0 ml, and read the color generated after about 2 hr at about 23° at 760 nm in a 1-cm cuvette. Provided appropriate standards and blanks are employed, considerable variation in these conditions may be permissi-

[9] V. L. Singleton and J. A. Rossi, Jr., *Am. J. Enol. Vitic.* **16,** 144 (1965).
[10] P. Cunniff, ed., "Official Methods of Analysis of AOAC International," 16th ed. AOAC International, Gaithersburg, MD, 1995.

ble. Important considerations are adequate FCR to react completely and rapidly with the oxidizable substances in the samples, sufficient time and mixing of the sample and the FCR solution before adding the alkali solution, and similar time/temperature conditions of color development.

Originally, saturated sodium carbonate was used for the alkaline reagent, which has obvious problems of temperature effects and so on. Sodium cyanide and sodium hydroxide have also been used successfully. It is important to have enough but not excessive alkalinity. About pH 10 is desired after combination with the acidic FCR and the samples. If the buffering capacity of the interconversion of carbonate and bicarbonate is not exceeded, evolution of free CO_2 bubbles to interfere with colorimetry is not ordinarily a problem.

The color may be developed more quickly at warmer temperature (Fig. 1), but, as discussed later, interferences may be greater. The blue color is relatively stable, and a standard, blank, and sample set read at 760 nm after 6 hr at room temperature gave slightly lower absorbance but similar analytical results to the 1-hr colorimetry, although with higher standard deviation. At higher temperatures, the loss of color with time is greater. Higher alkali levels also speed color development and its fading (Fig. 1).

The sample volume need not be 1.00 ml as long as the capacity of the linear range is not exceeded and conversion calculations and standards reflect the change. Microadaptation reduces costs. This procedure has been scaled down by a factor of five to a final volume of 20.0 ml.[9,11] For that procedure, 2.00 ml of a 1:10 diluted sample (compared to the 100-ml procedure), 10.0 ml of FCR diluted 1:10, and 8.00 ml of 75 g/liter sodium carbonate gave the final 20.0 ml. Use of semiautomatic manual pipettes and syringe diluters and dispensers at the same final volume (20 ml) was also satisfactory with only a very slightly increased standard deviation. With adequate equipment, further size reduction is certainly possible.

Flow automation also is quite successful and an automated flow adaptation gave, with low-sugar samples, essentially identical values and a slightly lower coefficient of variation, but either heating to develop maximum color in a reasonable flow time or color measurement when it is still developing is required. Singleton and Slinkard[11] used an air-segmented flow system that delivered 0.42 ml of sample or standard per 20 ml final volume into 9.00 ml of dilution water, 5.29 ml of 1:5 dilution of FCR, and 5.29 ml of 100 g/liter sodium carbonate. A short 7-turn coil mixed the diluted sample, another of 28 turns followed FCR addition to provide adequate intermediate reaction time, and a third of 14 turns mixed in the sodium carbonate. It is considered important, as already discussed, to mix in the alkali well

---

[11] K. Slinkard and V. L. Singleton, *Am. J. Enol. Vitic.* **28,** 49 (1977).

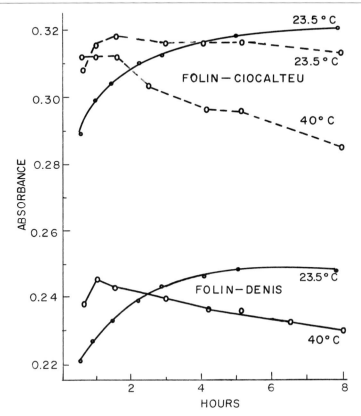

FIG. 1. Absorbance development from FCR and FDR with time at two temperatures and with 2.0 g/100 ml (solid lines) or 3.0 g/ml (dashed lines) sodium carbonate. Reproduced with permission from V. L. Singleton and J. A. Rossi, Jr., *Am. J. Enol. Vitic.* **16,** 144 (1965).

after the FCR to avoid premature alkaline destruction of the activity of FCR. The final mixture was passed through a 13-m coil in a 55° bath intended to provide about a 5-min delay and produce high color development similar to the manual method. Analyses were made at 40 samples or standards per hour. Absorbance was read at 760 nm in a flow cell with a 0.8-cm optical path.

Reading the absorbance manually while the color is still developing rather rapidly is impractical, but the reproducibility of flow automation makes it feasible. Celeste *et al.*[12] used this approach apparently at ambient

[12] M. Celeste, C. Tomas, A. Cladera, J. M. Estela, and V. Cerda, *Anal. Chim. Acta* **269,** 21 (1992).

temperature. Samples or standards (5–100 mg/liter of gallic acid) were injected, 60 samples per hour, into a flowing stream in 0.5-mm I.D. PTFE tubing. The sequence of addition was water, 0.7 ml/min; sample, 107 μl each; FCR, 1.0 ml/min of a 1:10 dilution; and 1.0 ml/min of 0.5 M sodium hydroxide. Mixing coils were not indicated, but a reaction coil of 1 m was followed by reading the absorbance at 760 nm in a 1-cm, 18-μl flow cell.

Standards and Blanks

A blank (phenol-free) solution fades rapidly from yellow to colorless unless the reagent has been partly reduced. The fact that the FCR is not stable under alkaline conditions emphasizes the importance of having sufficient excess present to react with all the phenols before it is destroyed.

Comparison standards as well as blanks are recommended to be included within each group of samples. Use of the absorbance produced under standard conditions as an "index" lacks the self-correction, easy transferability, and easy visualization built into the analysis if standards and blanks are used. Compounds used for standards have included tannic acid, gallic acid, catechin, tyrosine, and others. Most commonly, gallic acid has been used and the results are reported in milligram gallic acid equivalents (GAE) per liter. Earlier results on wines and spirits were considered "tannin" values because tannic acid was used as the standard. However, tannic acid from different preparations can vary, and other tannins cover a wide range of color yield per unit weight. Gallic acid is the significant phenolic unit in commercial tannic acid from oak galls. Gallic acid is equivalent on a weight basis if tannic acid is considered pentadigalloylglucose. The values in milligrams of tannic acid or gallic acid equivalents per liter on the same wine or spirit sample are very similar and relative values in a set of samples are directly comparable. Partly for this historical reason, gallic acid is widely used as the comparison standard. In addition, it is inexpensive, soluble in water, recrystallized easily from water, readily dried, and stable in the dry form. A stock solution is commonly made by dissolving 500 mg of gallic acid in a small amount of ethanol and diluting with distilled water to 1.00 liter. This will keep in a refrigerator for a day or two, but it is subject to slow oxidation and microbial attack. It is convenient to freeze portions suitable for the desired number of assays in oversized (so they do not break), screw-capped glass bottles. These may be held indefinitely and thawed as needed, taking care to mix any sublimed ice melt before opening.

The second most commonly used standard, (+)-catechin (mg CtE/liter), has advantages if flavonoid partitioning is being compared. As will be discussed under molar color yield, values can be interconverted.

The standard solution (500 mg GAE/liter) may be used as the top level in a series of standard dilutions, but its blue pigment production will be outside the absorbance range for satisfactory spectrophotometry in standard equipment. If standard and unknown samples prove to be too high, dilution of the blue color can indicate the proper phenol level of the samples, but they should also be diluted and reanalyzed. The measurable color yield should be linear or nearly so up to about 300 mg GAE/liter in the sample by the procedure described. Sample content is expressed most simply by direct computation from a plot prepared from standard dilutions with the same portion volumes as the samples rather than on the basis of the final reaction volume. The minimum detectable amount is of the order of 3 mg GAE/liter, especially if the sensitivity is extended by such techniques as cuvettes with longer light paths. Remember that oxidation, from air or otherwise, can alter stored samples, especially at low content.

Samples and Sample Preparation

Total phenol determination by FC (or FD) probably has been used most extensively with wines and spirits, but applications have been made with many kinds of samples, including fruit juices, plant tissues, sorghum grains, leather and antifeedant tannins, wood components, proteins, medicines, vanilla and other flavor extracts, olive oil, and water contaminated with phenols or treated with tannin to prevent boiler scale. Because of the potential for unusual problems or special interferences, some evaluation experiments should be conducted as new types of samples are analyzed.

Wines, brandies, whiskies, juices without appreciable insoluble pulp, and similar samples may be analyzed directly, with dilution if necessary, and consideration of potential interferences to be discussed shortly. Phenols from solid samples, of course, need to be converted to clear extracts suitable for colorimetry. An ethanol equivalent to 1 ml/100 ml of the final reaction mixture did not change the results in the normal assay with proper standards; however, the interference by free sulfur dioxide may be enhanced. Similarly, dilute aqueous solutions of other solvents unreactive in the assay may be usable (acetone, methanol, dimethylformamide have been reported), but testing and possibly preparing standards in the same solution are recommended.

An interesting application for intractable samples (e.g., solids present) is to conduct the reaction in suspension and measure the blue anionic pigment after its quantitative extraction into chloroform as a tetralkylammonium salt.[13]

---

[13] A. Cladera-Fortaza, C. Tomas-Mas, J. M. Estrela-Ripoll, and G. Ramis-Ramos, *Microchem. J.* **52**, 28 (1995).

Chemistry of Reaction

The pertinent chemistry of tungstates and molybdates is very complex.[14] The isopolyphosphotungstates are colorless in the fully oxidized 6^+ valence state of the metal, and the analogous molybdenum compounds are yellow. They form mixed heteropolyphosphotungstates–molybdates. They exist in acid solution as hydrated octahedral complexes of the metal oxides coordinated around a central phosphate. Sequences of reversible one or two electron reductions lead to blue species such as $(PMoW_{11}O_{40})^{4-}$. In principle, addition of an electron to a formally nonbonding orbital reduces nominal MoO^{4+} units to "isostructural" MoO^{3+} blue species.

Tungstate forms are considered to be less easily reduced, but more susceptible to one-electron transfer. In the complete absence of molybdenum, phophotungstates have been used to determine *ortho*-dihydric phenols selectively without including monophenols or *meta*-dihydric ones. Molybdates are considered to be reduced more easily to blue forms and electron migration is induced thermally. Mixed complexes as in the FC and FD reagents are intermediate, readily oxidizing monophenols and vicinal diphenols, but lacking in thermally enhanced electron delocalization. Detailed molecular and electronic structures of the blue reduction products are unclear and are likely to remain so in view of their complex nature. Long wavelength absorption maxima move from longer to short wavelengths and become more intense[14,15] with greater reduction. Tungstate analogs had maxima at shorter wavelengths and lower molar absorptivities than molybdates, but followed similar trends. In solutions of increasing pH, a series of one-electron reductions can occur.

Blue products of phosphomolybdate reduction can have Mo^{6+} to Mo^{5+} ratios of 9.0 to 0.6. The 4 e^- reduced species is the most stable blue form and develops readily from mixtures of Mo^{5+} and Mo^{6+} in the necessary heteropolyphosphate forms. Absorption peaks are rather broad for the purer species of blue product, and the occurrence of several species can account for the very broad peaks found from FC and FD reduction (Fig. 2). Because of the breadth of these peaks and the fact that other components in biological samples do not absorb in this region, analysis can be carried out at a wide range of wavelengths, 760 nm generally being chosen for FC.

Although it is possible to form complexes between phenols such as catechol and phosphotungstates and molybdates, the phenol being oxidized by the FCR appears to have no other effect than to supply electrons. Different substrates do not appear to become part of the blue chromophore

---

[14] M. T. Pope, *Prog. Inorg. Chem.* **39,** 181 (1991).
[15] E. Papaconstantinou and M. T. Pope, *Inorg. Chem.* **9,** 667 (1970).

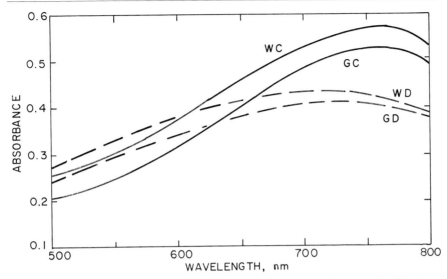

FIG. 2. Absorption spectra produced by wine (W) and gallic acid (G) with Folin–Ciocalteu (C) and Folin–Denis (D) reagents. Reproduced with permission from V. L. Singleton and J. A. Rossi, Jr., *Am. J. Enol. Vitic.* **16**, 144 (1965).

nor do the blue products appear different if generated by different substrates. This conclusion is based on the facts that the pertinent spectrum produced by different substrates is essentially the same (Fig. 2), gallic acid added to wine is recovered quantitatively,[9] and the absorbance produced from a mixture of natural phenols of different classes is equivalent to the sum of their individual contributions.[16]

Molar Color Yields

The blue color generated at room temperature calculated as molar absorptivity related to the reacting substance has been determined[16] for phenolic derivatives representing 29 monophenols, 22 catechols, 11 pyrogallols, 4 phloroglucinols, 9 resorcinols, 9 *para*-hydroquinols, 11 naphthols, 6 anthracenes, 17 flavonoid aglycones, 9 glycosides, 5 hydroxycoumarins, 7 aminophenols, and 19 nonphenolic substances. Selected data are shown in Table I.

[16] V. L. Singleton, *Adv. Chem. Ser.* **134**, 184 (1974).

TABLE I
MOLAR ABSORPTIVITY IN FC ASSAY FROM SELECTED PHENOLS AND POTENTIALLY INTERFERING SUBSTANCES[a]

Compound	Molar absorbance (÷1000)	Free phenolic hydroxyls	Reacting groups	Molar absorptivity per reactive group
Phenol	12.7	1	1	12.7
p-Coumaric acid	15.6	1	1	15.6
Tyrosine	15.7	1	1	15.7
Catechol	22.5	2	2	11.2
Chlorogenic acid	28.9	2	2	14.4
DOPA	24.6	2	2	12.3
Ferulic acid	19.2	1	1+	19.2
Vanillin	14.9	1	1	14.9
Pyrogallol	24.8	3	2	12.4
Gallic acid	25.0	3	2	12.5
Sinapic acid	33.3	1	2	16.6
Phloroglucinol	13.3	3	1	13.3
p-Hydroquinone	12.8	2	1	12.8
Resorcinol	19.8	2	1+	19.8
(+)-Catechin	34.3	4	3	11.5
Kaempferol	29.6	3	2	14.8
Quercetin	48.3	4	3+	16.1
Quercitrin	44.8	4	3	14.9
Malvin	40.5	3	2+	20.2
4-Methylesculetin	31.1	2	2	15.6
o-Aminophenol	21.8	1	2	10.9
p-Aminophenol	12.5	1	1	12.5
p-Methylaniline	11.6	0	1	11.6
o-Diaminobenzene	21.4	0	2	10.7
Flavone	0.1	0	0	0.1
Flavanone	1.9	0	0+	1.9
3-Hydroxyflavone	3.5	0	0+	3.5
4-Hydroxycoumarin	0.1	0	0	0.1
Acetylsalicylic acid	0.2	0	0	0.2
D-Fructose	0.0	0	0	0
Ascorbic acid	17.5	0	1+	17.5
Ferrous sulfate	3.4	0	1−	3.4
Sodium sulfite	17.1	0	1+	17.1

[a] Adapted with permission from V. L. Singleton, *Adv. Chem. Ser.* **134**, 184 (1974).

Other studies,[17,18] allowing for changed conditions, including the use of FDR, have generally agreed very well in relative terms, even if not in absolute molar absorptivity, and have added new compounds. Under standardized conditions of the FC analysis, phenol itself gave 12,700 molar absorbance.[16] Most other biologically likely monophenols tested gave similar or slightly higher (to 15,900) extinctions. Electron-attracting groups such as chloro, carbonyl, or nitro gave progressively less extinctions (4-chlorophenol, 12,100; salicylic acid, 8000; and picric acid, 0). Phloroglucinol and nearly all other *meta*-polyphenols reacted as monophenols, as did *para*-hydroquinols. Catechols and pyrogallol derivatives with free hydroxyls gave twice the color of monophenols in agreement with their *ortho*-quinone possibilities. Resorcinol behaved as a diphenol, but its derivatives generally tested as monophenols.

Flavonoids such as catechin closely approximated the sum of the color expected from the phloroglucinol A ring plus the reaction possibilities of their B ring (i.e., one plus two in this case). Flavonols such as quercetin, but not their 3-glycosides, gave more color than predicted, probably from participation of the enolic C ring. This idea is reinforced by the behavior of flavone and flavanones, which lack phenolic groups (Table I).

Generally, methoxyl substitution removed the reactivity of that phenolic group, but in some instances, particularly sinapic acid, there was indication of partial removal of a methyl group under assay conditions to generate additional free phenolic hydroxyls and additional blue pigment. Despite the alkaline conditions, ester and lactone formations involving phenolic hydroxyls appear to remain intact and inactive during the assay, at least at room temperature.

Reinforced with recovery from mixtures of phenols,[16] these data show that a good first approximation of the contribution of a given weight of a specific compound to total color by the FC assay can be made by calculation taking into account the number of predicted reacting groups and the molecular weight compared to the gallic acid standard. Even better estimate can be made if the color yield of the specific compound in question is compared experimentally to gallic acid under the specific reaction conditions in use. It is unreasonable to expect that the total UV-HPLC peak area will correlate with total phenol by FC unless such calculations have been made. A high UV absorber may be a low contributor to the phenolic total and vice versa.

---

[17] T. Swain and J. L. Goldstein, *in* "Methods in Polyphenol Chemistry" (J. B. Pridham, ed.), p. 131. Macmillan, New York, 1964.

[18] M. Haug, B. Enssle, M. H. Goldbach, and K. Gierschner, *Ind. Obst. Gemueseverwert.* **69,** 567 (1984).

Interferences

The inclusiveness of the oxidation due to the FCR makes it unsurprising that the analytical result can include "interfering" substances in many crude, natural samples. If the possible interfering substances and their likely concentrations are known, efforts to limit the FCR assay to phenols can often be successful. To a degree the assay should be considered a measure of oxidizable substrates, not just phenols. In any case, results can be very useful, if interpreted properly.

Interferences can be of three types: inhibitory, additive, and enhancing or augmenting. Conceivably, inhibition could be from oxidants competing with the FCR, but such a reaction in samples should have been completed in advance. Air oxidation after the sample is made alkaline can certainly decrease phenols oxidizable by FCR. This is a reason why FCR addition ahead of alkali has been emphasized. If this is done, exposure is limited and is the same for samples and standards. Efforts to sparge and blanket with nitrogen have shown no significant effect. Inclusion of solvents other than water in the samples has sometimes inhibited color formation, but in practical usage the effect has been small or avoidable by solvent change or correctable by matching standards and blanks with the samples.

Additive effects are to be expected if unanticipated phenols or enols (e.g., additives or microbial metabolites) are present, as will also be the case with nonphenol FCR reactants. Aromatic amines as well as aminophenols are included in assays for total phenol (Table I). Tryptophan and other indoles react quantitatively with FCR, as do some purines. This has been known for a long time.[5,8] Guanine (but not guanosine), xanthine, and uric acid reacted to give molar color yields with FCR equivalent to monophenols.[19,20] Adenine and other purines and the common pyrimidines gave little color formation (about 1/50th as much). Alkaloids such as caffeine do not appear to have been tested. In wines and many other samples, insufficient of these compounds is present to contribute importantly in the assay for total phenols.

The reaction of proteins with FCR sometimes has been considered an interference, but this is somewhat unfair as most of the reaction is from tyrosine and tryptophan content. Further confusion has arisen from the fact that the Lowry method of protein determination[21,22] uses FCR. However, in this method alkali is added and incubated with copper ions well before

---

[19] T. E. Myers and V. L. Singleton, *Am. J. Enol. Vitic.* **30**, 98 (1979).
[20] M. Ikawa, C. A. Dollard, and T. D. Schaper, *J. Agric. Food Chem.* **36**, 309 (1988).
[21] G. Legler, C. M. Müller-Platz, M. Mentges-Hettcamp, G. Pflieger, and E. Juelich, *Anal. Biochem.* **150**, 278 (1985).
[22] C. M. Stoscheck, *Methods Enzymol.* **182**, 50 (1990).

the addition of FCR. This biuret-type reaction causes the conversion of nonphenolic dipeptides and larger into reactive enolic compounds and cuprous ions. The FC total phenol analysis procedure presented here does not add copper and avoids this reaction. Cuprous and ferrous ions can interfere, but not significantly, at the levels found in biological samples.

Cysteine gives a molar absorptivity with FCR of about 3500 (one-half or less of a monophenol), and hydrogen sulfide or other sulfhydryls such as glutathione are reported to contribute in total "phenol" assays with FCR, but this has not been studied fully. It is uncertain if this is only from direct oxidation of FCR or from hydroquinone regeneration to allow further oxidation by FCR. Reaction of a mercaptan with a quinone ring to produce a thioether substituted hydroquinone has been shown with all sulfhydryl compounds tested except dithiothreitol (DTT), which appears to give internal disulfide rather than quinone substitution.[23] Dithiothreitol by itself gives high FC color, but Larson et al.[24] reported that in the Lowry protein analysis addition of DTT after alkaline destruction of all of the FCR produces an enhanced color yield still proportional to the original protein content. Furthermore, color development is much more rapid to the final measurable level. This procedure deserves investigation for the non-Lowry FC assay without copper salts.

On a molar basis, sugars alone do not react appreciably at room temperature (Table I), but they may interfere with phenol analysis if the sugar level is high.[11,25] At an elevated temperature, more interference is produced. This effect has been compensated for by applying standard corrections, separately determined corrections, or by preparing the standards in the same sugar concentration. Typical corrections[11] at room temperature and at 55° for various gallic acid levels are given in Table II.

The fructose effect was higher than that of glucose. Pectin at 5000 mg/liter added in equal volume to white wines gave only a 16-mg GAE/liter increment to the total in an automatic (heated) assay. Arabinose, galacturonic acid, and galactose had similarly low effects. The interference evidently is caused by the production in the strongly alkaline solution of enediol reductones from the sugar, a well-known reaction more intense with fructose. Under conditions of the assay and modest sugar levels, enediol production can be ignored at room temperature and is, even when heated, only a small part of the sugar present, but can be sizable compared to the content of phenols in sweet samples with low phenol levels. This may be why FC analysis of beers (made from cooked wort) has not been considered

---

[23] V. L. Singleton, M. Salgues, J. Zaya, and E. Trousdale, *Am. J. Enol. Vitic.* **36,** 50 (1985).
[24] E. Larson, B. Howlet, and A. Jagendorf, *Anal. Biochem.* **155,** 243 (1986).
[25] E. Donko and E. Phiniotis, *Szolez. Boraszat.* **1,** 357 (1975).

TABLE II
Approximate Corrections for Invert Sugar Content for FC Determination of Total Phenols[a,b]

Apparent phenol content (mg GAE/liter)	A Sugar content			B Sugar content		
	25 g/liter	50 g/liter	100 g/liter	25 g/liter	50 g/liter	100 g/liter
100	−5%	−10%	−20%	−20%	−30%	−60%
200	−5%	−8%	−20%	−20%	−25%	−38%
500	−4%	−6%	−10%	−17%	−24%	−38%
1000	−3%	−6%	−10%	−11%	−15%	−25%
2000	−3%	−6%	−10%	−10%	−13%	−17%

[a] Adapted with permission from K. Slinkard and V. L. Singleton, *Am. J. Enol. Vitic.* **28,** 49 (1977).

[b] FC conditions A = 25°, 120 min; B = 55°, 5 min. For instance, for a sample with 5.0% inverted sucrose under condition A, an apparent total phenol of 1000 mg GAE/liter should be corrected by 6% to give 940 mg GAE/liter.

as satisfactory as other methods. In the automated method,[11] 55° was chosen as the bath temperature because above that sugar began to participate over several days in browning reactions in white wine. Blouin et al.[26] nevertheless recommended 70° for 20 min after a careful statistical evaluation of FC assay conditions.

Ascorbic acid, an enediol, reacts readily with FCR and its presence must be considered. It reacts with polyphosphotungstate under acidic conditions (pH 3) in an assay that measures the blue color generated before the addition of alkali.[27] Verified with FCR, this procedure could be used to determine ascorbic acid before the phenols and its value then subtracted. In any case, appreciable blue formation from FCR before the addition of alkali indicates the presence of ascorbic acid or other very easily oxidized substance not requiring phenolate forms.

Ascorbic acid could have an augmentation effect on the amount of FCR reacting with the phenols present by reducing quinones as they form and prolonging the reaction. However, the FCR reaction with ascorbic acid appears sufficiently fast to prevent much of this effect. This may be one reason that a time lag is found desirable after combination of the sample and FCR and before the addition of alkali. Phenols oxidize little except as phenolates, and quinones should not form until after the ascorbic acid was already oxidized.

[26] J. Bloin, L. Llorca, F. R. Montreau, and J. H. Dufour, *Connaiss. Vigne Vin.* **4,** 403 (1972).
[27] F. W. Müller-Augustenberg and H. Kretzdorn, *Dtsch. Wein-Ztg.* **91,** 314 (1955).

The separate determination of ascorbic acid and correction of the total phenol accordingly are clearly important if enough is present to skew the results. In grape juice, little ascorbic acid remains after normal processing and less to none in wines or spirits unless it has been added.

The first oxidation product of ascorbic acid is dehydroascorbic acid. It evidently can accumulate as ascorbic acid reduces quinones from polyphenol oxidase action in grape juice/must production. Part of the interference in white wine analysis is from this source. Dehydroascorbic acid is not detected by usual ascorbic acid assays, but is enolic and can also react with FCR. Dehydroascorbic at 100 mg/liter has given FC values in heated flow automatic analysis equivalent to 45 mg GAE/liter. Separate analysis and subtraction of both forms are indicated, especially for white wines and other low phenol products.

Sulfites and sulfur dioxide react alone with FCR (Table I), but Somers and Ziemelis[28] showed that sulfite amplifies the reaction with phenols. This can be a serious problem in wines because not only is SO_2 often added for its antioxidant and antimicrobial effect, but yeast fermentation produces a small amount. Because the reaction appeared to be amenable to a correction formula, it was further investigated in our laboratories. The addition of sufficient acetaldehyde to fully bind the bisulfite present prevented this augmentation of the reactivity of phenols. Therefore it is believed the extra interference is caused by the rapid regeneration of oxidizable phenol by the sulfite, presumably by reduction or by substitution into the quinoid ring. Because duroquinone is not subject to augmentation by SO_2 and tetrabromocatechol and quinol are, the reduction of quinoids and not substitution is considered the likely mechanism.

Although sulfite in the aldehyde-bound form still reacts in the FC assay as it would alone, this reaction is reduced by about 50% to approximately a molar color production of 8000. In wines and most other modern food products, sulfites are low and are already in bound forms unless they have been freshly added, as was the case in the earlier work.[28]

Using procedures such as selective distillation, freed sulfur dioxide also can be removed before FC assay.[29] D'Agostino[30] reported that SO_2 could be removed even from highly treated juices and concentrates to give FC values comparable to untreated material. Sugar corrections for a higher range were also given. Moutounet[31] showed that interference by sugar and SO_2 is interactive and eliminated both by chromatographic treatment with

---

[28] T. C. Somers and G. Ziemelis, *J. Sci. Food Agric.* **31,** 600 (1979).
[29] G. Schlotten and M. Kacprowski, *Wein-Wiss.* **48,** 33 (1993).
[30] S. D'Agostino, *Vignevini* **13**(10), 17 (1986).
[31] M. Moutounet, *Connaiss. Vigne Vin.* **15,** 287 (1981).

a dextran derivative (Sephadex LH-20). All phenols were retarded (by adsorption not gel exclusion) enough to allow washing through, by a small amount of water, of the polar interference compounds. The retarded phenols were eluted in 60% acetone. After removal of the acetone, FC analysis matched the value obtained on a model portion of the sample without sugar or sulfite.

The augmentation effect of sulfur dioxide was very different for different phenols. Phloroglucinol (Fig. 3), all three dimethyl phenols, and resorcinol gave an additive but no augmentation effect. Phenol, hydroquinone, gallic acid, and catechol derivatives gave a large augmentation over what would be expected from the individual components separately. The addition of acetaldehyde in more than equimolar amounts to the free SO_2 decreased this augmentation and sufficient prevented it (Fig. 4). Therefore a "swamping" concentration of acetaldehyde should be used.

Unless there were unusual amounts of free sulfite present (>250 mg/liter), the addition of 1000 mg/liter of acetaldehyde to the samples removed the augmentation effect and allowed simple subtractive correction for the residual oxidizability of the total bound bisulfite (Fig. 5). About 30 min at about pH 3 and room temperature is allowed for SO_2 binding by the

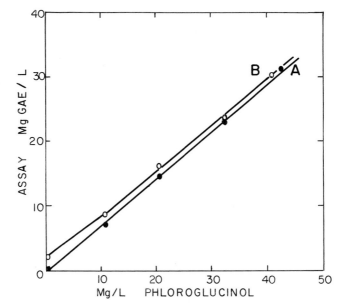

FIG. 3. Total phenol by Folin–Ciocalteu assay of phloroglucinol alone (A) and in the presence of 23 mg/liter of freshly added SO_2 (B).

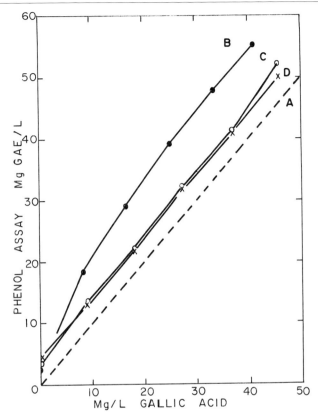

FIG. 4. Total phenol by Folin–Ciocalteu assay of gallic acid alone (A), with freshly added SO$_2$ at 24.2 mg/liter (B), 25.6 mg/liter SO$_2$ + 35 mg/liter acetaldehyde (C), and 25.3 mg/liter SO$_2$ + 79 mg/liter acetaldehyde (D).

acetaldehyde to occur. Under these conditions, precipitation caused by aldehyde bridging has not been a problem. The acetaldehyde itself added no FC color. Adjustments are made for dilution of course.

Automatic analysis (heated[11]) without any corrections of a series of 36 dry white table wines made with differences in pomace contact from seven grape varieties gave total phenols by FC of 135–817 mg GAE/liter. Separate analysis of the same wines by a less inclusive UV spectral shift method gave values invariably 33–69% lower than FC values, but with a correlation coefficient of 0.975 between the two methods. There was some difference among varieties: Thompson Seedless giving greater and Palomino smaller differences between the two methods. These and other studies, including

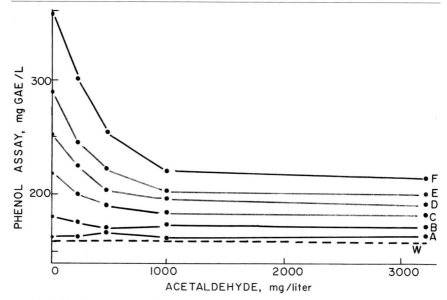

FIG. 5. Total phenol assay by Folin–Ciocalteu of a Riesling white wine alone (W) and with various levels of acetaldehyde following the addition of SO_2 in mg/liter of A = 25, B = 50, C = 100, D = 150, E = 200, and F = 300.

red wines at considerably higher total phenol levels, suggested total contributions of about 100 mg GAE/liter in typical light white wines without fermentable sugar. Such wines contain interfering substances in small but contributory amounts (bound sulfites, some purines, some dehydroascorbic acid, etc.) accounting for most, if not all, of the excess of the total phenol of FC over the sum of the readily identified individual phenols. Of course the comparison method may be at fault. Good relative values are possible routinely with FC and, with the consideration of potentially interfering substances and the levels present, more certainty is achieved.

Differential Methods and Variations

The total phenol by FC can be compared with values by other analyses on the same samples for subgroups such as tannins, flavonoids, anthocyanins, or phenolic acids. Among such other analyses are 520 nm absorbance for anthocyanins, leucoanthocyanidin conversion to cyanidin, and vanillin colorimetry for certain flavonoids. Especially for samples with high phenol content such as red wines, such comparisons have been useful, for example, in studies of sample characteristics due to origin,[32] storage changes,[33] and so on. Nevertheless, interpretation is not straightforward due to factors including overlap between substance categories and different assay characteristics. It is beyond the scope of this article to discuss these interrelationships. Several procedures make use of two FC analyses, one before and another after selective removal of some fraction. Only these will be discussed further.

Cliffe et al.[34] showed FC before and after laccase oxidation gave by difference the fraction oxidized by that enzyme (about 50% in apple juice). Jennings[35] compared samples analyzed by normal FC with those reacted 15 min in the presence of alkali before FCR addition. Molar extinctions were unchanged for compounds expected to act like monophenols except for quinol and considerably decreased for vicinal dihydroxyl phenols. The percentage decrease for specific phenols was up to 72%. It is unclear whether oxygen became limiting during the alkaline reaction. The method appears capable, with application over a wider series of controlled conditions, of clarifying the relative ease of oxidation among different phenols. Most of the other differential applications of two FC analyses have involved precipitation or adsorption reactions.

Flavonoids can be determined separately from other natural phenols by precipitation with formaldehyde. Phloroglucinol derivatives have electron-rich (nucleophilic) centers between the *meta*-hydroxyls. They react with formaldehyde and acetaldehyde in a strongly acid solution (pH <0.8) to produce first methylol substitution and then cross-linking leading to insoluble polymers. In natural samples the A rings of flavonoids are the only phloroglucinol derivatives likely to be present. This is the basis for a method to determine nonflavonoids separately that do not precipitate from flavonoids that do.[36]

The procedure is to add 5.0 ml of 1:2 concentrated HCl and 5.0 ml of 8 mg/liter formaldehyde in water to 10.0 ml of a sample (about 2000 mgGAE/liter or less). Mix well and let stand 24 hr (or more) at room temperature under nitrogen. The nitrogen is probably unnecessary considering the high acidity, but is a simple precaution. Precipitation increases slightly as long as 72 hr, but the differences are small and small amounts of other phenols can be entrained. If the formaldehyde is too low, precipita-

---

[32] E. Carruba, S. D'Agostino, B. Pastena, C. Alagna, and G. Torina, *Riv. Vitic. Enol.* **35,** 47 (1982).
[33] E. LaNotte and D. Antonacci, *Riv. Vitic. Enol.* **38,** 367 (1985).
[34] S. Cliffe, M. S. Fawer, G. Maier, K. Takata, and G. Ritter, *J. Agric. Food Chem.* **42,** 1824 (1994).
[35] A. C. Jennings, *Anal. Biochem.* **118,** 396 (1981).
[36] T. E. Kramling and V. L. Singleton, *Am. J. Enol. Vitic.* **20,** 86 (1969).

tion will be incomplete. If too high, the active sites (six and eight on catechin) can be excessively converted to methylol (soluble) derivatives and cross-linking decreased. Furthermore, the precipitate is soluble in 95% ethanol and appreciably in other aqueous solvents, including those from formaldehyde.

Formaldehyde at 40 mg per sample is about 15-fold maximum molar equivalent of the reactive phenols in higher samples in the indicated range and gave slightly more complete precipitation than did 120 mg per sample.[36] The formaldehyde solution was freshly prepared by dilution of 2.1 ml of 36% formalin to 100 ml. Note that the formaldehyde solution was listed erroneously as much too concentrated in some previous citations.[37,38]

After 24 hr, the reaction mixture is centrifuged or the supernatant decanted, and, if not perfectly clear, filtered through a 0.45-μm membrane filter. The residual nonflavonoid content is determined as usual by FC. Remember to compensate for the dilution compared to the original sample by either doubling the result or analyzing double the aliquot. Separately determine the total phenol by FC on the original, untreated sample. The total minus the nonflavonoid phenol (GAE or other matching units) equals the flavonoid content of the sample. The coefficient of variability of repeated nonflavonoid was about 3% and flavonoid values, because they involved two assays, were about double.

Obviously, if flavonoids are precipitated incompletely, the apparent nonflavonoids will be too high and if nonflavonoids are coprecipitated, too low. Less obviously, because sulfite augmentation would be swamped automatically by the formaldehyde, nonflavonoids by this procedure will be only subject to addition by sulfite, but unless the original total phenol was corrected to prevent any augmentation, the apparent flavonoid content could be too high. For a very low flavonoid content (<40 mg/liter), FC overestimated flavonoid content and had high variability, making it difficult to ascertain the completeness of removal of flavonoids by hyperoxidation of juices.[39] Also note that flavonoids with 4-carbonyl groups are deactivated and resistant to reaction with formaldehyde.

With some samples low in flavonoids, only a haze results from formaldehyde treatment. Conversely, anthocyanin glycosides, although flavonoids, are too soluble to precipitate completely unless sufficient condensed tannin or another less soluble phenol is available for cross-linking. Because phloroglucinol is precipitated completely by acidic formaldehyde, addition can

---

[37] V. L. Singleton, A. R. Sullivan, and C. Kramer, *Am. J. Enol. Vitic.* **22,** 161 (1971).
[38] V. L. Singleton, *in* "Methods of Plant Analysis" (H. F. Linskens and J. F. Jackson, eds.), p. 173. Springer Verlag, Berlin, 1988.
[39] V. Schneider, *Am. J. Enol. Vitic.* **46,** 274 (1995).

solve both problems. Faint pink may still remain in the strongly acidic solution even after phloroglucinol addition, but by spectral analysis this represents only a few milligrams of flavylium ions and an insignificant contribution to FC totals.

In general, phloroglucinol has not been added unless necessary for the two situations mentioned. When it is added, the procedure[37,38] is modified as follows: 10.0 ml of sample, 2.0 ml of concentrated HCl, and 5.0 ml of 8 mg/ml formaldehyde for low flavonoid samples and 12 mg/ml for high ones are well mixed and allowed to stand at room temperature for 2–3 hr. Then 3.0 ml of 10 mg/ml phloroglucinol is mixed into high anthocyanin samples or 1.0 ml of the same plus 2.0 ml of water into those with low flavonoids. After a total reaction time of about 24 hr, the FC analysis is completed as before.

As is the case with most such reactions, the formation of a high amount of precipitate can adsorb phenols that were otherwise soluble. Because of the high acidity, it may be necessary to raise the concentration of the sodium carbonate used in the FC analysis, which should be checked under the specific conditions employed.

This flavonid–nonflavonoid separation is not as precise as one would like. A few nonflavonoids precipitate slightly with formaldehyde,[16] fisetin (resorcinol A ring) did not, and flavonoid glycosides alone are incompletely precipitated. However, the analysis has led to important findings. Nonflavonoid phenols in grapes and wines, mostly hydroxycinnamates, are relatively similar regardless of processing, whereas the flavonoid content rises greatly as pomace extraction occurs.[36] However, in apple juices a similarly wide range of total phenol content can be found, but the nonflavonoid content maintains a relatively constant fraction of the total.[40] Tannins extractable from oak barrels are hydrolyzable (not flavonoid) whereas those present from the grape are condensed (flavonoid) tannins. Because grape phenols other than flavonoid are low and relatively constant, oak barrel extraction during wine (or spirit) aging or tannic acid addition can be quantitated.[37] Scalbert et al.[41] found good agreement with these results for a series of nonflavonoid tannins from various woods, but found that with added phloroglucinol these tannins and small, less soluble phenols were entrained appreciably.

Folin–Ciocalteu analysis has been applied before and after treatment with lead acetate, gelatin, hide powder, polyvinylpyrrolidone (PVP), polyamide, cinchonine, methylcellulose, sodium chloride, and ethyl acetate extraction. The first three can be passed over because lead acetate precipita-

[40] J. J. L. Cilliers, V. L. Singleton, and R. M. Lamuela-Raventós, *J. Food Sci.* **55,** 1458 (1990).
[41] A. Scalbert, B. Monties, and E. Janin, *J. Agric. Food Chem.* **37,** 1324 (1989).

tion is known to be nonspecific and incomplete, gelatin is a soluble protein and could add FC color (albeit low compared to other proteins), and hide powder, the classical agent for tannin removal, is difficult to obtain, standardize, and apply. Nevertheless, the reported successful application of FC analysis after such diverse treatments shows its relative insensitivity to extraneous substances.

The objective in all these cases is to selectively remove a specific fraction of FC total phenols. The separation commonly sought is larger polyphenols from smaller polyphenols. The tannin group is heterogeneous in size (molecular weights between about 500 and 3000, larger is generally insoluble) and in chemical nature. They are of interest as a group because of their astringency, ability to precipitate proteins, inhibit enzymes, tan hides, and so on. These properties depend on a sizable molecule displaying a large number of free phenolic groups capable of producing numerous hydrogen bonds with proteins. This condition is met by both hydrolyzable and condensed tannins. Hydrolyzable tannins release gallic acid, ellagic acid, and related phenols when deesterified from their central sugar or other polar groups. Condensed tannins are flavonoid polymers capable of partial conversion to cyanidin (anthocyanogens, proanthocyanidins, leucoanthocyanidins). Reactions, particularly oxidation, during processing and storage modify these compounds to decrease or more commonly increase their size and incorporate anthocyanins or other units not otherwise present.[6,42]

Saturation with sodium chloride is considered to precipitate from water the larger tannins (6–10 units). Those of lower polymeric size can be analyzed by FC in the salt solution and the lowest of all can be extracted into ethyl acetate. On a series of 27 red wines, plus 4 from heavily pressed pomace, and a red pigment extract, Margheri and co-workers [43–46] found sodium chloride invariably precipitated less phenol than did treatment with cinchonine. The difference averaged 18% and the salt precipitation was also less consistent.

Single precipitation or adsorption procedures inevitably do not have sharp cutoff points among such diverse sizes and types of polyphenols. Tannin adsorbents polyamide or insoluble PVP do not readily adsorb nontannins such as gallic acid or (+)-catechin and also pass sugar and SO_2, allowing potential interference in FC analysis to remain. This would be expected to occur with other adsorbents and precipitants, causing such

[42] H. P. S. Makkar and K. Becker, *J. Agric. Food Chem.* **44,** 1291 (1996).
[43] G. Margheri, D. Tonon, and F. Gottardi, *Vini Ital.* **18,** 337 (1976).
[44] G. Margheri and D. Tonon, *Riv. Vitic. Enol.* **30,** 376 (1977).
[45] G. Margheri, D. Tonon, and S. Inama, *Vini Ital.* **19,** 113 (1977).
[46] G. Margheri and G. Versini, *Vini Ital.* **21,** 83 (1979).

interference to be shown in unadsorbed or unprecipitated fractions. Particle size and method of preconditioning affect adsorption by these solids.[47]

Mitjavila et al.[48] used soluble PVP (25,000–30,000 molecular weight), which alone would not precipitate tannins from beers, fruit juices, wines, and spirits, but did so in the presence of sufficient trichloroacetic acid. This precipitate centrifuged and washed with trichloroacetic acid is then taken up in water. Folin–Ciocalteu analysis, allowing for the acid, can then be applied to the dissolved precipitate as well as to the supernatant and the original sample. Gallic acid and (+)-catechin were found in the supernatant and not in the precipitate. Both grape seed (flavonoid) tannin and gallotannic acid were in the precipitate and not in the supernatant. About half of a leucocyanidin preparation was found in each. In a series of 2 beers (light and dark), 11 wines (red, pink, and white), and 4 spirits, the sum of FC values for the supernatant and precipitated fractions was 95.1–105.2% of the original FC total (average 97.8%). Probably due to effects of ascorbic acid, the assay of juices was less successful. There appears to be no obstacle to application of flavonoid precipitation to either fraction allowing four categories: hydrolyzable tannins, flavonoid tannins, nontannin flavonoids, and phenolic acids and other small phenols.

Peri and Pompei[49] adapted the precipitation of tannins by cinchonine sulfate at pH 7.0–8.0 followed by separating the supernatant. The precipitate is dissolved in aqueous ethanol containing 10% HCl.[50] The two resultant solutions are then treated with acid formaldehyde to precipitate flavonoids. Folin–Ciocalteu analysis of the redissolved cinchonine precipitate gives hydrolyzable tannins in the supernatant from formaldehyde precipitation and condensed tannins by difference. Similarly, analysis for total phenol by FC on the supernatant from cinchonine precipitation after formaldehyde precipitation gives in the new supernatant a measurement of the simple nonflavonoids and by difference nontannin flavonoids. Testing of knowns in each group at about 500 mg GAE/liter and mixtures of them gave very good agreement between the actual assays and the calculated concentrations. With a white wine having had no barrel age at 813 mg GAE/liter total FC phenol, the distribution was 42% condensed tannin, 0% hydrolyzable tannin, 33% nonflavonoid phenols, and 25% nontannin flavonoids.[51]

Among 27 red wines, averaged for the 9 lowest in phenol (864–1720 mg CtE/liter), the middle third, and the highest 9 (4800–5900 mg CtE/

---

[47] C. Pompei, C. Peri, G. Montedoro, E. Miniati, and N. Pasquini, *Ann. Technol. Agric.* **20**, 21 (1971).
[48] S. Mitjavila, M. Schiavon, and R. Derache, *Ann. Technol. Agric.* **20**, 335 (1971).
[49] C. Peri and C. Pompei, *Phytochemistry* **10**, 2187 (1971).
[50] A. Brugirard and J. Tavernier, *Ann. Technol. Agric.* **1**, 311 (1952).
[51] C. Peri and C. Pompei, *Am. J. Enol. Vitic.* **22**, 55 (1971).

liter), the proportion of phenol precipitated by cinchonine increased and was, respectively, 32.1, 45.2, and 52.8% of the total FC phenol.[43-46] Conversely, the proportion of the FC total phenol extractable by ethyl acetate decreased and was, respectively, 36.1, 24.7, and 14.6% of the original total. If one assumes no overlap between those phenols extracted by ethyl acetate (nonpolar, dimers and smaller) and those precipitated by cinchonine (tannins equivalent to trimers and larger), there was 30–33% of the total phenol attributable to nontannin, nonextractable phenols, i.e., polar compounds such as caftaric acid (caffeoyltartaric acid), and glycosides.

Montedoro and Fantozzi,[52] with additions of highly methylated methylcellulose at double or more the tannin content, at pH 3–5, and with 100 g/liter of ammonium sulfate, did not precipitate hydroxy-substituted benzoates, cinnamates, simple phenols, flavonols, catechins, or phenolic glycosides, whereas cinchonine did precipitate portions of the last three. All three methods did precipitate gallotannic acid; methylcellulose more than PVP and slightly less than cinchonine. With model compounds[52] and a wide series of beverages,[53] recovery of added tannic acid was good and repeated analyses had coefficients of variation of 1.3–13.1%. Formaldehyde precipitation on the supernatant after methylcellulose adsorption gives nontannin nonflavonoids and nontannin flavonoids by difference.

As just discussed, the three most widely applied methods to fractionate the large, FC-active phenols are precipitation with polyvinylpyrrolidone,[48] cinchonine sulfate,[49] or methylcellulose.[52] Burkhardt[54] found on a must and red wines prepared from it with increasing degrees of pomace extraction that the three methods were equally suitable for laboratory analysis. For the three samples with the lowest phenol (530–780 mg CtE/liter), removal by cinchonine, PVP, and methylcellulose averaged, respectively, 1.3, 11.7, and 34.0% of the total FC phenol. For the three highest wines (2310–3280 mg CtE/liter), respective average removals were 48.7, 52.7, and 44.7%. From these data it appears that PVP had slightly greater affinity for the larger phenol molecules and methylcellulose a greater affinity in the dilute solution among mostly nontannin phenols. Unfortunately, formaldehyde precipitation was not included in these tests.

By a statistical study[55] of the analytical separation of total phenols by sodium chloride precipitation, formaldehyde precipitation, and comparison with other analyses and ratios, 93–100% of a group of red, pink, and wines made from a mixture of white and red grapes were classified correctly into

---

[52] G. Montodoro and P. Fantozzi, *Lebensm.-Wiss. Technol.* **7,** 155 (1974).
[53] E. Polidori, G. Montedoro, and P. Fantozzi, *Sci. Tecnol. Alimenti* **4,** 157 (1974).
[54] R. Burkhardt, *Lebensmittelchem. Gerichtl. Chem.* **30,** 206 (1976).
[55] G. Santa-Maria, G. L. Garrido, and C. Diez, *Z. Lebensm.-Unters. Forsch.* **182,** 112 (1986).

one of these three types. With combinations of FC with cinchonine and formaldehyde methods, White Pinot wine[56] had 275 mg GAE/liter total phenol, apportioned as 39 mg GAE/liter of tannin (all condensed, presuming no addition of hydrolyzable), 175 mg GAE/liter as simple phenols (neither tannins nor flavonoids), and 61 mg GAE/liter of flavonoids not tannins.

Forty white wines just after first racking[57] gave range and average (in milligrams GAE per liter by FC): total phenols 202–1075, 672; grape tannin 30–465, 216; nontannin flavonoids, 10–507, 208; and phenolic acids 122–465, 216. The sum of the means of the three fractions is exactly the mean of total phenols, even though the range in values is great. Coefficients of variation among all the *samples* (not within assays) were 36.7% for total phenol, 31.2% for phenolic acids, 60.3% for tannins, and 68.2% for nontannin flavonoids, yet relationships held. The methods used were FC and successive precipitation with methylcellulose and formaldehyde.

Conclusions and Extensions

Analyses of the Folin–Ciocalteu type are convenient, simple, require only common equipment, and have produced a large body of comparable data. Under proper conditions, the assay is inclusive of monophenols and gives predictable (but variable by reactive groups per molecule) reactions with the types of phenols found in nature. Because different phenols react to different degrees, expression of the results as a single number such as milligrams per liter gallic acid equivalence is necessarily arbitrary. Because the reaction is independent, quantitative, and predictable, analysis of a mixture of phenols can be recalculated in terms of any other standard.

The assay in fact measures all compounds readily oxidizable under the reaction conditions and its very inclusiveness allows certain substances to also react that are either not phenols or seldom thought of as phenols (e.g., proteins). Judicious use, with consideration of potential interferences in particular samples and prior study if necessary, can lead to very informative results. Aggregate analysis of this type is an important supplement to and often more informative than reems of data difficult to summarize from techniques such as HPLC that separate a large number of individual compounds.

The predictable reaction of components in a mixture makes it possible to determine a single reactant by other means and to calculate its contribution to the total FC phenol content. Relative insensitivity of the FC analysis

[56] D. Villa, *Vignevini* **12**(4), 37 (1985).
[57] A. M. Gattuso, M. C. Indovina, and L. Pirrone, *Vignevini* **13**(4), 35 (1986).

to many adsorbents and precipitants makes differential assay before and after several different treatments informative. A balance sheet to estimate the fraction of the total assay due to additional, still unknown components is a use capable of expansion. Because the reaction is quantitative and all the ingredients other than the added phenol are inorganic, investigation of the exact nature of the FCR oxidation product(s) from specific phenols appears feasible. Such a study would give new insight into the nature of phenol oxidation in general as well as oxidation by FCR specifically.

[15] Size Separation of Condensed Tannins by Normal-Phase High-Performance Liquid Chromatography

By VERONIQUE CHEYNIER, JEAN-MARC SOUQUET, ERWAN LE ROUX, SYLVAIN GUYOT, and JACQUES RIGAUD

Introduction

Condensed tannins, also called proanthocyanidins because they release anthocyanins when heated in acidic conditions, are ubiquitous plant components, consisting of chains of flavan-3-ol units (Fig. 1). Several classes can be distinguished on the basis of the hydroxylation pattern of the constitutive units. Among them, procyanidins, composed of (epi)catechin units (Fig. 1, $R_3 = H$), and prodelphinidins, deriving from (epi)gallocatechin (Fig. 1, $R_3 = OH$), are particularly widespread.

Within each class, monomeric units may be linked by C-4–C-6 and/or C-4–C-8 bonds (B type) or doubly linked, with an additional ether linkage (A type) and eventually substituted (e.g., glycosylated, galloylated). The structures of numerous oligomers have been elucidated.[1] However, the lower molecular weight proanthocyanidins are usually present in relatively low concentrations compared to polymers.[2] Besides, the degree of polymerization (DP) may vary greatly, as proanthocyanidins have been described up to 20,000 in molecular weight.[3]

Tannin properties, including radical scavenging effects and protein-binding ability, depend largely on their structure and particularly on the number of constitutive units (DP). Therefore, several methods have been

---

[1] L. J. Porter, in "The Flavonoids: Advances in Research since 1984" (J. B. Harborne, ed.), p. 23. Chapman and Hall, London, 1994.
[2] Z. Czochanska, L. Y. Foo, R. H. Newman, and J. L. Porter, *J. Chem. Soc. Perkin Trans I* 2278 (1980).
[3] E. Haslam and T. H. Lilley, *Crit. Rev. Food Sci. Nutr.* **27**, 1 (1988).

proanthocyanidins :

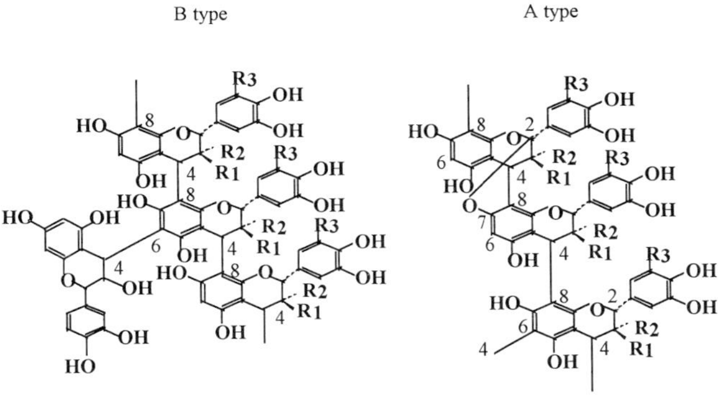

R_3 = H : procyanidin

R_3 = OH : prodelphinidin

constitutive units :

R_1 = OH, R_2 = R_3 = H : catechin

R_1 = R_3 = H, R_2 = OH : epicatechin

R_1 = R_3 = H, R_2 = O-G : epicatechin-3-O-gallate

R_1 = R_3 = OH, R_2 = H : gallocatechin

R_1 = H, R_2 = R_3 = OH : epigallocatechin

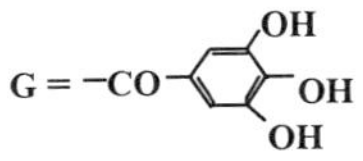

FIG. 1. Examples of proanthocyanidin structures.

developed to determine oligomeric and polymeric proanthocyanidins. Separation of lower molecular weight proanthocyanidins is usually achieved by reversed-phase high-performance liquid chromatography (HPLC), but the elution order is unrelated to the degree of polymerization. Various other techniques have been used to separate proanthocyanidins following this criterion. For instance, liquid chromatography on Sephadex LH-20[4] or Fractogel (Toyopearl)[5] allows fractionation of monomers and oligomers in increasing molecular weight order but becomes impractical beyond the tetramer level. Gel-permeation chromatography enables one to obtain a mass profile, but usually requires tedious derivatization procedures (acetylation, methylation)[6–8] that do not permit recovery of the samples. In contrast, the normal-phase HPLC method presented here, adapted from the TLC

[4] A. G. H. Lea and C. F. Timberlake, *J. Sci. Food Agric.* **25,** 1537 (1974).

[5] G. Derdelinckx and J. Jerumanis, *J. Chromatogr.* **285,** 231 (1984).

[6] V. Williams, L. J. Porter, and R. W. Hemingway, *Phytochemistry* **22,** 569 (1983).

[7] J. J. Karchesy, *in* "Chemistry and Significance of Condensed Tannins" (R. W. Hemingway and J. J. Karchesy, eds.), p. 197. Plenum Press, New York, 1989.

[8] H. D. Bae, L. Y. Foo, and J. J. Karchesy, *Holzforschung* **48,** 4 (1994).

method described by Lea,[9] allows separation of proanthocyanidins on a molecular weight basis, without derivatization.

Extraction and Purification of Proanthocyanidins

Fresh, frozen, or freeze-dried tissue can be extracted, although best results are generally obtained from fresh material. The most suitable solvents are ethanol, methanol, and acetone, which can be used pure or diluted with water.[10–13] Methanol or ethanol extract low molecular weight phenolics, including oligomeric proanthocyanidins, whereas larger polymers are better extracted with 60–70% aqueous acetone.

The sample is ground and/or homogenized in the extraction solvent [10 to 20 ml of either methanol or acetone–water, 60:40 (v/v) per g of plant material] for 30 min and centrifuged (10,000g, 15 min) at 4°. The extraction procedure is repeated two more times and the supernatants pooled. After removing the extraction solvent by rotary evaporation and replacing with ethanol–H_2O–trifluoroacetic acid (TFA) (55:45: 0.005, v/v/v), the tannins are separated from other phenolic compounds by chromatography on Toyopearl HW-40 (F) (TOSOHAAS). Phenolic acids, anthocyanins and flavanol monomers are eluted with ethanol-H_2O-TFA (55:45: 0.005, v/v/v) while monitoring absorbance at 280 nm. The tannin fraction is then recovered by elution with 60% acetone, taken to dryness, and stored at $-20°$ until used for further analysis.

Although this method should be suitable in most cases, it may require modifications, depending on the nature of tannins to be extracted and on the presence of other plant constituents. In particular, the extraction yield may be improved by sonicating and the samples may be protected against oxidation at this stage for instance by sulfiting or working under inert gas and/or at low temperature. Additional extraction steps using hexane or chloroform may also be applied prior to chromatography on Toyopearl to remove lipids and pigments such as carotenoids and chlorophyll.

Size Separation of Proanthocyanidins

Individual Analysis of Proanthocyanidin Oligomers

Normal-phase HPLC chromatograms are obtained by injecting 20 μl of the tannin extract dissolved in methanol (1 g/liter) and filtered through

[9] A. G. H. Lea, *J. Sci Food Agric.* **29,** 471 (1978).
[10] D. J. Cattell and H. E. Nursten, *Phytochemistry* **16,** 1269 (1977).
[11] L. Y. Food and J. J. Karchesy, *Phytochemistry* **28,** 1743 (1989).
[12] M. T. Escribano-Bailon, Y. Guttierez-Fernandez, J. C. Rivas-Gonzalo, and C. Santos-Buelga, *J. Agric. Food Chem.* **44,** 1731 (1992).
[13] I. McMurrough and D. Madigan, *J. Agric. Food Chem.* **44,** 1731 (1996).

a 0.45-μm membrane filter onto a Lichrospher Si 100 column (250 × 4 mm I.D., 5-μm particle size) (Merck, Darmstadt, Germany), protected with a guard column containing the same material (20 × 4 mm). The HPLC apparatus is a Waters system (Millipore, Bedford, MA) consisting of two M510 pumps, a U6K manual injector, an automated gradient controller, and a 440 absorbance detector, set at 280 nm and connected to a 746 integrator. The elution conditions are as follows: solvent A, CH_2Cl_2-MeOH-H_2O-TFA (10:86:2:0.005, v/v); solvent B, CH_2Cl_2–methanol–H_2O–TFA (82:18:2:0.005, v/v); elution with linear gradients from 0 to 40% A in 50 min, from 40 to 55% A in 5 min, from 55 to 100% B in 5 min, followed by washing with 100% A for 5 min and reconditioning the column, flow rate 1 ml min^{-1}, oven temperature 30°. The main individual peaks are collected in several runs and tentatively identified by thiolysis[14] and/or enzymatic hydrolysis and by coinjection with pure compounds previously isolated and identified in our laboratory,[15] when available.

Major peaks detected in the HPLC profile of a cacao bean extract corresponded to (−)-epicatechin and its C-4–C-8 linked oligomers, from dimer to pentamer, eluted in increasing molecular weight order.[16] Application of the normal-phase procedure to a cider apple pulp extract (*Malus domestica* var. Kermerrien), using slightly different elution conditions (linear gradients from 0 to 22% A in 55 min and from 22 to 100% A in 15 min), similarly allowed to elute successively (−)-epicatechin and its oligomers, from dimer to nonamer.

As well, although the chromatogram obtained from the grape seed extract appeared much more complex, due to the presence of catechin oligomers and mixed epicatechin–catechin derivatives in addition to the epicatechin oligomers found in the apple extract, grape seed procyanidins were eluted in successive groups corresponding to increasing *DP* up to the tetramer series.[16] Hydrolysis of the gallic esters with tannase improved the resolution of these successive groups as galloylated compounds eluted further than the corresponding nongalloylated compounds. Thus, although the elution of procyanidins in increasing molecular weight order was achieved by normal-phase HPLC, some overlapping may take place as the molecules become increasingly substituted. Besides, the separation of oligomeric procyanidins was rather poor compared to that obtained by reversed-phase HPLC; in particular, procyanidin dimers B5 and B7 coeluted with the corresponding C-4–C-8-linked procyanidins B2 and B1.

[14] J. Rigaud, J. Perez-Ilzarbe, J. M. Ricardo da Silva, and V. Cheynier, *J. Chromatogr.* **540**, 401 (1991).
[15] J. M. Ricardo da Silva, J. Rigaud, V. Cheynier, A. Cheminat, and M. Moutounet, *Phytochemistry* **30**, 1259 (1991).
[16] J. Rigaud, M. T. Escribano-Bailon, C. Prieur, J. M. Souquet, and V. Cheynier, *J. Chromatogr. A* **654**, 255 (1993).

Size Separation of Proanthocyanidin Oligomers and Polymers

Application of the following elution conditions allows improved separation of larger oligomers and polymers and minimized hydrolytic degradation during concentration of the collected fractions[17]: solvent A CH_2Cl_2-methanol-H_2O-TFA (10:86:2:0.005, v/v); solvent B CH_2Cl_2-methanol-H_2O-TFA (82:18:2:0.005, v/v); elution with linear gradients from 0 to 40% A in 50 min, from 40 to 55% A in 5 min, from 55 to 100% B in 5 min, followed by washing with 100% A for 5 min and reconditioning the column, flow rate 1 ml min^{-1}, oven temperature 30°, 25 μl injection volume.

Various tannin extracts, namely grape skin tannins containing both (epi)catechin and epigallocatechin units, along with small amounts (3–6%) of galloylated epicatechin units,[18] 30% galloylated B-type procyanidins extracted from grape seeds,[17] and A- and B-type epicatechin-based procyanidins from litchi pericarp,[19] were thus analyzed (Fig. 2). Successive fractions were collected throughout each elution, taken to dryness under vacuum, dissolved in 200 μl methanol, and used for thiolysis analysis.

Thiolysis characterization of the proanthocyanidin fractions collected after elution of the grape skin extract showed that the average degree of polymerization increased from 3.4 in the first fraction to 83.3 in the sixth one, confirming that proanthocyanidins are eluted, as expected, in increasing molecular weight order.[18] The seventh fraction collected gave no positive reaction when submitted to thiolysis or to Bate–Smith reaction,[20] meaning that the absorbance at 280 nm observed at the end of the chromatogram was not due to the presence of proanthocyanidins.

Identical analyses performed on extracts from grape seeds and from litchi pericarp similarly demonstrated that the mean degree of polymerization increases with the retention time for the first 60–70 min. However, a decrease of the mean DP was observed in the last fractions collected. This may be due to insolubilization of highly polymerized procyanidins and/or to their selective adsorption onto the glassware during the concentration step. In fact, larger molecular weight tannins found in the last fractions eluted from the grape skin extract were more concentrated and should be more soluble because they contain significant proportions of epigallocatechin units than tannins present in the corresponding fractions from litchi (A- and B-type procyanidins) and grape seed (galloylated procyanidins) extracts.

---

[17] C. Prieur, J. Rigaud, V. Cheynier, and M. Moutounet, *Phytochemistry* **36,** 781 (1994).

[18] J. M. Souquet, V. Cheynier, F. Brossaud, and M. Moutounet, *Phytochemistry* **43,** 509 (1996).

[19] E. Le Roux, T. Doco, P. Sarni-Manchado, Y. Lozano, and V. Cheynier, submitted for publication.

[20] L. J. Porter, L. N. Hrstich, and B. G. Chang, *Phytochemistry* **25,** 223 (1986).

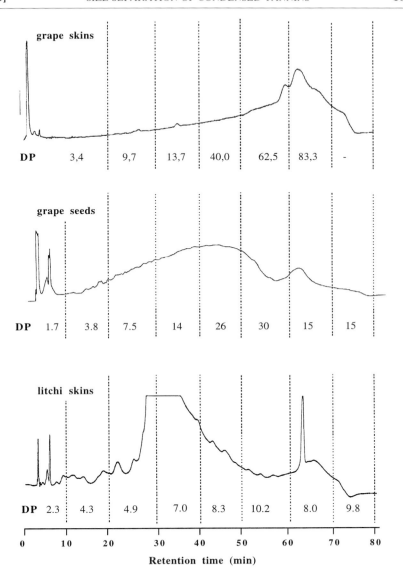

FIG. 2. Normal-phase HPLC profiles of tannin extracts from grape skins, grape seeds, and litchi pericarp and mean degree of polymerization (*DP*) of each fraction determined by thiolysis.

Although the DP increases with the retention time in each proanthocyanidin extract, no relationship between these two parameters could be established when comparing the three extracts studied. For example, the mean DPs calculated for the fractions eluted between 50 and 60 min were approximately 10 in the litchi pericarp extract, 30 in the grape seed extract, and 60 in the grape skin extract. This may be explained by differences in proanthocyanidin structures within the extracts. Besides, contamination of the proanthocyanidin extracts with other UV-absorbing material (e.g., oxidation products) may result in alteration of the elution profile, leading to misinterpretation of the molecular weight distribution. Therefore, a precise knowledge of the tannin composition is a prerequisite for applying the normal-phase HPLC separation method to a new plant extract. As a rule, the validity of the method should be checked carefully and the correspondence between retention times and average DP established in each particular case, especially when heterogeneous polymers are to be encountered.

[16] Resveratrol and Piceid in Wine

By ROSA M. LAMUELA-RAVENTÓS and ANDREW L. WATERHOUSE

Introduction

Resveratrol has been acclaimed for its preventive and curative effect against cardiovascular diseases and cancer, the principal causes of morbidity and mortality in developed countries. *trans*-Piceid, the 3β-glucoside of *trans*-resveratrol, and the *cis* isomers of resveratrol and piceid (see Fig. 1) have also been shown to have physiological activity. Piceid reduces the serum triglyceride and low-density lipoprotein–(LDL) cholesterol concentrations as well as the ratio of total cholesterol to high-density lipoprotein (HDL) cholesterol.[1]

Because resveratrol can inhibit LDL oxidation[2] and can block platelet aggregation[3,4] resveratrol may act to reduce the risk of cardiovascular diseases. As an antioxidant, it is more effective than butylhydroxytoluene

---

[1] H. Arichi, Y. Kimura, H. Okuda, K. Baba, M. Kozwa, and S. Arichi, *Chem. Pharm. Bull.* **30,** 1766 (1982).
[2] E. N. Frankel, A. L. Waterhouse, and J. E. Kinsella, *Lancet* **341,** 1103 (1993).
[3] A. A. E. Bertelli, L. Giovannini, D. Giannessi, M. Migliori, W. Bernini, M. Fregoni, and A. Bertelli, *Int. J. Tissue React.* **17,** 1 (1995).
[4] C. R. Pace-Asciak, S. E. Hahn, E. P. Diamandis, G. Soleas, and D. M. Goldberg, *Clin. Chim. Acta* **235,** 207 (1995).

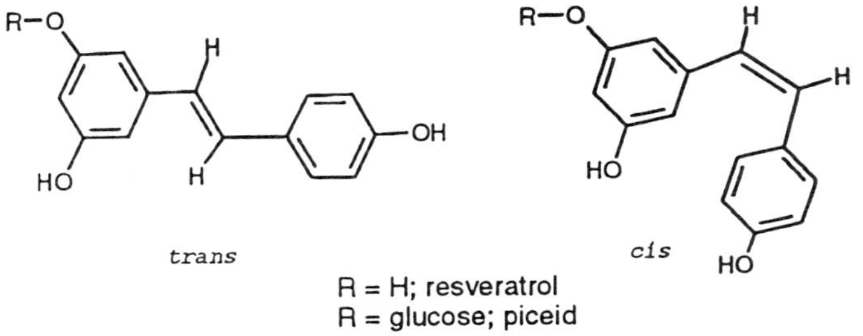

FIG. 1. Chemical structures of *trans*- and *cis*-resveratrol and *trans*- and *cis*-piceid.

R = H; resveratrol
R = glucose; piceid

(BHT), quercetin, or tocopherol on lipid peroxidation in liposomes and in rat liver microsomes[5]; however, other phenolic compounds have been shown to have more activity in other oxidation tests.[2]

Resveratrol also has antiinflammatory properties as exhibited by its ability to inhibit arachidonate metabolism,[4] specifically by inhibiting thromboxane A_2 formation (or thromboxane B_2 and hydroxyheptadecatrienoate [HHT]). These are stable products formed from thromboxane A_2 and lipoxygenase, products that could moderate thrombotic events.

Resveratrol may also have anticancer properties. It inhibits cellular events associated with tumor initiation, promotion, and progression. It inhibits free radical formation, which inhibits tumor initiation. It acts as an antimutagen as it induces the quinone reductase enzyme capable of detoxifying carcinogens. Moreover, it inhibits the hydroperoxidase activity of cyclooxygenase, so it will inhibit the arachidonic pathway to the formation of prostaglandins, which stimulates tumor cell growth and can activate carcinogenesis. It also acts by inhibiting, in a dose-dependent manner, preneoplastic lesions, slowing the progression of carcinogenesis.[6]

Which foods contain resveratrol? Curiously, it is not well distributed in foods; however, it is found in groundnuts, *Arachis hipogea*,[7] and grape products.[8,9] It also occurs in leaves, barks,[7] and roots of gymnosperms and dicotyledons.[10] In grape berries, resveratrol is located primarily in the skin cells and is absent or in low concentration in the fruit flesh.[9]

[5] J. P. Blond, M. P. Denis, and J. Bezard, *Sci. Aliments* **15**, 347 (1995).
[6] M. Jang, L. Cai, G. O. Udeani, K. V. Slowing, C. F. Thomas, C. W. W. Beecher, H. H. S. Fong, N. R. Farnsworth, A. Kinghorn, R. G. Mehta, R. C. Moon, and J. M. Pezzuto, *Science* **275**, 218 (1997).
[7] J. L. Ingham, *Phytochemistry* **15**, 1791 (1976).
[8] A. L. Waterhouse and R. M. Lamuela-Raventos, *Phytochemistry* **37**, 571 (1994).
[9] L. L. Creasy and M. Coffee, *J. Am. Soc. Horticult. Sci.* **113**, 230 (1988).
[10] J. Gorham, *Prog. Phytochem.* **6**, 203 (1980).

In grape products, not only is the aglycone of *trans*-resveratrol present, but *trans*-piceid can be quantified in grapes[8] and wines.[11,12] These compounds, resveratrol and piceid, are present in two isomeric forms, *cis* and *trans*; however, *cis*-resveratrol is a by-product of fermentation[12] and is found rarely in grapes.

In red wine vinification, maceration with skins and seeds during fermentation contributes to the extraction of the phenols present in these firm tissues.[13,14] Resveratrol requires a relatively long maceration time on the skins to be extracted.[14–16] However, the amount present also depends on enological techniques employed: yeast strains,[17] the fining agents used,[17] and the time of aging in oak.[18]

Red wines are the highest in *trans*-resveratrol content, with the average level of all four compounds—*trans*- and *cis*-resveratrol and *trans*- and *cis*-piceid—approximately 8 mg/liter, depending on the grape variety.[11] In rosé wines,[19] levels range between 1.4 and 3.0 mg/liter, whereas white wine is reported to have much lower resveratrol content than red ones, presumably due to minimal skin contact associated with white wine production, between 0.1 and to 1.2 mg/liter.[19] Moreover, because piceid is bound to a glucose, it seems that it may be absorbed more easily, as has been described for quercetin–glucoside.[20]

Assays of Resveratrol and Piceid

Gas Chromatography–Mass Spectrometry

Most gas chromatography (GC) methods required a prederivatization with bis[trimethylsilyl]trifluoroacetamide (BSTFA) prior to column appli-

[11] R. M. Lamuela-Raventos, A. I. Romero-Perez, A. L. Waterhouse, and M. C. de la Torre-Boronat, *J. Agric. Food Chem.* **43**, 281 (1995).
[12] F. Mattivi, F. Reniero, and S. Korhammer, *J. Agric. Food Chem.* **43**, 1820 (1995).
[13] B. J. Hector, J. B. Magee, C. P. Hegwood, and M. J. Coign, *Am. J. Enol. Vitic.* **47**, 57 (1996).
[14] R. Pezet and P. Cuenat, *Am. J. Enol. Vitic.* **47**, 287 (1996).
[15] P. Jeandet, R. Bessis, B. F. Maume, P. Meunier, D. Peyron, and P. J. Trollat, *J. Agric. Food Chem.* **43**, 316 (1995).
[16] R. M. Lamuela-Raventós, A. I. Romero-Pérez, A. L. Waterhouse, M. Lloret, and M. C. de la Torre-Boronat, *in* Wine: Nutritional and Therapeutic Benefits (T. R. Watkins, ed.), Vol. 661. American Chemical Society, Washington, D.C., 1997.
[17] U. Vrhovesk, S. Wendelin, and R. Eder, *Am. J. Enol. Vitic.* **48**, 214 (1997).
[18] P. Jeandet, R. Bessis, M. Sbaghi, P. Meunier, and P. Trollat, *Am J. Enol. Vitic.* **46**, 1 (1995).
[19] A. I. Romero-Pérez, R. M. Lamuela-Raventos, A. L. Waterhouse, and M. C. de la Torre-Boronat, *J. Agric. Food Chem.* **44**, 2124 (1996).
[20] P. C. H. Hollman, J. H. M. de Vries, S. D. van Leeuwen, M. J. B. Mengelers, and M. B. Katan, *Am. J. Clin. Nutr.* **62**, 1276 (1995).

cation with detection by flame ionization or mass spectrometry (MS). The disadvantages of GC methods compared to the liquid chromatography methods are as follows: time required for the extraction or for the derivatization (60 min); *trans* to *cis* isomerization may occur during derivatization[21]; and *trans*-polydatin is converted to the free isomers,[21] resulting in an overestimation of the aglycone.

High-Performance Liquid Chromatography

The first two methods reported for the analysis of *trans*-resveratrol by HPLC required multistep extraction procedures.[22,23] In 1994, McMurtey *et al.*,[24] using an electrochemical detector, and Pezet *et al.*,[25] using a fluorometric detector, accomplished analysis of *trans*-resveratrol by direct HPLC injection. Both methods were very sensitive to this compound and decreased the time needed for analysis. However, with these two methods, only *trans*-resveratrol was quantified. Later, the geometric isomers of the aglycones were separated and quantified.[26–28]

In 1995, Lamuela-Raventos *et al.*[11] and Mattivi *et al.*[12] published two different methods that allowed the quantification not only of resveratrol isomers, but also of both piceid isomers simultaneously. One method requires an extraction procedure using a RP-C18 cartridge,[12] whereas the other method avoids this procedure and its recovery issues.[11]

In the direct injection method, red and rosé wine samples can be chromatographed immediately after filtration. However, some filters, including nylon, PVDF, and polysulfone, retain more than 60% of *trans*-resveratrol. No filter adsorption loss is observed using Anopore (Maidstone, England) membrane filters (Anodisc, 0.2)[11] or poly(tetrafluoroethylene).[29]

Some white wines require an additional sample concentration step for the quantitation of the low levels of *cis* forms: 10 ml is concentrated to 3 ml by rotary evaporation (30°, *in vacuo*) and the concentrate is filtered through Whatman (Clifton, NJ) inorganic Anopore membrane filters, with

---

[21] G. J. Soleas, D. M. Goldberg, E. Ng, A. Karumanchiri, E. Tsang, and E. P. Diamandis, *Am. J. Enol. Vitic.* **48,** 169 (1997).

[22] E. H. Siemann and L. L. Creasy, *Am. J. Enol. Vitic.* **43,** 49 (1992).

[23] R. M. Lamuela-Raventos and A. L. Waterhouse, *J. Agric. Food Chem.* **41,** 521 (1993).

[24] K. D. McMurtrey, J. Minn, K. Pobanz, and T. P. Schultz, *J. Agric. Food Chem.* **42,** 2077 (1994).

[25] R. Pezet, V. Pont, and P. Cuenat, *J. Chromatogr. A* **663,** 191 (1994).

[26] G. J. Soleas, D. M. Goldberg, E. P. Diamandis, and A. Karumanchiri, *Am. J. Enol. Vitic.* **45,** 364 (1994).

[27] D. M. Goldberg, A. Karumanchiri, E. Ng, E. P. Diamandis, J. Yan, and G. J. Soleas, *J. Agric. Food Chem.* **42,** 1245 (1994).

[28] P. Jeandet, R. Bessis, B. F. Maume, and M. Sbachi, *J. Wine Res.* **4,** 79 (1993).

[29] B. C. Trela and A. L. Waterhouse, *J. Agric. Food Chem.* **44,** 1253 (1996).

a prefilter of glass microfiber to avoid plugging the membrane filter (Anotop 10 Plus, 0.2 mm). In previous work, we described a concentration step from 10 to 1 ml[19]; however, samples must be analyzed immediately after the concentration step or some precipitates are formed. Concentrating the sample to 3 ml ensures that the analyses are above the limit of quantitation and the sample is much more stable.

Chromatography

The HPLC system must have a UV–VIS detector that allows the collection of spectra on eluted peaks. The column is a Nucleosil, C_{18} 120 (25 × 0.4 cm), 5-mm particle size, with a precolumn of the same material; the column is maintained at 40°.

Separation is carried out with two solvents. Solvent A is glacial acetic in water, pH 2.40, and solvent B is 20% phase A with 80% acetonitrile, with a flow rate of 1.5 ml/min. The best resolution is obtained by applying the following elution program:

Time (min)	%A	%B
0	82	18
10	82	18
17	77	23
21	75.5	24.5
27	68.5	31.5
30	0	100

The HPLC conditions are critical for the separation of the four compounds from other interfering wine components. The pH of the solvent affects the elution. Solvent A must not have a pH higher than 2.40 because the separation of *cis*-resveratrol from an interfering peak at 21.9 min (see Fig. 2) will not be complete. The gradient conditions allow the separation of resveratrol and piceid from adjacent peaks. The eluent has to be monitored at 306 and 286 nm, the UV of optimum absorbancies of *trans* and *cis* isomers (Fig. 3). The glucosylated forms have spectra identical to their respective aglycons.

Standards. *trans*-Resveratrol can be purchased from commercial sources, including Sigma and PharmaScience Inc. *trans*-Piceid can be isolated from the dried roots of *Polygonum cuspidatum,* which is available from Asian herbal suppliers as Kojo-kon or Itadori-Kon, following the procedure described by Waterhouse and Lamuela-Raventós.[8]

Although it is not feasible to prepare pure *cis* standards because of light instability, solutions of *trans* compounds can be irradiated with light, and

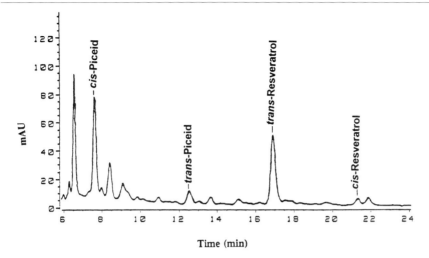

FIG. 2. Chromatogram of a red wine.

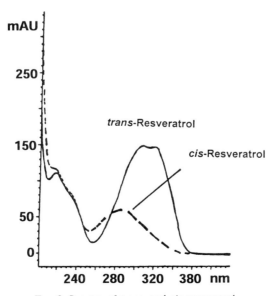

FIG. 3. Spectra of *trans*- and *cis*-resveratrol.

the lost *trans* is assumed to be converted quantitatively to the *cis*.[29] The quantitation of *trans*-piceid and *cis*-piceid can be performed based on the assumption of identical molar extinction coefficients of *trans*-resveratrol and *cis*-resveratrol at 306 and 285 nm.

[17] Antioxidant Activity by Headspace Gas Chromatography of Volatile Oxidation Products of ω-6 and ω-3 Polyunsaturated Lipids

By E. N. FRANKEL

Lipid Oxidation and Antioxidant Action

The oxidation of lipids containing polyunsaturated fatty acids (PUFA) and their oxidative susceptibility vary in humans, their red blood cells (RBC) and low-density lipoproteins (LDL) in health and disease. A large variety of conditions have been used to investigate lipid peroxidation in biological samples. Such terms used in the literature as oxidative stress, antioxidant capacity, lability, or susceptibility to lipid peroxidation or to autoxidative breakdown are often difficult to interpret because a large variation in experimental conditions and methods are used to determine the extent of peroxidation.

The chemistry of PUFA oxidation and antioxidant action is very complex in biological systems, and the main pathways are shown in Scheme 1 [Eqs. (1)–(8)]. Polyunsaturated lipids (LH) form alkyl radicals (L·) in the presence of an initiator generally composed of metals and hydroperoxides [Eqs. (1) and (2)]. Alkyl radicals react rapidly with oxygen to form peroxyl radicals (LOO·), which react with unsaturated lipids to produce hydroperoxides (LOOH), and propagate a chain reaction [Eqs. (3) and (4)]. Hydroperoxides decompose in the presence of metals to produce alkoxyl radicals (LO·), which cleave into a complex mixture of aldehydes [Eq. (5)]. These aldehydes are mainly responsible for causing damage in biological tissues.[1] In oxidized LDL, these aldehydes react with the apoprotein B (Apo B) of the LDL particles [Eq. (6)] to produce a new epitope that is recognized by macrophage scavenger receptors by producing foam cells and atherosclerotic plaques.

In the presence of radical stoppers such as antioxidants (AH), lipid peroxidation is retarded or inhibited by delaying the propagation stage

---

[1] H. Esterbauer, G. Jürgen, O. Quenhenberger, and E. Koller, *J. Lipid Res.* **28,** 495 (1987).

Initiation:

$$LOOH + Cu^+ \to LO\cdot + Cu^{2+} \quad (1)$$
$$LH + LO\cdot \to L\cdot \quad (2)$$

Propagation:

$$L\cdot + O_2 \to LOO\cdot \quad (3)$$
$$LOO\cdot + LH \to LOOH + L\cdot \quad (4)$$

Hydroperoxide Decomposition

$$LOOH + Cu^+ \to LO\cdot \to \text{Aldehydes} \quad (5)$$
$$\text{Aldehydes} \to \text{Modification and fragmentation of Apo B LDL} \quad (6)$$

Inhibited Peroxidation

$$LOO\cdot + AH \to LOOH + A\cdot \quad (7)$$
$$LO\cdot + AH \to LOH + \text{stable products} \quad (8)$$

SCHEME 1

of the autoxidation cycle [Eqs. (3) and (4), Scheme 1]. Several kinds of antioxidants can interrupt free radical lipid peroxidation by reacting with (1) the peroxyl radicals [Eq. (7)] to inhibit the formation of hydroperoxides and (2) the alkoxyl radicals [Eq. (8)] to inhibit the formation of aldehydes (Scheme 1).

Methodology

Different methods used to measure lipid peroxidation include analyses for lipid hydroperoxides, fluorescence, loss of PUFA, oxygen absorption, diene conjugation, thiobarbituric acid-reacting substances (TBARS), hydrocarbons and aldehydes by gas chromatography (GC), and chemiluminescence. The susceptibility of human serum, LDL, and RBC to lipid peroxidation increases by dietary deficiencies in vitamin E[2-5] or in the presence of various natural or artificial free radical initiators or oxidizing agents. The susceptibility of lipids toward this oxidative stress or damage is measured by determining TBARS,[6,7] fluorescent products,[8,9] volatile hydrocarbons,[10]

---

[2] B. Lubin and D. Chiu, *Pediatric Res.* **16,** 928 (1982).
[3] M.-L. Hu, E. N. Frankel, B. E. Leibovitz, and A. L. Tappel, *J. Nutr.* **119,** 1574 (1989).
[4] H. Esterbauer, J. Gebicki, H. Puhl, and G. Jürgen, *Free Radic. Biol. Med.* **13,** 341 (1992).
[5] J. F. Kearney and B. Frei, in "Natural Antioxidants" (B. Frei, ed.), p. 303. Academic Press, San Diego, 1994.
[6] J. Stocks and T. L. Dormandy, *Brit. J. Haematol.* **20,** 95 (1971).
[7] L. Vettore and C. J. G. Tedesco, *Haematologica* **60,** 250 (1975).
[8] J. J. M. Van den Berg, F. A. Kuypers, J. H. Qju, D. Chiu, B. Lubin, B. Roelofsen, and J. A. F. Op den Kamp, *Biochim. Biophys. Acta* **944,** 29 (1988).
[9] B. D. Goldstein and E. M. McDonagh, *J. Clin. Invest.* **57,** 1302 (1976).
[10] M. R. Clemens, H. Einsele, H. Frank, H. Remmer, and H. D. Waller, *Biochem. Pharmacol.* **32,** 3877 (1983).

and oxidized protein products.[11,12] Under the acid and heat conditions of the TBA assay, hydroperoxides and various aldehyde precursors decompose into TBARS,[13,14] which are related to the susceptibility of the tissue sample to peroxidation. The determination of TBARS in biological tissues can thus also be interpreted as a measure of oxidative susceptibility.

The TBA reaction has been used extensively in biological systems to determine malonaldehyde, which has been considered as a model for secondary products of lipid peroxidation. This colorimetric reaction is known, however, not to be specific for malonaldehyde.[15] Studies of the interactions of DNA with pure lipid oxidation products indicate that the biological importance of malonaldehyde may have been exaggerated in the literature.[16] Because of its convenience, the TBA method has become a common assay to determine the degree of peroxidation and oxidative susceptibility of a wide range of biological materials, including LDL. However, the validity of the TBA determination as an index of lipid peroxidation in biological samples has been a matter of considerable debate in the literature.[17]

The determination of TBARS inherently lacks specificity and is subject to interference by many compounds, including materials that are not due to lipid peroxidation. This method is also flawed by analytical artifacts and is affected by the same factors as lipid peroxidation.[17] Because many substances interfere with the TBA method, careful corrections are necessary by using appropriate blanks[15]; however, this practice is often neglected. Furthermore, the results of different modifications of the TBA test in human serum can vary as much as 10-fold.[18] With its lack of specificity, the TBA test is not recommended for human biological materials.[19]

Oxidation of Low-Density Lipoproteins

Modification of LDL by oxidation of its polyunsaturated lipid components has been implicated in the initiation of atherosclerosis.[4,20,21] Oxidation

---

[11] C. W. M. Haest, G. Plass, and D. Kamp, *Biochim. Biophys. Acta* **509**, 21 (1978).
[12] J. Palek and S. C. Liu, *Sem. Hematol.* **14**, 75 (1979).
[13] J. M. C. Gutteridge, J. Stocks, and T. L. Dormandy, *Anal. Chim. Acta* **70**, 107 (1974).
[14] S. Patton, *J. Am. Oil Chem. Soc.* **51**, 114 (1973).
[15] T. F. Slater, *Methods Enzymol.* **105**, 283 (1984).
[16] K. Fujimoto, W. E. Neff, and E. N. Frankel, *Biochim. Biophys. Acta* **795**, 100 (1984).
[17] D. R. Janero, *Free Radic. Biol. Med.* **9**, 515 (1990).
[18] H. Esterbauer and K. H. Cheeseman, *Methods Enzymol.* **186**, 407 (1990).
[19] B. Halliwell and J. M. C. Gutteridge, *Methods Enzymol.* **186**, 1 (1990).
[20] D. Steinberg, S. Parthasarathy, T. E. Carew, J. C. Khoo, and J. L. Witztum, *N. Engl. J. Med.* **320**, 915 (1989).
[21] H. Esterbauer, H. Puhl, M. Dieber-Rotheneder, G. Waeg, and H. Rabl, *Ann. Med.* **23**, 573 (1991).

proceeds by the interaction of PUFA components of LDL with active oxygen species or lipid peroxidation products. Protection against peroxidation comes from antioxidants, antioxidant enzymes, and cofactors.[21] *In vivo*, there is a steady competition between the oxidation and the protective processes that depend on the PUFA composition and on antioxidant levels.

Results of *in vitro* and *in vivo* studies are difficult to interpret because questionable methodology has been used to measure lipid peroxidation and the susceptibility of PUFA to oxidation. Determinations of aldehydes, oxysterols, hydroxyenals, fluorescence, and oxidized proteins or protein carbonyls provide useful information regarding oxidative damage most relevant to the modification of LDL. Aldehydes condense with lysine residues of Apo B to produce Schiff bases and enhance the affinity for scavenger receptors on macrophages.[22] Oxysterols are found in oxidized LDL and in the blood of fasting humans and in animals on a vitamin E-deficient diet. Hydroxyenals inhibit DNA, RNA, protein synthesis, and are cytotoxic at relatively high concentrations.[22] Fluorescence formation is largely due to modification of Apo B.[1] Measurements of oxidized proteins or protein carbonyls are most relevant to the oxidative modification of Apo B, but they are not sensitive and require significant oxidative changes; measures of protein carbonyls are also not specific and are influenced by factors other than peroxidation.

Other methods used for LDL oxidation provide less useful analytical information, including loss of linoleic (18:2) and arachidonic (20:4) acids, conjugated diene formation, TBARS, and electrophoretic mobility. The loss of 18:2 and 20:4 is a crude and insensitive measurement of lipid peroxidation because oxidative damage can be observed before significant changes in these PUFA can be measured. Determinations of conjugated dienes derived from PUFA hydroperoxides and several variations of the TBA test have been used most commonly to determine oxidation in LDL. Although diene conjugation is a useful method to measure lipid peroxidation in pure lipids, it reflects hydroperoxide formation but not the decomposition aldehyde products that actually cause oxidative damage to Apo B; this method cannot be used directly on human body fluids containing major UV-absorbing material. Thiobarbituric acid-reacting substances provide a nonspecific measure of lipid peroxidation in LDL that is flawed by many artifacts. Electrophoretic mobility is another nonspecific measure of LDL modification that may be confounded by other factors than lipid peroxidation.

Studies of dietary, genetic, and pathological modification of LDL require sensitive, chemically specific, and practical methods for lipid peroxidation.

[22] H. Esterbauer, R. J. Schaur, and H. Zollner, *Free Radic. Biol. Med.* **11,** 81 (1991).

Specific methodology is needed to clarify the question of whether LDL composition truly affects the oxidative susceptibility of lipoprotein particles. To elucidate how lipid peroxidation products act in the complex mechanism of lipid repair in biological materials, more useful and valid biochemical information may be obtained by using more specific assays, preferably using more than one assay for lipid peroxidation.[15,16]

The fatty acid composition of LDL is influenced by diets rich in oleic acid[23-25] and by supplementation of fish oil ω-3 PUFA.[26-28] Oleate-rich LDL particles from diets rich in oleic acid are highly resistant to oxidation, based on formation of TBARS,[23-25] conjugated dienes, and resistance to oxidation induced by endothelial cells.[23,24] However, the susceptibility to Cu^{2+}-catalyzed peroxidation, as measured by TBARS, is not changed in LDL from subjects fed fish oil and those fed corn oil diets.[27]

Effect of Antioxidants

Antioxidants are generally evaluated on the basis of their reactivity toward peroxyl radicals and their inhibition of hydroperoxide formation [Eqs. (4), (7), and Scheme 1]. However, their effects in scavenging alkoxyl radicals and in inhibiting aldehyde formation [Eq. (8)] have been overlooked, yet the production of aldehydes is likely to be most relevant to oxidative biological damage [Eq. (6)].

Diets can clearly promote lipid peroxidation when they include polyunsaturated and oxidized lipids and prooxidative metals. Dietary antioxidants can reduce the oxidative modification of LDL *in vitro* and can inhibit experimental atherosclerosis in animals.[4,5] The phenolic compounds in red wine and grape extracts have potent antioxidant activity in inhibiting the oxidation of human LDL *in vitro*.[29,30] This antioxidant activity is related to a wide range of phenolic compounds, including natural flavonoids that are widespread in the plant kingdom. Diets rich in fruits and vegetables

---

[23] S. Parthasarathy, J. C. Khoo, E. Miller, J. Barnett, J. L. Witztum, and D. Steinberg, *Proc. Natl. Acad. Sci. USA* **87**, 3894 (1990).
[24] P. Reaven, S. Parthasarathy, B. J. Grasse, E. Miller, F. Almazan, F. H. Mattson, J. C. Khoo, D. Steinberg, and J. L. Witztum, *Am. J. Clin. Nutr.* **54**, 701 (1991).
[25] E. M. Berry, S. Eisenberg, M. Friedlander, D. Harats, N. A. Kaufmann, Y. Norman, and Y. Stein, *Am. J. Clin. Nutr.* **56**, 394 (1992).
[26] A. Tripodi, P. Loria, M. A. Dilengite, and N. Carulli, *Biochim. Biophys. Acta* **1083**, 298 (1991).
[27] M. S. Nenseter, A. C. Rustan, S. Lund-Katz, E. Soyland, G. Maelandsmo, M. C. Phillips, and C. A. Drevon, *Arterioscl. Thromb.* **12**, 369 (1992).
[28] D. Harats, Y. Dabach, G. Hollander, M. Ben-Naim, R. Schwartz, E. M. Berry, O. Stein, and Y. Stein, *Atherosclerosis* **90**, 127 (1991).
[29] E. N. Frankel, J. Kanner, J. B. German, E. Parks, and J. E. Kinsella, *Lancet* **341**, 454 (1993).
[30] E. N. Frankel, A. L. Waterhouse, and J. E. Kinsella, *Lancet* **341**, 1103 (1993).

are also associated with lower risks of coronary heart disease. This benefit is usually related to the level of vitamins E and C and β-carotene. However, significant amounts of flavonoid compounds are also present in fruits and vegetables, and their potential nutritional effects also need to be considered seriously. The estimated daily intake of flavonoids ranges from 25 mg to 1 g,[31] yet very little is known about the absorption and metabolism of these compounds.

There is extensive literature on the evaluation of natural antioxidants in different unsaturated lipids and biological model systems. However, results of different investigators are difficult to interpret because of variations in test conditions and end points of lipid oxidation used. In the evaluation of natural antioxidants, varied results can be obtained by methods measuring products at different stages of lipid oxidation. The effects of antioxidants in inhibiting hydroperoxide formation [Eq. (7), Scheme 1] should be distinguished from their effects in preventing hydroperoxide decomposition [Eq. (8)]. We evaluated the effects of natural antioxidants by measuring both hydroperoxide formation, based on conjugated dienes, and hydroperoxide decomposition, based on hexanal formation, an important decomposition products of ω-6 PUFA peroxidation. Thus, we found that on the basis of conjugated diene formation, α-tocopherol exhibited prooxidant activity at high concentrations (500–1000 ppm) with bulk corn oil and corn oil-in-water emulsions, but was a good inhibitor of hexanal formation.[32] Assuming that hexanal determinations of hydroperoxide decomposition are related closely to biological damage, the effects of antioxidants in inhibiting hexanal formation may be thus more biologically relevant.

Several nonspecific assays were used to test the "antioxidant capacity" of biological tissues and fluids. In one study, the activity of water-soluble antioxidants in serum was measured by the suppression of chemiluminescence produced by luminol in the presence of free radicals generated by horseradish peroxidase and hydrogen peroxide.[33] In another study, the antioxidant activity of plasma and body fluids was based on the inhibition of the absorbance of an artificial radical cation [2,2'-azinobis(3-ethylbenzothiazoline-6-sulfonate)] formed in the presence of metmyoglobin and hydrogen peroxide.[34] These assays measure the relative reactivity of antioxidant toward artificial radicals, but they provide no quantitative information on what biological targets are protected by the antioxidants. Several studies

---

[31] J. Kühnau, *World Rev. Nutr. Diet* **24,** 117 (1976).
[32] S.-W. Huang, E. N. Frankel, and J. B. German, *J. Agric. Food Chem.* **42,** 2108 (1994).
[33] S. Maxwell, A. Cruickshank, and G. Thorpe, *Lancet* **344,** 193 (1994).
[34] C. A. Rice-Evans, N. J. Miller, and G. Paganga, *Free Radic. Biol. Med.* **20,** 933 (1996).

showed a marked variation in activity with different lipid systems and with different oxidizing conditions.[35,36] Different flavonoids showed widely different antioxidant activities and even prooxidant activities according to their abilities to scavenge hydroxyl or superoxide radicals or lipid peroxidation.[37] Therefore, the validity of radical trap methods may be questionable without knowing what specific biological substrate is protected. These nonspecific radical trap methods may be further confounded by many factors in body fluids, including effects that may not derive from lipid peroxidation.

To determine the real effects of antioxidants, it is important to obtain specific chemical information about what products of lipid peroxidation are inhibited. Several specific assays are needed to elucidate how lipid oxidation products act in the complex multistep process of oxidative damage in biological tissues (Scheme 1). The results of many complementary methods are required to determine oxidation products formed at different stages of the free radical chain. Because antioxidants show different activities toward hydroperoxide formation and decomposition, it is important that more than one method be used to monitor the oxidation process.

Analysis of Volatile Oxidation Products

To determine the decomposition products of lipid peroxidation, the technique of static headspace GC proved to be powerful and versatile in our laboratory for the quantitative analysis of volatile aldehydes in a large number of liquid, semiliquid, and solid samples. This technique was used to study quantitatively the oxidation of rat liver samples,[38] human red blood cell membranes,[39] and human LDL[40] at the molecular level. We also applied the headspace GC method in a nutritional study to test the effect of fish oil supplementation on the oxidative susceptibility of LDL fractions.[41] The method distinguished volatile oxidation products of ω-6 PUFA (pentane and hexanal) and ω-3 PUFA (propanal) in LDL samples from human subjects fed fish oil supplements. Specific volatile oxidation products were related to the fatty acid profiles and tocopherol content of LDL. The

---

[35] S.-W. Huang, A. I. Hopia, E. N. Frankel, and J. B. German, *J. Agric. Food Chem.* **44,** 444 (1996).

[36] A. I. Hopia, S.-H. Huang, K. Schwarz, J. B. German, and E. N. Frankel, *J. Agric. Food Chem.* **44,** 2030 (1996).

[37] Y. Hanasaki, S. Ogawa, and S. Fukui, *Free Radic. Biol. Med.* **16,** 845 (1994).

[38] E. N. Frankel, M.-L. Hu, and A. L. Tappel, *Lipids* **24,** 976 (1989).

[39] E. N. Frankel and A. L. Tappel, *Lipids* **26,** 479 (1991).

[40] E. N. Frankel, J. B. German, and P. A. Davis, *Lipids* **27,** 1047 (1992).

[41] E. N. Frankel, E. J. Parks, R. Xu, B. O. Schneeman, P. A. Davis, and J. B. German, *Lipids* **29,** 233 (1994).

method is suitable as a biological *in vitro* or *ex vivo* assay for lipid peroxidation. Because the headspace GC technique is rapid and does not require the workup of complex biological samples, it is ideal for the routine evaluation of antioxidants that require the processing of a large number of samples. This headspace assay of oxidative susceptibility has advantages over many other assays in permitting the specific analysis of multiple products of lipid peroxidation.

Headspace Capillary Gas Chromatography Method

The headspace GC analysis is a technique used to measure volatile compounds equilibrated with liquid or solid samples sealed in a closed system.[42,43] The method depends on the equilibrium between the gas and liquid with solutes distributed between the two phases. The content of solutes is determined by analyzing quantitatively samples of only the gas phase. The method is very sensitive to the equilibrium solute distribution between phases at the temperature selected for the analysis. Although this technique is simple, reliable quantitation requires stringent control of analytical conditions and careful calibrations with known compounds to be determined.

The oxidative susceptibility of liver preparations,[38] red blood cell membrane,[39] and LDL[40] is evaluated by headspace capillary GC to determine volatiles with a headspace sampler (Perkin-Elmer Sigma 3B gas chromatograph with an H-6 headspace sampler, Norwalk, CT), a capillary DB-1701 column (30 m × 0.32 mm, 1 μm thickness, J & W, Folsom, CA) heated isothermally at 70–80°. The GC conditions are helium carrier gas flow, 20 ml/min (helium pressure gauge of 60 psi); splitless injector temperature, 180°; detector temperature, 200°; and oven temperature, 75°. Volatile compounds derived from the oxidation of LDL are identified by the comparison of retention times with those of authentic reference compounds. The method is calibrated daily with standards of 5 and 10 μM hexanal.

Plasma LDL is prepared by sequential density ultracentrifugation in the presence of EDTA and is dialyzed exhaustively with phosphate-buffered saline, deoxygenated with nitrogen, and assayed immediately for oxidative susceptibility by headspace GC. Duplicate samples of 0.25 ml LDL (containing 1 mg LDL protein/ml) in phosphate-buffered saline are measured into special 6-ml bottles sealed and incubated for 2–4 hr at 37°

---

[42] H. Hachenberg and A. P. Schmidt, "Gas Chromatographic Headspace Analysis." Heyden and Son, London, 1977.

[43] B. Ioffe and A. G. Vitenberg, "Headspace Analysis and Related Methods." Wiley, New York, 1984.

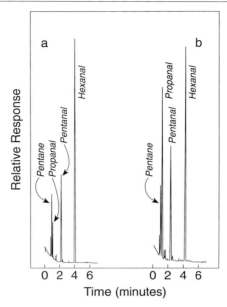

FIG. 1. Headspace gas chromatograms of copper-oxidized low-density lipoprotein isolated from an unsupplemented subject (a) and a fish oil-supplemented subject (b). [From E. N. Frankel, E. J. Parks, R. Xu, B. O. Schneeman, P. A. Davis, and J. B. German, *Lipids* **29,** 233 (1994), with permission from the AOCS Press.]

in the presence of 10–80 μM $CuSO_4$. After incubation, bottles are inserted into the headspace sampler heated at 40° and are pressurized with carrier gas for 30 sec, and an aliquot of headspace is injected directly into the gas chromatograph.

When a headspace sampler is not available, the same headspace technique is used with liver preparations[38] and RBCM[39] by sealing the samples in serum bottles and equilibrating under oxidizing conditions. After incubation, samples of 1-ml headspace are injected with a gas-tight syringe (Precision Sampling, Baton Rouge, LA) into a gas chromatograph.

Applications of the Gas Chromatography Headspace Assay

A typical chromatogram of a sample of human LDL oxidized with $CuSO_4$ shows four main components identified as pentane, propanal, pentanal, and hexanal (Fig. 1a).[41] Four main components are identified by comparison of retention times with authentic reference compounds, namely pentane, propanal, pentanal, and hexanal. Pentane and hexanal are derived from oxidation of ω-6 PUFA, and propanal from oxidation of ω-3 PUFA.[44] Pentanal is presumed to come from hexanal by oxidation in the presence of Cu^{2+}.

When hypertriglyceridemic subjects were supplemented with fish oil, LDL oxidation resulted in a significant increase in propanal formation. During *in vitro* oxidation of LDL with copper, propanal formation increased significantly in subjects supplemented with fish oil (Fig. 1b).[41] This increase correlated directly to increases in ω-3 PUFA in LDL. However, the oxidative susceptibility of LDL did not change, based on total volatile oxidation products. A previous study on LDL from human subjects consuming fish oil and corn oil also showed no change in susceptibility to copper-catalyzed oxidation, based on lipid peroxide determinations.[45] However, other studies reported significant increases in formation of TBARS resulting from fish oil diets.[46,47] Because TBARS are formed by oxidative decomposition of PUFA containing more than two double bonds,[48] the oxidation of any lipids containing ω-3 PUFA would be expected to produce high levels of TBARS. Therefore, TBARS determinations cannot be used as the sole determination of oxidative susceptibility to compare the effects of dietary ω-6 versus ω-3 PUFA on LDL oxidation. We need to use better techniques to provide the basic chemical information necessary to understand the mechanism of oxidation in LDL and other biological systems implicated in many diseases.

The effects of wine phenolics on the oxidative susceptibility of human LDL was investigated by measuring the amounts of hexanal and conjugated dienes formed by the Cu^{2+}-catalyzed oxidation of human LDL.[29] Hexanal formation was inhibited by 60 and 98% by the addition of diluted dealcoholized red wine (California Petite Syrah) containing 3.8 and 10 μM phenolic compounds (Fig. 2); conjugated diene was inhibited by 50 and 75% by diluted wine containing 2 and 4 μM phenolic compounds. Diluted wine containing 10 μM phenolics had the same antioxidant activity as 10 μM quercetin in inhibiting the oxidation of LDL. In contrast, 10 μM α-tocopherol inhibited hexanal formation only 60%.

The antioxidant activity of 20 California wines was evaluated by their abilities to inhibit the copper-catalyzed oxidation of human LDL *in vitro*.[49] The relative inhibition of LDL oxidation varied from 46 to 100% with red

---

[44] E. N. Frankel, *Prog. Lipid Res.* **22**, 1 (1983).
[45] M. S. Nenseter, A. C. Rustan, S. Lund-Katz, E. Soyland, G. Maelandsmo, M. C. Phillips, and C. A. Drevon, *Arterioscl. Thromb.* **12**, 369 (1992).
[46] A. Tripodi, P. Loria, M. A. Dilengite, and N. Carulli, *Biochim. Biophys. Acta* **1083**, 298 (1991).
[47] D. Harats, Y. Dabach, G. Hollander, M. Ben-Naim, R. Schwartz, E. M. Berry, O. Stein, and Y. Stein, *Atherosclerosis* **90**, 127 (1991).
[48] L. K. Dahle, E. G. Hill, and R. T. Holman, *Arch. Biochem. Biophys.* **98**, 253 (1962).
[49] E. N. Frankel, A. L. Waterhouse, and P. L. Teissedre, *J. Agric. Food Chem.* **43**, 890 (1995).

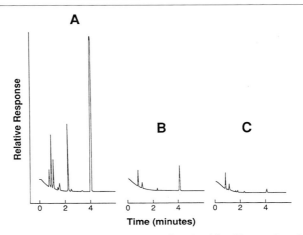

FIG. 2. Headspace gas chromatograms of samples of oxidized human low-density lipoprotein. (A) Control LDL, (B) plus 3.8 μM wine phenolics, and (C) plus 10 μM wine phenolics. [From E. N. Frankel, J. Kanner, J. B. German, E. Parks, and J. E. Kinsella, Lancet **341**, 454 (1993), with permission from The Lancet Ltd.]

wines and from 3 to 6% with white wines. The antioxidant activity of wines toward LDL oxidation was distributed widely among the principal phenolic compounds, including gallic acid, catechin, myricetin, quercetin, caffeic acid, rutin, epicatechin, cyanidin, and malvidin-3-glucoside. The antioxidant activities of extracts of different table and wine grapes in human LDL *in vitro* were comparable to those for wines.[50] Hexanal formation was inhibited by 62–91% with various grape extracts. The relative antioxidant activity of grape extracts toward LDL oxidation correlated with contents of anthocyanins, flavan-3-ols, flavonols, and hydroxybenzoates.

Pure phenolic compounds were evaluated for their activities in inhibiting human LDL oxidation by copper.[51] The antioxidant activities decreased in the following order: catechin, myricetin = epicatechin = rutin, gallic acid, quercetin, and cyanidin. Catechin oligomers and procyanidin dimers and trimers separated from red grape seeds had the same antioxidant activity as monomers catechin, epicatechin, and myricetin. Thus, the numerous phenolic compounds found in wine and grapes are potent antioxidants in inhibiting the *in vitro* LDL oxidation.

[50] A. S. Meyer, O.-S. Yi, D. A. Pearson, A. L. Waterhouse, and E. N. Frankel, *J. Agric. Food Chem.* **45**, 1638 (1997).
[51] P. L. Teissedre, E. N. Frankel, A. L. Waterhouse, H. Peleg, and J. B. German, *J. Sci. Food Agric.* **70**, 55 (1996).

Several studies have produced mixed results in showing *in vivo* antioxidant activity of phenolic compounds in red wine. However, the results of these studies are unconvincing because they used nonspecific assays for antioxidant action that may be inadequate due to the confounding and indirect effects of many serum and plasma components.[33,52,53] One study reported that after a washout period of white wine consumption for 2 weeks, the oxidizability of LDL isolated from subjects consuming red wine for 4 weeks did not change.[54] However, the effect of *in vivo* supplementation of polyphenols in red wine on the oxidizability of LDL cannot be tested *ex vivo* because these hydrophilic compounds behave like ascorbic acid[5] and other water-soluble antioxidants in being removed from LDL during isolation from plasma. The total antioxidant activity of polyphenols of grapes, wines, and green teas was evaluated by determining their scavenging ability toward an artificial radical cation system.[34] However, this approach for the evaluation of natural antioxidants by such an artificial radical model system provides no information on what lipid or protein is protected.

Conclusions

The technique of headspace GC is rapid and suitable for biological materials because no sample workup is necessary. This method can be used to determine the specific oxidation of ω-3 and ω-6 PUFA. It is ideal for the measurement of oxidative susceptibility as affected by diet and etiology of biological samples. This method may prove valuable for routine checking of antioxidant supplementation and clinical etiology.

A wide range of phenolic compounds were shown to inhibit the *in vitro* oxidation of human LDL, which plays an important role in the initiation of atherosclerosis. Plant phenolic antioxidants in fruits and vegetables may also have a protective effect on coronary heart disease and cancer, but the molecular basis of protection is not understood. A better knowledge of the mechanism of natural antioxidants will require more systematic studies with specific methods providing specific chemical information that can be related directly to oxidative modifications of biological systems. Pharmacokinetic data are needed to evaluate the effectiveness of the polyphenolic antioxidant compounds in fruits and beverages and their potential role in reducing coronary heart disease.

---

[52] K. Kondo, A. Matsumoto, H. Kurata, H. Tanahashi, H. Koda, T. Amachi, and H. Itakura, *Lancet* **344,** 1152 (1994).
[53] B. Fuhrman, A. Lavy, and M. Aviram, *Am. J. Clin. Nutr.* **61,** 549 (1995).
[54] Y. B. de Rijke, P. N. M. Demacker, N. A. Assen, L. M. Sloots, and M. B. Katan, *Am. J. Clin. Nutr.* **63,** 329 (1996).

[18] Determination of Tea Catechins by Reversed-Phase High Performance Liquid Chromatography

By PETER C. H. HOLLMAN, DINI P. VENEMA, SANTOSH KHOKHAR, and ILJA C. W. ARTS

Introduction

Catechins or 2-phenylbenzodihydropyrans belong to the class of flavonoids, polyphenolic compounds with antioxidant properties that occur ubiquitously in foods of plant origin. Over 4000 different naturally occurring flavonoids have been described and this list is still growing.[1] Major dietary sources of flavonoids are vegetables, fruits, and beverages such as tea and red wine.[2] Kühnau[3] estimated that the total flavonoid intake in the United States was 1 g/day expressed as glycosides, but most likely this estimate is too high. New, more specific food analyses suggested that the Dutch intake of flavonols and flavones, a subclass of flavonoids, was only one-fifth of the intake of flavonols and flavones in the United States estimated by Kühnau.[3]

The four major catechins of tea are (−)-epicatechin (EC), its galloyl ester (−)-epicatechin 3-gallate (ECg), (−)-epigallocatechin (EGC), and (−)-epigallocatechin 3-gallate (EGCg) (Fig. 1).[4,5] (+)-Catechin is only a minor compound in teas, but higher amounts of (+)-catechin are found in fruits.[4–6] *In vitro* data on the antioxidant activity of flavonoids show that they are better antioxidants than the antioxidant vitamins.[7] Within the group of flavonoids, tea catechins, especially gallates, are the most effective antioxidants.[8] An estimate of the daily intake of catechins has not been reported because quantitative data are lacking for many foods, particularly fruits and vegetables. In countries such as the United Kingdom and The

---

[1] E. Middleton and C. Kandaswami, *in* "The Flavonoids: Advances in Research since 1986" (J. B. Harborne, ed.), p. 619. Chapman & Hall, London, 1994.
[2] P. C. H. Hollman and M. B. Katan, *in* "Flavonoids in Health and Disease" (C. Rice-Evans and L. Packer, eds.), p. 483. Dekker, New York, 1997.
[3] J. Kühnau, *World Rev. Nutr. Diet.* **24,** 117 (1976).
[4] S. Khokhar, D. P. Venema, P. C. H. Hollman, M. Dekker, and W. M. F. Jongen, *Cancer Lett.* **114,** 171 (1997).
[5] Y.-L. Lin, I.-M. Juan, Y.-L. Chen, Y.-C. Liang, and J.-K. Lin, *J. Agric. Food Chem.* **44,** 1387 (1996).
[6] K. Herrmann, *Erwerbsobstbau* **32,** 4 (1990).
[7] C. A. Rice-Evans and N. J. Miller, *Biochem. Soc. Trans.* **24,** 790 (1996).
[8] C. A. Rice-Evans, N. J. Miller, and G. Paganga, *Free Radic. Biol. Med.* **20,** 933 (1996).

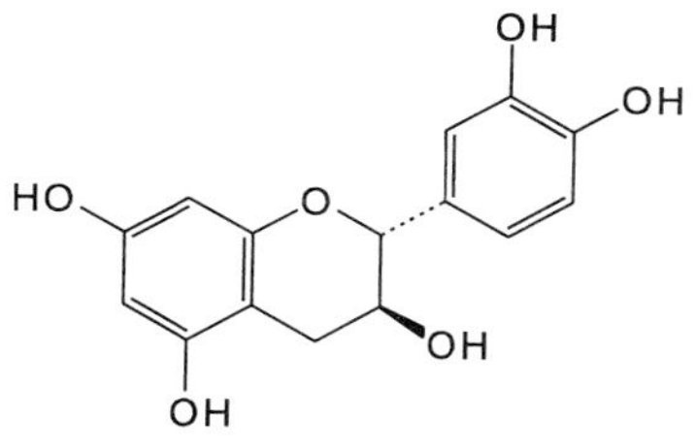

FIG. 1. Structures of tea catechins: (+)-catechin (I), (−)-epicatechin (II, R = H), (−)-epigallocatechin (II, R = OH), (−)-epicatechin 3-gallate (III, R = H), and (−)-epigallocatechin 3-gallate (III, R = OH).

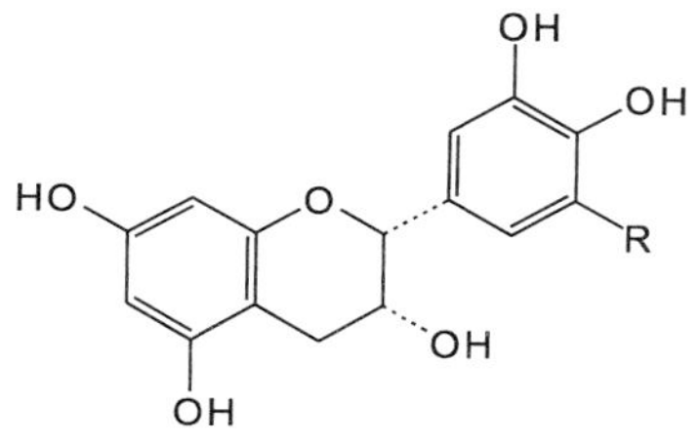

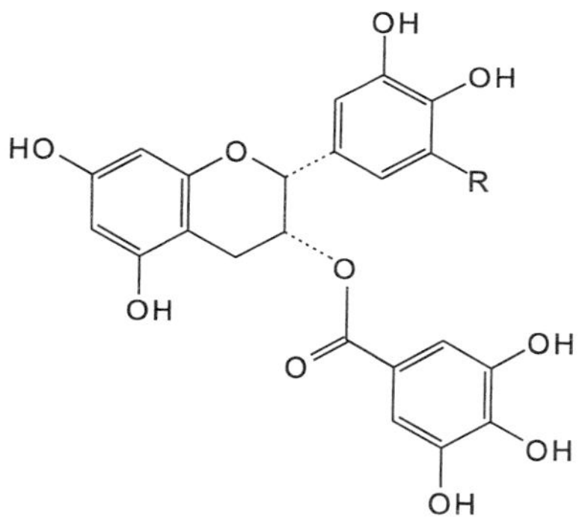

Netherlands with a high habitual tea consumption, tea is an important source of catechins. Because the concentration of catechins varies with type and brand of tea, it is important to determine catechins in a range of commonly consumed teas.

Methods and Materials

Chromatography

High-performance liquid chromatography (HPLC) equipment consists of an automatic injector, a Merck Hitachi L6000A and L6200 pump, an Inertsil ODS-2 column (150 × 4.6 mm, 5 μm; GL Sciences Inc., Tokyo, Japan) protected by an Opti-Guard PR C_{18} Violet A guard column (Optimize Technologies, Inc.), both placed in a column oven set at 30°, and a Kratos spectroflow 783 UV detector set at 278 nm. The detector output is sampled using a Nelson (PE Nelson, Cupertino, CA) Series 900 interface and Nelson integrator software (Model 2600, rev. 5). Solvents used for separation are 5% acetonitrile (eluent A) and 25% acetonitrile in 0.025 M phosphate buffer, pH 2.4 (eluent B). The gradient is as follows: 0–5 min, 15% B; 5–20 min, linear gradient from 15 to 80% B; 20–23 min, 80% B; and 23–25 min, 15% B. The flow rate is 1.0 ml min^{-1}. The sample injection volume is 10 μl.

Standards

Pure standards of (+)-catechin and EC are from Sigma (Sigma-Aldrich Chemie BV, Zwijndrecht, The Netherlands); EGC, ECg, and EGCg are from Apin (Apin Chemicals, Oxon, UK). Standards are dissolved in methanol containing citric acid (800 mg/liter). These stock solutions contain 1 mg of catechins/ml and are stored at 4°. The stability of the stock solutions is followed spectrophotometrically every other day for 1 week, and no deterioration occurred. Calibration solutions are freshly prepared for each series of analyses by diluting aliquots of the stock solutions in methanol. Calibration curves are constructed by linear regression of the peak area against concentration of the calibration solution (10–100 mg/liter). All calibration curves are linear through the origin in the measured range. Peaks are identified by comparing their retention time with the retention time of the standards.

Tea Extracts

Tea extracts are prepared as follows: pour 100 ml of boiling water onto 1 g of tea leaves. Stir the tea infusion occasionally for 5 min, decant, and allow to cool. Adjust the pH to 3.2 with citric acid, filter through a 0.45-μm filter (Acrodisc, CR PTFE; Gelman Sciences, Ann Arbor, MI), and inject into the HPLC system without any additional treatment.

Results and Discussion

Precision

Repeatability and reproducibility of the assay were determined by analysis in duplicate of five identical samples of black tea extract on 5 different days. The relative standard deviation of reproducibility was 4.0% for EC, 7.4% for EGC, 4.6% for ECg, and 5.6% for EGCg. The repeatability relative

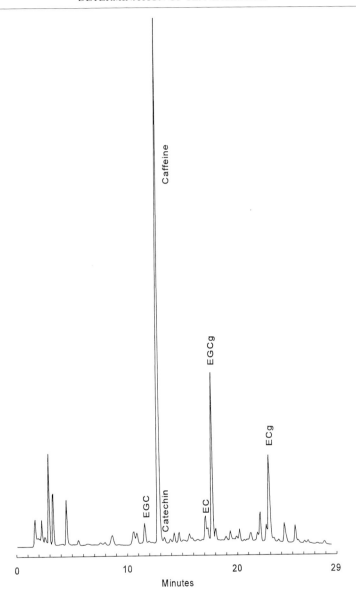

FIG. 2. Chromatogram of a black tea extract.

standard deviation was 2.9% for EC, 4.4% for EGC, 3.0% for ECg, and 3.4% for EGCg.

Chromatography

Gradient elution is necessary because of the large differences in retention of the catechins on this type of reversed-phase column. The separation chosen was a compromise between selectivity and speed of analysis. Figure 2 shows that the resolution between caffeine and (+)-catechin was not sufficient to quantify small amounts of (+)-catechins. However, (+)-catechin is only a minor catechin component in tea.[5] Resolution of epicatechin can be moderate in some samples (Fig. 2). Sample cleanup using solid-phase extraction[9] may be helpful, but will complicate sample pretreatment considerably. We found that fluorescence detection (280 nm excitation, 310 emission wavelengths) of (+)-catechin and (−)-epicatechin was more sensitive and selective than UV detection, thus obviating the need of sample cleanup. In addition, the interference of caffeine was minimized as the peak area of caffeine with fluorescence detection was only 100th of that of (+)-catechin.

Sample Stability

The addition of citric acid to the samples enhanced the stability of the catechins. No losses of catechins in tea extracts with citric acid were found after storage at 4° for at least 2 weeks. Without citric acid, catechins decreased after 1 day of storage.

Conclusions

With adequate precision, this method is simple and fast because sample pretreatment can be abandoned. The method can be applied for analyses of a large number of samples of all kinds of tea.

---

[9] S. Kuhr and U. H. Engelhardt, *Z. Lebensm. Unters. Forsch.* **192,** 526 (1991).

[19] Flavonoids as Peroxynitrite Scavengers *in Vitro*

By ANANTH SEKHER PANNALA, SURINDER SINGH, and
CATHERINE RICE-EVANS

Introduction

The use of antioxidants, both natural and synthetic, in the prevention and cure of many diseases is gaining wide importance in the medical field. For example, diets rich in fruit and vegetables are known to play a role in protection against coronary heart disease and certain types of cancer.[1] This has been attributed to a variety of cardioprotective and anticarcinogenic mechanisms of the individual constituents, including the free radical scavenging properties of antioxidant nutrients. Currently there is considerable interest in the antioxidant activities of dietary antioxidants, vitamins C and E, carotenoids, and plant phenolics, especially polyphonolic flavonoids and hydroxycinnamates.

Polyphenolic Flavonoids: Catechin and Its Gallate Esters

Polyphenolic flavonoids constitute a large class of compounds, ubiquitous in plants, containing a number of phenolic hydroxyl groups conferring the antioxidant activity.[2] Dietary sources of these compounds include green and black tea, red wine, grapes, and onions.[3] Polyphenols are reducing agents that function as antioxidants by virtue of their hydrogen-donating properties of their phenolic hydroxyl groups[4–7] as well as by their transition metal-chelating abilities.[8–10] In particular, epicatechin, epigallocatechin

---

[1] G. Block, *Nutr. Rev.* **50**, 207 (1992).
[2] J. B. Harborne, *in* "Plant Flavonoids in Biology and Medicine" (B. Cody, E. Middleton, and J. B. Harborne, eds.), p. 15. A. R. Liss, New York, 1986.
[3] C. A. Rice-Evans, N. J. Miller, and G. Paganga, *Free Radic. Biol. Med.* **20**, 993 (1996).
[4] C. A. Rice-Evans, *Biochem. Soc. Symp.* **61**, 103 (1995).
[5] W. Bors, W. Heller, C. Michel, and M. Saran, *Meth. Enzymol.* **186**, 343 (1990).
[6] C. A. Rice-Evans, N. J. Miller, and G. Paganga, *Free Radic. Biol. Med.* **20**, 933 (1996).
[7] S. V. Jovanovic, S. Steenken, M. Tosic, B. Marjanovic, and M. G. Simic, *J. Am. Chem. Soc.* **116**, 4846 (1994).
[8] G. Paganga, H. Al-Hashim, H. Khodr, B. C. Scott, O. I. Aruoma, R. C. Hider, B. Halliwell, and C. A. Rice-Evans, *Redox Rep.* **2**, 359 (1996).
[9] M. Thompson, C. R. Williams, and G. E. P. Elliot, *Anal. Chim. Acta* **85**, 375 (1976).
[10] J. E. Brown, H. Khodr, R. C. Hider, and C. A. Rice-Evans, *Biochem. J.* **330**, 1173 (1998).

(EGC), and their gallate esters (Fig. 1) have been shown to scavenge both aqueous and lipophilic radicals and to protect low-density lipoprotein (LDL) from oxidation by acting as chain-breaking antioxidants,[11,12] Plant polyphenols have also been shown to possess antimicrobial,[13] antiallergic,[14] anti-inflammatory,[15] and antimutagenic[16-20] properties.

Hydroxycinnamates

Hydroxycinnamates or phenylpropanoids are present in many dietary sources, such as wheat, corn, rice, tomatoes, white grapes, olives, white wine, apples, and pears. Hydroxycinnamates are present in dietary phytochemicals at higher concentrations than polyphenolic flavonoids.

Hydroxycinnamates (Fig. 2), which are distributed widely in plant tissues, are produced from L-phenylalanine and L-tyrosine via the shikimate pathway, forming *p*-coumaric acid, which undergoes further hydroxylation to caffeic acid and subsequent O-methylation to ferulic acid. These compounds occur in nature in various conjugated forms, resulting from enzymic hydroxylation, O-methylation, O-glycosylation, and esterification of *p*-coumaric acid to form quinic or carbohydrate esters.[21] Hydroxycinnamates have been reported to exhibit antimicrobial, antiallergic, and anti-inflammatory activities[22-24] as well as antimutagenic properties.[25,26]

[11] N. Salah, N. J. Miller, G. Paganga, L. Tijburg, G. P. Bolwell, and C. A. Rice-Evans, *Arch. Biochem. Biophys.* **322**, 339 (1995).
[12] S. Miura, J. Watanabe, M. Sano, T. Tomita, T. Osawa, Y. Hara, and I. Tomita, *Biol. Pharm. Bull.* **18**, 1 (1995).
[13] J. M. T. Hamiltonmiller, *Antimicrob. Agents Chemother.* **39**, 2375 (1995).
[14] N. Matsuo, K. Yamada, K. Yamashita, K. Shoji, M. Mori, and M. Sugano, *In Vitro Cell Dev. Biol.* **32**, 340 (1996).
[15] G. M. Shivji, E. Zielinska, S. Kondo, H. Mukhtar, and D. N. Sander, *J. Invest. Dermatol.* **106**, 787 (1996).
[16] G. C. Yen and H. Y. Chen, *J. Food Protect.* **57**, 54 (1994).
[17] J. Yamada and Y. Tomita, *Biosci. Biotech. Biochem.* **58**, 2197 (1994).
[18] A. Constable, N. Varga, J. Richoz, and R. H. Stadler, *Mutagenesis* **11**, 189 (1996).
[19] C. Han, *Cancer Lett.* **114**, 153 (1997).
[20] A. BuAbbas, E. Copeland, M. N. Clifford, R. Walker, and G. Ioannides, *J. Sci. Food Agric.* **75**, 453 (1997).
[21] R. Ibrahim and D. Barron, in "Methods in Plant Biochemistry" (P. M. Dey and J. B. Harborne, eds.), p. 75. Academic Press, London, 1989.
[22] A. Bell, in "The Biochemistry of Plants" (E. E. Conn, ed.), p. 1. Academic Press, New York, 1981.
[23] G. Surico, L. Varvaro, and M. Solfrizzo, *J. Agric. Food Chem.* **35**, 406 (1987).
[24] G. F. Sud'ina, O. K. Mirzoeva, M. A. Pushkareva, G. A. Korshunova, N. V. Sumbutya, and S. D. Varfolomeev, *FEBS Lett.* **329**, 21 (1993).
[25] A. W. Wood, M.-T. Huang, R. L. Chang, H. L. Newmark, R. E. Lehr, H. Yagi, J. M. Sayer, D. M. Jerina, and A. H. Cooney, *Proc. Natl. Acad. Sci. USA* **79**, 5513 (1982).
[26] M. Namiki, *Crit. Rev. Food Sci. Nutr.* **29**, 273 (1990).

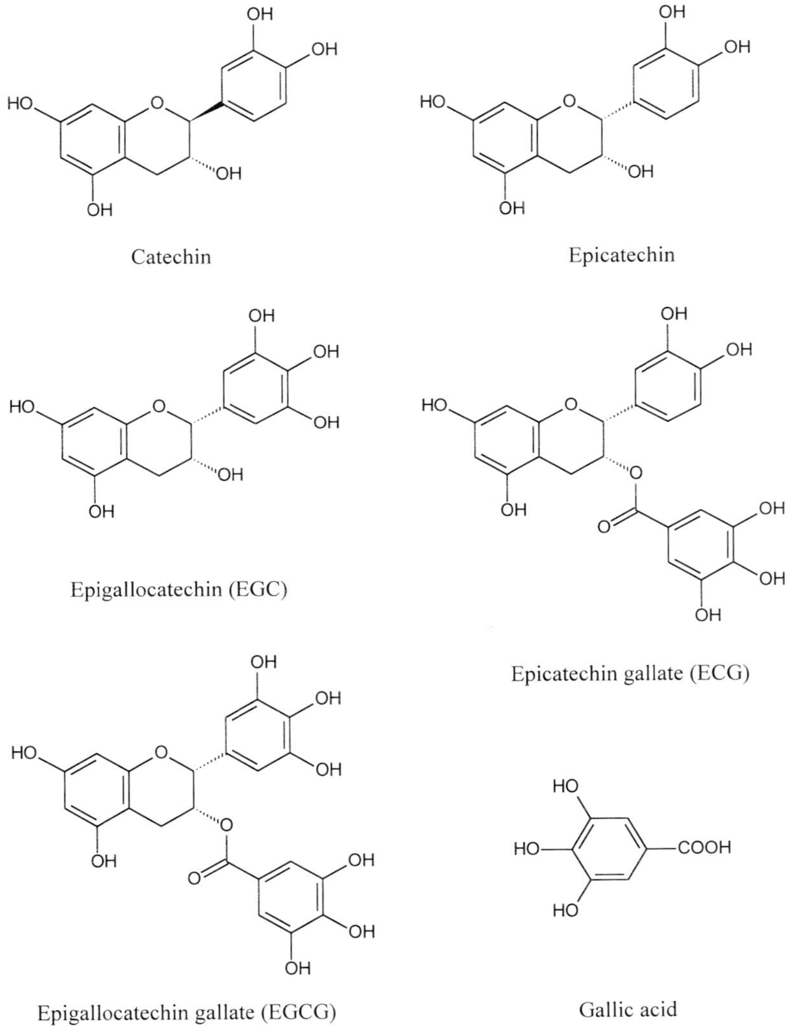

Catechin

Epicatechin

Epigallocatechin (EGC)

Epicatechin gallate (ECG)

Epigallocatechin gallate (EGCG)

Gallic acid

FIG. 1. Chemical structures of polyphenolic flavonoids.

Ferulic acid has also been shown to protect α_1-antiproteinase against inactivation by hypochlorous acid.[27] Hydroxycinnamates have also been shown to possess peroxyl radical-scavenging properties as measured by

[27] B. Scott, J. Butler, B. Halliwell, and O. I. Aruoma, *Free Radic. Res. Commun.* **19,** 241 (1993).

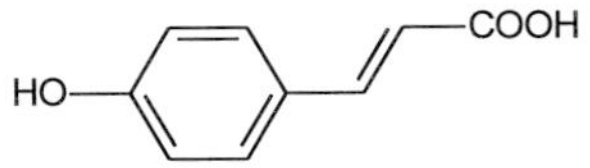

p-Coumaric acid

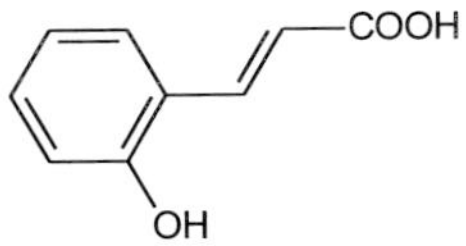

o-Coumaric acid

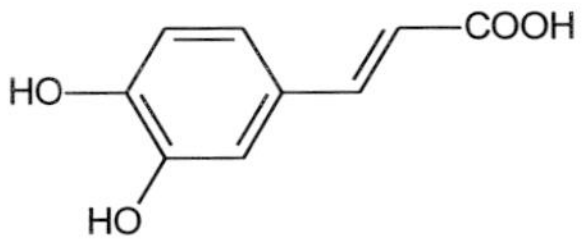

Caffeic acid

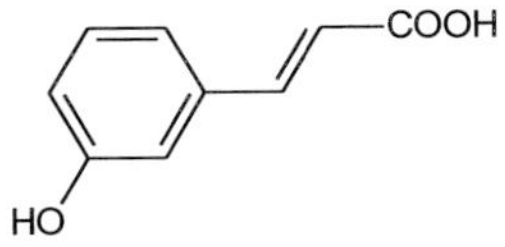

m-Coumaric acid

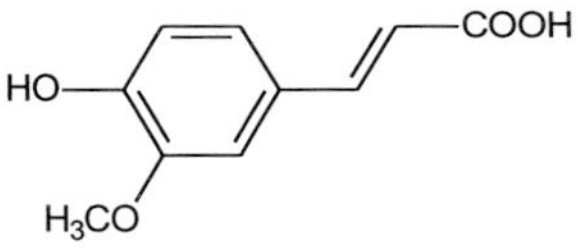

Ferulic acid

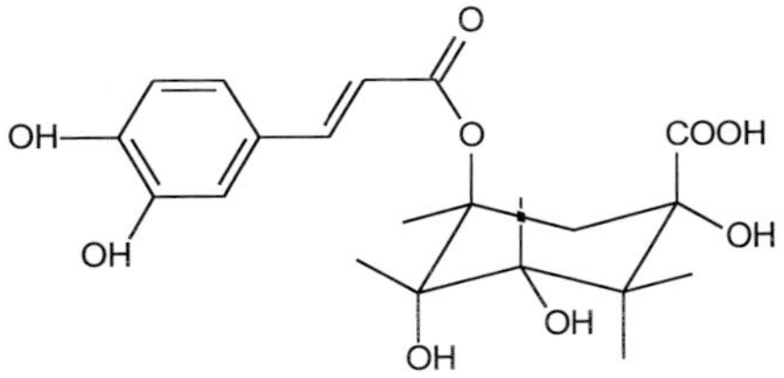

Chlorogenic acid

FIG. 2. Chemical structures of hydroxycinnamates.

their ability to prevent the lipid peroxidation of LDL mediated by heme proteins.[28,29]

Peroxynitrite and Tyrosine Nitration Assay

Peroxynitrite is a highly toxic oxidizing and nitrating species that can be produced *in vivo* by the rapid interaction of superoxide and nitric oxide. The second-order rate constant for this reaction is $6.7 \times 10^9\ M^{-1}\ \text{sec}^{-1}$.[30] Stimulated macrophages, neutrophils, and endothelial cells have all been

[28] C. Castelluccio, G. Paganga, N. Melikian, G. P. Bolwell, J. Pridham, J. Sampson, and C. A. Rice-Evans, *FEBS Lett.* **368,** 188 (1995).

[29] C. Castelluccio, G. P. Bolwell, C. Gerrish, and C. A. Rice-Evans, *Biochem. J.* **316,** 691 (1996).

[30] R. E. Huie and S. Padmaja, *Free Radic. Res. Commun.* **18,** 195 (1993).

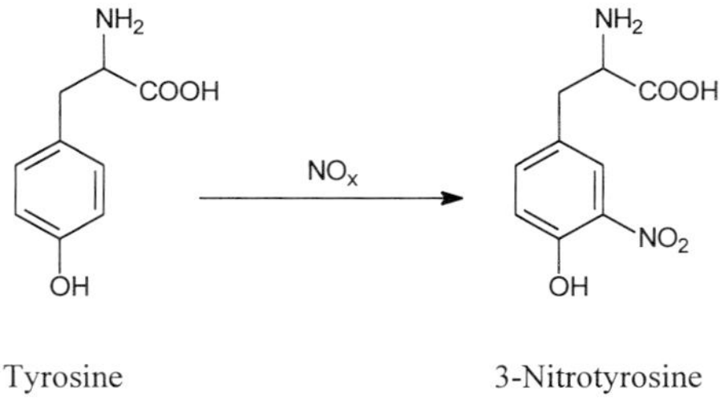

SCHEME 1. Tyrosine nitration on exposure to reactive nitrogen species (NO_x).

shown to generate peroxynitrite,[31-33] and data have provided evidence for the *in vivo* formation of peroxynitrite in, for example, human atherosclerotic coronary vessels, acute lung injury, and chronic inflammation.[34-37] Peroxynitrite at physiological pH (pK_a 6.8) undergoes protonation to form peroxynitrous acid that decays rapidly to form a mixture of reactive products; peroxynitrite and products derived from it have been reported to induce lipid peroxidation[38] and to modify amino acids in proteins. For example, tyrosine is especially susceptible to peroxynitrite-dependent nitration reactions forming 3-nitrotyrosine (Scheme 1). The underlying mechanism for the nitration of tyrosine is unclear. Nitration reactions can occur either via the nitronium ion (NO_2^+) or via the nitrogen dioxide radical ($NO_2\cdot$). Peroxynitrite could undergo a homolytic fission to generate nitrogen dioxide radical and the hydroxyl radical ($OH\cdot$) or a heterolytic fission to generate the nitronium ion and the hydroxide ion (OH^-).

The ability to inhibit peroxynitrite-dependent nitration provides a useful assay to screen various compounds for their ability to scavenge peroxynitrite and the nitrating species derived from it.[39,40] The aim of this study was to

[31] H. Ischiropoulos, L. Zhu, and J. S. Beckman, *Arch. Biochem. Biophys.* **298**, 446 (1992).
[32] M. C. Carreras, G. A. Pargament, S. D. Catz, J. J. Poderosso, and A. Boveris, *FEBS Lett.* **341**, 65 (1994).
[33] N. W. Kooy and J. A. Royall, *Arch. Biochem. Biophys.* **310**, 352 (1994).
[34] J. S. Beckman, J. Chen, H. Ischiropoulos, and J. P. Crow, *Meth. Enzymol.* **233**, 229 (1994).
[35] I. Y. Haddad, G. Pataki, P. Hu, Y. Ye, J. S. Beckman, and S. Matalon, *J. Clin. Invest.* **94**, 2407 (1994).
[36] N. W. Kooy, J. A. Royall, Y. Z. Ye, D. R. Kelly, and J. S. Beckman, *Am. J. Resp. Crit. Care Med.* **151**, 1250 (1995).
[37] H. Kaur and B. Halliwell, *FEBS Lett.* **350**, 9 (1994).
[38] R. Radi, J. S. Beckman, K. M. Bush, and B. A. Freeman, *Arch. Biochem. Biophys.* **288**, 481 (1991).
[39] M. Whiteman and B. Halliwell, *Free Radic. Res.* **25**, 275 (1996).
[40] A. S. Pannala, C. A. Rice-Evans, B. Halliwell, and S. Singh, *Biochem. Biophys. Res. Commun.* **232**, 164 (1997).

evaluate the abilities of plant polyphenolics (catechin, epicatechin, ECG, EGC, EGCG) and hydroxycinnamates (*p*-, *m*-, *o*-coumaric acids, chlorogenic acid, ferulic acid, and caffeic acid) to protect against damage by peroxynitrite as assessed by their abilities to inhibit tyrosine nitration.

Methods

Synthesis of Peroxynitrite

Peroxynitrite synthesis is carried out by modifying the method described by Beckman *et al.*[34] Acidified hydrogen peroxide (20 ml) and sodium nitrite (20 ml) solutions are drawn into two separate syringes, analogous to a stop flow setup. Simultaneous injection of the contents of both syringes into an ice-cooled beaker containing 1.5 M potassium hydroxide (40 ml) through a Y-shaped junction leads to rapid mixing, resulting in the formation of peroxynitrous acid followed by stabilization of the resulting peroxynitrite anion. Excess hydrogen peroxide is removed by passing the solution through a manganese dioxide column. The concentration of peroxynitrite is determined by measuring the absorbance at 302 nm ($\varepsilon = 1670\ M^{-1}\ cm^{-1}$). The typical yield of freshly prepared peroxynitrite ranges from 45 to 80 mM. Higher concentrations (>400 mM) of peroxynitrite can be obtained by freeze fractionation.

Peroxynitrite Scavenging Assay

Tyrosine Nitration on Exposure to Peroxynitrite. To determine the extent of nitration, a fixed concentration of tyrosine is reacted with increasing concentrations of peroxynitrite. A 50-μl aliquot of peroxynitrite (increasing concentrations: 0 to 1000 μM, final concentration) is added to a solution containing tyrosine (100 μM, final concentration) in 0.2 M phosphate buffer, pH 7. Concentrated buffer solution is utilized to ensure that the pH of the samples is not altered by the addition of alkaline peroxynitrite. A 100-μl aliquot of the internal standard, 4-hydroxy-3-nitrobenzoic acid (100 μM, final concentration) in 0.2 M phosphate buffer, pH 7, is added to give a final volume of 1 ml. Samples are then analyzed by high-performance liquid chromatography (HPLC) using a porous graphite column.

Peroxynitrite is added to 0.2 M phosphate buffer, pH 7, and allowed to stand for 10 min at room temperature. Tyrosine is subsequently added to this solution, which now contains the degraded peroxynitrite. This sample is analyzed by HPLC to observe the effect of degraded peroxynitrite on tyrosine.

Tyrosine Nitration Assay. The peroxynitrite-scavenging activity of hydroxycinnamates and catechin polyphenolics is determined by their ability to reduce peroxynitrite-induced tyrosine nitration. A 50-μl aliquot of peroxynitrite (500 μM) is added to a solution containing tyrosine (100 μM) in the presence of various concentrations (0–100 μM) of hydroxycinnamates or catechin polyphenolics in 0.2 M phosphate buffer, pH 7. The concentrated buffer solution is utilized to ensure that the pH of the samples is not altered by the addition of alkaline peroxynitrite. A 100-μl aliquot of the internal standard, 4-hydroxy-3-nitrobenzoic acid (100 μM, final concentration) in 0.2 M phosphate buffer, pH 7, is added to give a final volume of 1 ml. Appropriate controls, without antioxidants and the degraded peroxynitrite, are also carried out to estimate levels of tyrosine nitration. The peroxynitrite-scavenging ability of hydroxycinnamates and catechin polyphenolics is expressed as the percentage decrease in 3-nitrotyrosine formation compared to control samples. Statistical analysis is determined by Student's paired and unpaired t test (Microsoft Excel); $p \leq 0.05$ is considered to be statistically significant.

HPLC System. HPLC analysis is carried out on the samples to estimate amounts of tyrosine and 3-nitrotyrosine. A Hewlett Packard Model 1090M-II HPLC system with an autoinjector, auto sampler, and diode array detector linked to a HP 900-300 data station is used. Aliquots of samples (100 μl) are injected onto a porous graphite column (Hypercarb column, 100 × 4.6 mm. I.D.; 5-μm particle size). The mobile phase, which consists of (a) 50 mM phosphate buffer, pH 7, and (b) acetonitrile, is pumped in the following gradient system (min/% MeCN): 0/10, 4/10, 14/60, 15/10, 20/10 at a flow rate of 1 ml/min. Tyrosine is monitored at 275 nm whereas 3-nitrotyrosine formation and the internal standard are monitored at 430 nm.

Calibration Curves and Validation of HPLC Analysis. The amount of unreacted tyrosine and 3-nitrotyrosine formed is determined from calibration plots constructed using authentic standards. 4-Hydroxy-3-nitrobenzoic acid (100 μM) is used as an internal standard. Calibration plots of 3-nitrotyrosine and tyrosine are constructed over the range of 0–10 μM (low calibration) and 0–100 μM (high calibration). Known concentrations of tyrosine and 3-nitrotyrosine are spiked in pH 7 phosphate buffer to which 100 μl of the internal standard is added. Peak area ratios (PAR) of tyrosine : internal standard and 3-nitrotyrosine : internal standard are plotted against the spiked concentration of both species. Linear behavior with correlation coefficient values ≥0.995 are obtained over the calibration range of 0–10 and 0–100 μM.

Interaction of Hydroxycinnamates with Peroxynitrite

In order to establish whether hydroxycinnamates can themselves undergo nitration, their interaction with peroxynitrite in the absence of tyrosine was also investigated. This investigation is carried out in two phases. Initially the reaction is monitored by UV/visible spectrophotometry where the changes in the spectral characteristics are noted. These changes indicate the possible reaction products. To complement this study the samples are then analyzed by HPLC in a system similar to the tyrosine nitration assay to identify the products formed.

Spectrophotometric Study of Hydroxycinnamate Interaction with Peroxynitrite. A 50-μl aliquot of peroxynitrite (4 mM) is added to a solution containing the corresponding hydroxycinnamates (50 μM) in 0.2 M phosphate buffer, pH 7, giving a final volume of 4 ml. Samples are then analyzed by spectrophotometry on a Hewlett-Packard 8453 spectrophotometer. Peroxynitrite, allowed to degrade for 10 min in phosphate buffer pH 7 at room temperature, is used as the blank. Spectra of control samples of the antioxidants without the addition of peroxynitrite are also obtained for comparative purposes.

HPLC Study of Hydroxycinnamate Interaction with Peroxynitrite. To characterize the nature of the products formed in the reaction between hydroxycinnamates and peroxynitrite, hydroxycinnamate samples are reacted with peroxynitrite and analyzed by HPLC. A 50-μl aliquot of peroxynitrite (10 mM, final concentration 500 μM) is added to a solution containing 50 μM hydroxycinnamate in pH 7 phosphate buffer (0.2 M), giving a final volume of 1 ml. Samples are then analyzed by HPLC using analytical conditions described for the tyrosine nitration assay. An isocratic system with 15% acetonitrile is used for all samples except for ferulic acid, which is analyzed by the isocratic system at 15% for 10 min followed by an increase to 20% in the next minute and maintaining for a further 10 min.

pH-Dependent Analysis of Suspected Nitrated Compounds. In order to help identify peroxynitrite-modified products, peaks obtained from HPLC analysis are fraction collected. Each fraction collected is analyzed by spectrophotometry at two pH values (3 and 7) to observe the changes due to deprotonation of the phenolate moiety (pK_a 6.7 for 3-nitrotyrosine) of nitrated derivatives at higher pH value. The sample is diluted twofold in 0.2 M phosphate buffer either at pH 3 or at pH 7 and is subsequently scanned between 200 and 600 nm using a Hewlett-Packard 8453 spectrophotometer. Mobile phase with the appropriate buffer is used as the blank.

Mass Spectrometric Analysis of Suspected Nitrated Compounds. Mass spectrometric analysis is carried out on suspected nitrated hydroxycinnamates as a confirmation of the results obtained from the pH-dependent spectrophotometric study. Products obtained from the reaction between *p*-coumaric acid/ferulic acid and peroxynitrite are chosen based on the amounts of the suspected nitrated species formed. Quinones, the suspected oxidation products of catechols, are not selected due to their ability to undergo polymerization and subsequent breakdown. *p*-Coumaric acid and ferulic acid are exposed to peroxynitrite. Samples are prepared as described earlier. Analysis is carried out by HPLC, and products are fraction collected. Fractions are subsequently freeze dried and resolubilized in 5 ml of methanol. The solvent is evaporated, and the sample is analyzed by mass spectrometry. A small quantity of the sample is dissolved in ethyl acetate, injected onto the mass spectrometer, and analyzed using the chemical ionization mode. A Finnigan mass spectrometer is utilized for this purpose.

Results

Extent of Tyrosine Nitration by Peroxynitrite

When exposed to peroxynitrite at pH 7, tyrosine undergoes nitration to form predominantly 3-nitrotyrosine. The identity of 3-nitrotyrosine was confirmed by comparison with an authentic standard. Both the retention time and spectroscopic properties of the peroxynitrite-generated product and authentic 3-nitrotyrosine were identical. 3,4-Dihydroxyphenylalanine (DOPA), a possible hydroxylation product of the reaction between tyrosine and peroxynitrite, was not detected at all concentrations of peroxynitrite investigated. Exposure of tyrosine (100 μM) to increasing concentrations of peroxynitrite (0–1000 μM) resulted in an increase in the production of 3-nitrotyrosine and a subsequent decrease in the levels of tyrosine (Fig. 3). The total recovery of tyrosine and 3-nitrotyrosine in each sample was close to 100% at low concentrations of peroxynitrite (10–50 μM), but decreased to 80 and 60% at 500 and 1000 μM of peroxynitrite, respectively. No additional chromatographic peaks were detected at these higher concentrations of peroxynitrite.

Inhibition of Tyrosine Nitration by Catechins and Their Gallate Esters

The ability of flavanol antioxidants to decrease peroxynitrite-mediated tyrosine nitration was determined. The catechin polyphenols were coincubated with tyrosine prior to the addition of 500 μM peroxynitrite followed by the quantification of 3-nitrotyrosine formation. Results (Figs. 4 and 5) indicate that all the catechin polyphenols tested were potent scavengers of peroxynitrite due to their ability to prevent the nitration of tyrosine. None of the compounds interfered with the HPLC analysis of tyrosine and 3-nitrotyrosine. All compounds tested had a greater ability to reduce nitration of tyrosine than Trolox, which was used as a standard for comparative purposes. At higher concentrations (50 and 100 μM) of the polyphenols, the reduction of tyrosine nitration was close to 100%. At lower concentra-

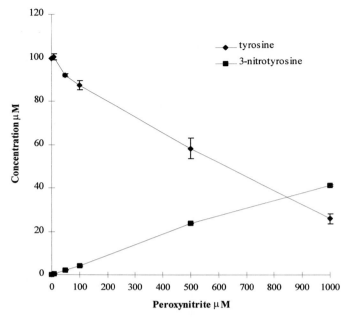

Fig. 3. Extent of tyrosine nitration at increasing concentrations of peroxynitrite. Unreacted and nitrated tyrosine were quantified as described in the methods section. Data points represent mean ± SD ($n = 3$).

tions (10 μM) the abilities of catechin polyphenols to minimize tyrosine nitration were ECG (38.1 ± 3.6%) ≥ EGCG (32.1 ± 7.5%) ≥ gallic acid (32.1 ± 1.9%) > catechin (23.9 ± 5.4%) ≥ epicatechin (22.9 ± 3.3%) ≥ EGC (19.9 ± 2.0%). Inhibition of tyrosine nitration by Trolox at 10 μM was 13.6 ± 2.9%.

Inhibition of Tyrosine Nitration by Hydroxycinnamates

The ability of hydroxycinnamates to decrease peroxynitrite-mediated tyrosine nitration was determined. Hydroxycinnamates were coincubated with tyrosine (100 μM final concentration) prior to the addition of peroxynitrite (500 μM final concentration), followed by the quantification of 3-nitrotyrosine formation. The results obtained (Figs. 6 and 7) indicate that the 3,4-disubstituted hydroxycinnamates were more potent inhibitors of peroxynitrite-mediated 3-nitrotyrosine formation compared to their monosubstituted counterparts. Products formed were distinct from tyrosine and 3-nitrotyrosine and were readily separable by HPLC. The disubstituted phenolics, caffeic, chlorogenic, and ferulic acids, were found to have a

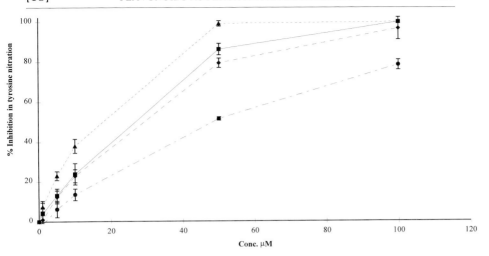

FIG. 4. Effect of catechin (■), epicatechin (◆), and ECG (▲) on peroxynitrite (500 μM)-mediated tyrosine (100 μM) nitration. Trolox (●) was used as the standard antioxidant for comparison. Data points represent mean ± SD ($n = 6$).

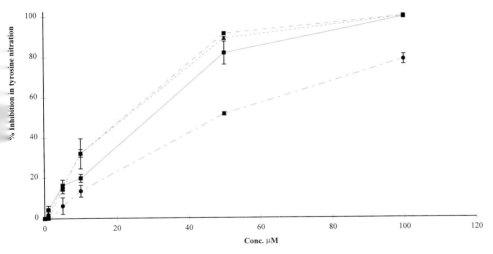

FIG. 5. Effect of EGC (■), EGCG (◆), and gallic acid (▲) on peroxynitrite (500 μM)-mediated tyrosine (100 μM) nitration. Trolox (●) was used as the standard antioxidant for comparison. Data points represent mean ± SD ($n = 6$).

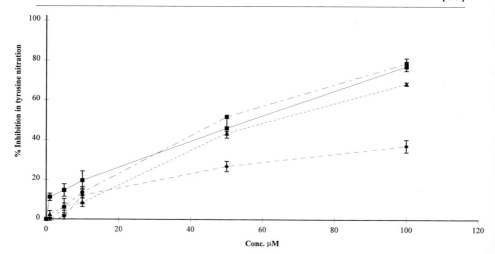

Fig. 6. Effect of *p*-coumaric acid (■), *m*-coumaric acid (♦), and *o*-coumaric acid (▲) on peroxynitrite (500 μM)-mediated tyrosine (100 μM) nitration. Trolox (●) was used as the standard antioxidant for comparison. Data points represent mean ± SD ($n = 6$).

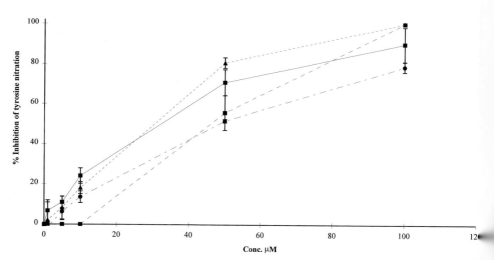

Fig. 7. Effect of chlorogenic acid (■), ferulic acid (♦), and caffeic acid (▲) on peroxynitrite (500 μM)-mediated tyrosine (100 μM) nitration. Trolox (●) was used as the standard antioxidant for comparison. Data points represent mean ± SD ($n = 6$).

greater ability to decrease the nitration of tyrosine compared to Trolox, which was used as a standard for comparative purposes, at all concentrations. Among the monosubstituted compounds, the extent of inhibition of tyrosine nitration by *p*- and *o*-coumaric acids was almost equal to that of Trolox, whereas *m*-coumaric acid was the least potent.

At equimolar concentrations of tyrosine to added phenolic (100 μM), an almost complete inhibition of tyrosine nitration was observed with caffeic acid, ferulic acid, and chlorogenic acid, whereas the extent of reduction by Trolox, *p*-coumaric acid, and *o*-coumaric acid was in the range of 68–78%. *m*-Coumaric acid only exhibited less than 40% activity under these conditions. At 50 μM antioxidant concentration, the abilities of hydroxycinnamates to minimize tyrosine nitration were caffeic acid (80.6 ± 2.6%) > chlorogenic acid (70.8 ± 6.5%) > ferulic acid (55.7 ± 8.6%) ≥ Trolox (51.7 ± 0.8%) > *p*-coumaric acid (45.9 ± 4.9%) > *o*-coumaric acid (43.5 ± 1.0%) > *m*-coumaric acid (26.8 ± 2.6%).

Changes in Spectral Characteristics of Hydroxycinnamates after Exposure to Peroxynitrite

In order to establish whether hydroxycinnamates can undergo nitration reactions, they were exposed to peroxynitrite in the absence of tyrosine. Spectrophotometric analysis of hydroxycinnamates after the addition of peroxynitrite revealed that there was a change in spectra of the samples in the visible region (Fig. 8). For ferulic, *p*-coumaric acid, *o*-coumaric, and, to a lesser extent, *m*-coumaric acid (Fig. 8), an increase in absorbance at approximately 430 nm was observed, suggesting the formation of a nitrated phenol. Although the catechol derivatives caffeic acid and chlorogenic acid exhibited spectral changes in the UV region when exposed to peroxynitrite, there was no specific change in the visible region, suggesting that nitration of the aromatic ring had not occurred (Fig. 8).

Analysis of Hydroxycinnamate and Peroxynitrite Reaction

Modification was observed in hydroxycinnamate spectra when exposed to peroxynitrite by spectrophotometric study. This was further complemented by HPLC analysis of the hydroxycinnamate samples exposed to peroxynitrite, which revealed that noncatechol hydroxycinnamates interact with peroxynitrite to yield products that differ from their catecholate counterparts. Analogous to the formation of 3-nitrotyrosine, it was observed that *p*-coumaric acid interacts with peroxynitrite to yield what appeared to be a single nitrated product (Fig. 9) with a retention time of 7.1 min. The identity of the product as a nitrated aromatic compound was confirmed by collection of the peak followed by spectral analysis at pH 3 and pH 7

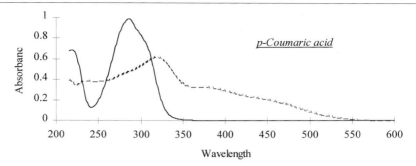

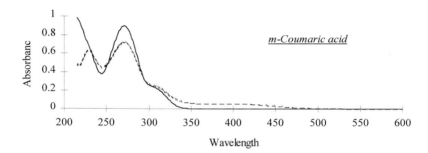

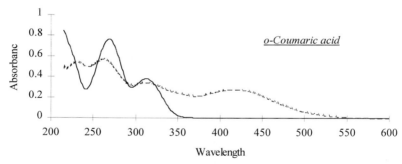

FIG. 8. Spectral changes in hydroxycinnamates (50 μM) on exposure to peroxynitrite (500 μM). —, spectra of the original sample; ---, spectra of samples after being exposed to peroxynitrite. [Reproduced with permission from A. S. Pannala et al., Free Radic. Biol. Med. Elsevier, Oxford, UK.[59]]

and also by mass spectral analysis. Figure 10 shows the characteristic shift in the spectrum of a nitrated species due to a change in pH. The presence of the nitro group on the aromatic ring dramatically increases the acidity of the phenolic hydroxyl group such that it exists predominantly as the

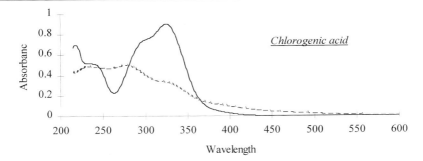

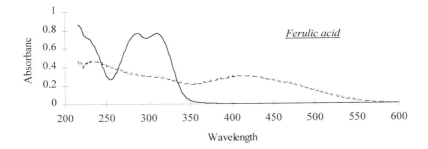

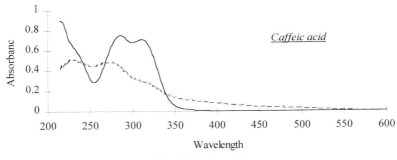

FIG. 8. (*continued*)

phenolate species at pH 7 (pK_a 6.7). Protonation of the phenolate at pH 3 leads to the spectral change observed in Fig. 10 with a reduction in absorbance at 430 nm. The identity of the peak was further confirmed as a nitrated species by mass spectral analysis. The mass spectrum (Fig. 11a) showed a molecular ion peak at 209 (M +) that corresponds to the mass of *p*-coumaric acid with the addition of a single nitro group. Loss of an

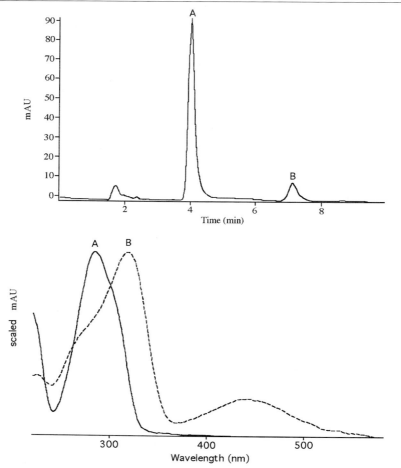

Fig. 9. Chromatogram and corresponding spectra of p-coumaric acid (50 μM) after exposure to peroxynitrite (500 μM). Peak A: p-coumaric acid; peak B: nitrated p-coumaric acid. [Reproduced with permission from A. S. Pannala et al., Free Radic. Biol. Med. Elsevier, Oxford, UK.[59]]

oxygen gave a peak at 193 and another peak at 179, which could possibly be the loss of a nitroso group. A further peak at 123, which corresponds to the loss of the hydroxyl group and the alkyl side chain, confirmed the identity of the compound as a nitro-p-coumaric acid.

Analysis of o-coumaric acid (Fig. 12) exposed to peroxynitrite revealed the formation of two products with retention times of 10.3 and 17 min,

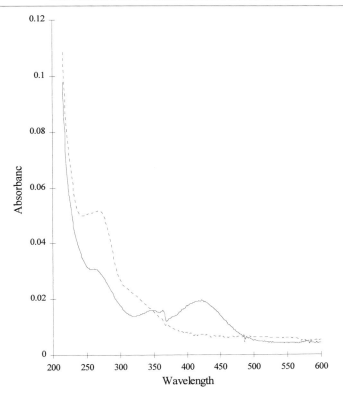

FIG. 10. Spectrophotometric analysis of the nitrated species observed by HPLC. Typical spectra of the nitrated sample at pH 7 (—) and 3 (---). Spectral change due to the protonation of the phenolate moiety (pK_a 6.7). [Reproduced with permission from A. S. Pannala et al., Free Radic. Biol. Med. Elsevier, Oxford, UK.[59]]

respectively. Spectra of these peaks displayed characteristics typical of a nitrated aromatic species. This was confirmed spectrophotometrically by the change in spectral shift, which results from protonation/deprotonation of the nitrophenolate species at pH 3/7, analogous to that of p-coumaric acid. The HPLC chromatogram of m-coumaric acid exposed to peroxynitrite (Fig. 13) showed the presence of an additional peak at 8 min. The identity of this peak was verified as a nitrated species in a similar manner to that of o- and p-coumaric acid. The chromatogram of ferulic acid exposed to peroxynitrite showed the formation of two major products (Fig. 14). The peak with a retention time of 17.5 min is believed to be a nitrated species. pH-dependent analysis of this peak suggested that it might be a nitrated species with the characteristic spectral change at low pH. Mass spectral

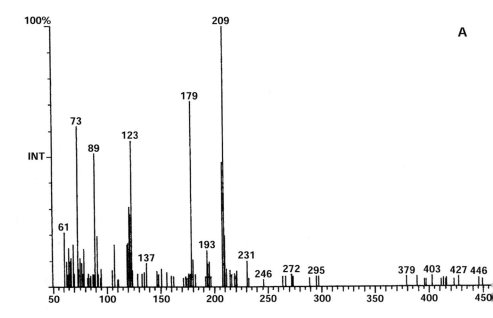

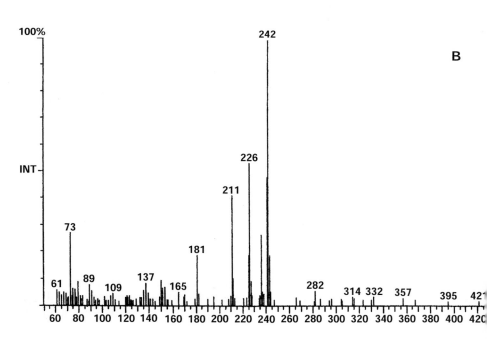

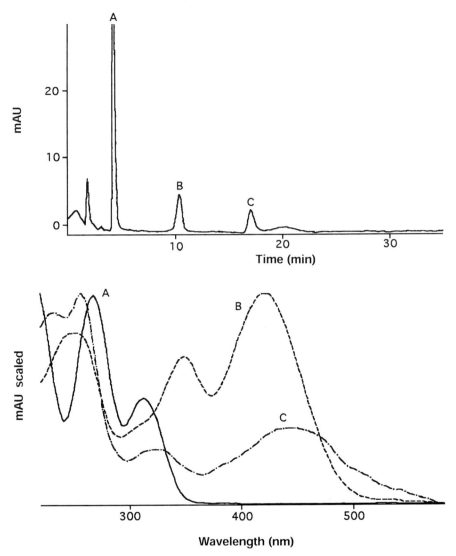

FIG. 12. Chromatogram and corresponding spectra of o-coumaric acid (50 μM) after exposure to peroxynitrite (500 μM). Peak A: o-coumaric acid; peaks B and C: nitrated o-coumaric acid. [Reproduced with permission from A. S. Pannala et al., Free Radic. Biol. Med. Elsevier, Oxford, UK.[59]]

FIG. 11. Mass spectral analysis of suspected (A) p-coumaric acid-nitrated species and (B) ferulic acid-nitrated species. [Reproduced with permission from A. S. Pannala et al., Free Radic. Biol. Med. Elsevier, Oxford, UK.[59]]

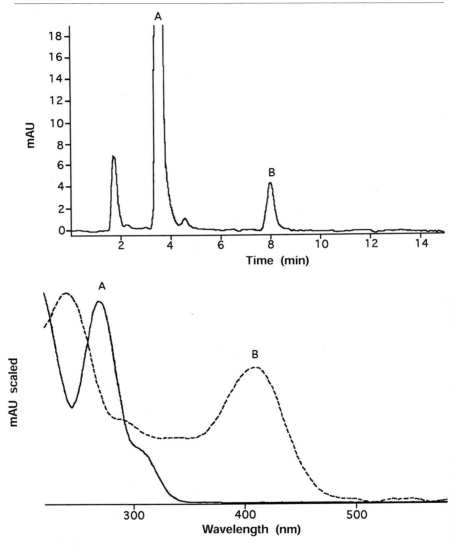

FIG. 13. Chromatogram and corresponding spectra of m-coumaric acid (50 μM) after exposure to peroxynitrite (500 μM). Peak A: m-coumaric acid; peak B: nitrated m-coumaric acid. [Reproduced with permission from A. S. Pannala et al., Free Radic. Biol. Med. Elsevier, Oxford, UK.[59]]

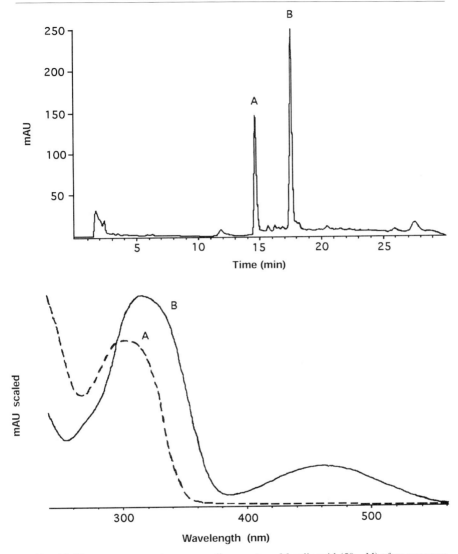

FIG. 14. Chromatogram and corresponding spectra of ferulic acid (50 μM) after exposure to peroxynitrite (500 μM). Peak A: ferulic acid-like species; peak B: nitro ferulic acid. The peak with a retention time of 11.8 min corresponds to that of unreacted ferulic acid. [Reproduced with permission from A. S. Pannala et al., Free Radic. Biol. Med. Elsevier, Oxford, UK.[59]]

analysis of this compound confirmed that it is indeed a nitrated species. The mass spectrum (Fig. 11b) showed a molecular ion peak at 242 (M + 3) that corresponds to the mass of ferulic acid with the addition of a single nitro group. Subsequent signals at 226 and 211, which correspond to the loss of an oxygen and a methoxyl group, provided further confirmation. The identity of the compound was confirmed to be mononitrated ferulic acid with the presence of an additional peak at 181 that corresponds to the loss of the carboxylic acid along with a methylene group. The peak with the retention time of 14.7 min has a spectrum similar to that of ferulic acid. The precise identity of this peak is unknown.

When exposed to peroxynitrite, the catecholate hydroxycinnamate, caffeic acid, gave rise to the formation of one major product with a retention time of 5.5 min (Fig. 15). This product, like most of the other peroxynitrite-modified hydroxycinnamate derivatives, absorbs in the visible region. However, the spectral profile of this product differs from that of other nitrated derivatives that do not show a marked difference in the visible region (Fig. 16). The small shift seen may be due to protonation/deprotonation of the carboxylate moiety (Fig. 16). Spectral data suggest that this product is not a nitrated species and could possibly be a quinone. An observation that supports this assumption is the relative instability of this derivative, which is characteristic of quinones.

Discussion

When exposed to peroxynitrite, tyrosine undergoes nitration to form 3-nitrotyrosine. It has been suggested that this reaction will generate the formation of hydroxylated (DOPA) and dimerized products (dityrosine) in addition to nitrated products.[31,34,39,41,42] However, from the reaction we have observed, the predominant product is 3-nitrotyrosine.[40] The total recovery of tyrosine and 3-nitrotyrosine from the reaction of tyrosine and peroxynitrite was close to 100% at low concentrations of peroxynitrite (10–50 μM), but decreased to 80% at 500 μM and to 60% at 1 mM peroxynitrite concentration. The remaining could possibly be accounted for by the formation of other oxidized products such as the hydroxylated product, DOPA, and the phenoxyl radical-coupled product, dityrosine. It has been reported that the formation of dityrosine is only to an extent of 2% of the total reaction products.[42] Hence, the assay measures the antioxidant activity of the compounds with respect to their ability to scavenge peroxynitrite in

---

[41] G. L. Squadrito and W. A. Pryor, *Am. J. Physiol.* **268**, L699 (1995).

[42] J. P. Eiserich, C. E. Cross, A. D. Jones, B. Halliwell, and A. Van der Vliet, *J. Biol. Chem.* **271**, 19199 (1996).

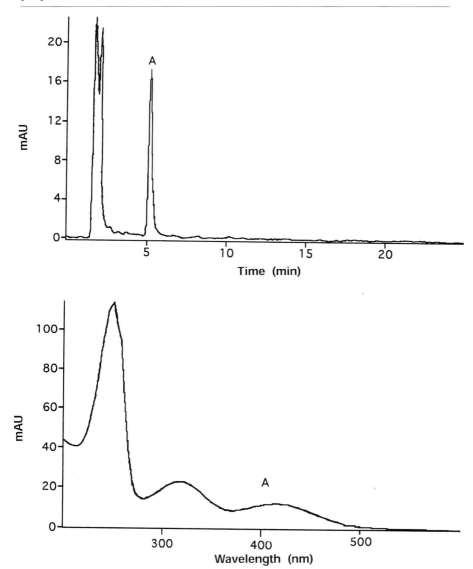

FIG. 15. Chromatogram and corresponding spectra of caffeic acid (50 μM) after exposure to peroxynitrite (500 μM). Peak A is believed to correspond to the quinone derivative of caffeic acid. [Reproduced with permission from A. S. Pannala et al., Free Radic. Biol. Med. Elsevier, Oxford, UK.[59]]

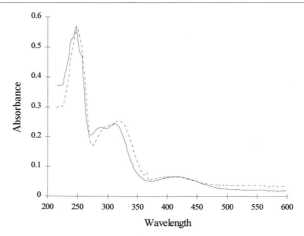

FIG. 16. Spectral analysis of the putative quinone at two different pH conditions: pH 3 (—) and 7 (---). No change was observed in the visible region, but a minor change can be seen in the UV region due to protonation of the carboxylic acid. [Reproduced with permission from A. S. Pannala et al., Free Radic. Biol. Med. Elsevier, Oxford, UK.[59]]

preventing nitration reactions. However, the underlying mechanism for peroxynitrite-mediated nitration is unclear. Nitration reactions can occur either via the nitronium ion (NO_2^+) or via the nitrogen dioxide radical ($NO_2 \cdot$). Peroxynitrite could undergo a homolytic fission to generate a nitrogen dioxide radical and the hydroxyl radical ($\cdot OH$) or a heterolytic fission to generate nitronium ion and the hydroxide ion (OH^-). Irrespective of the nitrating species, the end product 3-nitrotyrosine has been characterized.

Polyphenols are reported to be effective antioxidants. They have been reported to act as superoxide radical scavengers,[43–46] peroxyl radical scavengers,[7,47] inhibitors of lipid peroxidation,[48–50] scavengers of $ABTS^{\cdot+}$ radical[11] and metal chelators,[8,10] and as inhibitors of LDL oxidation induced by

[43] Y. Hanasaki, S. Ogawa, and S. Fukui, Free Radic. Biol. Med. **16,** 845 (1994).
[44] N. Cotelle, J. L. Bernier, J. P. Henichart, J. P. Catteau, E. Gaydou, and J. C. Wallet, Free Radic. Biol. Med. **13,** 211 (1992).
[45] C. Yuting, Z. Rongliang, J. Zhingjian, and J. Yong, Free Radic. Biol. Med. **9,** 19 (1990).
[46] Y. C. Zhou and R. L. Zheng, Biochem. Pharmacol. **42,** 1177 (1991).
[47] M. Erben-Russ, C. Michel, W. Bors, and M. Saran, J. Phys. Chem. **91,** 2362 (1987).
[48] C. G. Fraga, V. S. Martino, G. E. Ferraro, J. D. Coussio, and A. Boveris, Biochem. Pharmacol. **36,** 717 (1987).
[49] A. Negre-Salvayre, Y. Alomer, M. Troly, and R. Salvayre, Biochim. Biophys. Acta **1096,** 291 (1991).
[50] J. Terao, M. Piskula, and Q. Yao, Arch. Biochem. Biophys. **308,** 278 (1994).

redox active metals and macrophages.[51,52] The reduction potentials of polyphenolic antioxidants are lower than those of alkyl peroxyl radicals and the superoxide radical, possibly due to the extended cojugation and the number of hydroxyl groups on the ring system.[53] In addition, their abilities to chelate transition metal ions such as iron and copper can retard free radical production, thereby preventing any biological damage.[8] Polyphenols can also act as prooxidants as they can undergo oxidation by metal ions.[10] Other studies have shown the sequence of reactivities of the flavanols as H-donating antioxidants against the ABTS$^{·+}$ radical generated in the aqueous phase as ECG ≈ EGCG > EGC > epicatechin ≈ catechin,[11] suggesting that incorporation of the gallate moiety via an ester linkage at the 3-position enhances the antioxidant activity of the catechins, as does increasing the number of hydroxyl groups on the phenolic B ring. A similar pattern of reactivity is observed for the DPPH$^·$ radical and the superoxide radical[7]-scavenging activity. Theaflavins are another class of naturally occurring antioxidants, and incorporation of the gallate moiety via an ester linkage has reported to increase their antioxidant activity.[54]

The ability of catechin and their gallate esters to prevent peroxynitrite-induced tyrosine nitration was found to be the following order: ECG ≥ EGCG ≥ gallic acid > catechin ≥ epicatechin ≥ EGC > trolox. From this series of antioxidants tested, ECG and EGCG are the most effective inhibitors of peroxynitrite-mediated tyrosine nitration. The mechanism by which these compounds prevent tyrosine nitration could be one of two possible mechanisms: either competing directly with tyrosine for the nitrating species, thereby undergoing nitration themselves, or neutralizing the nitrating species by donating a pair of electrons. Nitration reactions on the catechin polyphenol can essentially occur at the 6- and 8-positions of the chromanol ring due to activation of these sites by the hydroxyl groups at 5- and 7-positions and the pyrone ring, the activation being more at the 6-position.

Electron donation, from the catechol side chain or the gallate moiety, could also be a mechanism by which catechin polyphenols act as peroxynitrite scavengers. When oxidized, catechols donate a pair of electrons forming the corresponding quinone (Scheme 2). The catechol side chain on the flavonoid can donate electrons to the nitrating species, thereby forming a quinone. Gallate esters of the catechins, ECG and EGCG, can donate

---

[51] H. Mangiapane, J. Thomson, A. Salter, S. Brown, G. P. Bell, and D. A. White, *Biochem. Pharmacol.* **43**, 445 (1992).
[52] C. De Whalley, S. M. Rankin, J. R. S. Hoult, W. Jessup, and D. Leake, *Biochem. Pharmacol.* **39**, 1743 (1990).
[53] S. V. Jovanovic, I. Jankovic, and L. Josimovic, *J. Am. Chem. Soc.* **114**, 9018 (1992).
[54] N. J. Miller, C. Castelluccio, L. Tijburg, and C. A. Rice-Evans, *FEBS Lett.* **392**, 40 (1996).

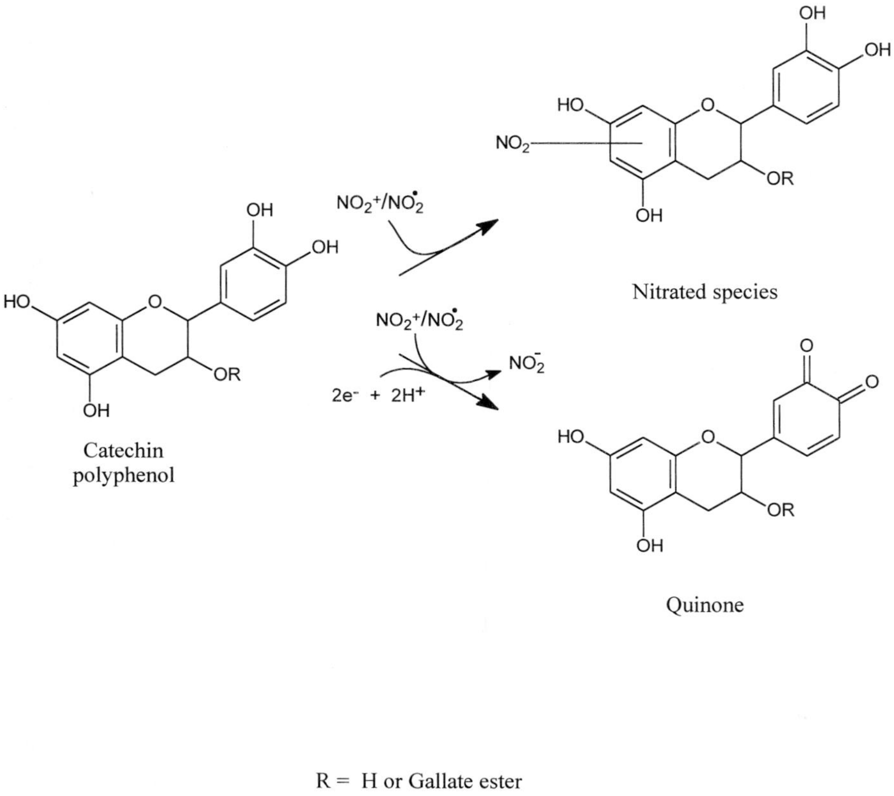

SCHEME 2. Mechanism of action of catechin and its gallate esters.

electrons from both the catechol and the gallic acid moiety to form the corresponding quinones. Hence it could possibly be the reason for their higher peroxynitrite-scavenging activity when compared to other catechin polyphenolics.

Among the hydroxycinnamate group, disubstituted compounds were found to be more effective scavengers of peroxynitrite than monosubstituted compounds. The ability of hydroxycinnamates to protect against damage by peroxynitrite, as detected by a decrease in the nitration of tyrosine, is in the following order: caffeic acid ≈ chlorogenic acid ≈ ferulic acid > Trolox > p-coumaric acid > o-coumaric acid > m-coumaric acid. It is possible that once again there are two possible mechanisms by which hydroxycinnamates can inhibit the peroxynitrite-mediated nitration of tyrosine. Monohydroxycinnamates (ferulic acid and the coumaric acids) can

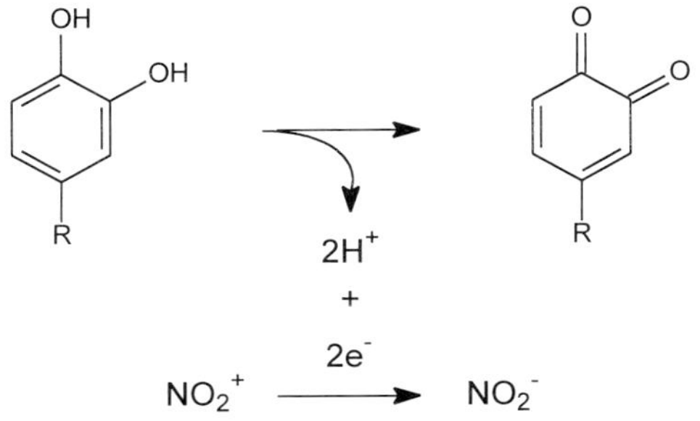

SCHEME 3. Suggested electron donation reaction for catecholates.

undergo nitration reactions and protect tyrosine from undergoing nitration. However, for caffeic and chlorogenic acids (both catecholates) the mechanism of action appears to be via electron donation to form the corresponding quinones. Electron transfer reactions involving monohydroxycinnamates and catecholates can, in principle, occur by either a single or a two-electron transfer, respectively. For example, a single electron transfer from a monohydroxycinnamate to the nitronium ion is predicted to generate nitrogen dioxide gas and the alkoxyl radical. Alternatively, when a two-electron transfer occurs from a catecholate to the nitronium ion, the corresponding quinone and the nitrite ion are likely to be formed (Scheme 3).

In lipid systems, phenolic compounds have been reported to act as antioxidants by virtue of their ability to donate hydrogen atoms, leading to the formation of phenoxyl radicals. It has also been reported that monophenolates (e.g., p-coumaric acid) are less effective than catecholates (e.g., caffeic acid) in protecting lipids against peroxidation.[28,55] Due to the presence of an aromatic ring, hydroxylated coumaric acid derivatives can undergo nitration when exposed to peroxynitrite. This was confirmed by spectrophotometry that revealed an increase in absorbance at approximately 430 nm when p-coumaric acid was exposed to peroxynitrite. This was further substantiated by HPLC, which showed an additional peak exhibiting a characteristic nitrated phenol spectrum. The presence of the hydroxyl group at the 4-position increases the likelihood of nitration occurring at the 3-position on the aromatic ring, analogous to the formation of 3-nitrotyrosine. This was further confirmed by mass spectral analysis where a peak for a mononitrated species was observed.

o-Coumaric acid, which has a hydroxyl group at the 2-position, can potentially undergo nitration at either the 3- or the 5-position on the aromatic ring. Figure 12 shows the chromatogram of o-coumaric acid exposed

---

[55] H. Chimi, J. Cillard, P. Cillard, and M. Rahmani, *J. Am. Oil Chemists Soc.* **68,** 307 (1991).

to peroxynitrite. The presence of two additional peaks, at 10.3 and 17 min, which showed characteristic nitrated phenol spectra, confirmed this hypothesis. *m*-Coumaric acid, which has a hydroxyl group at the 3-position, is predicted to mainly undergo nitration at the 4-position, as both the 2- and the 6-positions are sterically hindered. The limited ability of *m*-coumaric acid to undergo nitration and its predicted low antioxidant activity are reflected in the present study where it was found to be the least active compound. Spectral comparison indicated that there was only a small increase in absorbance in the visible region after *m*-coumaric acid was exposed to peroxynitrite. Analysis by HPLC revealed the formation of a nitrated species, but the extent of nitration was minimal.

Both the catechol derivatives, caffeic acid and chlorogenic acid, show maximal peroxynitrite-scavenging activity. From spectral data it was further observed that neither compound showed an appreciable change in the visible region, suggesting that nitration of these compounds does not occur. Analysis by HPLC of caffeic acid exposed to peroxynitrite revealed only the presence of what appeared to be a quinone. It is therefore likely that catecholate compounds act as antioxidants by donating electrons, possibly leading to the formation of nitrite and the corresponding quinones (Scheme 3). Trolox and α-tocopherol are also known to be oxidized by peroxynitrite to give the appropriate quinones in aqueous solutions.[56,57]

With its electron-donating methoxyl group at the 3-position, ferulic acid has a greater ability to stabilize the phenoxyl radical, accounting for its higher activity.[58] It can also undergo nitration reactions due to partial activation of the 5-position by the 4-hydroxy group. This was further confirmed by spectrophotometry demonstrating that a major change occurs in the appropriate position in the visible region and by HPLC that showed the formation of a peak consistent with nitration when ferulic acid is exposed to peroxynitrite. The increase in absorbance at approximately 430 nm suggests that ferulic acid has undergone nitration. This observation was further supported by mass spectral data that confirmed the identity of the compound to be a mononitrated ferulic acid species.

Conclusion

In summary, the present investigation has revealed that catechin polyphenols and hydroxycinnamates can inhibit peroxynitrite-induced nitration

---

[56] N. Hogg, V. M. Darley-Usmar, M. T. Wilson, and S. Moncada, *FEBS Lett.* **326,** 199 (1993).
[57] N. Hogg, J. Joseph, and B. Kalyanaraman, *Arch. Biochem. Biophys.* **314,** 153 (1994).
[58] E. Graf, *Free Radic. Biol. Med.* **13,** 435 (1992).
[59] A. S. Pannala, R. Razaq, B. Halliwell, S. Singh, and C. A. Rice-Evans, *Free Radic. Biol. Med.* **24,** 594 (1998).

of tyrosine. The scavenging of peroxynitrite occurs by two possible mechanisms: preferential nitration for monohydroxycinnamates and electron donation for catecholates.

Acknowledgments

We thank the Ministry of Agriculture, Fisheries and Food, United Kingdom (CR-E), and Arthritis and Rheumatism Council, United Kingdom (SS), for the financial support. The authors thank Dr. H. Khodr, King's College London, for measuring the pK_a values of 3-nitrotyrosine and Mr. D. Sahota and Mr. C. Walker, Drug Control Centre, United Kingdom, for the mass spectral analysis.

Section IV

Thiols

[20] Determination of Oxidized and Reduced Lipoic Acid Using High-Performance Liquid Chromatography and Coulometric Detection

By CHANDAN K. SEN, SASHWATI ROY, SAVITA KHANNA, and LESTER PACKER

Introduction

α-Lipoic acid, also known as thioctic acid, 1,2-dithiolane-3-pentanoic acid, 1,2-dithiolane-3-valeric acid, or 6,8-thioctic acid, has generated considerable clinical interest as a cellular thiol-replenishing and redox-modulating agent.[1-5] Biologically, lipoate exists as lipoamide in at least five proteins where it is linked covalently to a lysyl residue. Four of these proteins are found in the α-keto acid dehydrogenase complex, the pyruvate dehydrogenase complex, the branched chain keto acid complex, and the α-keto dehydrogenase complex. The fifth lipoamide moiety is in the glycine cleavage system.[6] The mitochondrial E3 enzyme, dihydrolipoyl dehydrogenase, reduces lipoate (LA) to dihydrolipoate (DHLA) at the expense of NADH. The enzyme shows a marked preference for the naturally occurring R-enantiomer of LA.[7] Lipoate is also a substrate for the NADPH-dependent enzyme glutathione reductase.[8] It has also been found that thioredoxin reductase from calf thymus and liver, human placenta, and rat liver reduces both LA and lipoamide efficiently.[9]

To develop an understanding of the mechanisms involved in the biological functions of supplemented LA, much of the current interest is focused on the fate of exogenous LA in cultured cells, as well as in animal and

---

[1] C. K. Sen, *J. Nutr. Biochem.* **8**, 660 (1997).
[2] C. K. Sen, *Biochem. Pharmacol.* **55**, (1998).
[3] C. K. Sen, S. Roy, and L. Packer, *in* "Therapeutic Potential of the Antioxidant and Redox Properties of α-Lipoic Acid" (L. Montagnier, R. Olivier, and C. Pasquier, eds.), p. 251. Dekker, New York (1997).
[4] L. Packer, S. Roy, and C. K. Sen, *Adv. Pharmacol.* **38**, 79 (1997).
[5] C. K. Sen and L. Packer, *FASEB J.* **10**, 709 (1996).
[6] K. Fujiwara, K. Okamura-Ikeda, and Y. Motokawa, *FEBS Lett.* **293**, 115 (1991).
[7] N. Haramaki, D. Han, G. J. Handelman, H. J. Tritschler, and L. Packer, *Free Radic. Biol. Med.* **22**, 535 (1997).
[8] U. Pick, N. Haramaki, A. Constantinescu, G. J. Handelman, H. J. Tritschler, and L. Packer, *Biochem. Biophys. Res. Commun.* **206**, 724 (1995).
[9] E. S. J. Arner, J. Nordberg, and A. Holmgren, *Biochem. Biophys. Res. Commun.* **225**, 268 (1996).

human tissues. Both LA and its reduced form DHLA have remarkable antioxidant properties.[10] Both of these compounds have been reported to regulate a number of critical cell functions such as biosynthesis of reduced glutathione (GSH), recycling of other oxidized antioxidants, regulation of cellular reducing equivalent homoeostasis, and regulation of agonist-induced activation of equivalent homeostasis, and regulation of agonist-induced activation of transcription factors, gene expression, and apoptosis.[11] To understand how LA and DHLA regulate various aspects of cell function, it is important to be able to detect LA and DHLA accurately in biological samples. Conventional high-performance liquid chromatography (HPLC) detection using ultraviolet or fluorescence detection methods are not feasible because both LA and DHLA lack adequate chromophores. Another detection method that has been used widely is based on gas chromatography and mass spectrometry. This method is highly sensitive; however, the major limitation of this method is that it cannot distinguish between oxidized and reduced forms of LA. The only method available to detect LA and DHLA reliably was reported previously from this laboratory using HPLC and electrochemical (EC) detection with dual Au|Hg electrodes.[12] The method was based on the reduction of disulfides in LA at electrode 1 followed by oxidation of thiols at the downstream electrode 2. To avoid oxidation of DHLA to LA during the measurement, oxygen must be excluded rigorously from the system. One of the major drawbacks of the reported assay method is that the Au|Hg electrodes lose sensitivity after a few (30–50) injections and must be reconditioned. This loss of sensitivity of the electrodes is caused mainly by the consumption of Hg at the electrode surface. Reaction of thiols with electrodes slowly depletes the Hg amalgam plating. In addition, the preparation of the Au|Hg amalgam electrode requires cautious handling as is usually recommended for work with mercury. This article reports on an improved, highly sensitive, and more convenient method for the detection of oxidized and reduced LA in biological samples using HPLC-EC coulometric detection.

Principles of HPLC-EC Coulometric Detection

Electrochemistry involves heterogeneous chemical reactions between a compound and an electrode in which an electron is transferred from the solution to the electrode, or vice versa, and a measurable current is formed as a result. For such oxidation–reduction reactions to occur, energy in the

[10] L. Packer, E. H. Witt, and H. J. Tritschler, *Free Radic. Biol. Med.* **19,** 227 (1995).
[11] S. Roy and L. Packer, *Biofactors* **7,** 263 (1998).
[12] D. Han, G. J. Handelman, and L. Packer, *Methods Enzymol.* **251,** 315 (1995).

form of an electric potential is required. In a traditional electrochemical detector, the potential is held constant (DC mode) and the current is measured as a function of time. When an electroactive species flows through the electrode, current is formed. The magnitude of this current is proportional to the concentration of the compound in solution on the electrode. Most electrochemical detectors for HPLC operate in the amperometric mode. In such a mode the solution of the compound only passes over a flatbed of a working electrode. Under such conditions only a small flat surface area of the electrode is available for interaction with the analyte. As a result, only 5–15% of the electroactive species is oxidized or reduced by the electrode. In contrast, coulometric detectors use flow-through or porous graphite electrodes. The surface area of such electrodes is large, allowing almost 100% of the analyte to react with it. Thus, the efficiency in coulometric detection is approximately 100% compared to the conventional amperometric detection that has only 5–15% detection efficiency.

HPLC Apparatus and Analytical Cells

The HPLC system consists of a ESA (ESA Inc., Chelmsford, MA) Model 580 solvent delivery module, a rheodyne injector (Cotati, CA), an on-line degassing system (Altech, Deerfield, IL), and a ESA Coulochem II multielectrode detector (Chelmsford, MA). The detector contains the following analytical cells: (i) a Model 5010 analytical cell (electrode 1), (ii) a Model 5011 high sensitivity analytical cell (electrode 2), and (iii) a Model 5020 guard cell, which is placed between the pump and the injector (ESA Inc., Coulochem II, Chelmsford, MA). The electrodes of the coulometric detector for LA/DHLA and GSH/glutathione disulfide (GSSG) assays are set as follows: electrode 1, 0.40 V; electrode 2, 0.85 V; and guard cell, 0.90 V. Data are collected using a PE Nelson 900 Series Interface and analyzed using software Turbochrom 3 (Perkin Elmer, San Jose, CA).

Column and Mobile Phases

A C_{18} column (150 mm long $\times$ 4.6 mm I.D., 5-μm pore size; Alltech, Deerfield, IL) is used for the separation of LA/DHLA and GSH/GSSG. The mobile phase for LA/DHLA separation consists of 50% of 50 mM NaH_2PO_4, (pH 2.7), 30% acetonitrile, and 20% methanol. For the separation of GSH/GSSG, the mobile phase consists of 98% 50 mM NaH_2PO_4 (pH 2.7) and 2% acetonitrile. For both assays, the flow rate is maintained at 1 ml/min throughout the analysis.

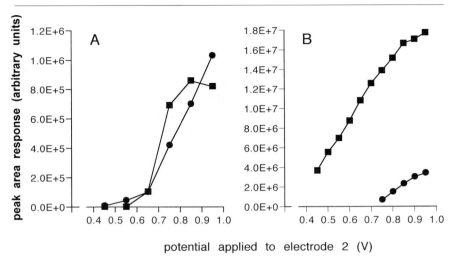

FIG. 1. Current–voltage response curves for lipoate (5 μM LA, ■) and dihydrolipoate (5 μM DHLA, ●) (A) and glutathione (12.5 μM GSH, ■) and glutathione disulfide (5 μM GSSG, ●) (B). As the potential applied to electrode 2 is increased, the peak area response increases for DHLA and GSH. For LA and GSSG the peak area response plateaus at about 0.8 and 0.9 V, respectively.

Current–Voltage Response Curve

The optimal use of an electrochemical detector for liquid chromatography requires knowledge of the appropriate potentials to drive the desired electrochemical reaction. This potential is dependent on a large number of factors, including the nature of the electrode surface, pH, composition of the mobile phase, and chemistry of the compound of interest. A plot of current generated (peak height) versus applied potential difference is commonly referred to as a hydrodynamic (HDV) voltammogram or a current–voltage $(C-V)$ curve. The optimum potential for the oxidation of LA, DHLA, GSH, and GSSG was determined by injecting the compounds onto the column and adjusting the potential difference across electrode 2 from 0.40 to 0.95 V (Fig. 1). The optimum potential for the quantitative measurement of LA/DHLA and GSH/GSSG is 0.85–0.9 V. This range of potential difference allows to obtain maximum peak area with minimal background in response to a given injection of analyte.

Standards and Standard Curve

Lipoate and DHLA standard compounds are from either ASTA-Medica (Frankfurt-am-Main, Germany) or Sigma Chemical Company (St. Louis,

MO). Dihydrolipoate is highly unstable and oxidizes readily at room temperature. To avoid oxidation, DHLA standards are prepared in the mobile phase (pH 2.7) of LA/DHLA and stored at $-80°$ or in liquid nitrogen. Concentrations of LA or DHLA versus the corresponding peak area response are plotted in Fig. 2. The linear range of both compounds is shown as inserts of Fig. 2. Injection of large amounts of the compounds may overwhelm the redox capacity of the electrode and may cause a deviation from the linear relationship between peak area and sample quantity. Repeated regression analysis of standards over a several day period result in high correlations ($R = 0.99$), indicating low variability of analysis in different days. Figure 3 shows standard curves for GSH and GSSG assays. The HPLC-EC coulometric method for the detection of GSH and GSSG has been reported.[13] Concentrations of GSH and GSSG used to prepare the standard curve were kept in a range that matches the concentration of these compounds in biological cell samples (described later).

Detection of LA/DHLA and GSH/GSSG in Biological Samples

To demonstrate that the method is applicable for biological samples, LA/DHLA (Fig. 4) and GSH/GSSG (Fig. 5) were analyzed from human T lymphocyte and rat skeletal muscle-derived L6 cells.

Cells and Cell Culture

Human Jurkat T cells clone E6-1 [American Type Culture Collection (ATCC), Rockville, MD] are grown in RPMI 1640 medium (GIBCO-BRL, Life Technologies, Gaithersburg, MD) supplemented with 10% fetal calf serum (FCS), 100 U/ml penicillin/100 μg/ml streptomycin, 110 mg/liter sodium pyruvate, and 2 mM L-glutamine (University of California, San Francisco, CA). Rat skeletal muscle-derived L6 myoblasts are also from ATCC. For experiments, cells are seeded at a concentration of 0.5×10^6 cells per well in a 6-well, flat-bottom tissue culture-treated polystyrene plate (Falcon, Becton-Dickinson Labware, NJ). Cultures are grown in Dulbecco's modified Eagle's medium (DMEM, GIBCO-BRL) supplemented with 10% FCS, 5 mM glutamine, 0.3% D-glucose, 50 U/ml of penicillin, and 50 μg/ml of streptomycin. Cells are maintained in a standard culture incubator with humidified air containing 5% (v/v) CO_2 at $37°$.

Extraction of Thiol/Disulfides from Biological Samples

For HPLC analyses, cells are pelleted (400g, 5 min) at $4°$, and the pellet and medium are deproteinized separately by treatment with 4% (final

[13] J. Lakritz, C. G. Plopper, and A. R. Buckpitt, *Anal. Biochem.* **247,** 63 (1997).

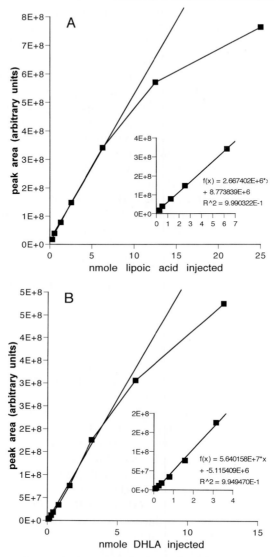

FIG. 2. Representative standard curves for lipoic acid (LA) (A) and dihydrolipoate (DHLA) (B). At high concentrations the peak area response is not linear to the amount of LA or DHLA injected. (Inset) The working range where concentration to the peak area response is linear. The conditions of chromatography are described in the text. Guard cell, 950 mV; screening electrode E1, 400 mV; analytical electrode E2, 850 mV.

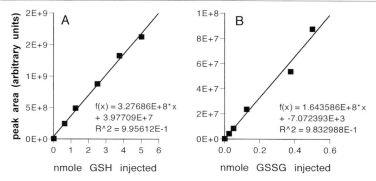

FIG. 3. Representative standard curve for reduced glutathione (GSH) (A) and glutathione disulfide (GSSG) (B). The concentration range plotted for GSH and GSSG was selected to match concentrations that are expected in biological samples. The conditions of chromatography are described in the text. Guard cell, 950 mV; screening electrode E1, 400 mV; analytical electrode E2, 850 mV.

concentration) monochloroacetic acid (Fisher Scientific, Springfield, NJ) for GSH and GSSG measurements. For the measurement of LA and DHLA the cell pellet and supernatant culture medium are deproteinized by treatment with the LA/DHLA HPLC mobile phase, the composition of which is described in a previous section. Following deproteinization, the mixtures

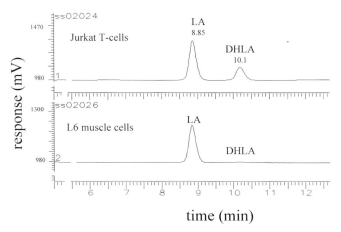

FIG. 4. Reduction of lipoic acid (LA) to dihydrolipoate (DHLA) by human Jurkat T cells grown in suspension culture and rat L6 myoblasts grown in monolayer. Cells were treated with 0.5 mM LA for 30 min. LA and DHLA were detected in cells using the HPLC coulometric detector as described in the text. Individual LA and DHLA peaks in each chromatogram are labeled.

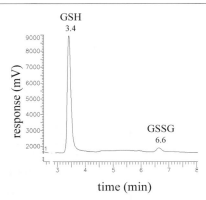

FIG. 5. Chromatogram of glutathione (GSH) and glutathione disulfide (GSSG) detected in a Jurkat T-cell extract. Following acid extraction of cells, GSH and GSSG were detected using the HPLC coulometric detector as described in the text. Individual GSH and GSSG peaks are labeled.

are snap-frozen in liquid nitrogen and stored at −80° for HPLC determinations. The HPLC assay is done within 1 week of storage. Immediately before the assay, samples are thawed, vortexed, and then centrifuged at 15,000g for 2 min. The clear supernatant is filtered using microfilterfuge tubes (Rainin, Woburn, MA) fitted with a 0.45-μm nylon filter and used for injection.

Cellular Reduction of LA to DHLA

Exogenous free LA added to Jurkat T cells in culture is taken up rapidly by the cells and reduced to DHLA.[12,14] Figure 4 illustrates the presence of DHLA in human Jurkat T cells and traces of it in L6 skeletal muscle cells that were treated with 0.5 mM of LA for 30 min. No DHLA is detected in the cells immediately after the addition of LA (not shown), suggesting that the DHLA detected after 30 min of treatment is formed in the cells. The reduction of LA to DHLA is more marked in Jurkat T cells compared to that in L6 cells where only trace amounts of DHLA could be detected.

In summary, the HPLC-EC coulometric detection method described here sensitively detects both oxidized and reduced species of lipoate (Fig. 4) and glutathione (Fig. 5) from biological samples. Detection limits for LA and DHLA for this HPLC method are 1–5 pmol, which is at least 10 times more sensitive for the detection of LA compared to the Au|Hg electrode-based detection reported previously.

[14] G. J. Handelman, D. Han, H. Tritschler, and L. Packer, *Biochem. Pharmacol.* **47,** 1725 (1994).

[21] Flow Cytometric Determination of Cellular Thiols

By CHANDAN K. SEN, SASHWATI ROY, and LESTER PACKER

Introduction

The oxidation–reduction (redox) state of cellular thiols plays a central role in antioxidant defense and in the regulation of a large number of signal transduction pathways.[1–4] Vicinal dithiols have been identified to serve key cellular signaling and metabolic functions.[5] Redox signaling is implicated in the modulation of a number of signal transduction processes that are known to be associated with the pathogeneses of human disease. Oxidative stress results in the oxidation of intracellular thiols to disulfides. For example, in oxidative stress situations, cellular glutathione (GSH) is oxidized rapidly to glutathione disulfide (GSSG). When produced in excess, all of the intracellular GSSG cannot be recycled to GSH by NADPH-dependent glutathione reductase activity, and GSSG is effluxed from oxidatively challenged cells. Such efflux lowers cellular GSH levels. The intracellular redox state and amount of GSH in the cell may be influenced by a large number of factors.[6]

Flow cytometric determination of cellular thiols is a powerful tool with considerable clinical potential. In contrast to conventional biochemical assays where cell extracts are studied to obtain mean results, flow cytometry allows the collection of results from individual cells. As a result, cells may be immunostained on the surface for any specific marker and information may be obtained for a subpopulation of cells. For example, using appropriate antibodies, results can be obtained from $CD4^+$ and $CD4^-$ subpopulations of a lymphocyte population of cells.[7] In such cases, appropriate fluorophores should be selected and the instrument should be properly set up to ensure that the compensation is well adjusted such that emission from one fluorophore does not interfere with the emission of the other.

The heterocyclic molecule bimanes or 1,5-diazabicyclo[3.3.0]octadienediones react preferentially with thiol groups.[8] Halo (e.g., chloro or

---

[1] C. K. Sen and L. Packer, *FASEB J.* **10,** 709 (1996).
[2] J. M. Muller, M. R. A. Rupec, and P. A. Baeuerle, *Methods* **11,** 301 (1997).
[3] C. K. Sen, *Biochem. Pharmacol.* **55,** 1747 (1998).
[4] C. K. Sen, *Curr. Top. Cell Regul.* **36,** in press.
[5] C. Gitler, M. Mogyoros, and E. Kalef, *Methods Enzymol.* **233,** 403 (1994).
[6] C. K. Sen, *J. Nutr. Biochem.* **8,** 660 (1997).
[7] C. K. Sen, S. Roy, D. Han, and L. Packer, *Free Radic. Biol. Med.* **22,** 1241 (1997).
[8] E. M. Kosower and N. S. Kosower, *Methods Enzymol.* **251,** 133 (1995).

bromo) derivatives of bimane are nonfluorescent in their native forms but emit strong fluorescence when reacted with thiols. This property of bimanes has been utilized to establish several types of thiol-detecting methods based on high-performance liquid chromatography (HPLC),[9] electrophoresis,[10] and flow cytometry.[11] Monochlorobimane (MCB) has been used commonly in clinical studies for the detection of cellular GSH.[12-15] The reaction of MCB with GSH requires glutathione S-transferase activity. Because of a low affinity of human glutathione S-transferase isoenzymes for MCB, this thiol probe has been evaluated not to be ideal for human cell GSH measurements.[16-18] Thus, the focus has been shifted to probes that would not be dependent on the activity of glutathione S-transferase for their thiol reactivity. Replacement of the Cl atom of MCB by Br renders the bimane much more chemically reactive than MCB. Monobromobimane (MBB) specifically labels both protein and nonprotein thiols by a glutathione S-transferase-independent mechanism. The reaction of bromobimanes with thiols are second order and are dependent on the pH, the active nucleophile being the thiolate anion such as GS^-. The reaction of bromobimane with a thiolate converts the nonfluorescent agent into water-soluble fluorescent products[8] that can be detected flow cytometrically. Two other properties of the thiol-reacted MBB fluorophore have made it a quantitative analytical tool that has been used for the electrophoretic analysis of proteins. First, the quantum yield of each fluorescent adduct is identical and independent of the protein species. Second, MBB labels a known number of cysteine residues for each protein.[10]

Preparation of Cells

This technique is most suited for cells that grow in suspension. This article describes the measurement of cellular thiols from cultured Jurkat T cells and human peripheral blood lymphocytes.

[9] R. C. Fahey, G. L. Newton, R. Dorian, and E. M. Kosower, *Anal. Biochem.* **111,** 357 (1981).
[10] D. O. O'Keefe, *Anal. Biochem.* **222,** 86 (1994).
[11] D. W. Hedley and S. Chow, *Methods Cell Biol.* **42,** 31 (1994).
[12] F. J. T. Staal, M. Roederer, L. A. Herzenberg, and L. A. Herzenberg, *Proc. Natl. Acad. Sci. U.S.A.* **87,** 9943 (1990).
[13] M. Roederer, F. J. T. Staal, M. Anderson, R. Rabin, P. A. Raju, L. A. Herzenberg, and L. A. Herzenberg, *Ann. N.Y. Acad. Sci.* **677,** 113 (1993).
[14] F. J. T. Staal, S. W. Ela, M. Roederer, M. T. Anderson, and L. A. Herzenberg, *Lancet* **339,** 909 (1992).
[15] L. A. Herzenberg, S. C. De Rosa, J. Gregson Dubs, M. Roederer, M. T. Anderson, S. W. Ela, S. C. Deresinski, and L. A. Herzenberg, *Proc. Natl. Acad. Sci. U.S.A.* **94,** 1967 (1997).
[16] J. A. Cook, S. N. Iype, and J. B. Mitchell, *Cancer Res.* **51,** 1696 (1991).
[17] D. W. Hedley and S. Chow, *Cytometry* **15,** 349 (1994).
[18] G. A. Ublacker, J. A. Johnson, F. L. Siegel, and R. T. Mulcahy, *Cancer Res.* **51,** 1783 (1991).

Jurkat T Cells

Human Jurkat T cells were obtained from American Type Culture Collection (ATCC, Rockville, MD). Wurzburg T cells, a subclone of the Jurkat T cells, was a kind gift of Dr. L. A. Herzenberg of Stanford University, California.[19] Both types of cell are grown in RPMI 1640 medium containing 10% fetal calf serum (FCS), 1% (w/v) penicillin/streptomycin, 110 mg/liter sodium pyruvate, and 2 mM L-glutamine (University of California, San Francisco) in humidified air containing 5% (v/v) CO_2 at 37°.

Human Peripheral Blood Lymphocytes

Blood drawn from an antecubital vein is collected in heparinized tubes. Peripheral blood lymphocytes (PBL) are isolated by a standard density gradient separation on Ficoll–Hypaque (Pharmacia, Sweden). Peripheral blood lymphocytes are seeded at 1×10^6/ml in RPMI 1640 culture medium containing 10% heat-inactivated FCS and maintained in humidified air containing 5% CO_2 at 37°. Peripheral blood lymphocytes isolated and seeded as just described are stimulated mitogenically with phytohemagglutinin P (2 μg/ml; grade B, Pharmacia, Sweden). Four hours after such stimulation, PBL are either treated or not (as mentioned in the respective figure legend) with supplements. The supplement treatment time for PBL is 48 hr.

Differential Assessment of Cellular Thiols

As is clear from the principle of the reaction between MBB and thiols, the interaction is not specific for any particular cellular thiol. In the assay system reported here, the specificity of fluorescence signals for distinct thiol pools was ensured by using defined cellular thiol regulatory agents such as buthionine sulfoximine, phenylarsine oxide, and N-ethylmaleimide. This method exploits the ability of buthionine sulfoximine to selectively deplete cellular GSH, phenylarsine oxide to block vicinal dithiols, and N-ethylmaleimide to block all cellular free thiols. The differential assessment of MBB fluorescence of cells indicates that only around half of the total emission signal is GSH dependent (Fig. 1). The difference of results obtained from GSH-adequate cells and GSH-depleted cells provides a specific estimation of the intracellular GSH level.

Cells in culture are resuspended (10^6/ml) in the standard culture medium as described earlier and seeded in 24-well plates at 10^6 cells/well. For each treatment group of cells, four separate wells containing 1 ml of cell suspension are prepared. Cells from each of the four wells of a treatment

---

[19] C. K. Sen, S. Roy, and L. Packer, *FEBS Lett.* **385,** 58 (1996).

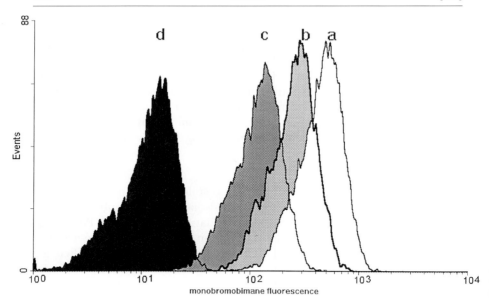

FIG. 1. Differential estimation of cellular thiols. Jurkat T cells were treated with 40 μM monobromobimane (MBB) for 15 min and then fluorescent emission from UV excited cells was collected. In this histogram, the y axis represents the relative cell number and the x axis represents the intensity of thiol-reacted bimane fluorescence in a four-decade log scale. (a) total fluorescence, (b) buthionine sulfoximine (BSO)-treated cells, (c) phenylarsine oxide (PAO)-treated cells, and (d) N-ethylmaleimide (NEM)-treated cells. a minus b estimates intracellular GSH, a minus c estimates intracellular vicinal dithiols, and a minus d estimates total cellular thiols. Analysis of data showed that almost all (97%) of the MBB emission was sensitive to the treatment of NEM. PAO and BSO treatment of cells blocked 73 and 44% of the total emission from MBB-reacted cells, respectively. Reprinted with permission from C. K. Sen et al., Free Radic. Biol. Med. **22,** 1241 (1997).

group are pelleted (400g for 5 min) and resuspended (10^6 cells/ml) in phosphate-buffered saline (PBS) at room temperature. Each treatment group is subjected to the following four types of treatment: (a) the control is not treated with any thiol regulatory agent so that the total emission from these cells can be read and (b) GSH-depleted cells are treated with 150 μM buthionine sulfoximine (BSO; aqueous solution) for 18 hr in culture. Pretreatment of cells with BSO decreased intracellular GSH level by ∼95% as estimated by HPLC electrochemical detection (not shown). The concentration of BSO and time required to obtain maximum depletion of intracellular GSH without loss of viability should be optimized for each cell type. (c) Vicinal dithiol-blocked cells are treated with 20 μM phenylarsine oxide [PAO; in dimethyl sulfoxide (DMSO)] for 10 min before treatment of cells with bimane and (d) total thiol-blocked cells are treated with 250 μM

N-ethylmaleimide (NEM; aqueous solution) for 10 min before bimane treatment.

Monobromobimane Loading

Monobromobimane (Molecular Probes, Eugene, OR) is dissolved in acetonitrile to obtain an 8 mM stock concentration. Cells are pelleted (400g, at room temperature for 5 min) and resuspended in PBS (pH 7.4) at 10^6/ml and, if required, treated with PAO or NEM for 10 min as described in the previous section. Following this interval, the MBB stock solution is added to the cell suspension at room temperature such that the final concentration of the bimane reagent is 40 μM. Bimane-loaded cells are analyzed using a flow cytometer.

Flow Cytometry

Bimane-loaded cells are excited using a 20-mW powered UV line of a Innova 90-4 argon ion laser (Coherent, Palo Alto, CA) set of 350 nm in a EPICS Elite (Coulter, Miami, FL) flow cytometer. Fluorescent emission from cellular sulfhydryl-reacted bimane is recorded using cytometer settings, and filter arrangements as shown in Fig. 2. A forward scatter/side-

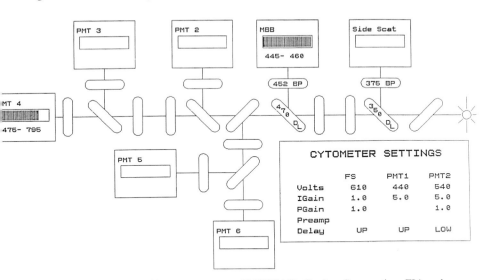

FIG. 2. Flow cytometer filter arrangements (OPTICAD, Coulter Corporation, FL) and a typical example of cytometer settings used for the measurement of emission from monobromobimane (MBB)-reacted cells. A 20-mW powered UV line of a Innova 90-4 argon ion laser (Coherent, Palo Alto, CA) set at 350 nm was used for excitation. The open circle at the right represents the laser source. PMT, photomultiplier tube. Generated using the standard Coulter Elite software version 4.02 (Coulter Corporation).

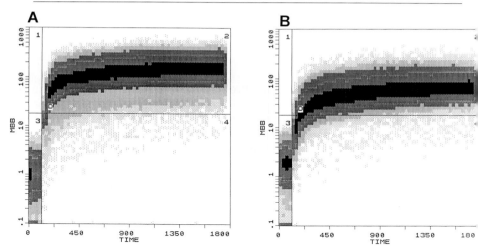

FIG. 3. Kinetics of monobromobimane (MBB) fluorescence development in Jurkat T cells. Autofluorescence (see quadrant D3, bottom left) was collected for 2 min from cells (10^6/ml) suspended in phosphate-buffered saline (pH 7.4). After this, monobromobimane (40 μM) was added to the cell suspension, and the kinetics of fluorescence development was followed for a total of 30 min. Time plotted on the x axes is represented in seconds. The y axes represent the MBB fluorescence plotted in log scale. Data were collected at a flow rate of 300–400 cells/sec for 1800 sec. (A) Control nontreated cells and (B) GSH-depleted cells (treated with 150 μM buthionine sulfoximine for 18 hr).

scatter dot plot is obtained from which a morophometrically homogeneous viable cell population is gated. Data are collected from at least 10,000 cells at a flow rate 250–300 cells/sec.

Multiparameter flow cytometric analyses may be performed with surface immunostained cells using fluorochrome-coupled monoclonal antibodies. For example, T cells may be immunotyped for the presence or absence of CD3 (mature T cells), CD4 (helper T cells), and CD8 (cytotoxic suppressor T cells), and thiol data can be collected from a subpopulation of cells split on the basis of their immunotyping.[7] Precaution should be taken to ensure that the antibodies used for the staining process do not influence cellular thiols. For example, treatment of T cells with certain antibodies may result in GSH efflux from the cell. This must be avoided or controlled for.

Fluorescence Emission from Monobromobimane-Treated Cells

Emission from UV laser-excited cells that were loaded with MBB was followed for 30 min. Mean fluorescence was collected continuously from 300 to 400 cells/sec. The kinetics of increase in fluorescent emission is illustrated in Fig. 3. The sensitivity of the emission to the different thiol regulatory agents is shown in Fig. 1. Almost all (97%) of the emission from MBB-reacted cells was quenched when thiols were blocked by 250 μM NEM treatment for 10 min before bimane treatment. This suggests that almost the entire emission originated from the reaction of MBB with thiols and that the thiol nonspecific response is minimal. Analysis of data obtained from Jurkat T cells showed that 44% of the total emission is contributed by cellular GSH and that vicinal dithiols in the cell account for 73% of the total emission. Based on these results, the different pools of intracellular thiol may be quantitated as indicated in the legend of Fig. 1.

Low Concentration of α-Lipoic Acid Increases Cellular Thiols

From concentrations as low as 10 μM, α-lipoic acid is effective in increasing the cellular GSH level. A dose-dependent effect of α-lipoic acid in the concentration range of 10–125 μM was observed (Fig. 4). Using HPLC electrochemical detection we observed that 10 μM lipoate resulted in a 16 ±13% increase of Jurkat T cell GSH (not shown). We have observed that in cultured cells there is almost always a small population (up to 10%) of cells that are either nonviable or smaller in size compared to the main population (>90% of all cells). If these cells are studied jointly with the main population of cells, the obtained results may be expected to have a

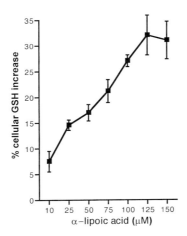

FIG. 4. Flow cytometric determination of the effect of α-lipoic acid on cellular GSH content. Intracellular GSH was estimated from monobromobimane-reacted Jurkat T cells as described in Fig. 1. Cells were treated with a racemic mixture of α-lipoic acid for 18 hr either in the presence or in the absence of 150 μM of buthionine sulfoximine. Results are expressed as percentage change compared to respective α-lipoic acid nontreated controls. Data are mean ± SD.

high standard deviation. Elimination of these cells by gating of the main population allows more consistent results to be obtained. During biochemical measurements, these outlying cells contribute to total protein measurements, but because their thiol status is much lower than that of the main population, the net result suffers from high standard deviation. For results that are expressed on a per cell basis it should be noted that manual cell-counting efficiency is certainly much less than the precise flow cytometric estimation. The effect of short-term (18 hr) and long-term (16 weeks) α-lipoic acid treatment on different intracellular thiol pools is illustrated in Fig. 5. Results show consistently that the effect of α-lipoic acid on intracellular thiols is more pronounced in response to chronic treatment compared to overnight treatment.

The lymphocyte population isolated from human peripheral blood may consist of a heterogeneous mixture of cells as indicated by the forward/side scatter dot plot (Fig. 6a). Typically, this mixture consists predominantly of lymphocytes, but may also contain some other cell types. The homogeneous lymphocyte population is gated as shown in Fig. 6a. That this cell population actually consists of lymphocytes may be verified separately by immunostaining. Monobromobimane emission detected from this gated cell population showed that the thiol content of the majority (>90%) of nontreated resting cells spanned across the M3 domain, indicating that the majority of the population had thiols ranging over the upper 90 percentile. One small subpopulation was clearly separated from the majority population with respect to thiol content. The thiol content of these cells in the M1 domain was in the lowest 1 percentile range. It would be predicted that this cell population should be more vulnerable to oxidative stress and be functionally impaired. α-Lipoic acid, lipoamide, or N-acetyl-L-cysteine (NAC) treatment corrected the thiol pool of this subpopulation. Treatment of lymphocytes with excess glutamate decreased the thiol pool of the main cell population[7,20,21] and also resulted in a shift of cells from the regular thiol M3 domain to the low thiol M1 domain population. α-Lipoic acid, lipoamide, or NAC prevented such a shift. This effect should be of remarkable significance with respect to improving the physiological functioning of cells that are thiol deficient under physiological or pathological conditions.[15]

Summary

Several biochemical techniques are based on chromatography or electrophoresis for the determination of thiols from biological samples. These

[20] C. K. Sen, S. Roy, and L. Packer, in "Biological Oxidants and Antioxidants: Molecular Mechanisms and Health Effects" (L. Packer and A. S. H. Ong, eds.), 5, 13. AOCS Press, Champaign, 1998.

[21] D. Han, C. K. Sen, S. Roy, M. S. Kabayashi, and L. Packer, Am. J. Physiol. 273, R1771 (1997).

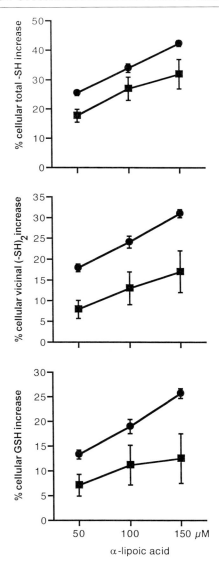

FIG. 5. Effect of treatment duration on α-lipoic acid-dependent changes in different pools of intracellular thiols. Wurzburg cells were maintained in culture medium containing 0 (control), 50, 100, or 150 μM α-lipoic acid for 18 hr (■) or 16 weeks (●). Cellular total protein sulfhydryls (top), vicinal dithiols (middle), and GSH (bottom) were estimated from monobromobimane-treated cells as described in Fig. 1. Results are expressed as percentage change compared to respective α-lipoic acid nontreated controls. Data are mean ± SD.

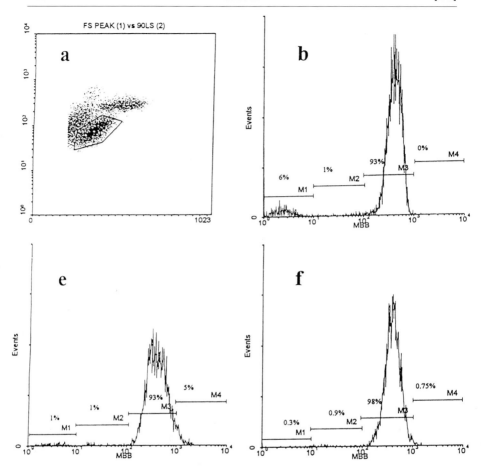

FIG. 6. Effect of lipoate, lipoamide, and N-acetylcysteine on human peripheral blood lymphocyte subpopulations. (a) Forward scatter/side scatter dot plot profile of PBL after 48 hr of mitogen-activated culture. A morphometrically homogeneous population of the predominant type of cells was gated for the study of cellular thiols. For the rest of the histograms (b–h) the x axis is a four-decade-log scale representing total monobromobimane fluorescent emission (estimates total cellular thiols, see Fig. 1). Each decade of the log scale is marked as follows: $10^0 \rightarrow 10^1 = $ M1; $10^1 \rightarrow 10^2 = $ M2; $10^2 \rightarrow 10^3 = $ M3; $10^3 \rightarrow 10^4 = $ M4. The fraction of the total population of cells present in each of the four marked domains is annotated just adjacent to the corresponding zone. The y axis represents the relative cell number (events) in each zone of the x axis, (b) nontreated control, (c) 50 μM lipoate treated, (d) 100 μM lipoate treated, (e) 100 μM lipoamide treated, (f) 100 μM N-acetylcysteine treated, (g) 5 mM L-glutamate treated, and (h) cotreated with 5 mM glutamate and 100 μM lipoate. Reprinted with permission from C. K. Sen et al., Free Radic. Biol. Med. **22**, 1241 (1997).

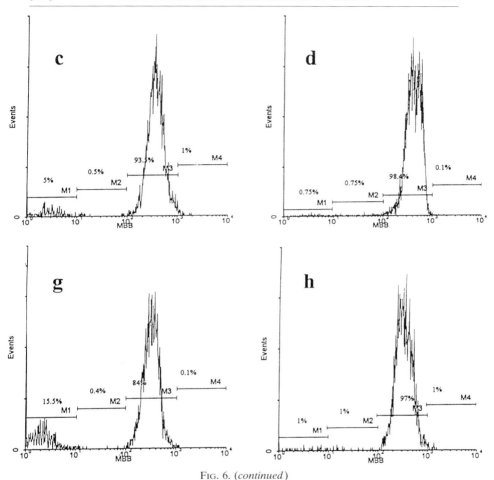

FIG. 6. (continued)

techniques are indispensable for the accurate and sensitive detection of specific thiols. Flow cytometric determination of cellular thiols is a powerful technique that is perhaps best suited for clinical application, particularly for cells in blood or other body fluids. Information can be obtained from a small sample amount with a relatively little and quick sample treatment. This technique offers an unique advantage to study the thiol status of a subset of cells because data are collected from individual cells. Multiparameter flow cytometry allows the study of different subsets of immunotyped cells. A major drawback of the flow cytometric method is the lack of specificity for the determination of distinct thiols. The reaction between

MBB and thiols is not specific for any particular intracellular thiol, although almost all of the entire thiol-reacted bimane emission is specific for thiols in general. This limitation can be partly overcome by the treatment of cells with known thiol regulatory agents as described in the section on the differential assessment of cellular thiols.

Acknowledgments

This work was supported by NIH DB 50430 and by a research grant from the Finnish Ministry of Education.

[22] Measurement of Glutathione, Glutathione Disulfide, and Other Thiols in Mammalian Cell and Tissue Homogenates Using High-Performance Liquid Chromatography Separation of N-(1-Pyrenyl)maleimide Derivatives

By Lisa A. Ridnour, Roger A. Winters, Nuran Ercal, and Douglas R. Spitz

Introduction

As glutatione and cellular thiol-related research progresses, the role of these thiols has been extended from acting only in the maintenance of steady-state redox equilibrium, to a proposed involvement in many stress-induced metabolic and bioregulatory functions, including signal transduction and gene expression.[1-3] Glutathione depletion has been identified in various disease processes, including Parkinson's disease and human immunodeficiency virus (HIV) infection.[4,5] Thiols are also known to protect against cytotoxicity associated with exposure to radiation and chemotherapeutic agents.[6,7] Therefore, the evaluation of cellular thiol status may pro-

[1] A. Meister and M. E. Anderson, *Annu. Rev. Biochem.* **52,** 711 (1983).
[2] R. M. Clancy, D. Levartovsky, J. Leszczynska-Piziak, J. Yegudin, and S. B. Abramson, *Proc. Natl. Acad. Sci. U.S.A.* **91,** 3680 (1994).
[3] J. S. Stamler, *Curr. Top. Microbiol.* **196,** 19 (1995).
[4] T. L. Perry, D. V. Godin, and S. Hansen, *Neurosci. Lett.* **33,** 305 (1982).
[5] R. Buhl, H. A. Jaffe, F. B. Wells, A. Mastrangeli, C. Saltini, A. M. Cantin, and R. G. Chrystal, *Lancet* **2,** 1294 (1989).
[6] B. A. Arrick and C. F. Nathan, *Cancer Res.* **44,** 4224 (1984).
[7] T. Wang, X. Chen, R. L. Schecter, S. Baruchel, M. Alaoui-Jamali, D. Melnychuk, and G. Batist, *J. Pharmacol. Exp. Ther.* **276,** 1169 (1996).

vide valuable information associated with stress-induced alterations in both signal transduction and metabolic pathways, as well as insight regarding the levels of oxidative stress a cell or tissue system may be experiencing. To this end, the growing interest in thiol metabolism has warranted the development of methods that provide rapid and easy thiol detection and quantification.

Although several methods are currently available for the evaluation of glutathione levels, there are some disadvantages to many of these methods, including complicated and time-consuming protocols and analysis, insufficient sensitivity of thiol detection in small amounts of sample, inability to measure oxidized glutathione (GSSG), and inability to determine several relevant thiols other than glutathione in a single-step process. Winters et al.[8] have described a method for the fluorescence detection of several cellular thiols using high-performance liquid chromatography (HPLC) that employs a rapid and simple thiol derivatization with N-(1-pyrenyl)maleimide (NPM). Derivatization is accomplished when a reduced thiol interacts with the maleimide moiety of NPM.[8–10] Under neutral to basic pH conditions,[11] this reaction is complete within 1 min.[8] Once formed, the fluorescent derivatives are stabilized by acidification (pH 2.5), allowing for sample storage up to 6 weeks prior to HPLC analysis (data not shown). The NPM assay utilizes a simple isocratic elution system where all of the analytes of interest elute from the column within 35 min. In addition, total GSH values from both Tietze spectrophotometric[12] and monobromobimane HPLC[13] protocols were found to be in excellent agreement with those obtained using the NPM assay.[8] This observation shows that the NPM assay provides reliable glutathione measurements when compared to other widely accepted methods.

Because oxidized glutathione comprises a small fraction of the total glutathione pool under steady-state conditions the ability to detect changes in GSSG is advantageous in providing information regarding intracellular levels of oxidative stress. Using the NPM assay, GSSG may be quantified following incubation with 2-vinylpyridine, which binds reduced glutathione, making it unavailable for the derivatization reaction.[14] Oxidized glutathione

---

[8] R. A. Winters, J. Zukowski, N. Ercal, R. H. Matthews, and D. R. Spitz, *Anal. Biochem.* **227**, 14 (1995).

[9] J. K. Weltman, R. P. Szaro, A. R. Frackelton, R. M. Dowben, J. R. Bunting, and R. E. Cathou, *J. Biol. Chem.* **248**, 3173 (1973).

[10] C. W. Wu, L. R. Yarbrough, and F. Y. H. Wu, *Biochemistry* **15**, 2863 (1976).

[11] M. Johansson and D. Westerlund, *J. Chromatogr.* **385**, 343 (1987).

[12] F. Tietze, *Anal. Biochem.* **27**, 502 (1969).

[13] G. L. Newton, R. Dorian, and R. C. Fahey, *Anal. Biochem.* **114**, 383 (1981).

[14] O. W. Griffith, *Anal. Biochem.* **106**, 207 (1980).

is then reduced by the addition of NADPH and GR and derivatized as described earlier.

Analytes originally standardized by this assay include both oxidized and reduced glutathione (GSSG, GSH), cysteine (Cys), homocysteine (H-Cys), γ-glutamylcysteine (γ-GC), and cysteinylglycine (Cys-Gly). In a later report, Ercal et al.[15] have shown that N-acetylcysteine (NAC) may also be quantified using the NPM assay. When using this method, spike experiments have shown that 94.2 and 97.2% of GSH and GSSG, respectively, are recoverable.[8] Moreover, all analytes are easily detectable at the level of 100 fmol per 20 μl injection volume.

Because some inconsistencies were encountered in the chromatographic features following the reduction of oxidized glutathione, we have now optimized the assay conditions for the evaluation of GSSG. This modified protocol is described in detail in this article.

Materials

HPLC-grade acetonitrile, methanol, acetic acid, and 85% O-phosphoric acid are from Fisher Scientific (St. Louis, MO). N-(1-Pyrenyl)maleimide, 2-vinylpyridine, and DL-homocysteine are from Aldrich (Milwaukee, WI). Ultrapure oxidized (GSSG) and reduced (GSH) glutathione, L-cysteine, cysteinylglycine, γ-glutamylcysteine, N-acetylcysteine, and NADPH are from Sigma Chemical Co. (St. Louis, MO). Glutathione reductase (GR) is from Boehringer-Mannheim (Indianapolis, IN). Thirteen-millimeter, 0.45-μm nylon Acrodisc filters are from Gelman (Ann Arbor, MI).

The HPLC system is from Shimadzu Scientific Instruments, Inc. (Columbia, MD). The system consists of an LC-10AT pump, an SIL-10A autoinjector, and an RF-1501 spectrofluorophotometer (operating at 330 nm excitation and 375 nm emission), which are all operated by an SCL-10A system controller. The 100 × 4.6 mm column and 1.0 cm × 4.6 mm guard column are packed with 3-μm particle size C_{18} material and are from Astec (Whippany, NJ). Thiols are resolved by isocratic elution using a mobile phase consisting of 65% acetonitrile, 35% nanopure H_2O, 1 ml/liter acetic acid, and 1 ml/liter O-phosphoric acid. The pH of this solvent system is approximately 2.5. The pH of the mobile phase is found to dramatically affect retention times and must be carefully optimized for each system. Integration of chromatograms is accomplished using EZChrom software, also purchased from Shimadzu.

---

[15] N. Ercal, S. Oztezcan, T. C. Hammond, R. H. Matthews, and D. R. Spitz, *J. Chromatogr. Biomed. Appl.* **685**, 329 (1996).

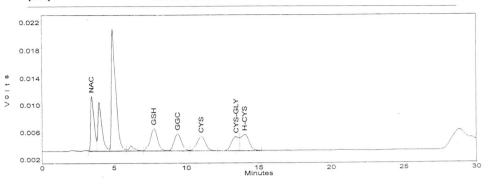

FIG. 1. Chromatogram showing 25 pmol per 20-μl injection of NAC, GSH, γ-GC, Cys, Cys-Gly, and H-Cys run at low sensitivity.

Method

All reduced standards are prepared using serial dilutions of 1 mM thiol diluted in 50 mM phosphate buffer containing 1.34 mM diethylenetriaminepentaacetic acid (DETAPAC), pH 7.8, or 100 mM Tris containing 5 mM serine, 10 mM borate, and 1 mM DETAPAC, pH 7.5. The Tris serine borate buffer should be used when the inhibition of γ-glutamyltranspeptidase activity is desired.[16] Standard dilutions are prepared in a total of 20-μl volumes (standard and buffer), then adjusted to 250 μl with nanopure H_2O. At this point, the pH of the solution should be 7 or greater in order to facilitate the derivatization reaction.[11] Diluted thiols are derivatized by the addition of 750 μl of 1.0 mM NPM dissolved in acetonitrile. Solutions are incubated for 5 min at room temperature, then acidified by the addition of 1 μl of 1:6 diluted HCl. Following acidification, the pH of the solution should be approximately 2.5. Figure 1 shows a chromatogram of 25 pmol on column injection of NAC, GSH, Cys, γ-GC, Cys-Gly, and H-Cys. Unlabeled peaks to the left of GSH and to the far right of H-Cys are thought to arise from impurities in the NPM compound and are also found in the NPM blank.[8] Using this method, linear regression correlation coefficients (R^2) of 0.999 are commonly achieved for standard curves of thiols in the range of 200 fmol to 200 pmol per 20-μl injection. In addition, using the high-sensitivity spectrofluorophotometer setting, 100 fmol of all analytes is easily detected. Retention times, linear regression correlation coefficients, and limits of quantification at high sensitivity detector settings are summarized

[16] S. S. Tate and A. Meister, *Proc. Natl. Acad. Sci. U.S.A.* **75**, 4806 (1978).

TABLE I
Standard Curves for Thiol Analytes

Analyte	Retention time (min)	Limit of quantification (fmol)[a]	Correlation coefficient (R^2)
NAC	3.12	100	0.999
GSSG[b]	7.71	100	0.998
GSH	8.12	100	0.999
γ-GC	9.73	100	0.999
Cys	11.32	100	0.998
Cys-Gly	13.98	100	0.999
H-Cys	14.49	100	0.999

[a] Limit of quantification determined at high-sensitivity detector settings.
[b] GSH following conversion from GSSG.

in Table I. Retention times may vary between runs depending on the age and use of the column as well as slight fluctuations in temperature or pH of the mobile phase.

To optimize the GSSG protocol, we began by employing the assay conditions as described originally.[8] In doing so, we found that the addition of 5 μl of undiluted GR results in a significant shift of the GSH peak following the reduction of GSSG or when employed with GSH standard (Figs. 2A and 2B). In addition, the GSH peak is not resolved completely from NPM blank peaks. Glutathione reductase concentrations are then optimized by testing the effect of serial dilutions of the enzyme on the GSH peak retention time. We found that GR diluted 1:50 in nanopure H_2O provides a rapid and complete conversion of GSSG to GSH while retaining excellent chromatographic characteristics, as demonstrated in Figs. 2C and 2D. Similarly, NADPH and 2-vinylpyridine concentrations and acidification conditions are also optimized for the oxidized glutathione NPM assay. To verify improvements in the modified assay, GSSG standards are compared to GSH standards. Serial dilutions of the GSSG standard are prepared in 100 mM Tris containing 5 mM serine, 10 mM borate, and 1 mM DETAPAC, pH 7.5. Standard dilutions are prepared in a total volume of 40 μl standard and buffer. Standard volumes are adjusted with 44 μl nanopure H_2O, 16 μl of 6.25% 2-vinylpyridine in absolute ethanol is added, and mixtures are incubated at room temperature for 1 hr. This reaction enables the determination of oxidized glutathione following reduction by NADPH/GR and NPM derivatization. Following the 1-hr incubation, 95 μl of a 2-mg/ml solution of NADPH dissolved in nanopure H_2O is added. Five microliters of 1:50 GR in nanopure H_2O is then aliquoted individually to each mixture followed by the resuspension of the solution

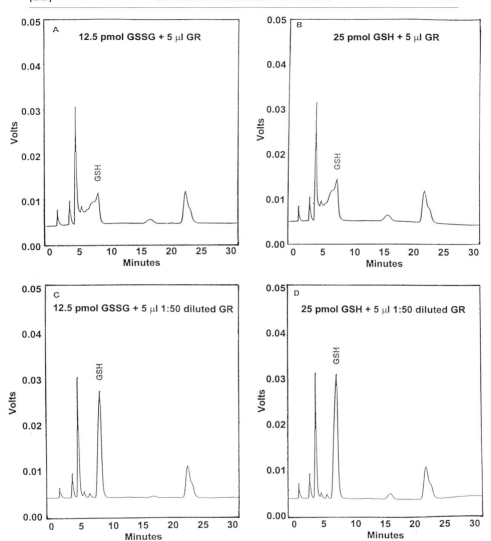

FIG. 2. Chromatogram showing the effects of concentrated vs 1:50 diluted glutathione reductase on the retention time of the GSH peak using GSSG and GSH standards.

four to five times (approximately 5 sec) with a 100-μl pipette. At this point, 100 μl of the reaction is added immediately to a separate tube containing 150 μl nanopure H_2O and 750 μl of 1.0 mM NPM in acetonitrile. Solutions are allowed to incubate for 5 min at room temperature and then are acidified

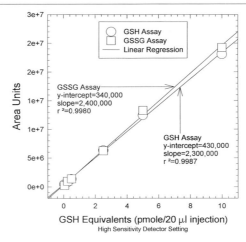

FIG. 3. Standard curves for GSSG and GSH run at the high-sensitivity detector setting.

with 2 μl of 1:6 HCl in nanopure H_2O, resulting in a pH of approximately 2.5. Figure 3 shows standard curves at high sensitivity for both oxidized and reduced glutathione when measured by the modified NPM method. Table II summarizes the ptotocol for preparation of reduced thiol and oxidized glutathione standards.

For sample preparation, cell monolayers are rinsed with 3–4 ml of ice-cold PBS, scrape harvested in 250 μl of ice cold buffer, and collected in 1.5-ml Eppendorf tubes. Homogenization of cell samples is accomplished by resuspending the cells vigorously in buffer with a small-bore pipette tip. Similarly, tissue is prepared by mincing the sample finely in ice-cold buffer followed by a brief (3 × 1-sec burst) homogenization while on ice using a polytron homogenizer with a microtip. Sonication is not advised as it could result in the artifactual oxidation of thiols. Homogenates are diluted in a final volume of 20 μl such that they fall within the standard curve of interest (2–10 mg/ml protein) and then are derivatized immediately to minimize the oxidation of glutathione. Samples for reduced thiol analyses may be prepared in either phosphate or Tris buffer. However, it is advised that Tris buffer be used for the measurement of GSSG as 50 mM phosphate buffer results in insufficient resolution of the glutathione peak. Cell and tissue samples are filtered prior to HPLC analysis to enhance chromatographic features and to increase the life of the column. All samples are normalized per milligram protein using the method of Lowry.[17] Figure

[17] O. H. Lowry, N. J. Rosebrough, A. L. Farr, and R. J. Randall, *J. Biol. Chem.* **193,** 265 (1951).

TABLE II
Reduced and Oxidized Protocols for Preparation of Standard Curves

Thiol analyzed (pmol)	Buffer + thiol stock (μl)	Stock concentration	Volume (μl) H₂O	Final volume (ml)
\multicolumn{5}{c}{Reduced thiol standards}				
0.00	20.0 + 0.0	—	230	1.0
0.10	15.0 + 5.0	1 μM	230	1.0
0.25	7.5 + 12.5	1 μM	230	1.0
0.50	17.5 + 2.5	10 μM	230	1.0
2.50	7.5 + 12.5	10 μM	230	1.0
5.00	17.5 + 2.5	100 μM	230	1.0
10.00	15.0 + 5.0	100 μM	230	1.0
25.00	7.5 + 12.5	100 μM	230	1.0
50.00	17.5 + 2.5	1 mM	230	1.0
100.00	15.0 + 5.0	1 mM	230	1.0

GSSG analyzed (pmol)	Buffer + stock (μl)	Stock concentration	H₂O (μl)	6.25% 2-vp (μl)[a]	NADPH (μl)[b]	GR (μl)[c]
		Oxidized glutathione standards				
0.0	20.0 + 0.0	—	44	16	95	5
0.1	15.0 + 5.0	1 μM	44	16	95	5
0.25	7.5 + 12.5	1 μM	44	16	95	5
0.50	17.5 + 2.5	10 μM	44	16	95	5
2.5	7.5 + 12.5	10 μM	44	16	95	5
5.0	17.5 + 2.5	100 μM	44	16	95	5
10.0	15.0 + 5.0	100 μM	44	16	95	5
25.0	7.5 + 12.5	100 μM	44	16	95	5
50.0	17.5 + 2.5	1 mM	44	16	95	5
100.0	15.0 + 5.0	1 mM	44	16	95	5

[a] 2-Vinylpyridine (2-vp) was prepared in absolute ethanol at a concentration of 6.25% (v:v). After adding 2-vp, samples were incubated at room temperature for 1 hr prior to the addition of NADPH.
[b] NADPH was prepared in nanopure H₂O at a concentration of 2 mg/ml.
[c] GR was diluted 1:50 in nanopure H₂O. After adding GR to the mixture, the sample was mixed by resuspending four to five times and was then quickly aliquoted to a tube containing 150 μl H₂O and 750 μl of 1.0 mM NPM in acetonitrile.

4A shows a chromatogram of reduced thiol peaks obtained following the analysis of glucose-deprived MCF-7[18] cell homogenates, whereas Fig. 4B demonstrates the modulation of reduced thiol pools in the same cells treated with 1 mM N-acetylcysteine. Comparable results are obtained when moni-

[18] X. Liu, A. K. Gupta, P. M. Corry, and Y. J. Lee, *J. Biol. Chem.* **272**, 11690 (1997).

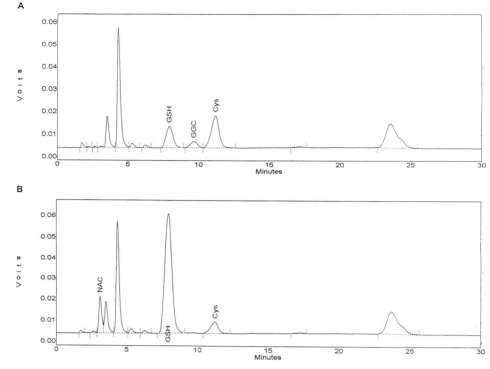

FIG. 4. Chromatograms from 20-μl injections of MCF-7 cell homogenates run at low-sensitivity detection.

toring thiol pools in whole liver homogenates (data not shown). These data demonstrate the utility of this assay in assessing alterations in reduced thiol pools caused by shifts in metabolism.

Conclusion

In summary, the NPM assay provides a rapid and simple method for analyzing both oxidized and reduced glutathione, as well as other thiols, including cysteine, γ-glutamylcysteine, homocysteine, cysteinylglycine, and N-acetylcysteine. The NPM assay provides an excellent method for determining modulations in intracellular thiols caused by oxidative stress. Because changes in redox potential have been associated with alterations in metabolism, signal transduction, and gene expression, the NPM assay provides a sensitive and accurate means of correlating thiol status with these biological processes.

Acknowledgments

This work was supported by NIH Grants RO1 HL51469 (DRS) and F32ES05781 (LAR).

[23] Ratio of Reduced to Oxidized Glutathione as Indicator of Oxidative Stress Status and DNA Damage

By MIGUEL ASENSI, JUAN SASTRE, FEDERICO V. PALLARDO, ANA LLORET, MARTIN LEHNER, JOSE GARCIA-DE-LA ASUNCION, and JOSE VIÑA

Introduction

Glutathione (GSH) is a tripeptide (γ-Glu-Cys-Gly) present at a high level (millimolar range) in all living cells and which participates in numerous cellular functions,[1] including protection against oxidative damage caused by free radicals. Oxidative stress is defined as a disturbance between the prooxidant and the antioxidant balance in favor of the former.[2] Thus, the glutathione status (GSH/GSSG ratio) is a good indicator of oxidative stress.

Several methods have been proposed for the determination of glutathione status in biological samples.[3] Accurate determination of this status is largely dependent on the prevention of GSH autoxidation during sample processing. As the disulfide form (GSSG) is present only in minimal amounts with respect to the reduced form,[4] a small GSH autoxidation during sample processing can give erroneously high GSSG levels.[5]

In order to prevent GSH autoxidation during sample preparation, GSH can be trapped with suitable agents such as N-ethylmaleimide (NEM), 2-vinyl pyridine (2-VP), or iodoacetic acid (IAA). N-Ethylmaleimide is preferred because of its rapid reaction rate (completed within 1 min), in

[1] J. Viña, "Glutathione: Metabolism and Physiological Functions." CRC Press, Boca Raton, FL, 1990.
[2] H. Sies, *Angewandte Chem.* **25,** 1058 (1986).
[3] F. A. M. Redegeld, A. S. Koster, and W. P. van Bennekom, *in* "Glutathione: Metabolism and Physiological Functions" (José Viña, ed.,), p. 11. CRC Press, Boca Raton, FL, 1990.
[4] N. S. Kosower and E. M. Kosower, *Int. Ref. Cytol.* **54,** 109 (1978).
[5] M. Asensi, J. Sastre, F. V. Pallardo, J. García de la Asunción, J. Estrela, and J. Viña, *Anal. Biochem.* **217,** 323 (1994).

contrast to 2-VP (20–60 min) or IAA (5–15 min). Furthermore, alkylation with NEM can be achieved at 4°, whereas alkylation with 2-VP or IAA occurs at room temperature.

Trichloroacetic acid (TCA), metaphosphoric acid, sulfosalicylic acid, picric acid, and perchloric acid (PCA) are used to precipitate proteins. Perchloric acid is preferred because it can be removed as a precipitate of the potassium salt at neutral pH. In blood samples, oxyhemoglobin-derived oxidants are formed during acid precipitation of hemoglobin,[6] and these compounds can oxidize GSH and other substrates such as pyruvate during sample processing. In the case of glutathione, this oxidation does not take place when the acid precipitant of proteins contains NEM[5,7] because it penetrates the red cell membrane quicker than protons and reacts with GSH before these have access to intracellular thiols.[7]

Thus GSSG can be determined using the high-performance liquid chromatography (HPLC) method described in this article. Glutathione cannot be determined in the same chromatogram because the GSH–NEM adduct cannot be measured readily. An ideal method to determine GSH content is to measure total glutathione by HPLC as described by Reed *et al.* (adding the value for GSH + twice the value found for GSSG) or enzymatically, as described by Tietze,[8] and to subtract the value found for GSSG. Reed's method can be used to determine *total* glutathione because the GSH that is oxidized is recovered as GSSG, but as shown below, it cannot be used to determine GSSG or the GSH/GSSG ratio.

GSSG Determination

As stated earlier, the accurate measurement of GSSG in the presence of GSH relies on rapid and effective GSH quenching. In order to prevent GSH oxidation during sample preparation, we recommend N-ethylmaleimide as a GSH quenching agent and perchloric acid to precipitate proteins.

Sample Preparation of Blood Samples

Blood samples are treated with PCA (6% final concentration) containing NEM (20 mM final concentration) and bathophenanthrolinedisulfonic acid (BPDS) (1 mM final concentration) as a metal chelator. Blood samples are then derivatized and analyzed by HPLC.

[6] D. Galleman and P. Eyer, *Anal. Biochem.* **191,** 347 (1990).
[7] D. Galleman and P. Eyer, *Biol. Chem.* **371,** 881 (1990).
[8] F. Tietze, *Anal. Biochem.* **27,** 502 (1969).

Reagents

12% PCA containing 40 mM NEM and 2 mM BPDS

Procedure

1. Add 0.5 ml of whole blood to 0.5 ml of ice-cold 12% PCA containing 40 mM NEM and 2 mM BPDS. Blood samples must be treated with PCA immediately after extraction from the animal or subject. Mix thoroughly.
2. Centrifuge at 15,000g for 5 min at 4°.
3. Take 0.5 ml of acidic supernatant and keep in ice until derivatization. Samples can also be stored frozen at −20° for up to 1 week.

Sample Preparation of Tissue Samples

Tissue samples must be freeze-clamped[9] immediately after obtaining them.

Reagents

6% PCA containing 20 mM NEM and 2 mM BPDS

Procedure

1. Homogenize (1/10, w/v) the tissue sample in ice-cold 6% PCA containing 20 mM NEM and 2 mM BPDS.
2. Centrifuge at 15,000g for 5 min at 4°.
3. Take 0.5 ml of acidic supernatant and keep in ice until derivatization. Samples can also be stored frozen at −20° for up to 1 week.

Derivatization

Reagents

1 mM γ-Glutamyl glutamate (Glu-Glu) prepared in 0.3% perchloric acid

2 M potassium hydroxide (KOH) containing 0.3 M 3-[N-morpholino]-propanesulfonic acid (MOPS)

1% 1-fluoro-2,4-dinitrobenzene (FDNB) dissolved in ethanol

---

[9] A. Woolemberg, O. Ristau, and G. Schoffa, *Pflueger's Arch. Ges. Physiol. Menschen Tiere* **270**, 399 (1960).

Procedure

1. Add 50 µl of 1 mM glutamyl glutamate and 10 µl of a pH indicator solution (1 mM *m*-cresol purple) to 500 µl of acidic supernatant.
2. Adjust pH to 8.0–8.5 with 2 M KOH containing 0.3 M MOPS to prevent excessive alkalinization. Check pH after neutralization with a pH meter. It is important not to reach pH 10, as hydrolysis of the adduct GSH–NEM and GSH autoxidation may occur in this case.[10,11]
3. Centrifuge samples at 15,000g for 5 min.
4. Add an aliquot of 25 µl of each supernatant to 50 µl of 1% FDNB in a small glass tube. After a 45-min incubation in the dark at room temperature, derivatized samples are desiccated under vacuum and stored at $-20°$ in the dark until injection. Samples processed in this way are stable for several weeks.

HPLC Analysis

Reagents

Mobile phase A: 80% methanol (HPLC grade), 20% water (HPLC grade)

Mobile phase B: Prepared by adding 800 ml of a stock sodium acetate solution to 3.2 liters of solvent A. The stock sodium acetate solution is prepared by adding 1 kg sodium acetate (HPLC grade) and 448 ml of water (HPLC grade) to 1.39 liters of glacial acetic acid (HPLC grade).

Procedure. This is carried out as described previously.[5] Samples processed as described earlier are dissolved in 50 µl of 80% methanol (mobile phase A) and injected into the HPLC system. An NH_2 Spherisorb column (20 × 0.4 cm, 5-µm particles) is used. An NH_2 µBondapak column is also suitable for this method. The flow rate is 1.0 ml/min during all the procedure. Mobile phases and the gradient are the following:

Solvent A: 80% methanol
Solvent B: 0.5 M sodium acetate in 64% methanol

After a 25-µl injection of the derivatized solution, the mobile phase is held at 80% A, 20% B for 5 min, followed by a 10-min linear gradient up to 1% A, 99% B. The mobile phase is held at 99% B until GSSG has eluted. The GSH–NEM adduct decomposes and

---

[10] P. Sacchetta, D. Di Cola, and G. Federici, *Anal. Biochem.* **154**, 341 (1986).
[11] E. Beutler, S. K. Srivastava, and C. West, *Biochem. Biophys. Res. Commun.* **38**, 341 (1970).

appears as three peaks. The proportion of the total adduct that appears in each peak is not constant and thus the total amount of GSH cannot be determined in this fashion.

GSH Determination

Glutathione measurement is performed by a modification of the glutathione S-transferase method of Brigelius et al.[12] This is based on the conjugation of chlorodinitrobenzene with GSH catalyzed by glutathione S-transferase. The adduct formed, S-(2,4-dinitrophenyl)glutathione, exhibits a maximum of absorbance at 340 nm.

The precipitation of proteins is carried out by acid treatment. Perchloric acid causes an autoxidation of GSH during sample processing,[5] which is especially important when assaying blood samples. Indeed we have found oxidation of up to 25% of the GSH present when PCA was used. This oxidation does not occur when NEM is used as a thiol-trapping agent. Thus, PCA can be used to determine GSSG (as described earlier). To determine GSH, we use trichloroacetic acid as a deproteinizing agent. The final TCA concentration in the deproteinizing solution must be 15%. Lower concentrations result in a loss of GSH upon storage, even at $-20°$. Under these conditions the samples can be stored for 1 week.

Preparation of Blood Samples

Reagents

30% TCA containing 2 mM ethylenediaminetetraacetic acid (EDTA) as ion chelator

Procedure

1. Add 0.5 ml of whole blood to 0.5 ml of ice-cold 30% TCA containing 2 mM EDTA. Blood samples must be treated with TCA immediately after extraction from the animal or subject. Mix thoroughly. Keep samples on ice until centrifugation.
2. Centrifuge at 15,000g for 5 min at 4°.
3. Take 0.5 ml of acidic supernatant and keep it in ice until it is used for spectrophotometric determination. Samples can also be stored frozen at $-20°$ for up to 1 week.

Preparation of Tissue Samples

Reagents

15% TCA containing 1 mM EDTA as ion chelator

---

[12] R. Brigelius, C. Muckel, T. P. M. Akerboom, and H. Sies, *Biochem. Pharmacol.* **32,** 2529 (1983).

Procedure

1. Obtain the sample by the freeze-clamp technique.
2. Homogenize (1/10, w/v) the tissue sample in 15% TCA containing 1 mM EDTA as an ion chelator.
3. Centrifuge at 15,000g for 5 min at 4°.
4. Take 0.5 ml of acidic supernatant and keep it on ice until spectrophotometric determination. Samples can also be stored frozen at −20° for up to 1 week.

Spectrophotometric Determination

Reagents

0.5 M potassium phosphate buffer, pH 7, containing 1 mM EDTA
1-Chloro-2,4-dinitrobenzene (CDNB) (2 mg/ml of ethanol)
Glutathione *S*-transferase solution prepared by dissolving 500 U/ml of phosphate buffer. This solution is dialyzed in 100 ml of phosphate buffer at 4° for 6 hr changing the buffer every 2 hr. The enzyme solution can be stored at −20° until utilization.

Procedure

1. Add the following reaction mixture to a microcuvette: 825 μl of 0.5 M potassium phosphate buffer, pH 7, containing 1 mM EDTA;

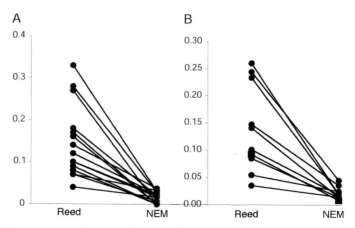

FIG. 1. Measurement of GSSG levels by Reed's and NEM methods in peripheral blood mononuclear cells. (A) Levels of GSSG (nmol/10^6) in peripheral blood mononuclear cells measured by both methods. (B) GSSG/GSH ratio in peripheral blood mononuclear cells using both methods. Each point represents a different experiment determined by both methods.

25 μl of the acidic supernatant of the sample; and 10 μl of CDNB solution.
2. Record the absorbance at 340 nm as a baseline.
3. Add 10 μl of glutathione S-transferase solution (prepared as indicated earlier) to start the reaction.
4. Record the absorbance at 340 nm until the end point of the reaction ($\varepsilon = 9.6 \text{ m}M^{-1} \text{ cm}^{-1}$).

Values of GSSG/GSH Ratio Obtained Establish Relationships between Glutathione Redox Status and Other Metabolic Parameters

When glutathione is measured by the method just described, values for GSSG are obtained that are consistently lower than those found using previously described methods. When values of GSSG and of the GSH/

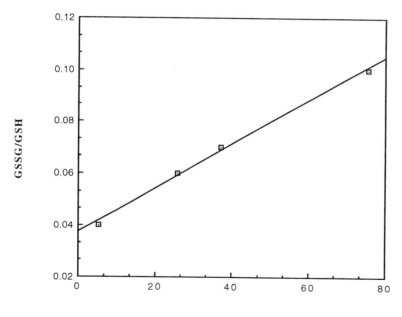

FIG. 2. Relationship between blood glutathione oxidation and blood lactate/pyruvate levels in exhaustive exercise. Values from blood of human volunteers are shown. Subjects performed exhaustive physical exercise on a treadmill, and their blood lactate, pyruvate, GSH, and GSSG levels were determined before and immediately after exercise.

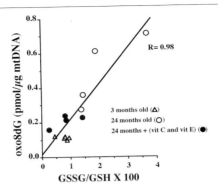

Fig. 3. Relationship between age-induced mitochondrial glutathione oxidation and mitochondrial DNA damage in rats. Relationship between values of mitochondrial GSSG/GSH ratio and levels of oxo-8-deoxyguanosine in mitochondrial DNA from livers of rats. Line of regression and correlation coefficient (r) are shown. Aging causes an increase in glutathione oxidation and DNA damage that is protected by dietary supplementation with antioxidants.

GSSG ratio obtained by a standard method used widely in the past[13] were compared with values found using the method described here, GSSG values were lower and the GSH/GSSG ratio was higher for each sample measured with this method (see Fig. 1).

Glutathione can be used to indicate oxidative stress.[1] In the past some attempts were made to correlate glutathione oxidation with other metabolic parameters. For instance, Gohil et al.[14] tried to correlate glutathione oxidation in exercise with lactate levels in blood. They did not find such a relationship. However, using a method for glutathione determination similar to the one described here, we found that such a relationship does indeed exist.[15] Thus, we were able to establish that glutathione oxidation occurs only when exercise is exhaustive (Fig. 2). Furthermore, we have been able to establish a relationship between glutathione oxidation and mitochondrial DNA damage both in aging and in apoptosis. This is described next.

Relationship between Values of Mitochondrial GSSG/GSH Ratio and Levels of oxo8dG in mtDNA

Oxidative damage to DNA can be estimated by determining the occurrence of oxidized bases such as 8-oxo-7,8-dihydro-2′-deoxyguanosine

---

[13] D. J. Reed, J. R. Babson, P. W. Beatty, A. E. Brodie, W. W. Ellis, and D. W. Potter, *Anal. Biochem.* **106**, 55 (1980).

[14] K. Gohil, C. Viguie, W. C. Stanley, G. Brooks, and L. Packer, *J. Appl. Physiol.* **64**, 115 (1988).

[15] J. Sastre, M. Asensi, E. Gascó, F. V. Pallardó, J. A. Ferrero, T. Furukawa, and J. Viña, *Am. J. Physiol.* **263**, R992 (1992).

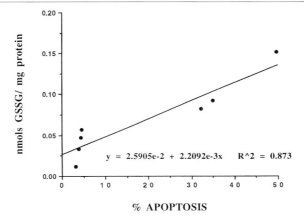

FIG. 4. Increased GSSG concentration during apoptosis in fibroblasts. GSSG was determined in fibroblasts in which apoptosis was induced by incubation in a medium devoid of fetal calf serum.

(oxo8dG). We studied the effect of aging and of antioxidant treatment on the levels of oxo8dG in mitochondrial DNA of mice and rats and confirmed that indeed they are higher in old animals than in young ones.[16] Dietary supplementation with antioxidants protects against the age-related increase in the levels of oxo8dG.

We also measured the levels of GSH and GSSG in mitochondria from young and old animals and found that the GSSG/GSH ratio increases with aging and that this increase is prevented by dietary supplementation with antioxidants. Furthermore, we determined the relationship between the glutathione redox ratio (GSSG/GSH) and the levels of oxo8dG in mitochondrial DNA. Figure 3 shows that such a relationship exists.

We have also investigated the possible relationship between glutathione oxidation and apoptosis in fibroblasts and have found that such a relationship does indeed exist and that the proportion of apoptotic fibroblasts *in vivo* correlates with the level of GSSG in those cells. Figure 4 shows such a relationship. Again, as was the case with exercise described earlier, other authors[17] studied whether glutathione oxidation occurs in apoptosis,

[16] J. García de la Asunción., A. Millán, R. Plá, L. Bruseghini, A. Esteras, F. V. Pallardo, J. Sastre, and J. Viña, *FASEB J.* **10,** 333 (1996).
[17] D. J. Van den Dobbelsteen, C. Stefan, Y. Nobel, J. Schlegel, Y. A. Cotgreave, S. Orrenius, and A. F. G. Slater, *J. Biol. Chem.* **271,** 15420 (1996).

but they could not find an oxidation of glutathione in apoptotic cells. When GSSG was measured as described here, a relationship was indeed found between glutathione oxidation and apoptosis.

Concluding Remarks

The cellular levels of GSH are at least two order of magnitude higher than those of GSSG.[18] Thus a 2% oxidation of GSH leads to an error in the estimation of GSSG of about 100%. Oxidation of GSH must be lowered to levels of about 0.1%. This can be achieved using the method described in this article. Use of other methods leads to GSSG values that are consistently higher than those found using this method (see Fig. 1).

When using the present method the glutathione redox ratio (GSSG/GSH) accurately reflects the redox ratio of cells or of subcellular organelles and relationships can be found between glutathione oxidation and other metabolic parameters such as lactate levels in exercise (Fig. 2), oxidized DNA bases in aging (Fig. 3), or apoptosis (Fig. 4).

Methods for GSSG determination in samples that also contain GSH (i.e., virtually all biological samples) must prevent autoxidation of GSH to levels lower than 0.1%. The method described here meets such a requirement and may be used to determine the glutathione redox ratio in various physiological and pathophysiological situations.

[18] J. Viña, R. Hems, and H. A. Krebs, *Biochem. J.* **170,** 627 (1978).

[24] Nonenzymatic Colorimetric Assay of Glutathione in the Presence of Other Mercaptans

By JEAN CHAUDIERE, NADIA AGUINI, and JEAN-CLAUDE YADAN

Introduction

Many methods have been described for the measurement of glutathione (GSH). Direct assays of total mercaptans are based on chromogenic reactions of sulfhydryl groups with electrophilic reagents. Such assays have obvious limitations in terms of specificity. Other methods are time-consuming and are based on glutathione reductase-coupled assays, with interferences of mixed disulfides of glutathione and enzyme inhibitors, or on chromatographic techniques.[1-7] Reliable chromatographic methods are relatively complex and require expensive materials. Such observations underline the need for faster and easier methodology, especially when large series of biological samples must be processed on a daily basis.

A colorimetric procedure has been developed that takes advantage of a GSH-specific reaction in the absence of a coupling enzyme.[8] This methodology has been adapted in the form of a reagent kit that is currently available to research laboratories through various distributors. Other investigators have confirmed that this assay compared favorably with other procedures.[9]

This article summarizes information that can be used to understand the advantages and limitations of this methodology. The chromogenic reagent 1 used in our assays is 7-trifluoromethyl-4-chloro-N-methylquinolinium. In aqueous solution, reagent 1 absorbs ultraviolet (UV) light with a maximal absorbance wavelength of 316 nm and a molar extinction coefficient of approximately 10,500 M^{-1} cm^{-1}.

Assay of Total Mercaptans

In an aqueous reaction medium whose pH is in the range of 7 to 8, reagent 1 reacts rapidly and quantitatively at ambient temperature with any alkylmercaptan RSH to yield a stable chromophoric thioether. The resulting solution of thioether adduct absorbs visible light strongly, with a maximal absorbance wavelength in the range of 340–360 nm, without interference of reagent absorbance.

As shown in Fig. 1 (reaction 1), the thioether group is in the 4-position on the quinolinium ring, which corresponds to a classical addition–elimination reaction in which the chloride ion is the leaving group. The pH profile of this reaction (data not shown) suggests that the initial addition step requires the basic or thiolate form of mercaptans, i.e., RS$^-$. As expected, the kinetics of the addition–elimination reaction are first order in reagent 1. Reaction

[1] F. Tietze, *Anal. Biochem.* **27,** 502 (1969).
[2] O. W. Griffith, *Anal. Biochem.* **106,** 207 (1980).
[3] M. E. Anderson, *in* "Glutathione: Chemical, Biochemical and Medical Aspects" (D. Dolphin, R. Poulson, and O. Avramovic, eds.), p. 339. Wiley, New York, 1989.
[4] N. K. Burton and G. W. Aherne, *J. Chromatogr.* **382,** 253 (1986).
[5] R. C. Fahey and G. L. Newton, *Methods Enzymol.* **143,** 85 (1987).
[6] R. C. Fahey, *in* "Glutathione: Chemical, Biochemical and Medical Aspects" (D. Dolphin, R. Poulson, and O. Avramovic, eds.), p. 303. Wiley, New York, 1989.
[7] D. Gérard-Monnier, S. Fougeat, and J. Chaudière, *Biochem. Pharmacol.* **43,** 451 (1992).
[8] J. C. Yadan, M. Antoine, and J. Chaudière, Patent PCT/FR 92 10184; WO 93 13071 (1992).
[9] M. Floreani, M. Petrone, P. Debetto, and P. Palatini, *Free Radic. Res. Commun.* **26,** 449 (1997).

FIG. 1. Chromogenic reactions of reagent 1 with mercaptan RSH in aqueous medium. Reaction (1) involves the thiolate form of RSH and yields stable thioethers at pH 7.8. Their maximal absorbance wavelengths are in the range of 350–360 nm. Reaction (2) is a β elimination reaction that is only observed with the GSH-derived thioether obtained from reaction 1. It yields a stable 7-trifluoromethyl-4-thioquinolone whose maximal absorbance wavelength is 400 nm.

times can therefore be shortened by means of a large excess of reagent 1, as well as by increasing the pH of the reaction medium.

This addition–elimination reaction can be used in a colorimetric assay of total RSH at pH 7.8. As shown in Fig. 2, if the initial RSH concentration in the reaction medium is in the range of 5–200 μM, a linear calibration

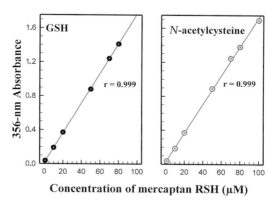

FIG. 2. Examples of calibration curves obtained in the 356-nm assay of total sulfhydryl groups at pH 7.8. Final absorbances were measured after 10-min incubations of 0.6 mM reagent 1 with standard samples at 25°. In mixtures of mercaptan RSH, the slope of the calibration curve obtained with GSH can be used as an average estimation of the apparent molar absorbance of monothiol-derived adducts of reagent 1.

TABLE I
STABLE CHROMOPHORIC PRODUCTS OBTAINED FROM REACTIONS OF REAGENT 1 WITH VARIOUS MERCAPTAN RSH[a]

Mercaptan RSH	Thioether adduct, pH 7.8 λ_{max} (nm), abs $(M^{-1} \cdot cm^{-1})$[b]	Alkalinization, pH ≥ 12.7 λ_{max} (nm)
Glutathione	356, 17400	4-Thioquinolone: 400[c]
Coenzyme A	359, 16000	4-Quinolone: 329
Homocysteine	352, 18000	Unchanged adduct
Cysteine	352, 24270	(S ± N) adduct: 356[d]
Cysteinylglycine	350, 15100	(S ± N) adduct: 355[d]
N-Acetylcysteine	358, 16700	4-Quinolone: 329
Captopril	359, 16000	4-Quinolone: 329
1-Thio-β-D-glucose	341, 14000	4-Quinolone: 329
Mercaptosuccinate	362, 15000[e]	4-Quinolone: 329
DL-penicillamine	354, 16000[e]	Unchanged adduct
α-Mercaptopropionylglycine	353, 17000	Hemithioketal: 300
Diethyl dithiocarbamate	279, 13000	4-Quinolone: 329
Dithiothreitol	355, 32200[f]	4-Quinolone: 329

[a] At pH 7.8 and on subsequent increase in pH.
[b] Apparent molar absorbances may differ from true molar extinction coefficients.
[c] 7-Trifluoromethyl-4-thioquinolone is only produced from the GSH-derived thioether as a result of a specific β-elimination reaction.
[d] A nitrogen adduct (4-iminoquinoline) is produced through intramolecular displacement of the sulfur substituent by the proximal amino group.
[e] Formation of thioether is slow at pH 7.8; reaction completion requires 30 min.
[f] The two sulfhydryl groups react with reagent 1.

curve is obtained. At the 356-nm wavelength, the molar extinction coefficient of the GSH-derived thioether product is about 17,400 M^{-1} cm^{-1}.

As shown in Table I, the maximal absorbance wavelength of thioethers derived from various alkylmercaptans is in the range of 340–360 nm, with the exception of dithiocarbamates. Inspection of apparent molar absorbances per sulfhydryl unit leads to the conclusion that a good estimation of total sulfhydryl groups can be obtained in a mixture of low molecular weight mercaptans if the assay is performed at the maximal absorbance wavelength of the GSH-derived thioether, i.e., 356 nm.

In biological samples where GSH predominates, using the apparent molar absorbance of the GSH-derived thioether as an average value for RSH will usually result in less than 10% error. In practice, the slope of the calibration curve obtained with known concentrations of pure GSH should be used instead of a theoretical value.

One should note that the two sulfhydryl groups of dithiol compounds, such as lipoic acid or dithiothreitol, will react with reagent 1 (data not shown).

In our hands, the chromogenic reaction of all biological mercaptans of low molecular weight is complete in less than 10 min at pH 7.8. However, some exogenous mercaptans may require longer reaction times at this pH due to steric hindrance or to a higher sulfhydryl pK_a. We have not found any significant interference due to nucleophiles other than mercaptans with this procedure.

The detection limit of the assay is 5 μM if a 1-cm optical path length is used for absorbance measurement. Accuracy is high as individual standard errors on the mean are lower than 2% in the 10–200 μM range. Day-to-day reproducibility is better than 98%.

Glutathione-Specific Assay

Because the addition of mercaptans to reagent 1 involved the thiolate form RS^-, exclusively, i.e., the conjugate base of RSH, we studied the stability of the previously described thioethers in more alkaline conditions.

In principle, the addition of hydroxide ion should produce hemithioketal groups in the 4-position. If subsequent elimination of the sulfur substituent occurs, it should produce a stable 4-quinolone. Such reactions are indeed observed around pH 11, but their yields are not quantitative, and some of the starting thioethers are left unchanged.

More interestingly, however, when the pH is raised above 12.7, the GSH-derived thioether undergoes a β elimination that is not observed with other mercaptans (see Table I). As shown in Fig. 1 (reaction 2), this reaction yields a chromophoric 4-thioquinolone. The latter strongly absorbs visible light, with a maximal absorbance wavelength of 400 nm and a molar extinction coefficient of approximately 18,000 M^{-1} cm^{-1}.

Its structure was confirmed by comparison of its nuclear magnetic resonance and UV/VIS spectra with those of authentic 7-trifluoromethyl-4-thioquinolone, produced on reaction of reagent 1 with Na_2S at pH 9.

At pH $\geq$12.7, other thioether derivatives of reagent 1 are left unchanged (homocysteine and penicillamine), transformed into other adducts through the intramolecular reaction of amino groups (cysteine and cysteinylglycine), or hydrolyzed into 7-trifluoromethyl-4-quinolone, with the exception of α-mercaptopropionylglycine (α-MPG), which yields a fairly stable hemithioketal derivative.

As shown in Table I (pH $\geq$ 12.7), the maximal absorbance wavelength of such products is in the range of 300–358 nm, which is at least 40 nm below the maximal absorbance wavelength of 7-trifluoromethyl-4-thioquinolone. Interestingly, the maximal absorbance of 4-quinolone is approximately 10,800 M^{-1} at 329 nm, and hydrolysis of reagent 1 in such alkaline conditions will only yield a minor residual absorbance at 400 nm.

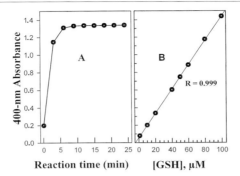

FIG. 3. Examples of kinetics and calibration curves obtained in the 400-nm assay of the GSH-derived 4-thioquinolone at pH 13.5. (A) Examples of kinetics observed with 80 μM GSH and 0.6 mM reagent 1. (B) Example of calibration curve. Final absorbances were measured at least 10 min after the addition of 30% NaOH to the GSH-derived thioether obtained from reaction 1 (see Figs. 1 and 2) at 25°. Other contaminating mercaptans do not affect this calibration curve. The latter is also obtained from GSH itself if the initial pH of the reaction medium is fixed to 13.5, as the kinetics of reaction 1 (Fig. 1) are then very fast.

These observations enabled us to develop a glutathione-specific assay that is based on the colorimetric measurement of 7-trifluoromethyl-4-thio-quinolone at the 400-nm wavelength. In our optimal procedure, the pH of the thioether-containing solution is raised above 13.4 by adding a small volume of 30% (v/w) sodium hydroxide, and the β-elimination reaction is complete in less than 15 min, as shown in Fig. 3A.

In biological samples, the reproducibility of such kinetics was improved greatly in the presence of 0.025% (v/w) Lubrol as a neutral detergent. The 4-thioquinolone produced in such conditions is stable for several hours in the dark. The linear calibration curve, which is obtained with known concentrations of GSH, is shown in Fig. 3B.

The detection limit of the assay is 17 μM if a 1-cm optical path length is used for the absorbance measurement. The accuracy is similar to that of the previous procedure, with individual standard errors on the mean lower than 2% in the 20–200 μM range. Day-to-day reproducibility is better than 97%.

Interferences Observed in Glutathione-Specific Assay

Endogenous compounds that may interfere with the assay include cysteine and free amines, disulfides of glutathione, and compounds containing the γ-glutamylcysteine structure.

The amino groups of cysteine, cysteinylglycine, and free amines react with reagent 1 or thioether adducts in the strongly alkaline conditions of the assay. If such compounds are present in large concentrations, they will induce a contaminating absorbance near 400 nm, which is due to the formation of stable 4-iminoquinolines. This is due to an intramolecular substitution of thioether by proximal amino groups in cysteine or cysteinylglycine and to intermolecular substitution by free amines. The molar absorbance of cysteine- or cysteinylglycine-derived iminoquinolines is much smaller than that of 7-trifluoromethyl-4-thioquinolone.

With very few exceptions, the steady-state concentration of intracellular cysteine or cysteinylglycine, as well as that of free amino groups, is less than 1% of that of GSH in the cytosol and mitochondrial matrix, and such interferences will be negligible if amino buffers are avoided.

When interferences are suspected, they are easy to detect as free amino groups do not react with reagent 1 at pH 7.8: One should conclude that there are significant interferences if $[GSH]_{400} > [RSH]_{356}$ when the two assay procedures are used sequentially.

Glutathione disulfide (GSSG) produces about 1.5 equivalent of thioquinolone, which is compatible with the following reactional sequence, in which 1.5 equivalent of GS(H) would be produced at pH > 13.4:

$$GSSG + OH^- \rightarrow GSOH + GS^-$$
$$2\ GSOH \rightarrow GSO_2H + GSH$$

Similar interferences should be expected with mixed disulfides of glutathione. This problem is redhibitory in some extracellular compartments, such as blood plasma or biliary fluid, or in cell-free systems where strongly oxidizing conditions are imposed. As a rule, however, the ratio [GSH]/[GSSG + RSSG] is higher than 50 within aqueous intracellular compartments, where such interferences will therefore be negligible.

γ-Glutamylcysteine structures possess sulfhydryl groups that behave as that of glutathione in this assay. They include trypanothione, γ-glutamylcysteine itself, or synthetic carboxylic esters of glutathione.[10,11] This should be a minor limitation, as trypanothione is produced exclusively by trypanosomes, whereas the steady-state concentration of γ-glutamylcysteine is very small, i.e., less than 1% of the GSH concentration within most animal cells.

In addition, we have observed that α-MPG, a synthetic mercaptan sometimes used in pharmacology, slows down the kinetics of β elimination markedly. Apparently, a 4-dithioketal, which is the mixed adduct of reagent 1 with both GSH and α-MPG, is formed irreversibly. Its alkaline decomposi-

---

[10] M. E. Anderson, E. J. Levy, and A. Meister, *Methods Enzymol.* **234**, 492 (1994).
[11] E. J. Levy, M. E. Anderson, and A. Meister, *Methods Enzymol.* **234**, 499 (1994).

tion into 4-thioquinolone is quantitative and highly reproducible, but it takes more than 90 min at pH 13.5.

Overall, such interferences should not be significant in most cellular extracts or homogenates of solid tissues, and it is easy to rule out or to anticipate problems with cell-free systems. However, interferences due to GSSG or related mixed disulfides preclude the use of blood plasma samples in this assay.

Recommended Procedure

Instrumentation

Assays should be performed in a thermostatted spectrophotometer equipped for measurement of absorbance in the 300- to 500-nm region. Measurements should be performed at a fixed temperature ±2° in the range of 20–30°. Spectrophotometric cuvettes made of glass or plastic are suitable.

Chemicals and Reagents

Buffer: 200 mM potassium phosphate, pH 7.8 (25°), containing 0.025% (w/v) Lubrol and 0.2 mM diethylenetriaminepentaacetic (DTPA). Amine buffers must be avoided. Reagent 1: 7-trifluoromethyl-4-chloro-N-methylquinolinium is synthesized as described previously[8]; a stock solution of 12 mM of the pure trifluoromethyl sulfonate or methyl sulfate salt is prepared and stored in 0.2 N aqueous HCl. This stock solution is stable for 6 months when it is stored at 4°.
Sodium hydroxide solution: 30% (w/v) aqueous NaOH
Physiological saline: 0.9% aqueous NaCl
Metaphosphoric acid (MPA): 5% (w/v) aqueous solution. This solution is stable for 3 days at 4°.

Sample Preparation

The following procedure has been used successfully with cell suspensions of erythrocytes, isolated hepatocytes, lymphocytes, endothelial cells, and tumor cells such as HL-60, Jurkat and MCF-7 cells, and also with solid tissues such as liver, heart, kidney, and lung.

a. All steps should be performed at 4°.
b. Whenever possible, biopsies of solid tissues should be obtained from organs that have been perfused *in situ* with physiological saline.
c. Intact cells or solid tissues should be washed with physiological saline before cell disruption or tissue homogenization.

d. Aqueous cell lysates or homogenates of solid tissues should be obtained in 5% MPA to prevent fast sample degradation due to oxidation or to enzyme reactions.

e. Proteins are then precipitated by centrifugation at 3000 g for 10 min, and the clear supernatant is collected and kept at 4° until it is used in the assays.

Such MPA extracts are usually stable for at least 1 hr at 4°, but appropriate controls of stability should be performed if longer storage times are envisaged before the assays. Alternatively, intact and perfused organs can be frozen in liquid nitrogen and then lyophilized directly. The resulting dry sample can be homogenized and deproteinized in 5% MPA as described earlier. In our hands, intact lyophilates of preperfused rat heart can be stored for 4 days at −70° without a significant loss of endogenous GSH (data not shown).

Colorimetric Assay of Total RSH at 356 nm

Adjust the 356-nm absorbance baseline of the spectrophotometer to zero with buffer only. Perform three independent measurements of blank absorbance (A_0 mesaured for [RSH] = 0) at 356 nm. The resulting mean value will have to be subtracted from absorbance values (A) obtained in the presence of sample.

For each colorimetric measurement, the reaction medium is prepared as follows:

1. Take an initial volume of sample (V_i) of 20–300 μl.
2. Complete to 900 μl with buffer (buffer volume = 900 μl − V_i).
3. Add 50 μl of reagent 1 and mix thoroughly.
4. Incubate for 10 min at 4° or at the selected working temperature.
5. Measure the final absorbance (A) at 356 nm.

The resulting solution of thioether adducts is stable for at least 1 hr.

The concentration of total RSH in samples introduced in the spectrophotometer cuvette will be derived from the equation $[RSH]_{tot} = (A - A_0)/(Sl)$, where S is the slope of the linear calibration curve, expressed in M^{-1} cm^{-1}, and l is the optical pathlength expressed in centimeters.

Colorimetric Assay of GSH at 400 nm

Adjust the 400-nm absorbance baseline of the spectrophotometer to zero with buffer only. For each colorimetric measurement, the reaction medium is prepared as follows:

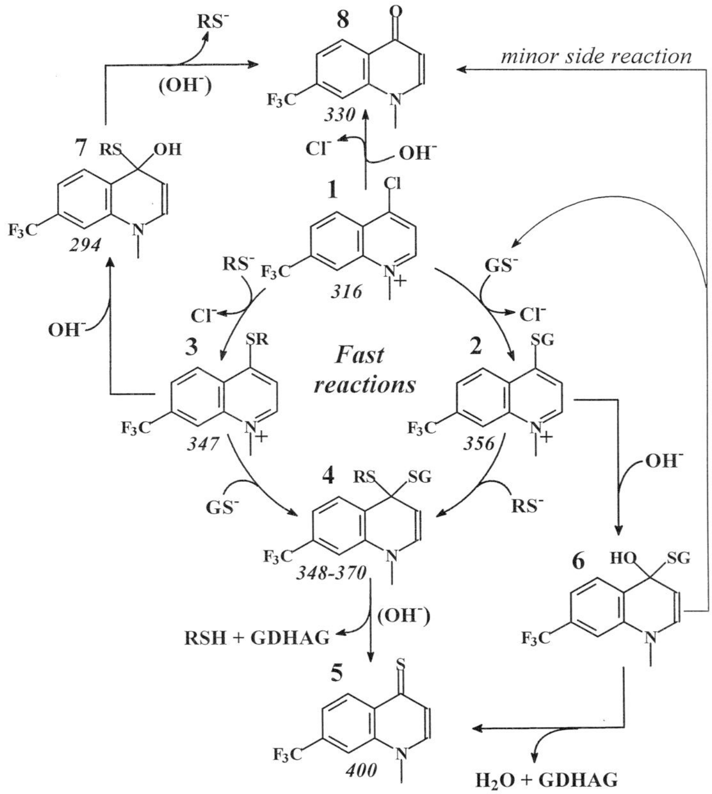

FIG. 4. Reactional steps involved in the production of chromophoric products of reagent 1 and mercaptans. Maximal absorbance wavelengths of intermediates and products are indicated in italics, with the exception of the putative intermediate 6, whose lifetime is too short for spectrophotometric visualization. At pH 7.8, thioethers are produced through reactions [1 → 2] and [1 → 3]. At pH > 12.7, the predominant GSH-specific reactional pathway is [1 → 2 → 6 → 5]. In the presence of excess α-mercaptopropionylglycine (RSH = α-MPG), another reactional pathway of GSH involves irreversible and very fast reactions [2 → 4] and [3 → 4], followed by the slow decomposition [4 → 5]. Products 2 and 3 are stable at pH 7.8. Products 5 and 8 are stable at the higher pH values required for their production. GDHAG, γ-glutamyldehydroalanylglycine.

1. Take an initial volume of sample (V_i) of 20–300 μl.
2. Complete to 900 μl with buffer (buffer volume = 900 μl − V_i).
3. Add 50 μl of reagent 1 and mix thoroughly.
4. Add 50 μl of sodium hydroxide solution and mix thoroughly.
5. Incubate for at least 10 min.
6. Measure the final absorbance (A) at 400 nm.

Solutions of 4-thioquinolone are stable for at least 1 hr if they are kept in the dark.

In this GSH-specific assay, the measurement of blank values requires this sequence in the presence of a sample of buffer. Three independent measurements of blank absorbance (A_0 measured for [GSH] = 0) should be obtained at 400 nm. The resulting mean value will have to be subtracted from absorbance values (A) obtained in the presence of sample.

The GSH concentration of samples introduced in the spectrophotometer cuvette are derived from the equation [GSH] = $(A - A_0)/(Sl)$, where S is the slope of the linear calibration curve, expressed in M^{-1} cm^{-1}, and l is the optical pathlength expressed in centimeters. One should check that the initial concentration of reagent 1 was in sufficient excess over GSH, i.e., ≥ 4 times higher than the [GSH] value obtained.

Summary of Reactions Involved and Conclusion

A tentative description of the reactions involved in the formation of chromophoric products of mercaptans and reagent 1 can be found in Fig. 4.

In agreement with this scheme, the overall yield of alkaline β elimination of GSH-derived adducts ranges from 92 to 97% at pH $\geq$ 13.4 because some residual GSH is recycled in step [**6** → **8**], whereas excess reagent 1 is eventually hydrolyzed entirely to the quinolone (compound **8**). The GSH specificity of the alkaline β-elimination steps may be explained by six-membered concerted reactions of the corresponding hemi- and/or dithio-ketal intermediates.

In conclusion, the assay, which is based on the colorimetric measurement of 7-trifluoromethyl-4-thioquinolone at the 400-nm wavelength, has the unique advantage to be glutathione specific in the absence of enzyme. Another advantage is that a colorimetric estimation of total sulfhydryl groups can be obtained at the 356-nm wavelength, before the addition of sodium hydroxide to the mixture of thioethers produced at pH 7.8.

The sensitivities of the two assays are very similar, and given the stability of chromophoric products, large series of samples can be processed within 20–30 min for subsequent absorbance measurement.

[25] Protozoological Method for Assaying Lipoate in Human Biologic Fluids and Tissue

By HERMAN BAKER, BARBARA DEANGELIS, ELLIOTT R. BAKER, and SEYMOUR H. HUTNER

Introduction

A procedure for assaying lipoic acid concentration in biologic fluids and tissues is described using a eukaryotic protozoan *Tetrahymena thermophila* (ATCC 30008, Rockville MD). *Tetrahymena thermophila* has a specific and sensitive (30 pg/ml) requirement for lipoic acid. Unlike humans and other microorganisms, *T. thermophila* cannot synthesize lipoic acid, hence its requirement for exogenous lipoic acid is specific. The lipoic acid supplied to *T. thermophila* by processing biologic fluids and tissues during the assay procedure permits the derivation of a practical assay by turbidometrically assessing its growth response to various lipoate concentrations.

Review of Procedures for Assay

The maintenance medium (Table I) for growing the inoculum is dispensed in 10-ml amounts in 25 × 125-mm screw-capped borosilicate tubes, autoclaved for 20 min at 15 lbs psi, 118–121° then allowed to cool; the medium can be stored at 4° for months. Transfers are made biweekly into fresh maintenance medium; one drop of the culture is used for transfer. The organism reaches full growth in 3 days at 29–32°. The growth medium for assaying lipoate is shown in Table II. This medium is stored in a glass-stoppered bottle at 4°, with a few milliliters of volatile preservative to prevent contamination by bacteria and molds. The volatile preservative, 1 part (by volume) chlorobenzene, 1 part 1,2-dichloroethane, and 2 parts 1-chlorobutane, is removed during autoclaving by the steam distillation effect. On cold storage, the soluble starch added after boiling in distilled water and added to the medium (Table II) causes some turbidity; turbidity is lost after final autoclaving for assay purposes. After dispensing 2.5 ml of basal medium into 25-ml borosilicate micro-Fernbach flasks (Kimble Glass), the standards and solution to be assayed are added and the volume brought to 5 ml with distilled water (Table III); flasks are then covered with autoclavable polypropylene caps (Pioneer Plastics, Jacksonville, FL). The covered flasks are placed into 2-quart Pyrex baking trays and autoclaved for 20 min and then are covered with another inverted baking tray

TABLE I
MAINTENANCE MEDIUM FOR
Tetrahymena thermophila

Constituent	g/dl
Trypticase BBL[a]	0.5
Yeast autolyzate[b]	0.2
Glucose	0.2
pH 6.5	

[a] From Becton-Dickenson Microbiology Systems, Cockeysville, Missouri.
[b] From Sigma Chemical, St. Louis, Missouri.

and allowed to cool at room temperature. For inoculation, 2 ml of a 3- to 4-day-old maintenance culture is diluted with 10 ml of sterile (autoclaved) distilled water; a drop of this solution serves as an inoculum for each flask in the tray. After inoculation, the flasks are recovered with the baking tray, the combination, sealed with masking tape, is incubated at 29–32° for 3 days. Microbial growth is expressed in turbidometric units, as measured with a turbidometer equipped with a red sensitive probe. A Coleman, Jr. or a spectrophotometer at 650 nm can be used.

Standards and Assay Use

A lipoate stock solution is prepared by dissolving 100 mg of the readily available racemate DL-α-lipoamide (Sigma Chemicals, St. Louis, MO) in 100 ml of a mixture containing 50 ml absolute ethanol and 50 ml of distilled water. Although *T. thermophila* can use DL-α-lipoic acid, lipoamide is used as a standard because it is believed to be the active moiety found in eukaryotes.[1]

Tenfold dilutions of the stock solution are made serially in distilled water to obtain working standards of 0.1, 1.0, and 10 ng/ml. This procedure dilutes out any growth inhibition for *T. thermophila* effected by the ethanolic stock solution. Standards are stored in glass-stoppered bottles; they are prepared monthly. Some volatile preservative is added to standards to prevent microbial contamination. A control flask consisting of 2.5 ml basal medium and 2.5 ml water, without lipoamide, is always included in the standard curve to estimate lipoate introduced by extraneous contamination. The standard curve (Table III) is prepared by making additions to 2.5 ml basal medium in individual 25-ml borosilicate micro-Fernbach flasks.

[1] L. Packer, R. Sashwati, and C. K. Sen, *Adv. Pharmacol.* **38**, 79 (1997).

TABLE II
BASAL MEDIUM FOR LIPOATE ASSAY USING *Tetrahymena thermophila*[a,b]

Constituent	Amount	Constituent	Amount
Casamino acids[c]	6000 mg	Calcium pantothenate	1 mg
Citric acid	100 mg	Sodium guanylate	30 mg
Potassium phosphate monobasic	100 mg	Adenine	20 mg
Magnesium sulfate heptahydrate	400 mg	Uracil	30 mg
Metal mixture[d]	100 mg	Cytidine	10 mg
Calcium ion[e]	50 mg	Thymidine	10 mg
Sodium acetate (anhydrous)	500 mg	L-Tryptophan	150 mg
Diacetin	0.4 ml	DL-Methionine	200 mg
Thiamin hydrochloride	500 mg	Glycine	100 mg
Biotin	5 mg	DL-asparagine	1000 mg
Folic acid	100 mg	Glycerol	10 ml
Disodium riboflavin-5-phosphate	500 mg	Glucose	1500 mg
Pyridoxal hydrochloride	1 mg	Soluble starch[f]	5000 mg
Pyridoxamine dihydrochloride	1 mg	Distilled water	(to) 500 ml
Nicotinamide	1 mg		

[a] From H. Baker, B. DeAngelis, E. R. Baker, and S. H. Hutner, *Free Rad. Biol. Med.* **25** (in press), with permission from Elsevier Science.

[b] Boil the medium in about 200 ml of distilled water. After cooling, adjust pH to 6.6–6.8 with 2.5 M potassium hydroxide and add starch (see footnote f).

[c] Acid-hydrolyzed casein (Difco).

[d] $ZnSO_4 \cdot 7H_2O$, 6.9 g; $MnSO_4 \cdot H_2O$, 4.2 g; $Fe(NH_4)_2(SO_4)_2 \cdot 6H_2O$, 7.8 g; $CoSO_4 \cdot 7H_2O$, 730 mg; $CuSO_4 \cdot 5H_2O$, 170 mg; $(NH_4)_6Mo_7O_{24} \cdot 4H_2O$, 120 mg; $Na_3VO_4 \cdot 16H_2O$, 80 mg; and H_3BO_3, 80 mg. Prepared as a finely ground mixture (triturate), these amounts suffice for 200 liters of assay medium supplying essential trace elements (Zn, Mn, Fe, Co, Cu, Mo, V, B).

[e] Twenty-five grams of calcium carbonate dissolved in minimal concentrated hydrochloric acid and brought to 100 ml with distilled water yields a solution containing 100 mg of calcium ion per 1 ml of solution. May be stored indefinitely at room temperature with a volatile preservative.

[f] Soluble starch is autoclaved separately in 200 ml distilled water to ensure solubility; it is added while hot to the medium before volume adjustment to 500 ml. Upon storage of the medium in the cold, the starch causes some turbidity; this does not affect the assay medium, as the turbidity is lost after the final autoclaving for the assay.

Standard solutions in the flasks are brought to a final volume of 5 ml with distilled water (Table III) and prepared for autoclaving (*vide supra*). The standard growth curve is plotted from flasks containing 10, 30, 100, 300, 1000, and 3000 pg of lipoamide per milliliter of final growth medium. The growth (ordinate) of the standard lipoate concentration per milliliter (abscissa) is plotted on semilogarithmic paper (Keuffel and Esser No. 46-5490 three cycle), and lipoate concentration of the unknown is calculated from the curve.

TABLE III
PREPARATION OF STANDARD DL-α-LIPOAMIDE CURVE[a]

Flask no.[b]	Concentration		Lipoamide standard per flask[c]	Distilled water (ml)
	pg/ml	pg/5 ml		
1	—	—	—	2.5
2	10	50	0.5 ml of 0.1 ng/ml	2.0
3	30	150	1.5 ml of 0.1 ng/ml	1.0
4	100	500	0.5 ml of 1.0 ng/ml	2.0
5	300	1,500	1.5 ml of 1.0 ng/ml	1.0
6	1000	5,000	0.5 ml of 10 ng/ml	2.0
7	3000	15,000	1.5 ml of 10 ng/ml	1.0
Prepared samples, e.g., fluids and tissue extracts				
	Sample addition (ml)			
8	1.0			1.5
9	1.5			1.0
10	2.0			0.5
11	2.5			—

[a] From H. Baker, B. DeAngelis, E. R. Baker, and S. H. Hutner, *Free Rad. Biol. Med.* **25** (in press), with permission from Elsevier Science.
[b] All flasks must contain 2.5 ml basal medium.
[c] DL-α-Lipoamide concentration (ng/ml) × 4.9 = pmol/ml.

Biologic Fluids and Tissue

The procedure for extracting, measuring, assaying, and calculating lipoate in biologic fluids and tissues is given in Tables III and IV. To prepare a 0.02 M phosphate buffer, add 3 g of $NaH_2PO_4 \cdot H_2O$ to 800 ml of distilled water. Adjust pH to 5.5 with 0.02 M Na_2HPO_4 and then add distilled water to 1 liter. To prepare 0.01 M phosphate buffer, dilute 1 liter of the 0.02 M buffer with 1 liter distilled water and add some volatile preservative for storage; buffers are stable for months at room temperature. The crude proteolytic enzyme used to digest bound lipoate in specimens is protease, type II (Sigma); directions are in Table IV. Lipoate is freed from protein binding by autoclaving in pH 5.5 buffer; further liberation of lipoate is facilitated by the use of this crude protease (Table IV). This enzyme enables liberation of lipoate from various lengths of lipoylpeptides. Because *T. thermophila* responds to lipoamide, lipoamidase is not needed to convert the amide to lipoic acid for analyses. The protease solution, when used, is assayed as a plasma sample, using water in place of plasma (Tables III and IV) to document the contaminating lipoate content; it usually contains negligible lipoate concentration.

TABLE IV
LIPOATE ASSAY[a]

Preparation of biologic fluids and liver tissue for lipoate assay

Blood, whole blood, plasma, red blood cells
1. Dilute 1 ml of specimen with 1 ml of 0.02 M phosphate buffer, pH 5.5.
2. Autoclave for 20 min.
3. Stir debris to make a suspension; add 8 ml of 0.01 M phosphate buffer, pH 5.5, containing 4 mg of protease Type II enzyme (Sigma)
4. Incubate mixture for 18–20 hr at 37°.
5. Autoclave sample for 20 min; centrifuge off debris and save supernatant for assay.
6. Assay No. 5 as in Table III; final dilution of original biologic fluid is 1 : 10.

Urine
1. Dilute 1 ml urine with 1 ml of 0.02 M phosphate buffer, pH 5.5.
2. Autoclave for 20 min.
3. Add 8 ml of distilled water.
4. Assay No. 3 as in Table III; final dilution of urine is 1 : 10.

Cerebrospinal fluid (CSF)
1. Take 1 ml of CSF and add 1 ml of 0.02 M phosphate buffer.
2. Autoclave mixture, centrifuge off debris, and save supernatant.
3. Add 1 ml of 0.01 M phosphate buffer to 1 ml of supernatant.
4. Assay No. 3 as in Table III; final dilution is 1 : 3.

Liver
1. Homogenize enough liver tissue in distilled water to contain 20 mg wet-weight tissue per milliliter of mixture. Separately save 1 ml for protein analysis.
2. Add 1 ml of sample No. 1 to 1 ml of 0.02 M phosphate buffer, pH 5.5.
3. Autoclave for 20 min.
4. Add 3 ml of 0.02 M phosphate buffer, pH 5.5, containing 0.5 ng/ml of protease Type II enzyme (Sigma) per milliliter.
5. Incubate mixture at 37° for 3 days.
6. Autoclave sample for 20 min, centrifuge off debris, and save supernatant. Take 1 ml of supernatant, and add 4 ml distilled water.
7. Assay diluted supernatant in No. 6 as in Table III; supernatant now contains 0.8 mg treated liver per milliliter of supernatant used for assay.

Calculations for lipoate concentration

Fluids

a is the concentration of lipoate standard (ml) as derived from sample growth turbidity
b is the volume of sample when diluted with buffer
c is the total volume in growth flasks
d is milliliter (volume) of sample used for test
e is milliliter (volume) of diluted sample (sample + diluent) added to assay flask.

$$\frac{a:b:c}{d:e} = \text{concentration of lipoate in sample per milliliter}$$

(continued)

TABLE IV (*continued*)

Calculations for lipoate concentration
Urine
Multiply concentration of lipoate in sample per milliliter by total 24-hr volume (ml) and express per 24-hr sample
Liver tissue
a is the concentration of lipoate standard (per ml) as derived from sample growth turbidity
b is the total volume in growth flasks, e.g., 5 ml
c is the concentration of tissue in milligrams per milliliter after diluting with buffer or water
d is milliliter (volume) of diluted sample (sample and diluent) added to assay flask
$$\frac{a:b}{c:d} = \text{concentration of lipoate per milligram of sample}$$

[a] From H. Baker, B. DeAngelis, E. R. Baker, and S. H. Hutner, *Free Rad. Biol. Med.* **25** (in press), with permission from Elsevier Science.

Blood

Blood is drawn from an antecubital vein into Vacutainers (Becton-Dickenson, Sunnyvale, CA) containing EDTA, as anticoagulant. Red blood cells (RBC) and plasma are obtained from EDTA-treated whole blood; plasma is pipetted away from RBC after the RBC are centrifuged off. RBC are then washed three times with saline (0.85% NaCl); the washings are discarded and 1 ml of packed RBC is used for assay. Enzyme treatment of specimens and subsequent autoclaving deproteinizes lipoate and destroys thermolabile drugs; further dilutions (Table IV) render thermostable drugs harmless to *T. thermophila*. The supernatant, obtained by centrifuging off the debris, is saved for assay as in Table III.

Urine

An aliquot of a 24-hr specimen is used for assay (Table IV); morning urine alone is not reliable for assay. Urine is autoclaved in 0.02 M phosphate buffer to coagulate any protein present; water is added to the supernatant (Table IV) to dilute out components that may be toxic to *T. thermophila*. No enzyme treatment is necessary, as we found lipoate in urine is free. The assay is carried out as detailed in Table III.

Cerebrospinal Fluid

Cerebrospinal fluid is obtained by lumbar puncture from subjects before local anesthesia for surgical procedures; traumatic or xanthochromic taps were discarded. The assay method is shown in Table III.

Liver Tissue

A specimen of liver, obtained by needle biopsy or at autopsy, is freed from coagulum and adhering tissue, sliced, washed three times with saline, suspended in water, and homogenized in a blender or, if the sample is small, in a tissue grinder. A separate portion of the tissue specimen is used to determine protein content to serve as a standard of reference for expressing tissue results.[2,3] This avoids confabulating results with wet-weight fluctuations and those due to fatty infiltration of the tissue. The tissue is prepared for assay as in Table IV and assayed as shown in Table III.

[2] O. H. Lowry, N. J. Rosebrough, A. L. Farr, and R. J. Randall, *J. Biol. Chem.* **193,** 265 (1951).
[3] O. Frank, A. Luisada-Opper, M. F. Sorrell, A. D. Thomson, and H. Baker, *Exp. Mol. Pathol.* **15,** 191 (1971).

[26] Antioxidant Activity of Amidothionophosphates

By OREN TIROSH, YEHOSHUA KATZHENDLER, YECHEZKEL BARENHOLZ, and RON KOHEN

Introduction

In general, reactive oxygen species (ROS) can be divided into two major oxidant categories. The first of these includes reactive free radicals, which can serve as damage causing agents. An example of this is the recurrent production of peroxyl radicals that leads to rapid accumulation of oxidative damage in the propagation process of lipid peroxidation. The second group consists of the relatively more stable ROS, such as lipid hydroperoxides and hydrogen peroxide, which serve as radical precursors. These more stable ROS can become free radicals by accepting an electron from a metal or a reducing agent. Examination of the spectrum of activity of the low molecular weight antioxidants showed that most of them react via H· donation and are transformed into stable radicals. These antioxidants can therefore react with free radicals with great efficiency, but cannot react with nonradical ROS species without producing secondary free radicals. Thus, in addition, such agents display prooxidant properties.

The mechanism of controlling oxidation by removing hazardous but relatively stable ROS, thereby preventing them from transforming into more damaging free radicals, is relatively unexplored. By evaluating the spectrum of low molecular weight antioxidants, which are capable of reaching lipophilic compartments and reacting with lipid hydroperoxides, we found that the number of such available compounds is extremely small. One of them is ebselen, which mimics glutathione peroxidase.[1] We concentrated on the development of such compounds on the assumption that with sufficient reactivity these molecules can overcome the drawbacks of conventional antioxidants and be complementary to them. Such an approach may open a way to prevent oxidation by inhibiting initiation. The proposed molecules may allow for the repair of oxidative damage that has already occurred.

We developed a family of novel antioxidants possessing a chemical structure analogous to that of the thiourea family of molecules, which are well-known antioxidants and potent hydroxyl radical scavengers also capable of removing hydrogen peroxide and superoxide radicals.[2,3] The therapeutic use of these molecules, however, is minimal due to their high toxicity.[4] In order to design less toxic reagents without losing antioxidant properties we replaced the thiocarbonyl group of thiourea, $(H_2N)_2-C=S$, with a more electronegative $P=S$ group. The thiourea toxicity is thought to be due to its conversion into a positively charged alkylating agent by the loss of the sulfur atom during metabolism, hence the change of the central carbon to a phosphorus atom. The center of the new molecule and the active site are based on a primary amidothionophosphate (AMTP) chemical bond (Fig. 1).[5]

These antioxidants may serve as scavengers of free radicals. They can, however, react with nonradical ROS without any prooxidative effects. The reactive part of the molecule (referred to as X) determines its potential to react as an antioxidant. The mechanism of protection of the other part of the molecule (referred to as R) is determined by other physicochemical properties such as the balance between hydrophilicity and hydrophobicity, which determine penetration through biological membranes, as well as the partition and exact location of the reactive group in hydrophobic and amphipathic environments, such as in biological membranes and lipoproteins. Only the correct combination of X and R will enable optimization of the antioxidant effect. Because AMTPs can break down peroxides in biological lipid assemblies and can also react strongly with sodium hypochlorite, they can thus protect against oxidation of biologically important com-

---

[1] H. Sies, *Methods Enzymol.* **234,** 477 (1994).
[2] M. J. Kelner, R. Bagnell, and K. J. Welch, *J. Biol. Chem.* **265,** 1306 (1990).
[3] G. R. Dey, D. B. Naik, K. Kishore, and P. N. Moorthy, *J. Chem. Soc. Perkin Trans.* **2,** 1625 (1994).
[4] M. R. Boyd and R. A. Neal, *Drug Metab. Dispos.* **4,** 314 (1976).
[5] O. Tirosh, Y. Katzhendler, Y. Barenholz, I. Ginsburg, and R. Kohen, *Free Radic. Biol. Med.* **20,** 421 (1996).

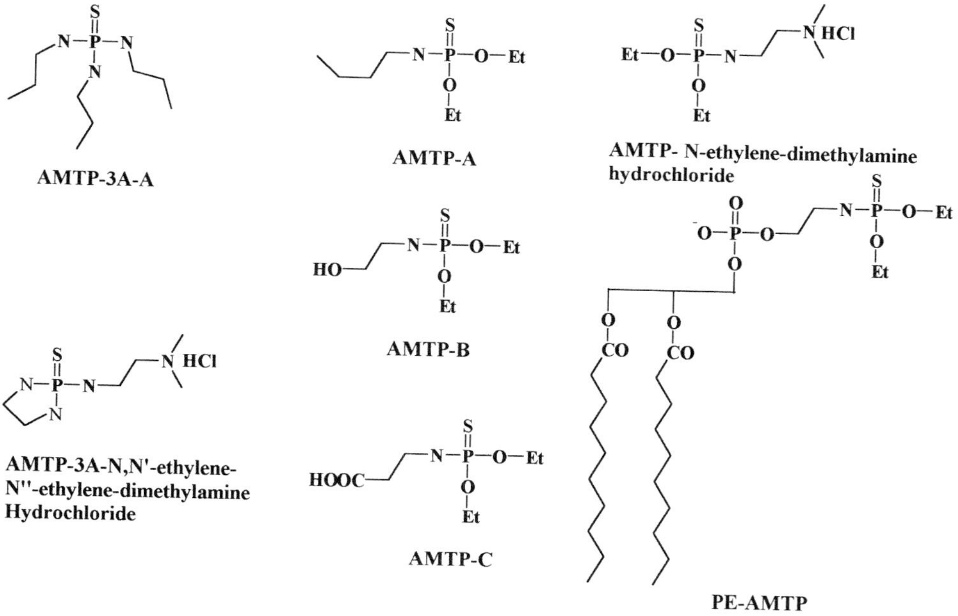

FIG. 1. Chemical structure of the AMTP family of antioxidants.

ponents such as lipids, proteins, plasma lipoproteins, low-density lipoproteins (LDL), membranes, and cells.

Synthesis: Examples of AMTP Preparation

Procedure for Synthesis of AMTP-B
(2-Hydroxyethylamidodiethyl Thionophosphate)

Diethyl chlorothiophosphate (2 g/10.6 mmol) is dissolved in 50 ml dry dichloromethane, and the solution is added dropwise with stirring over 30 min to an ice-cold solution of ethanolamine (3.2 g = 53 mmol) in 50 ml of dry dichloromethane. After 1 hr of stirring, 25 ml of acidified water (HCl), pH 2.0, is added and the organic layer is washed three times to extract all the free amine. The organic layer is dried on anhydrous magnesium sulfate, filtered, and evaporated to dryness. A clear liquid is obtained: Yield: 1.8 g (79%) ^{31}P NMR (72 ppm single peak). IR: 1400, 1490, 2900, 3500 (a broad peak). ^{1}H NMR (300 MHz, CDCl$_3$) d 1.25–1.35 (t, 6H), d 3.03–3.13 (m, 2H), d 3.6–3.68 (t, 2H), d 3.95–4.15 (bq, 4H).

Synthesis of N,N'N''-Tripropylamidothionophosphate (AMTP-3A)

Trichlorothiophosphate (1 equivalent) is dissolved in 50 ml dry dichloromethane. The solution is added dropwise with stirring to a solution of propylamine (9 equivalents) in 50 ml dichloromethane. After stirring for 1 hr, 25 ml acidified water (HCl), pH 2.0, is added and the organic layer is washed five times to extract all the free amine. The organic layer is dried on anhydrous magnesium sulfate and evaporated to dryness. The white powder residue is collected and dried under reduced pressure for 5 hr. Yield: 85%. ^{31}P NMR (CDCl$_3$, 121.42 MHz)-singlet 63 ppm. ^{1}H NMR (CDCl$_3$, 299.9 MHz)-9H (t, 0.8–1 ppm), 6H (m, 1.4–1.6 ppm), 6H (dt, 2.8–3 ppm).

Reactivity against Sodium Hypochlorite and Hydrogen Peroxide

Hydrogen Peroxide Decomposition. AMTP-B, which possesses only one primary amino group, is compared to AMTP-3A, with three amino groups (100 mM final concentration). The activity of these compounds is tested by their rate of reaction with 1 M hydrogen peroxide in a solution of dioxane : water, 1 : 9 (v/v), having a dielectric constant of 9.6. The reaction rates are monitored by following the disappearance of the thiophosphate using ^{31}P NMR. ^{31}P NMR spectra are used to evaluate the disappearance of the AMTP compounds and to elucidate the oxidation product following exposure to hydrogen peroxide. The rate of oxidation of AMTP thionophosphate to phosphate by hydrogen peroxide is followed by ^{31}P NMR. AMTP-3A is consumed 15 times faster than AMTP-B. In 3 hr, 20 and 95% of AMTP-3A and AMTP-B, respectively, are left, and after 30 hr, 30% of AMTP-B remains. The decomposition products of the original AMTPs have chemical shifts between 10 and -5 ppm, which means that sulfur is replaced by oxygen.[5] In the control mixtures of dioxane and water, both AMTPs are stable.

Reactivity against Sodium Hypochlorite: Cell Cultures. Human skin fibroblasts are grown in Dulbecco's modified Eagle's medium (DMEM) supplemented with glutamine, penicillin, streptomycin, and 10% fetal calf serum. Radiolabeled cell monolayers are prepared by the addition of 10 mCi/ml of [^{51}Cr]NaCrO$_4$ (New England Nuclear) to 100 ml of trypsinized cells grown in 75-ml tissue culture bottles. Cells are then dispensed into 24-well tissue culture plates (Nunc, Roskilde, Denmark) and grown to confluency in a CO$_2$ incubator. Before exposing the cells to sodium hypochlorite the DMEM medium is replaced by Hanks' balanced salt solution (HBSS) buffer. AMTP-B and ascorbic acid are compared as protecting agents for the cells (Fig. 2). To assess cytotoxicity, supernatant fluids are removed and centrifuged at 2000 rpm for 2 min (600 g). Solubilized radioac-

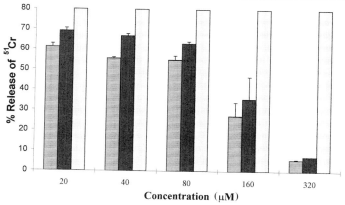

Fig. 2. Protection of fibroblast cells in culture against exposure to oxidative stress of 1 mM HOCl for 1 hr. First column, AMTP-B at various concentrations; second column, ascorbic acid at various concentrations; third column, control. The reaction medium was HBSS buffer.

tivity is then determined in a Kontron gamma counter. The total radioactivity associated with untreated controls is solubilized by the addition of 1 ml of 1% Triton X-100. AMTP-B and ascorbic acid show a strong protective effect on the cells, probably by scavenging the sodium hypochlorite in the medium and thereby preventing damage to essential biological molecules on the cell membranes (Fig. 2). We speculate that AMTP-B, which is an unionized semilipophilic molecule [partition coefficient P_B(oil/water) = 0.25, P_B(octanol/water) = 3.22] is also effective in scavenging intracellular hypochlorous acid.

AMTP as Effective Organoperoxide Reducer and as Antioxidant Lacking Prooxidative Properties

AMTP-B, 10 mM, pH 7.2, is incubated at 37° in an aqueous solution of Cu^{2+}, 10 mM. Oxygen consumption from the solution is measured with a biological oxygen monitor. No oxygen consumption is observed from the Cu^{2+} solution with and without AMTP. However, with 10 mM ascorbic acid or cysteine there is significant oxygen consumption. To further test the lack of prooxidant effects in the antioxidant activity of the AMTPs, small unilamellar liposomes (vesicles) (SUV) are prepared from 25 mM egg phosphatidylcholine/7 mM cholesterol in 50 ml HEPES (20 mM) buffer, pH 7.2, using the high-pressure homogenizer Model Minlab type 8.30 H (APV Rannie, Albertslund, Denmark) according to a published procedure.[6]

[6] Y. Barenholz and S. Amselem, in "Liposome Technology" (G. Gregoriadis, ed.), 2nd ed., Vol. I, p. 501. CRC Press, Boca Raton, FL, 1993.

TABLE I
ABILITY OF AMTP-3A IN CONTRAST TO NaS$_2$O$_4$ TO PROMOTE DECOMPOSITION OF LIPID HYDROPEROXIDES IN LIPOSOMES WITHOUT PROOXIDATIVE SIDE EFFECTS (LOSS OF PUFA)

Liposome	Time (hr)	Lipid hydroperoxides (mM)	Total PUFAs lost (%)
Without AMTP-3A[a]	0	0.078	75
	96	0.42	
With AMTP-3A[a]	0	0.069	9.2
	96	0.022	
Without dithionite	0	0.24	25
	168	0.25	
With dithionite	0	0.24	55
	168	0.0	

[a] At 37°.

Egg phosphatidylcholine (EPC2) is from Lipoid KG, Ludwigshafen, Germany, and cholesterol is from Sigma (St. Louis, MO). The liposomes are used following storage for 1 year at 4° in vacutainer tubes. These storage conditions result in the accumulation of hydroperoxides; however, most of the polyunsaturated fatty acids (PUFA) in the liposomes are not oxidized. The liposomes are exposed to oxidation conditions under air at 37° for 96 hr in the presence and absence of 2 mM AMTP-3A. Lipid hydroperoxides are monitored using a spectroscopic method that is modified to the micromolar range as follows: 50 μl of liposome dispersion is dissolved in 1 ml of ethanol. Fifty microliters of a 50% KI solution is added and the mixture is incubated for 20 min in the dark. Absorbance at 400 nm is measured. Liposome acyl chain composition is evaluated by GC at the beginning and the end of the incubation period. Lipid hydroperoxides in the liposomes constantly accumulated following incubation at 37° (from 0.07 to 0.42 mM after 96 hr) (Table I). However, in the presence of AMTP-3A no accumulation is observed. The content of lipid hydroperoxide is even reduced to the lowest detection threshold of the method (Table I). In order to evaluate whether the decomposition of the lipid hydroperoxides observed is due to an oxidative process (such as metal-induced decomposition) or due to a nonoxidative decomposition process, we measured the acyl chain composition of the liposomes. Results show that control liposomes (without AMTP) lose 66% of linoleic acid, 100% of arachidonic acid, and 100% of docosahexaenoic acid. In contrast, liposomes containing AMTP-3A lose only 9% of linoleic acid, 10% of arachidonic acid, and 10% of docosahexaenoic acid. These results indicate that the decomposition of the lipid hydroperoxides is not accompanied by an oxidative damage to the PUFA.

In another experiment we evaluated the same effect of peroxide decomposition, but this time by another compound. We used sodium dithionite ($Na_2S_2O_4$) on partially oxidized liposomes. Dithionite, in contrast to the AMTP, is a strong reducer agent that can reduce metals and oxygen. Interaction of this compound with peroxides or plain oxygen reduces them to water. The experiment included incubation of the liposomes with 10 mM of $Na_2S_2O_4$ for a 1-week period at room temperature. The liposomes are then analyzed for their acyl chain composition and peroxide content. Results show that although the lipid hydroperoxide content is reduced dramatically in the presence of $Na_2S_2O_4$, the acyl chain composition is strongly altered and most of the PUFAs are completely oxidized (Table I).

Lack of Ability of AMTP to Scavenge Peroxyl Radicals

AMTP might serve as the only selective antioxidant to separate between the contribution of lipid hydroperoxide and peroxyl radicals to the accumulation of oxidative damage in an oxidizing lipidic system. We demonstrated that AMTPs are unable to interact with peroxyl radicals. The reaction mixture consists of 1.5×10^{-8} M R-phycoerythrin (R-PE) in PBS, pH 7.4, at 37°. The oxidation reaction is started by adding 2,2′-diazobis(2-amidinopropane) dihydrochloride (AAPH) (temperature-dependent peroxyl radical generator) to a final concentration of 4.0 mM, and the decay of R-PE is monitored every 30 sec for 30 min. Scavenging of the peroxyl radicals is done by 5 μM uric acid and by AMTP-B at 10 and 100 μM.

AMTPs are antioxidants that can simultaneously donate two electrons.[5] Therefore, it is expected that when interacting with peroxyl radicals (which react preferentially with one electron at a time, as is the case with tocopherol) these compounds will have very low reactivity. This was proven using a flux of peroxyl radicals produced by AAPH that induced constant loss of R-PE fluorescence due to fluorophore oxidation. In the presence of 5 μM uric acid, an inhibition lag of 8 min is observed in the oxidation of R-PE. AMTP-B at concentrations of 10 and 100 μM does not induce any lag period in the loss of fluorescence, indicating that it does not scavenge peroxyl radicals.

AMTPs do not show any ability to scavenge peroxyl radicals or to inhibit the propagation phase of the lipid peroxidation chain reaction process. Measurements are conducted by three different methods. Peroxyl radicals are initiated by AAPH, an azo compound that decomposes at a temperature-dependent rate to produce peroxyl radicals. ^{31}P NMR of AMTP-B does not show any change after reaction with AAPH. No inhibition of oxygen consumption is observed when lipid peroxidation is induced in an emulsion by AAPH, and no protection against bovine serum albumin oxidation by peroxyl radicals is observed.

Conclusion

AMTP compounds may represent a new approach in the design of antioxidants. They are capable of reacting with hydroxyl radicals, sodium hypochlorite, hydrogen peroxide, superoxide, and lipid hydroperoxide, but are unable to inhibit free radical chain reactions or to scavenge peroxyl radicals. The mechanism by which they prevent lipid oxidation implies that oxidative damage to lipids might be inhibited by decomposition of unstable ROS such as lipid hydroperoxides instead of reaction with free radicals, e.g., peroxyl radical, in an attempt to break the radical chain. The unique qualities of AMTP might serve as a tool that will allow us to gain a better understanding of the factors involved in biological oxidation processes.

[27] Quantitation of Anethole Dithiolthione Using High Performance Liquid Chromatography with Electrochemical Detection

By KATRINA TRABER and LESTER PACKER

Introduction

It has been known since the 1950s that cabbage, broccoli, and other cruciferous vegetables all contain dithiolthiones, especially 1,2-dithiol-3-thione.[1] These vegetables have been used as medicines since ancient times and, more recently, have been reported to have an anticancer effect. This anticancer effect has been shown to be related to the sulfur-containing dithiolthiones.[2] ADT (Anethole dithiolthione, Sulfarlem) is a synthetic dithiolthione that has been used clinically for many years as a choleretic and in the treatment of xerostomia.

More recently, ADT has been shown to have chemoprotective and antioxidant-like effects.[3] Based on these data, the effects of ADT on NF-κB, a redox-sensitive transcription factor, were examined. In Würzburg T cells, ADT (i) inhibits the H_2O_2-induced activation of NF-κB, (ii) inhibits lipid peroxidation induced by H_2O_2, and (iii) increases cellular glutathione.[4]

---

[1] L. Jirousek and J. Starka, *Nature* **45,** 386 (1958).
[2] M. Albert-Pielo, *J. Ethnopharmocol.* **9,** 261 (1983).
[3] M. Christen, "Proceedings of the International Symposium on Natural Antioxidants, Molecular Mechanisms and Health Effects" (L. Packer, M. Traber, and W. Xin, eds.), p. 236. ACOS Press, Champaign, IL, 1996.
[4] C. K. Sen, K. Traber, and L. Packer, *Biochem. Biophys. Res. Commun.* **218,** 148 (1996).

This indicates that ADT is modifying cellular responses to oxidative stress and that ADT is a redox active agent. Similarly, dihydrolipoic acid (DHLA), a thiol dithiolane antioxidant, has also been shown to inhibit peroxide-mediated activation of NF-κB and has many of the same effects on cellular response to oxidative stress that are shown by ADT.[4-6] Thus, redox-sensitive agents can have marked effects on cell processes.

Because of its widespread use as a therapeutic drug and its potential use as an antioxidant, it is useful to measure ADT concentrations *in vivo* as well as *in vitro*. To this end, a method was developed to extract ADT from phospholipid liposomes as well as Würzburg cells. In addition, a highly sensitive high-performance liquid chromatography (HPLC) method was developed to measure the ADT concentrations found in these extracts. These methods can be used in the future not only to further study ADT as a potential antioxidant, but also as a clinical method to determine plasma concentrations of ADT.

Methods

Anethole dithiolthione is a kind gift of Solvay Pharma-LTM (42 rue Rouget de Lisle-92151 Suresnes, Cedex, France)

HPLC Analysis

The HPLC electrochemical detection system consists of a Beckman 144M solvent delivery module and a BAS amperometric detector with a BAS gold electrode plated with mercury. Samples are separated by an Alltech Altima C_{18} column (150 mm, I.D. 4.6 mm). The mobile phase is 70% methanol and 0.1 M monochloroacetic acid (pH 2.9). The mobile phase flows through the system at 1.0 ml/min. Samples are injected onto a 50-μl loop and are separated by the column. The ADT is detected by a BAS gold electrode essentially as in Han *et al.*[7] The upstream electrode is set to −1.001 V and the downstream electrode is set to 0.050.

ADT Extraction

ADT is extracted from cells, liposomes, or buffer prior to injection onto the HPLC. The extraction method is a variation of the chloroform/methanol method developed by Folch *et al.*[8] To maximize the recovery of ADT from

---

[5] V. E. Kagan, A. Shvedova, E. Serbinova, S. Khan, C. Swanson, R. Powell, and L. Packer, *Biochem. Pharmacol.* **44,** 1637 (1992).
[6] L. Packer, E. Witt, and H. J. Tritschler, *Free Radic. Biol. Med.* **19,** 227 (1995).
[7] D. Han, G. J. Handelman, and L. Packer, *Methods Enzymol.* **251,** 315 (1995).
[8] J. Folch, M. Lees, and G. H. Sloane-Stanley, *J. Biol. Chem.* **226,** 497 (1957).

the extraction, the amounts of methanol and buffer added to the chloroform are varied. The method is discussed in more detail under Results.

To a cell pellet, liposomes, or 20 μl phosphate-buffered saline (PBS), 1 ml of PBS and 1 ml chloroform are added. The sample is shaken well and centrifuged in a Fisher centrifuge at 500 g for 5 min. The bottom (chloroform) layer is removed and a 500-μl aliquot is dried down under nitrogen. The residue is resuspended in 200 μl of fresh mobile phase and injected onto the HPLC for measurement.

Cell Culture

Human lymphoma Würzburg T cells [a clone of Jurkat T cells, developed by Dr. Patrick Baeuerle (Frieburg, Germany)] are a kind gift of Dr. Leonard Herzenberg of Stanford University, California. The cells are grown in RPMI 1640 medium supplemented with 10% fetal calf serum (FCS), 1% (w/v) penicillin–streptomycin, 1% sodium pyruvate, and 1% L-glutamine (University of California, San Francisco Cell Culture Facility) in humidified air containing 5% (v/v) CO_2.

Liposome Formation

Liposomes are made from phosphatidylcholine diolyl (DOPC, Sigma Chemical Co., St. Louis, MO) 1 mg/ml in hexane. In a glass conical tube, 2 ml of DOPC is dried down under nitrogen. To the residue, 800 μl PBS is added. The solution is sonicated in a bath sonicator for 5 min until the solution becomes uniformly cloudy. The phospholipid solution is extruded through a French press. The solution is then filter sterilized before using with cells.

Results

HPLC Method

Thus far, our studies concerning the antioxidant properties of ADT have been limited in scope because we did not have an accurate and reliable method to measure ADT concentrations. An HPLC method to measure ADT with fluorescence detection was reported by Masoud and Bueding,[9] but ADT detection limits were above those that could be found in cells. To remedy this problem, an electrochemical method was developed that is more sensitive than the existing methods.

Because ADT has vicinal thiol groups similar to lipoic acid, the HPLC

---

[9] A. Masoud and E. Bueding, *J. Chromatogr.* **276**, 111 (1983).

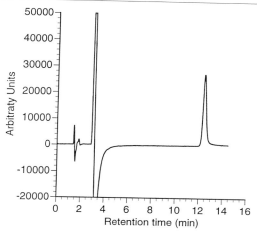

FIG. 1. Chromatogram of ADT.

separation method developed for lipoic acid by Han et al.[7] was used as a starting point. Because ADT is highly insoluble in water, acetonitrile was eliminated from the mobile phase and the proportion of methanol was increased to 70%. With this mobile phase, the ADT was eluted from the column after approximately 12 min. A typical chromatogram of 250 pmol of ADT is shown in Fig. 1. The large peak occurring at approximately 3.5 min is a result of residual oxygen in the sample and is commonly found during electrochemical detection.

This ADT detection method is highly sensitive. At amounts ranging from 0.5 pmol to 1 nmol injected, ADT has been quantitated accurately (Fig. 2). Within this range, the standard curve is linear and extremely reproducible.

Extraction of ADT from Liposomes and Cells

An extraction method was developed in order to measure the concentration of ADT in cells, liposomes, and buffer. During chloroform/MeOH extraction,[8] because of the insolubility of ADT in water, the ADT concentrated in the nonpolar chloroform layer of the extract. This layer was dried down, resuspended in mobile phase, and injected onto the HPLC. Results from this method were relatively reliable, but had recovery rates of <20%. Figure 3 shows the standard curves obtained from extracting known amounts of ADT from samples containing ADT added to PBS buffer or a Würzburg cell pellet. The standard curves obtained from this experiment are both linear, but the recovery was between 10 and 20% for both PBS and cell extractions.

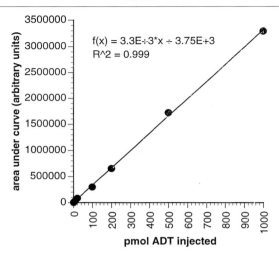

FIG. 2. ADT detection.

Because of these results, it was apparent that some of the ADT remained with the methanol in the aqueous layer and therefore did not concentrate in the chloroform as we had originally hypothesized. To investigate this, a known amount of ADT was added (2.0 nmol) to a cell pellet or PBS and was extracted with a constant amount of chloroform and varying ratios of

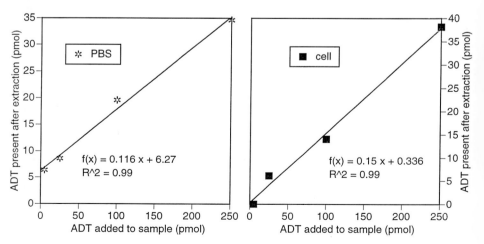

FIG. 3. Standard curves for ADT extraction.

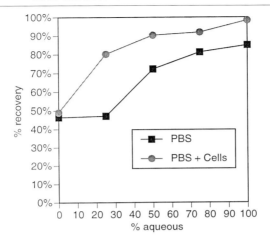

FIG. 4. Effect of PBS on recovery of ADT.

PBS and methanol (Fig. 4). The results of this experiment showed that as the proportion of PBS increased to 100%, the recovery of ADT increased to near 100%. This indicated that a high concentration of PBS in the extraction solution forces the ADT to concentrate in the chloroform layer.

The previous extraction experiment was then repeated, this time using 100% PBS instead of the 50:50, PBS:methanol mixture used previously. This modified extraction method gave clean, reproducible chromatograms

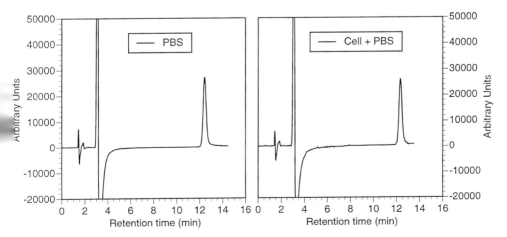

FIG. 5. Modified extraction method.

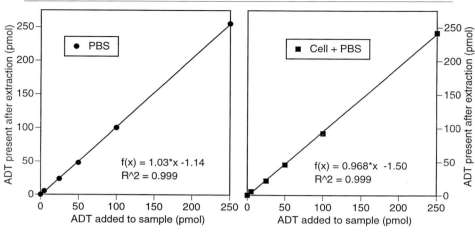

Fig. 6. Recovery of ADT.

(Fig. 5) and, contrary to the previous extraction experiment, the recovery of ADT was near 100% (Fig. 6).

Discussion

This article demonstrates a highly reproducible and effective method to measure ADT in cells as well as in aqueous solutions. The present method is 50-fold more sensitive than previously reported for fluorescence detection of ADT following HPLC separation.[9] In addition, the extraction method described herein removes nearly 100% of the ADT from cellular or aqueous samples.

In addition to the increased sensitivity of the EC method, this new method will also allow us to measure any redox products of ADT. Because ADT has a structure similar to lipoic acid, ADT may interact with other cellular thiols in a manner similar to lipoic acid. When lipoic acid enters the cell, the disulfide bond is reduced to form dihydrolipoic acid. It is not yet known if ADT may be metabolized in a similar manner when it is incubated with Würzburg cells.

Acknowledgment

We thank Dr. M. O. Christen of Solvay Pharma-LTM for valuable discussions.

Section V

Vitamin E and Coenzyme Q_{10}

[28] Gas Chromatography–Mass Spectrometry Analysis of Vitamin E and Its Oxidation Products

By Daniel C. Liebler, Jeanne A. Burr, and Amy J. L. Ham

Introduction

Vitamin E (α-tocopherol, TH[1]) is the principal inhibitor of lipid peroxyl radical propagation in biological membranes and lipoproteins.[1,2] α-Tocopherol exerts antioxidant effects primarily by trapping peroxyl radicals. Products of this reaction include two groups of unstable primary products (Fig. 1).[3–5] The first are 8a-substituted tocopherones derived from radical addition to the 8a position of the tocopheroxyl radical or from hydrolysis of the tocopherone cation.[5] The second product group includes isomeric epoxy(hydroperoxy)tocopherones (epoxytocopherones), which are formed by an unknown mechanism.[3,6] The 8a-substituted tocopherones hydrolyze readily to α-tocopherolquinone (TQ),[7] whereas the isomeric epoxytocopherones hydrolyze to epoxyquinones TQE1 and TQE2.[3] A two-electron reduction of TQ yields α-tocopherolhydroquinine (THQ).[8]

A variety of chromatographic and spectroscopic methods have been employed to analyze TH (reviewed in Sheppard *et al.*[9]). The most sensitive of these employ reverse-phase high-performance liquid chromatography) (HPLC) with electrochemical or fluorescence detection.[10–14] Strong ultraviolet (UV) absorbance at 265 nm permits UV-based detection of TQ in the picogram range,[15] whereas the weaker THQ chromophore permits

---

[1] G. W. Burton and U. K. Ingold, *Acc. Chem. Res.* **19**, 194 (1986).
[2] L. Packer and J. Fuchs, eds., "Vitamin E in Health and Disease." Dekker, New York, 1993.
[3] D. C. Liebler, P. F. Baker, and K. L. Kaysen, *J. Am. Chem. Soc.* **112**, 6995 (1990).
[4] D. C. Liebler, K. L. Kaysen, and J. A. Burr, *Chem. Res. Toxicol.* **4**, 89 (1991).
[5] D. C. Liebler and J. A. Burr, *Biochemistry* **31**, 8278 (1992).
[6] M. Matsuo, S. Matsumoto, Y. Itaka, and E. Niki, *J. Am. Chem. Soc.* **111**, 7179 (1989).
[7] D. C. Liebler, K. L. Kaysen, and T. A. Kennedy, *Biochemistry* **28**, 9772 (1989).
[8] T. Hayashi, A. Kanetoshi, M. Nakamura, M. Tamura, and H. Shirahama, *Biochem. Pharmacol.* **44**, 489 (1992).
[9] A. J. Sheppard, J. A. T. Pennington, and J. L. Weihrauch, *in* "Vitamin E in Health and Disease" (L. Packer and J. Fuchs eds.). Dekker, New York, 1993.
[10] S. K. Howell and Y. M. Wang, *J. Chromatogr.* **227**, 174 (1982).
[11] G. A. Pascoe, C. T. Duda, and D. J. Reed, *J. Chromatogr.* **414**, 440 (1987).
[12] M. E. Murphy and J. P. Kehrer, *J. Chromatogr.* **421**, 71 (1987).
[13] L. J. Hatam and H. J. Kayden, *J. Lipid Res.* **20**, 639 (1979).
[14] G. T. Vatassery and D. F. Hagen, *Anal. Biochem.* **79**, 129 (1977).
[15] G. T. Vatassery and W. E. Smith, *Anal. Biochem.* **167**, 411 (1987).

somewhat less sensitive detection at 280 nm.[8] Electrochemical detection also offers a sensitive means of detecting TQ and THQ in reverse-phase HPLC analyses.[11,12] TQE1 and TQE2, however, lack a strong UV chromophore and appropriate redox chemistry for electrochemical detection.

We have developed a stable isotope dilution capillary gas chromatography–mass spectrometry (GC-MS) method for analyzing TH and its principal end products, TQ, TQE1, TQE2, and THQ. The method permits simultaneous analysis of TH and its oxidation products at subpicomole levels in a single sample and provides a comprehensive assessment of vitamin E status and antioxidant reactions in biological systems.

Preparation of Deuterium-Labeled α-Tocopherol and Oxidation Products

Deuterium-labeled α-tocopherol and oxidation products are synthesized from 5,7-dimethyl-[C^2H_3]-(R,R,R)-α-tocopheryl acetate, which was a gift of the Natural Source Vitamin E Association (Kingsport, TN). Deuterium-labeled α-tocopherol can also be synthesized by a one-pot procedure described previously.[16] Standards of d_6-TH and its oxidation products are prepared from d_6-α-tocopherol acetate. d_3-TH for preparation of d_3-THQ (see later) is prepared from d_3-α-tocopherol acetate. In a screw cap tube flushed with nitrogen, 25 μmol of the acetate ester is treated with 500 μmol lithium aluminum hydride in 2 ml tetrahydrofuran at 0° for 30 min. The excess lithium aluminum hydride is quenched at 0° with 0.2 ml methanol, followed by 0.2 ml water. After acidification of the mixture with 0.5 ml 1 N HCl, the product is extracted with 3 × 2 ml hexane. Evaporation of the hexane in vacuo affords 5,7-dimethyl-[C^2H_3]-(R,R,R)-α-tocopherol (d_6-TH). Deuterated standards of TQ, TQE1, TQE2, and THQ are prepared by oxidation of d_6-TH or d_3-TH as described previously[3,5] and purified by reversed phase HPLC on a 4.6 × 250-m Spherisorb ODS-2, 5 μm column eluted with methanol at a flow rate of 1.5 ml min^{-1}.

THQ and d_3-THQ for calibration curves or for internal standard use are prepared from TQ or d_3-TQ. It is necessary to prepare THQ immediately prior to use, as this compound oxidizes readily to TQ. TQ (5 nmol) is dissolved in 0.5 ml ethanol at room temperature, to which 10 mg sodium borohydride is added. After 5 min, excess borohydride is quenched with 0.2 ml 0.1 N acetic acid, 0.3 ml water is added, and the reaction mixture is extracted with 1 ml hexane. Aliquots of this hexane solution containing THQ or d_3-THQ are removed immediately for preparing calibration curves or for use in samples (see later).

[16] S. Urano, Y. Hattori, S. Yamanoi, and M. Matsuo, *Chem. Pharm. Bull.* **28**, 1992 (1980).

Because HPLC-purified samples of d_6-TQ typically accumulate detectable amounts of d_6-THQ on storage, it is necessary to treat d_6-TQ stock solutions with $FeCl_3$ to reoxidize any residual d_6-THQ to d_6-TQ. A solution nominally containing 5 nmol d_6-TQ in 1 ml hexane is stirred vigorously with 1 ml water containing 10 mg $FeCl_3$ for 30 sec. A 20-ml aliquot of the hexane layer (500 fmol) is then taken for the preparation of calibration curves or samples.

Unlabeled TQ, TQE1/TQE2, and THQ are synthesized from TH as described previously.[3,5]

Preparation of Calibration Curves

Calibration curves are prepared from purified standards of TH, TQ, TQE1, and TQE2. Standard solutions containing 25, 50, 100, 500, 1000, or 2000 fmol each of unlabeled TH, TQ, TQE1, and TQE2 are added to 1 ml silanized autosampler vials. To these are added 500 fmol each of d_6-TH, d_6-TQ, d_6-TQE1, and d_6-TQE2 and the solvent is evaporated *in vacuo*. Samples are then derivatized and analyzed as described later. Calibration curves for all analytes should be prepared either daily or for each batch of samples to be analyzed.

Sample Extraction and Derivatization

All glassware used in sample preparation and analysis is silanized with dichlorodimethylsilane (Pierce, Rockford, IL) in toluene, rinsed with methanol, and dried. Procedures for analysis of TH and its oxidation products from rat liver microsomes are described here. Possible modifications of these procedures for analyses of different tissue samples are discussed later. Samples corresponding to 1 mg microsomal protein (typically 0.5–2.0 nmol TH) in 1 ml aqueous buffer are added to silanized glass tubes containing 1 ml 1 N sodium dodecyl sulfate and 1 ml ethanol containing 50 nmol butylated hydroxytoluene. To this is added 0.5 pmol each of d_6-TH, d_6-TQ, d_6-TQE1, and d_6-TQE2 as internal standards. The samples are then extracted twice with 1 ml hexane. Brief (15 sec) microprobe sonication is used to disperse the mixture, and the organic and aqueous layers are then separated by centrifugation at 3000 rpm. The combined hexane extracts are evaporated to dryness in a Speed-Vac concentrator (Savant Instruments, Farmingdale, NY).

The extraction procedure described here is a modification of that reported by Burton *et al.*[17] The method entails denaturation of sample proteins

---

[17] G. W. Burton, A. C. Webb, and K. U. Ingold, *Lipids* **20,** 29 (1985).

with SDS in aqueous ethanol, followed by extraction with heptane or hexane. α-Tocopherol recovery is most reproducible if the sample is allowed to sit for 30 min following the addition of SDS and ethanol and prior to extraction with hexane.

Because some oxidized TH may be present in the form of 8a-substituted tocopherones or epoxytocopherones (Fig. 1), samples are treated to convert these acid-labile species to TQ and TQE1/2, respectively.[5] The evaporated extract residue is redissolved in 2 ml ethanol/1 N HCl (1:1, v/v) for 30 min at room temperature. The mixture is then reextracted with 3×1 ml hexane, and the hexane extracts are transferred to crimp-top autosampler vials and evaporated. The residue is redissolved in 100 μl dimethylformamide and then treated with 100 μl BSTFA/TMCS (10:1, v/v) at room temperature for 2 hr.

Complete derivatization is achieved in either pyridine or dimethylformamide, although the derivatives appear to be considerably less stable in pyridine. This is particularly important, for example, when multiple samples are left at room temperature overnight for automated analyses. Samples in dimethylformamide appear to be stable for 24 hr. Omission of TMCS from the derivatization results in incomplete derivatization of the tertiary side-chain hydroxyl group of TQ and TQE1/TQE2. Fresh ampules of BSTFA-TMCS are opened daily.

GC-MS Analysis

GC-MS analyses are done with a Fisons-VG MD-800 instrument equipped with a Carlo Erba 8000 Series gas chromatograph and a CTC A-200S autosampler (Fisons Instruments, Beverly, MA). However, other comparable GC-MS instruments are suitable for these analyses, provided that an on-column injector is aviable. Samples are introduced by on-column injection at 150° and separated on a 30 m $\times$ 0.25 mm DB-5ms column (J & W Scientific, Folsom, CA). The initial column temperature is 150°. After 1 min, the column is programmed at 25° min^{-1} to 260°, then at 5° min^{-1} to 280°, and then held at that temperature for 20 min. Compound detection is by electron ionization mode at 70 eV. Full scan spectra are obtained over a 100- to 600-amu mass range. Selected ion monitoring (SIM) is done at nominal mass for fragments of interest with a 1-amu mass window. Dwell time for each ion is 12 sec with a 0.6-sec interchannel delay. Peak areas in selected ion chromatograms are obtained by manual integration using Fisons-VG Lab-Base data system software.

The derivatives elute in the order TQE1–TMS, TQE2–TMS, and TQ–TMS, followed by TH–TMS. Both epoxyquinones elute as a major and minor peak, which correspond to epoxide diastereomers.

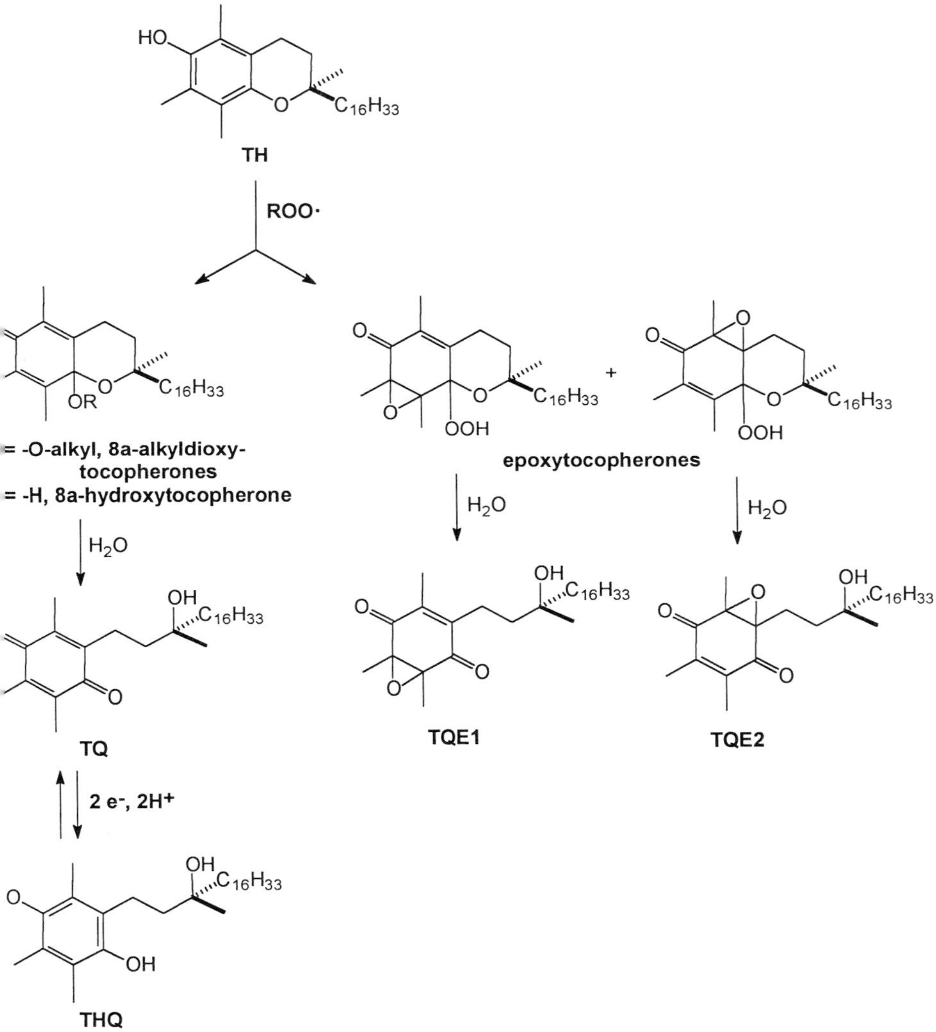

FIG. 1. Oxidation pathways for TH in peroxyl radical scavenging reactions. [From A. J. L. Ham and D. C. Liebler, *Arch. Biochem. Biophys.* **339**, 157 (1997). Used with permission.]

Selected Ion Monitoring

α-Tocopherol is detected by SIM of the molecular ion at m/z 502. Because the monoisotopic mass is 502.42, this ion frequently appears at either m/z 502 or 503 in full-scan analyses, depending on instrument calibra-

TABLE I
IONS USED FOR GC-MS SIM ANALYSIS OF TH,
TQ, TQE1, TQE2, AND THQ

Compound	Ions $(m/z)^a$
TH–TMS	503/509
TQ–TMS	293/299
TQE1–TMS	309/315
TQE2–TMS	309/315
THQ–TMS	309/312

^{a} Ion pairs correspond to m/z for the unlabeled fragment and the corresponding fragment from labeled standard. [From D. C. Liebler et al., Anal. Biochem. **236,** 27 (1996). Used with permission.]

tion. Thus, SIM acquisition is done at m/z 502 ± 0.5. Monitoring of TH via the more intense fragment ion at m/z 237 is complicated by a fragment ion from the TMS derivative of cholesterol, which elutes near TH–TMS. Because the cholesterol content of most biological samples greatly exceeds that of TH, this near-coelution effectively precludes the quantitation of TH through the m/z 237 fragment ion.

Selected ion monitoring detection of TQ is achieved by monitoring the fragment ion at m/z 293. Similarly, detection of TQE1 and TQE2 is accomplished by monitoring the analogous fragment ion at m/z 309. Although TQE1 and TQE2 each can be detected selectively by monitoring characteristic ions at m/z 267 and m/z 211, this would appear to be generally unnecessary. Moreover, the need to monitor two ions rather than one to detect TQE1/TQE2 would reduce the analyzer time devoted to signal acquisition for these compounds and therefore reduce the sensitivity of the analysis. A summary of the ions used for SIM of TH, TQ, TQE1, and TQE2 and their deuterated analogs is given in Table I. Calibration curves for TH, TQ, TQE1, and TQE2 are typically linear from 25 fmol to 2000 pmol. Calibration curves for TH are linear to 4000 fmol. Correlation coefficients typically range from 0.995 to 0.999.

Oxidation of α-Tocopherol in Rat Liver Microsomes and in Perfused Rat Liver

Use of this GC-MS method to analyze the fate of TH in antioxidant reactions is illustrated by our previously published studies of TH oxidation in rat liver microsomes and perfused rat liver.[18] Microsomes prepared from

[18] D. C. Liebler, J. A. Burr, L. Philips, and A. J. L. Ham, Anal. Biochem. **236,** 27 (1996).

TABLE II
OXIDATION OF TH TO TQ, TQE1, AND TQE2 IN RAT LIVER MICROSOMES

Time (min)[a]	TQE1	TQE2	TQ	TH	Total
0	3 ± 2	31 ± 8	9 ± 12	288 ± 24	330 ± 24
15	11 ± 3	61 ± 4	159 ± 30	104 ± 18	334 ± 30

[a] Microsomes were incubated with 0.15 mM cumene hydroperoxide and 0.1 mM Fe(NH$_4$)$_2$(SO$_4$)$_2$ for 15 min at 37°. Product amounts are in pmol mg^{-1} microsomal protein and represent mean ± SD (n = 3). [From D. C. Liebler et al., Anal. Biochem. **236,** 27 (1996). Used with permission.]

livers of male Sprague-Dawley rats are incubated at 1 mg ml^{-1} microsomal protein in 50 mM Tris–acetate, pH 7. Oxidation is initiated by adding cumene hydroperoxide (0.15 mM) followed immediately by Fe(NH$_4$)$_2$(SO$_4$)$_2$ (0.1 mM). After incubation for 15 min at 37°C under an air atmosphere, oxidations are terminated by the addition of 1 mM deferoxamine mesylate, and 0.5-ml samples containing 0.5 mg microsomal protein are extracted as described earlier and the hexane extracts are evaporated under nitrogen. The residue is then dissolved in 2 ml of ethanol/1 N HCl (1:1, v/v). After 30 min at room temperature, the mixture is extracted with 3 × 1 ml of hexane and the extracts are again evaporated. The products then are silylated for GC-MS analysis as described earlier.

Results reported for cumene hydroperoxide- and iron-catalyzed oxidations in microsomes are typical of TH oxidations studied in other *in vitro* systems.[3,5] Oxidation for 15 min consumed approximately 64% of the initial TH (Table II). TQE1 and TQE2 accounted for 4 and 16%, respectively, of the TH consumed, whereas TQ accounted for 80%. Because the extracts containing TH oxidation products were treated acid prior to analysis, the amount of TQ measured represents the total of TQ and all 8a-substituted tocopherones formed in the microsomes. Essentially all of the TH consumed can be accounted for by GC-MS as TQ and TQE1/TQE2, within experimental error.

For experiments in perfused rat liver, livers from male Sprague–Dawley rats are perfused in a nonrecirculating system with Krebs–Henseleit bicarbonate buffer (pH 7.4, 37°) saturated with an oxygen/carbon dioxide mixture (95:5) as described.[19] After a 30-min perfusion with buffer only, the liver is perfused for 10 min with buffer only (control) or buffer containing 2 mM *tert*-BuOOH. After perfusion for an additional 5 min with buffer only, the liver is perfused with 60 ml ice-cold saline and homogenized,

[19] A. J. L. Ham and D. C. Liebler, *Arch. Biochem. Biophys.* **339,** 157 (1997).

TABLE III
Levels of TH and Its Oxidation Products in Mitochondria and Liver Homogenate of Rats Fed Standard Diet[a]

	Control		tert-BuOOH	
Compound	Mitochondria	Homogenate	Mitochondria	Homogenate
Neutral extract[b]				
TH	168 ± 5[c]	166 ± 10	137 ± 5	150 ± 4
TQE1	ND[d]	ND	5 ± 0	4 ± 0
TQE2	ND	ND	5 ± 0	3 ± 0
T(H)Q	4 ± 2	2 ± 1	37 ± 2	14 ± 1
HCl extract[e]				
TH	158 ± 3	183 ± 21	130 ± 2	143 ± 6
TQE1	ND	ND	6 ± 1	5 ± 1
TQE2	4 ± 1	1 ± 1	17 ± 1	20 ± 1
T(H)Q	16 ± 1	11 ± 2	50 ± 2	46 ± 6

[a] Livers from rats fed a standard chow diet were perfused for 10 min either with perfusion medium (control) or with medium containing 2 mM tert-BuOOH. At the end of the perfusion, homogenate and mitochondria wre prepared from the livers and TH and its oxidation products. [From A. J. L. Ham and D. C. Liebler, *Arch. Biochem. Biophys.* **339**, 157 (1997). Used with permission.]
[b] Extracted at neutral pH.
[c] Values are in pmol mg^{-1} mitochondrial or homogenate protein.
[d] Not detected.
[e] The sample was treated with HCl prior to extraction to convert 8a-substituted tocopherones and epoxytocopherones to T(H)Q and TQE1/TQE2, respectively.

and mitochondria and whole homogenate are isolated. α-Tocopherol and products are extracted from homogenate and from isolated mitochondria as described earlier. The homogenate is filtered through two layers of cheesecloth prior to extraction. To detect oxidized TH present as 8a-substituted tocopherones and epoxytocopherones, extracts from mitochondria and homogenate are treated with dilute HCl as described previously,[5] and products of 8a-substituted tocopherone hydrolysis (TQ and THQ) and of epoxytocopherone hydrolysis (TQE1 and TQE2) are then analyzed.

Table III summarizes the distribution of TH and its oxidation products in mitochondria and homogenate after perfusion. In livers perfused without tert-BuOOH, only TH and traces of T(H)Q are detected in both mitochondria and homogenate. However, treatment with HCl releases additional T(H)Q and TQE2, presumably from tocopherone and epoxytocopherone precursors. After perfusion with tert-BuOOH, levels of products present in both homogenate and mitochondria are elevated over those in control perfusions. Levels of products detected after treatment with HCl are increased two- to threefold, again indicating that tocopherone and epoxytocopherone products are present.

Additional Notes on Sample Preparation and Analysis

The sample preparation protocol described here provides high recoveries and excellent reproducibility for analyses of subcellular membrane fractions (e.g., microsomes and mitochondria), liver homogenates, and fresh cell suspensions, where the biological material present can be dispersed uniformly prior to sampling. However, poor reproducibility can be a problem in analyses of samples where the biological material is not dispersed uniformly, such as previously frozen, pelleted cells, or samples containing significant amounts of connective tissue. Suspension of previously frozen cell pellets in 0.1% Triton X-100 prior to sampling achieves a more uniform cell dispersion and eliminates problems with assay reproducibility. In samples containing connective tissue, protease treatment may improve dispersion of the tissue material prior to sampling.

An important consideration in the analysis of antioxidants is the possibility of artifactual oxidation during sample workup. We routinely employ the synthetic antioxidant butylated hydroxytoluene in our sample extraction. The amount of butylated hydroxytoluene used is 1000-fold in excess of the TH in most samples. Although this antioxidant carries through the entire assay, it does not intenfere with GC-MS analyses. In samples where significant transition metal ion contamination is expected (e.g., in iron initiated microsomal oxidations, see later), 0.1 mM deferoxamine may be added to the sample immediately prior to extraction to chelate iron and prevent metal-catalyzed TH oxidation.

On-column injection of the TMS derivatives of TH and its oxidation products was preferable to spitless injection. Although satisfactory analysis of TH—TMS via splitless injection has been reported,[20] analytical sensitivity and reproducibility for splitless injection of TMS derivatives of TQ and TQE1/TQE2 were poor.

The method also permits quantification of both TQ and THQ, which exist in a redox equilibrium [Eq. (1)].

$$TQ + 2H^+ + 2e^- \rightleftharpoons THQ \tag{1}$$

The distribution between reduced and oxidized forms is easily affected by reductants or oxidants during sample workup. This may be manifested as an apparent absence of TQ and a complete reduction of the added d_6-TQ

---

[20] K. U. Ingold, G. W. Burton, D. O. Foster, L. Hughes, D. A. Lindsay, and A. Webb, *Lipids* **22,** 163 (1987).

internal standard to d_6-THQ. However, redox conversion of the internal standards during workup does not invalidate the method, as a d_6-TQ standard is used for TQ measurement and a d_3-THQ standard is used for THQ measurement. If measurement of both TQ and THQ is intended, the possibility that TQ and THQ can undergo redox interconversion during workup prior to addition of internal standards must be considered. It is possible that the measured distribution of these products may not accurately reflect real tissue TQ/THQ ratios. The reproducibility of d_3-THQ and d_6-TQ internal standard recoveries should provide an indication of the degree to which artifactual perturbations of the TQ/THQ redox ratio occur during workup.

Acknowledgments

This work was supported in part by USPHS Grant CA 59585 and Southwest Environmental Health Sciences Center Grant ES 06694.

[29] Determination of Vitamin E Forms in Tissues and Diets by High-Performance Liquid Chromatography Using Normal-Phase Diol Column

By JOHN K. G. KRAMER, ROBERT C. FOUCHARD, and KRISHNA M. R. KALLURY

Introduction

Vitamin E is a generic term that includes a mixture of tocopherols (T) and tocotrienols (T_3) that differ in the number and position of methyl groups on the fused chromanol ring and the absence or presence of three double bonds in the isoprenoid side chain (see Fig. 1).

Current methods for the analysis of vitamin E in tissues involve mechanical homogenization of a fresh or thawed tissue sample, followed by the addition of ethanol, and extraction of the homogenate with hexane or acetone.[1-14] The hexane extract is saponified to remove coextracted neutral

---

[1] J. N. Thompson and G. Hatina, *J. Liq. Chromatogr.* **2**, 327 (1979).
[2] B. J. Zaspel and A. S. Csallany, *Anal. Biochem.* **130**, 146 (1983).
[3] J. L. Buttriss and A. T. Diplock, *Methods Enzymol.* **105**, 131 (1983).
[4] G. W. Burton, A. Webb, and K. U. Ingold, *Lipids* **20**, 29 (1985).
[5] E.-L. Syvaoja, K. Salminen, V. Piironen, P. Varo, O. Kerojoki, and P. Kiovistoinen, *J. Am. Chem. Soc.* **62**, 1245 (1985).

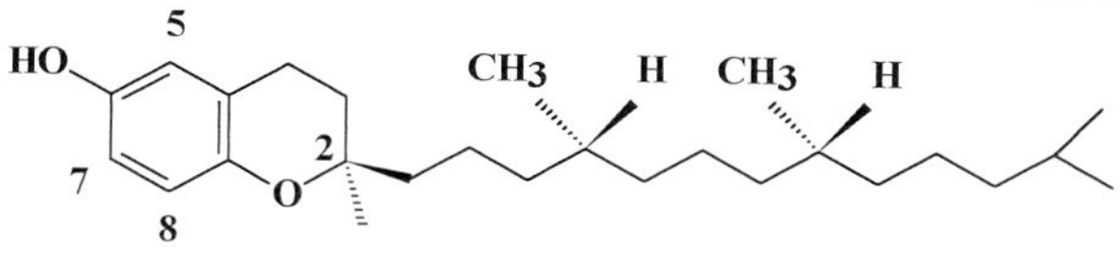

Fig. 1

Tocopherol and **Tocotrienol** structures

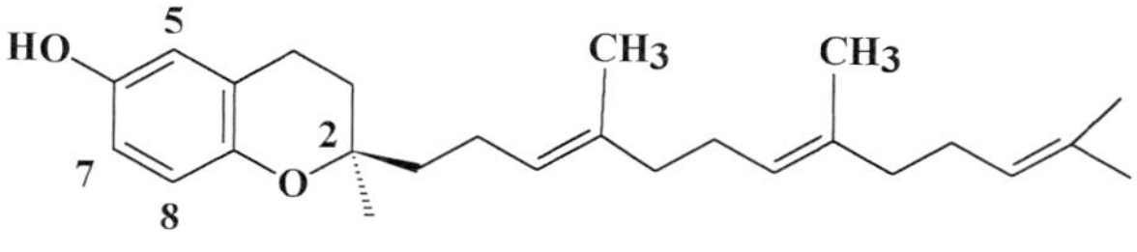

Position of CH₃ groups	Tocopherol	Tocotrienol
5, 7, 8	α-Tocopherol (α-T)	α-Tocotrienol (α-T$_3$)
5, 7	β-Tocopherol (β-T)	β-Tocotrienol (β-T$_3$)
7, 8	γ-Tocopherol (γ-T)	γ-Tocotrienol (γ-T$_3$)
8	δ-Tocopherol (δ-T)	δ-Tocotrienol (δ-T$_3$)
No CH₃ groups	Tocol	

lipids,[3,5-8,12,14] or is analyzed directly.[1,2,4,9-11,13] Saponification is still recommended, despite extensive loss of all T[7,15] and T$_3$.[16]

The analysis of vitamin E in foods and feeds,[1,8,9,14,15,17-20] oils and seeds,[1,19,21-27] plasma,[2,4,6-10,13-15,28-32] platelets,[30] and red cells[4,7,28,32] differs from tissues only in the prior homogenization step. Foods and feeds are

---

[6] S. N. Meydani, A. C. Shapiro, M. Meydani, J. B. Macauley, and J. B. Blumberg, *Lipids* **22**, 435 (1987).
[7] T. Ueda and O. Igarashi, *J. Micronutr. Anal.* **3**, 15 (1987).
[8] H. E. Indyk, *Analysis* **113**, 1217 (1988).
[9] C. G. Rammell, A. B. Pearson, and G. R. Bentley, *N. Z. Vet. J.* **36**, 133 (1988).
[10] Y. K. Chung, D. C. Mahan, and A. J. Lepine, *J. Anim. Sci.* **70**, 2485 (1992).
[11] S. G. Kaasgaard, G. Hølmer, C.-E. Høy, W. A. Behrens, and J. L. Beare-Rogers, *Lipids* **27**, 740 (1992).
[12] E. Berlin, D. McClure, M. A. Banks, and R. C. Peters, *Comp. Biochem. Physiol. A* **109**, 53 (1994).
[13] D. W. Alexander, S. O. McGuire, N. A. Cassity, and K. L. Fritsche, *J. Nutr.* **125**, 2640 (1995).
[14] Y. H. Wang, J. Leibholz, W. L. Bryden, and D. R. Frazer, *Br. J. Nutr.* **75**, 81 (1996).
[15] V. Piironen, P. Varo, E.-L. Syvaoja, K. Salminen, and P. Koivistoinen, *Intern. J. Vit. Res.* **54**, 35 (1983).
[16] C. K. Chow, H. H. Draper, and A. S. Csallany, *Anal. Biochem.* **32**, 81 (1969).

ground and extracted[1,9] or saponified first,[1,8,14,15,17–20] fats and oils are directly extracted[21–25,27] or saponified first,[19,25,26] and plasma, platelets, and red cells are treated with ethanol[2,6–10,13–15,28–31] or detergents[4,32] to remove protein by precipitation.

The vitamin E is then analyzed by high-performance liquid chromatography (HPLC) on normal-phase[1,3–5,8–11,15,18,19,21–23,25–27] or reversed-phase columns.[2,6–8,12–14,17,19,20,24,28–32] Generally, reversed-phase columns do not separate the β and γ isomers,[8,20] and normal-phase silica columns show poor reproducibility.[8,15,19,23,25] A better separation is obtained with a normal-phase column bonded with amino–cyano groups.[9,19,23] However, the amino group ionizes when polar organic solvents are used, which results in peak broadening and increases in retention times. This problem can be overcome by adding formic acid to the solvent, but that affects the stability of the column.[23]

The present method for the determination of all the vitamin E forms in tissues eliminates the thawing step by pulverizing the tissue at dry ice temperature.[33] Total vitamin E is then extracted with hexane and analyzed directly by HPLC using a diol column and fluorescence detection, without prior saponification of the sample.[34] The method is modified for the analysis of diets to determine both free and esterified vitamin E. Also discussed are the influence of coextracted neutral lipids, the response factors of the

---

[17] C. H. McMurray and W. J. Blanchflower, *J. Chromatogr.* **176**, 488 (1979).
[18] W. M. Cort, T. S. Vicente, E. H. Waysk, and B. D. Williams, *J. Agric. Food Chem.* **31**, 1330 (1983).
[19] W. D. Pocklington and A. Dieffenbacher, *Pure Appl. Chem.* **60**, 877 (1988).
[20] C. J. Hogarty, C. Ang, and R. R. Eitenmiller, *J. Food Comp. Anal.* **2**, 200 (1989).
[21] P. Tayor and P. Barnes, *Chem. Ind.* Oct. 17, 722 (1981).
[22] W. Müller-Mulot, G. Rohrer, G. Oesterhelt, K. Schmidt, L. Allemann, and R. Maurer, *Fette Seifen Anstrichm.* **85**, 66 (1983).
[23] C. G. Rammell and J. J. L. Hoogenboom, *J. Liq. Chromatogr.* **8**, 707 (1985).
[24] K. Warner and T. L. Mounts, *J. Am. Oil Chem. Soc.* **67**, 827 (1990).
[25] G. W. Chase, Jr., C. C. Akoh, and R. R. Eitenmiller, *J. Am. Oil Chem. Soc.* **71**, 877 (1994).
[26] T.-S. Shin and J. S. Godber, *J. Chromatogr. A* **678**, 49 (1994).
[27] J. T. Budin, W. M. Breene, and D. H. Putnam, *J. Am. Oil Chem. Soc.* **72**, 309 (1995).
[28] J. G. Bieri, T. J. Tolliver, and G. L. Catignani, *Am. J. Clin. Nutr.* **32**, 2143 (1979).
[29] C. H. McMurray and W. J. Blanchflower, *J. Chromatogr.* **178**, 525 (1979).
[30] J. Lehmann and H. L. Martin, *Clin. Chem.* **28**, 1784 (1982).
[31] G. R. Gutcher, W. J. Raynor, and P. M. Farrell, *Am. J. Clin. Nutr.* **40**, 1078 (1984).
[32] A. Garrido, M. Gárate, R. Campos, A. Villa, S. Nieto, and A. Valenzuela, *J. Nutr. Biochem.* **4**, 118 (1993).
[33] J. K. G. Kramer and H. W. Hulan, *J. Lipid Res.* **19**, 103 (1978).
[34] J. K. G. Kramer, L. Blais, R. C. Fouchard, R. A. Melnyk, and K. M. R. Kallury, *Lipids* **32**, 323 (1997).

different vitamin E forms, the use of internal standards to quantitate vitamin E, and the structure and properties of the diol column.

Materials

All solvents are HPLC grade, except absolute ethanol. The tocopherols (α, β, γ, δ, and 5,7-T) were obtained from Matreya, Inc. (Pleasant Gap, PA). Tocol, the preferred internal standard for vitamin E analyses, is now available from Matreya Inc. α-T_3 was a gift from Dr. Bill Collins (Center for Food and Animal Research, Ottawa, Ontario, Canada). Red palm oil (nutrolein, golden palm olein) is an enriched fraction of T and T_3 and was a gift from the Malaysian Palm Oil Council of America, Inc. (Chicago, IL). Triolein was obtained from Nu-Chek-Prep, Inc. (Elysian, MN).

Pulverization of Tissue and Extraction of Vitamin E

The animals are anesthetized and tissues are removed, rinsed with saline solution, and immediately frozen between two blocks of dry ice (or placed into liquid nitrogen). Tissues are stored at $-70°$ until analyzed. All subsequent steps are carried out in a low-light environment.

A stainless steel mortar and pestle (prepared in a machine shop, Fig. 2) and spatulas are cooled by placing them on dry ice.[33] A representative

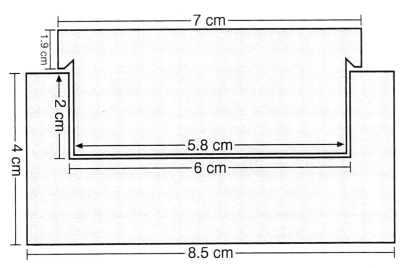

FIG. 2. A stainless steel mortar and pestle used to pulverize tissues at $-70°$.

sample of frozen tissue is accurately weighed (about 1 g), pulverized using the steel mortar and pestle at dry ice temperature, and then transferred quantitatively into a 50-ml conical tube with cap (Falcon polyethylene tube, Becton Dickinson Labware, Franklin Lakes, NJ) containing 2 ml of water cooled in an ice bath. Five milliliters of cold ethanol and 4 μg of a chosen internal standard of vitamin E are added to the conical tube. The sample is sonicated in an ice bath at 50% power for 45 sec (Model W-225; Heat Systems-Ultrasonic Inc., Farmingdale, NY). Add 5 ml of cold ethanol and 10 ml of cold hexane. The tube is mixed with a vortex for 30 sec and centrifuged at 2500 rpm for 15 min at 0°. Five milliliters of the hexane layer (to avoid any transfer of the lower layer) is taken to dryness using a rotary evaporator and immediately taken up in 1 ml of hexane and transferred into an amber HPLC vial. The vials are stored at $-70°$ until analyzed by HPLC the same day.

Extraction of Vitamin E from Oils and Diets

Vitamin E is extracted from oils (about 1 g) after the addition of cold water (2 ml), ethanol (10 ml), hexane (10 ml), and a known amount of internal standard. After mixing and centrifugation, an aliquot of the hexane layer is analyzed directly by HPLC.

If tocopherol esters are present in the diet, a modified procedure is used. To about 1 g of diet, cold water (2 ml), ethanol (10 ml), hexane (10 ml), and a known amount of internal standard are added. The sample is sonicated in an ice bath. As much as possible of the hexane is removed and the remainder is extracted with two additional 10-ml volumes of hexane. The hexane layers are combined, reduced using a rotary evaporator, taken up in 1 ml of hexane, and analyzed directly by HPLC. This extract represents the total free vitamin E in the diet. To the aqueous–ethanolic residue, ascorbic acid (100 mg) and 60% KOH/H_2O (w/w) (2 ml) are added, but no internal standard, and the mixture is heated for 10 min at 70°. The mixture is then extracted with three portions of 10 ml hexane. The combined hexane layers are washed with water to remove traces of KOH, and the solvent is removed and taken up in 1 ml hexane for HPLC analysis. This fraction represents esterified vitamin E forms. To quantitate the vitamin E esters, α-T is used as an external standard because this is the usual form in which vitamin E is added to the diet. The average recovery of α-T measured externally was found to be 95 $\pm$ 7% (n = 10).

Choice of Internal Vitamin E Standard

The internal standard should be a vitamin E form that does not occur in the sample. The synthetic vitamin E form 5,7-T was considered as an internal standard. However, 5,7-T coelutes with α-T_3 on the diol column and is therefore of limited use. Tocol is an excellent choice for internal standard because it has not been found to occur naturally and it elutes ahead of all other vitamin E forms on both normal-phase[7] and reversed-phase[3] HPLC columns. Tocol was not available commercially at the time of this study, but is now available from Matreya Inc. 2,2,5,7,8-Pentamethyl-6-hydroxy-chroman (similar to α-T, with a methyl instead of the isoprenoid side chain) was used as a suitable internal standard for vitamin E analysis.[35] This compound is now also available commercially (Wako Pure Chemical Industries Ltd., Richmond, VA).

HPLC Conditions

Vitamin E is measured using an HPLC (Model 1090, Hewlett-Packard, Palo Alto, CA) equipped with an autosampler (25-μl syringe), a fluorescence detector (Model FL2000; Spectra-Physics Analytical, Fremont, CA), and a data acquisition system (HP 3365 Series II ChemStation). The excitation is set at 296 nm and the emission at 330 nm. A normal-phase Supelcosil LC-Diol column (25 cm $\times$ 4.6 mm, 5 μm particle size; Supelco Inc., Bellefonte, PA) is used, operated at room temperature, with hexane/2-propanol (99:1) as the mobile phase at a flow rate of 1 ml/min. The solvent mixture is prepared fresh each day and sparged with helium for 10 min before use. Do not sparge continuously because this will selectively remove 2-propanol from the solvent mixture and increase the retention times of the vitamin E forms. These two liquids form an azeotropic mixture (77:23) that boils at 62.7°.[36]

Vitamin E Separations

A typical separation of all four tocopherols and four tocotrienols is shown in Fig. 3. The mixture in Fig. 3 was prepared by adding a hexane extract of vitamin E from red palm oil to a standard mixture of tocopherols. The elution order of the vitamin E forms on the diol column is similar to that observed with a normal-phase silica column.[1,3,5,8,15,18,19,21–23,25,27] Silica columns were reported to be unstable and occasionally yield poor reproducibility,[8,15,19,23,25] which was also experienced in our laboratory. Good separations were reported using an amino–cyano polar-phase column, but the amino group ionizes when polar organic solvents are used.[23] However, the

[35] I. Ikeda, Y. Imasato, E. Sasaki, and M. Sugano, *Intern. J. Vit. Nutr. Res.* **66**, 217 (1996).
[36] Z. M. Kurtyka, in "CRC Handbook of Chemistry and Physics" (D. R. Lide, ed.), p. 6. CRC Press, Boca Raton, FL (1990).

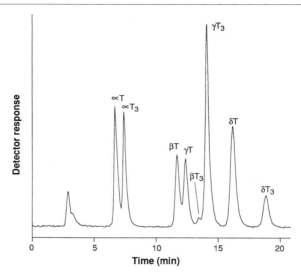

FIG. 3. Separation of a mixture of four tocopherols (T) and four tocotrienols (T_3) by HPLC on a normal-phase Supelcosil LC-Diol column (Supelco Inc.), using the solvent hexane/2-propanol (99:1) as the mobile phase at a flow rate of 1 ml/min. [From J. K. G. Kramer, L. Blais, R. C. Fouchard, R. A. Melnyk, and K. M. R. Kallury, *Lipids* **32**, 323 (1997).]

diol column gives both reproducible and quantitative results. The column was stable over hundreds of analyses.

Diol Column

The diol column is prepared commercially by reacting 5-μm spherical silica particles containing surface silanol (Si-OH) groups with 3-glycidoxypropyltrimethoxysilane to produce an epoxysilica bond phase with a ligand density of 3.8 μmol/m^2 on the silica.[34] Intentional hydrolysis of the epoxide ring, by refluxing with 0.1 M HCl for 2 hr, did not produce any significant cleavage of the epoxide ring, as evidenced by the solid-state ^{13}C cross-polarization, magic angle spinning, nuclear magnetic resonance (CP/MAS NMR) spectrum of the packing material (Fig. 4). For details regarding the conditions of the ^{13}C CP/MAS NMR, see Kramer *et al.*[34] If the epoxide ring had cleaved, the signal at δ 42.3 ppm (O-CH$_2$-CH of the epoxide ring) should have disappeared, and the signal at δ 70.9 ppm (O-CH$_2$-CH of the epoxide ring) should have been reduced to half its size. Two new signals should have registered at δ 64.2 ppm (CH$_2$-OH) and δ 72.1 ppm (CH-OH) corresponding to the diol. This column should therefore be considered an "epoxide" rather than a "diol" column. The epoxide structure would ex-

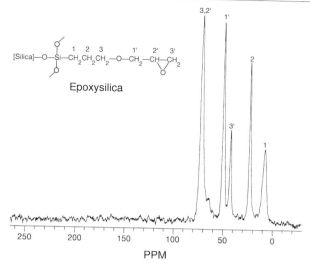

FIG. 4. The ^{13}C cross-polarization, magic angle spinning, nuclear magnetic resonance of the "Supelcosil LC-Diol" packing material. Signals were observed at δ 7.8 (C_1), 23.4 (C_2), 42.3 (C_3'), 49.8 (C_1'), and 70.9 ppm (C_3 and C_2'). The expected signal corresponding to the diol structure was not observed at δ 64.2 (C_3') and 72.1 ppm (C_2').

plain the stability of the column to polar and nonpolar solvents, its incapability to form strong hydrogen-bonded structures with proton-donor analytes, its moderate polarity, and its observed superior performance.

Fluorescence Detector Response for Different Vitamin E Forms

The calibration curves for the four tocopherols and α-T_3 are shown in Fig. 5. The fluorescence response is linear ($r > 0.99$) over the range from 10 to 100 ng for each of the vitamin E forms tested. As shown in Fig. 5, the response is least for α-T and most for δ-T, and the following correction factors were calculated for α-T (2.30), β-T (1.52), γ-T (1.32), δ-T (1.00), and α-T_3 (1.19). The need for determining response factors for the different vitamin E forms was recognized by others,[24,25] contrary to an earlier report that claimed that the responses were similar for all vitamin E forms.[1]

Influence of Triacylglycerols on the Quantitation of Vitamin E Using Fluorescence Detection

The coextraction of neutral lipids with hexane was a major concern when UV detection was used in the analysis of tocopherols.[19,21] For this

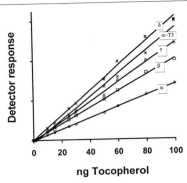

FIG. 5. Calibration curves for the fluorescence detector responses of tocopherols (T) and α-tocotrienols (α-T$_3$). [From J. K. G. Kramer, L. Blais, R. C. Fouchard, R. A. Melnyk, and K. M. R. Kallury, *Lipids* **32**, 323 (1997).]

reason, the hexane extract was saponified to remove the neutral lipids, mainly triacylglycerols. Saponification is known to destroy all the vitamin E forms, particularly δ-T[7,8,15,16,20,23] and α-T$_3$.[16] With the introduction of fluorescence detectors, the need to remove the neutral lipids by saponification is not necessary, as the coextracted neutral lipids do not appear to interfere with fluorescence detection.[21] Furthermore, fluorescence detection is at least 10 times more sensitive than UV detection.[19]

Ueda and Igarashi[7] systematically investigated the effect of adding increasing amounts of corn oil on the recovery of tocopherols after saponification. They showed that the recoveries of all tocopherols, except α-T,

TABLE I
FLUORESCENCE DETECTOR RESPONSE OF TOCOPHEROLS (T) AND TOCOTRIENOLS (T$_3$)[a]

Triolein (mg)	Relative response (%)					SD
	α-T	β-T	γ-T	δ-T	α-T$_3$	
0	100[b]	100	100	100	100	—
10	97[c]	98	99	98	96	2.2
25	94	95	94	97	93	1.4
50	91	91	91	91	90	1.7
100	87	87	87	87	87	1.8
200	86	84	85	86	83	3
400	81	81	81	82	80	3.5

[a] 50 ng each, in the presence of different amounts of triolein.
[b] The fluorescence response of each vitamin E form in the absence of triolein was arbitrarily set at 100%.
[c] Each value represents the mean of three determinations.

TABLE II[a]

TOCOPHEROL (T) AND TOCOTRIENOL (T$_3$) CONTENT OF LIVER AND HEART FROM NEWBORN PIGLETS FED MILK REPLACER DIETS OR SOW MILK FOR 4 WEEKS[b]

Diets	α-T	β-T	γ-T	α-T$_3$
Liver				
Sow milk	19.5	0.1	0.4	1.0
Soybean oil	21.2	0.9	2.1	1.4
Canola oil	18.3	1.1	1.1	1.5
Mixture of oils[c]	10.3	1.0	1.6	1.2
HEAR oil[d]	9.9	0.7	1.0	1.3
Pooled SED[e]	3.2	0.33	0.5	0.16
Analysis of variance (mean squares)[f]				
Litters (L)	288*	0.83	2.1	0.09
Diets (D)	281**	1.33	12.1	0.31
L × D	36	0.49*	4.6**	0.15
Among piglets	60***	0.14***	0.4***	0.10**
Among determinations	1.5	0.009	0.005	0.02
Heart				
Sow milk	8.9	0.0	0.2	1.0
Soybean oil	23.3	0.3	4.0	1.2
Canola oil	16.7	0.1	1.2	1.1
Mixture of oils	15.4	0.1	2.5	1.1
HEAR oil	9.5	0.1	1.3	0.9
Pooled SED	3.5	0.05	0.5	0.33
Analysis of variance (mean squares)				
Litters	270*	0.022	1.0	0.11
Diets	297*	0.087***	18.3***	0.13
L × D	57	0.010	1.4	0.32
Among piglets	38***	0.011***	0.7***	0.67***
Among determinations	1.6	0.0002	0.1	0.04

[a] Reprinted with permission from Kramer et al.[34]
[b] Each value represents the mean of two analyses/piglet and six piglets/diet. Data given as μg/g wet weight.
[c] Mixture of high oleic acid sunflower oil, soybean oil, and linseed oil (55:31:14) to mimic canola oil in fatty acid composition.
[d] Mixture of canola oil (0.8% 22:ln-9) and high erucic acid rapeseed (HEAR) oil (42.9%, 22:ln-9) to give a final oil containing 20% erucic acid (22:ln-9).
[e] SED, pooled standard error of the difference. Means greater than SED are significant ($p < 0.05$).
[f] Analysis of variance. Mean squares and significance: * $p < 0.05$; ** $p < 0.01$; *** $p < 0.001$.

were severely reduced when 100 and 200 mg of oil were added. We used the same approach to test the influence of increased triolein on the fluorescence response of all four tocopherols and α-T$_3$ (50 ng each), but without saponifying the samples (Table I). No difference appears to exist in response between the different tocopherol forms and α-T$_3$ to the amount of coexisting

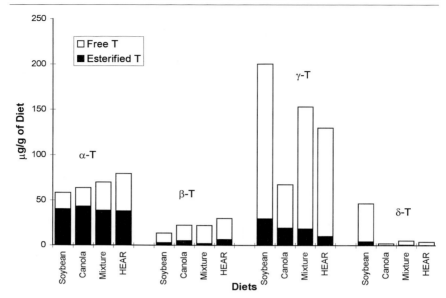

FIG. 6. Free and esterified tocopherols in the four milk replacer diets fed to the piglets. α-T succinate was added at 45.5 μg/g of diet.

triolein. This made it convenient to use any of the tocopherols as internal standard.

The amount of neutral lipids in tissues is generally about 1–3%, or 10–30 mg/g of tissue. Because only half the sample is used in the analysis (only 5 of the 10 ml of hexane was used), this represents 5–15 mg of neutral lipids per vial. Therefore, the recoveries of the tocopherols for tissue analysis should be better than 93–97% based on the results of Table I. However, the analysis of 1 g of diet that may contain 20% fat/oil contains 200 mg of neutral lipids per gram of diet. If half the sample is used, this represents 100 mg neutral lipids per vial, which will lead to an average recovery of about 87% for the different vitamin E forms. The lowest recoveries of about 81% are obtained for oil samples when half of a 1-g sample is concentrated in the vial, i.e., 400 mg of neutral lipids per vial. However, oils are generally rich in vitamin E and appropriately less oil can be used to obtain good results. Correction factors should be applied based on the neutral lipid content of the sample investigated.

Application of Analysis of Vitamin E to Liver and Heart Tissues and Diets

Table II shows the results of vitamin E analysis of liver and heart from piglets fed milk replacer diets for 4 weeks. Because δ-T was not present in these tissues, it was selected as the internal standard. Results showed that α-T is the major vitamin E form in liver and heart, with smaller amounts of β-T, γ-T, and α-T$_3$. The total vitamin E content in these tissues was greater than 25 μg/g wet weight, which is higher than reported previously for similar tissues,[7,9–11,13,14] an indication that there was a greater recovery using this method. The results showed a significant among animal variation, quite common when larger animals are investigated. However, the reproducibility between duplicate determinations performed on different days using separate 1-g tissue samples from the same animal are very high, showing no significant differences. This method was shown to be reproducible. The recovery of the internal standard δ-T in the present study was 95 ± 7% ($n = 60$).

Figure 6 shows the results of free and esterified vitamin E in the piglet milk replacer diets. α-T succinate was added to the diets at 45.5 μg/g diet. Based on this addition, the recovery of the added α-T succinate ranged from 83 to 94% as determined by the esterified vitamin E form.

General Comments

The present method takes advantage of the pulverization technique at $-70°$ that was used successfully to avoid lipolysis of tissue lipids,[33] and now appears also to reduce the destruction of vitamin E. We do not recommend saponification for the analysis of vitamin E in tissues or oils, except for diets to which vitamin E esters are added. In the latter case, the free vitamin E forms should be removed first to avoid losses during saponification that is required to hydrolyze the vitamin E esters. The diol column appears to combine the superior separation of normal-phase silica columns with the added stability and reproducibility that the epoxide structure of the packing material provides.

Acknowledgment

Contribution number S006 from the Southern Crop Protection, Food Research Center, Agriculture and Agri-Food Canada, Guelph, ON, N1G 2W1.

[30] Sensitive High-Performance Liquid Chromatography Techniques for Simultaneous Determination of Tocopherols, Tocotrienols, Ubiquinols, and Ubiquinones in Biological Samples

By Maurizio Podda, Christine Weber, Maret G. Traber, Rainer Milbradt, and Lester Packer

Introduction

Lipophilic antioxidants are responsible for the protection of cell membranes against lipid peroxidation. There is increasing evidence that the cell membrane is an important initiation site of cell signaling induced by oxidative perturbation.[1]

Vitamin E is the major lipophilic antioxidant in plasma, membranes, and tissues[2] and is the collective name for the eight naturally occurring molecules (four tocopherols and four tocotrienols) that exhibit α-tocopherol activity. Tocotrienols differ from tocopherols in that they have an isoprenoid instead of a phytyl side chain. The four forms of tocopherols and tocotrienols differ in the number of methyl groups on the chromanol nucleus (α- has three, β- and γ- have two, and δ- has one). Tocotrienols are present in tissues at low concentrations but may exert substantial effects due to their antioxidant and hydroxymethylglutaryl (HMG)-CoA reductase inhibiting activity.[3–5]

Coenzyme Q functions as an electron carrier in mitochondria. The coenzyme Q redox couple, ubiquinol/ubiquinone, is formed from the oxidation of NADH or NADPH between the NADH–Q reductase complex and cytochrome reductase.[6] However, coenzyme Q is also present in considerable amounts in endoplasmic reticulum, Golgi apparatus, lysosomes, peroxisomes, and plasma membranes.[7] This broad distribution in intra-

---

[1] Y. Devary, C. Rosette, J. A. DiDonato, and M. Karin, *Science* **261,** 1442 (1993).
[2] M. G. Traber and H. Sies, *Annu. Rev. Nutr.* **16,** 321 (1996).
[3] E. Serbinova, V. Kagan, D. Han, and L. Packer, *Free Radic. Biol. Med.* **10,** 263 (1991).
[4] R. A. Parker, B. C. Pearce, R. W. Clark, D. A. Gordon, and J. J. Wright, *J. Biol. Chem.* **268,** 11230 (1993).
[5] A. A. Qureshi, W. C. Burger, D. M. Peterson, and C. E. Elson, *J. Biol. Chem.* **261,** 10544 (1986).
[6] F. L. Crane, Y. Hatefi, R. L. Lester, and C. Widmer, *Biochim. Biophys. Acta* **25,** 220 (1957).
[7] A. Kalen, B. Norling, E. L. Appelkvist, and G. Dallner, *Biochim. Biophys. Acta* **926,** 70 (1987).

cellular organelles emphasizes that coenzyme Q may play an antioxidant role in membranes.[8,9] Indeed, ubiquinol protects against lipid peroxidation more efficiently than α-tocopherol in low-density lipoproteins[10-12] and is the most sensitive antioxidant in ultraviolet (UV)-irradiated skin. Indeed, lipid hydroperoxides are formed in UV-irradiated skin when ubiquinol is consumed, but before tocopherol levels are reduced appreciably.[13]

The simultaneous quantitation of these lipophilic antioxidants is therefore of general interest for understanding their regulatory and protective effects in tissues on oxidative stress.

Extraction of Vitamin E and Coenzyme Q from Tissues

Vitamin E is usually extracted from biological samples, after thorough dissolution, either directly or after a saponification step.[14] In order to determine total vitamin E in samples containing tocopherol acetate, the saponification step is necessary for liberating the free tocopherol. However, precautions in order to protect the vitamin E against destruction during heating in alkaline solution, e.g., addition of ascorbate or pyrogallol, and shielding from light are important.[14] In early studies, coenzyme Q was extracted also after saponification, but this procedure was found to lead to ubiquinol oxidation, ethoxy substitutions in the quinone ring, and formation of ubichromanol.[15]

Vitamin E extraction with or without a saponification step is typically performed using hexane. Ethanol/hexane mixtures (often 2:5),[16] chloroform/methanol,[17] or 2-propanol[18] have been used for direct extraction of coenzyme Q. To achieve an enhanced sample solubilization, sodium dodecyl sulfate (SDS) treatment with direct ethanol/hexane extraction has

---

[8] L. Ernster and G. Dallner, *Biochim. Biophys. Acta* **1271**, 195 (1995).
[9] R. E. Beyer, *Free Radic. Biol. Med.* **8**, 545 (1990).
[10] B. Frei, M. C. Kim, and B. N. Ames, *Proc. Natl. Acad. Sci. U.S.A.* **87**, 4879 (1990).
[11] R. Stocker, V. W. Bowry, and B. Frei, *Proc. Natl. Acad. Sci. U.S.A.* **88**, 1646 (1991).
[12] R. Takayanagi, K. Takeshige, and S. Minakami, *Biochem. J.* **192**, 853 (1980).
[13] M. Podda, M. G. Traber, C. Weber, L.-J. Yan, and L. Packer, *Free Radic. Biol. Med.* **24**, 55 (1998).
[14] C. Bourgeois, "Determination of Vitamin E: Tocopherols and Tocotrienols." Elsevier, London, 1992.
[15] F. Crane and R. Barr, *Methods Enzymol.* **18**, 137 (1971).
[16] B. Finckh, A. Kontush, J. Commentz, C. Hubner, M. Burdelski, and A. Kohlschutter, *Anal. Biochem.* **232**, 210 (1995).
[17] M. E. Gotz, A. Dirr, R. Burger, B. Janetzky, M. Weinmuller, W. W. Chan, S. C. Chen, H. Reichmann, W. D. Rausch, and P. Riederer, *Eur. J. Pharmacol.* **266**, 291 (1994).
[18] J. Lagendijk, J. B. Ubbink, and W. J. Vermaak, *J. Lipid Res.* **37**, 67 (1996).

been recommended[19] for vitamin E extraction and validated also for coenzyme Q extraction by Lang et al.[20] We have used a modification of these procedures to extract tocopherols, tocotrienols, ubiquinols, and ubiquinones.

Extraction Procedure

Two milliliters of phosphate-buffered saline (PBS), on ice and bubbled with nitrogen, are added to 50–100 mg of tissue in a 10-ml Potter–Elvehjem tube containing 50 μl butylated hydroxytoluene (BHT) (1 mg/ml). In the case of skin, the samples must first be ground under liquid nitrogen using a mortar and pestle and then directly transferred into the Potter–Elvehjem tube. The tissue is then homogenized for 1 min using a Teflon pestle. The homogenate is transferred to an amber, screw-cap tube with 1 ml SDS (0.1 mM) and vortexed for 30 sec, sonicated for 15 sec, chilled in ice water, and vortexed for 15 sec. Two milliliters of ethanol (on ice and bubbled with nitrogen) is added and again vortexed for 30 sec, sonicated for 15 sec, chilled, and vortexed for 15 sec. Using a solvent safe tip and a positive displacement pipette, 2 ml of hexane (on ice) is added. The tube is vortexed for 90 sec and subsequently centrifuged at 1000× g for 3 min. Transfer 1.75 ml of the upper hexane layer to a conical vial. This vial is placed in a water bath (30°), and the solvent is evaporated under a light nitrogen stream. The residue is dissolved in 100–400 μl of methanol/ethanol (1:1) by vortexing for 10 sec, sonicating for 5 sec, chilled in ice water, and vortexed for 20 sec before injection into the high-performance liquid chromatography (HPLC) setup.

Storage

These sample vials should be kept in the dark and under nitrogen at least at −20°. Despite the common belief of fast degradation of such samples, only a 5.7 ± 1.1% decrease of ubiquinol is found after 4 hr at 4° in an autosampler in different tissue samples. Ubiquinone, tocopherols, and tocotrienols show no change after 4 hr. When a human skin sample, physiologically containing ubiquinone 10, is spiked with ubiquinol 9, barely detectable levels of ubiquinone 9 result after 4 hr, confirming that only a small amount of oxidation occurs.

Saponification

The described method has also been used to determine the content of lipophilic antioxidants in mouse chow. This requires a saponification step

---

[19] G. W. Burton, A. Webb, and K. U. Ingold, *Lipids* **20**, 29 (1985).
[20] J. K. Lang, K. Gohil, and L. Packer, *Anal. Biochem.* **157**, 106 (1986).

as mentioned earlier. One gram of chow is mixed vigorously with 5 ml of 1% (w/v) aqueous ascorbic acid. After addition of 10 ml of ethanol and 1.5 ml of saturated KOH, the samples are saponified at 70° for 30 min, carefully protected from light. After cooling on ice and addition of 5 ml of water, the samples are extracted with 10 ml of hexane. Five milliliters of the upper hexane layer is collected and taken to dryness under nitrogen. The residue is diluted 25-fold and injected onto the HPLC.

Chromatography and Detection

Early separations of vitamin E were performed by column chromatography, paper chromatography, and thin-layer chromatography. Gas chromatography has been used successfully for vitamin E detection either of the free tocopherol forms or as acetyl/butyl or silyl esters.[14] Since 1971 HPLC methods have been used for vitamin E detection. In general, reversed-phase systems resolve different tocopherols and tocotrienols less efficiently than normal-phase systems. Reversed-phase systems, however, have the advantage of being more robust. In increasing order of selectivity and sensitivity, UV spectrophotometers, fluorescence detectors, and electrochemical detectors have been used for vitamin E detection.[14]

A very sensitive method of detection is required for the quantitation of lipophilic antioxidants in tissues, as sample size is the limiting variable in several experimental designs. This is especially important for tocotrienols because they are present at very low concentrations. HPLC separation is required for analysis of vitamin E, but UV detection does not provide sufficient sensitivity. The simultaneous determination of tocopherols and ubiquinol in biological tissue using electrochemical (EC) detection with in-line UV detection for ubiquinone determination has been described previously by our laboratory.[20] However, separation of the different tocopherols and tocotrienols was not accomplished. In contrast, an HPLC/EC method that efficiently separates and detects various vitamin E homologs[21] is not suitable for ubiquinol/ubiquinone separation due to the extremely prolonged retention times of these highly lipophilic compounds.

Procedure

The following method is very sensitive and allows the simultaneous determination of individual tocopherols, tocotrienols, ubiquinols, and ubiquinones using gradient HPLC and electrochemical detection for vitamin E homologs and ubiquinols with in-line UV detection for ubiquinones. The

[21] C. Suarna, R. L. Hood, R. T. Dean, and R. Stocker, *Biochim. Biophys. Acta* **1166**, 163 (1993).

TABLE I
PHYSICOCHEMICAL DATA OF TOCOPHEROLS, TOCOTRIENOLS, UBIQUINOLS, AND UBIQUINONES[a]

Substance	Molecular weight	λ_{max} [nm]	$E_{1cm}^{1\%}$	ε	Ref.
α-Tocopherol	430.7	292	75.8	3270	b
β-Tocopherol	416.7	296	89.4	3730	b
γ-Tocopherol	416.7	298	91.4	3810	b
δ-Tocopherol	402.7	298	87.3	3520	b
α-Tocotrienol	424.7	292	91.0	3870	b
β-Tocotrienol	410.7	295	87.5	3600	b
γ-Tocotrienol	410.7	298	103.0	4230	b
δ-Tocotrienol	396.7	292	83.0	3300	b
Ubiquinol 9	796.6	290	51.7	4120	c
Ubiquinol 10	864.7	290	46.4	4010	c
Ubiquinone 9	794.6	275	185	14,700	c
Ubiquinone 10	862.7	275	165	14,240	c

[a] Extinction coefficients are given for ethanol solutions. From M. Podda et al., J. Lipid Res. **37**, 893 (1996).
[b] M. Kofler, P. F. Sommer, H. R. Bolliger, B. Schmidli, and M. Vecchi, Vit. Horm. **20**, 407 (1962).
[c] Y. Hatefi, Adv. Enzymol. **25**, 275 (1963).

method is exemplified for the measurement of these lipophilic antioxidants in several tissues, including liver, heart, and skin.

Materials

Highest purity solvents and reagents are used. The reagent alcohol consists of 5/95% (v/v) 2-propanol/ethanol (Fisher Scientific, PA). Lithium perchlorate was obtained from Johnson Matthey (Deptford, NJ). Sodium dodecyl sulfate and sodium dithionite were obtained from Fisher, and butylated hydroxytoluene (BHT) is from Sigma (St. Louis, MO). Ubiquinone 9 and 10 standards were a gift from Nisshin Flour Milling Co., Ltd. (Tokyo, Japan). Tocopherol standards were kindly provided from Henkel (La Grange, IL). Tocotrienols were purified from palm oil tocotrienol-rich fraction (TRF) by Dr. Asaf A. Qureshi.

Animals

All experiments and animal handling procedures were approved by the Animal Care and Use Committee of the University of California, Berkeley. Female hairless mice (SKH1,8-12 weeks old) were purchased from Charles River Laboratories (Wilmington, MA) and were kept under standard light

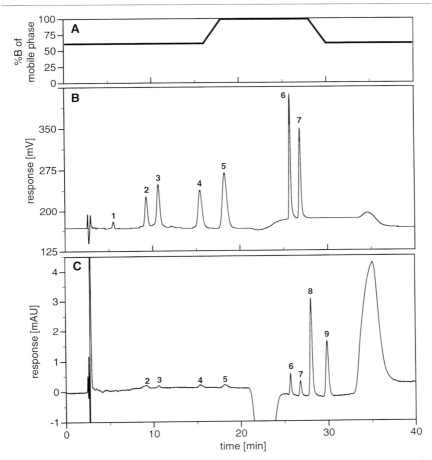

FIG. 1. Chromatogram of method with standard compounds. (A) Gradient scheme. (B) EC detection of BHT (1), γ-tocotrienol (2), α-tocotrienol (3), γ-tocopherol (4), α-tocopherol (5), ubiquinol 9 (6), and ubiquinol 10 (7). (C) UV detection at 275 nm of ubiquinone 9 (8) and ubiquinone 10 (9). The standards of γ-tocotrienol (2), α-tocotrienol (3), γ-tocopherol (4), α-tocopherol (5), ubiquinol 9 (6), and ubiquinol 10 (7) are also visible at this wavelength. Note: peaks of compounds detected by both UV and EC are present in the UV tracing 2 sec prior to detection by EC. From Podda M. et al., J. Lipid Res. **37,** 893 (1996).

and temperature conditions. Food (Harlan Teklad Rodent Diet #8656, Madison, WI) and water were provided *ad libitum*. Four mice were anesthetized, killed by cervical dislocation, and the tissues removed, rinsed with PBS saline, blotted dry, and an aliquot frozen in liquid nitrogen.

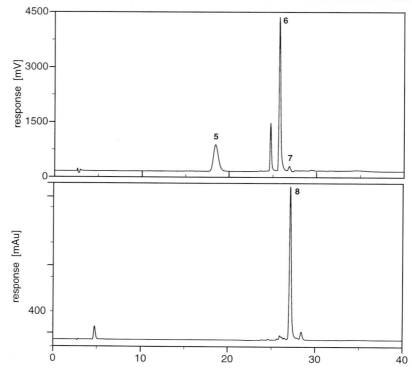

FIG. 2. Chromatogram of liver; peak identities are shown in Fig. 1 legend. *Top:* EC; *bottom:* UV 275 nm.

Preparation of Standards

Individual standards of α- and γ-tocopherols, α- and γ-tocotrienols, ubiquinol 9 and 10, and ubiquinone 9 and 10 are diluted in ethanol to obtain stock solutions between 50 and 200 μM; the exact concentration is determined spectrophotometrically (see Table I). Ubiquinol standards are prepared as described by Lang et al.[20] A stock solution of ubiquinone (50–150 μM) in 1 ml ethanol is reduced with 100 mg of sodium dithionite in 3 ml of water at room temperature in the dark for 30 min. Ubiquinols are extracted with hexane, dried under nitrogen, and resuspended in ethanol. The concentration of the resulting stock solution is determined spectrophotometrically using the molar extinction coefficient for ubiquinol 9 or ubiquinol 10 (Table I). The stock solution should be kept at −200° (stable for months).

Working standards are obtained by mixing appropriate amounts of the stock solutions. Working standards used for repeated determinations are

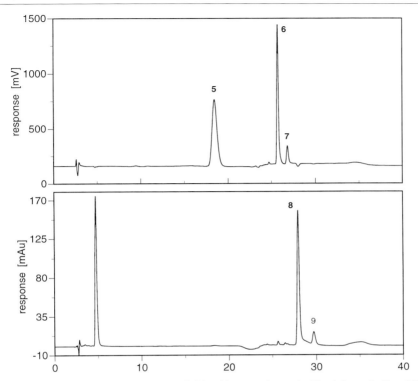

FIG. 3. Chromatogram of heart; peak identities are shown in Fig. 1 legend. *Top:* EC; *bottom:* UV 275 nm.

stored in a cooled autoinjector (4°) for a maximum of 4 hr. As in the tissue samples, the gradual oxidation of solutions of ubiquinols 9 and 10 in ethanol is found to be less than 6% after 4 hr under these conditions.

HPLC Analysis

The HPLC system we use consists of a Hewlett Packard 1050 series gradient pump, a SCL-10A Shimadzu system controller with a SIL-10A autoinjector with sample cooler, a Beckman Ultrasphere ODS C_{18} column, 4.6 mm I.D., 25 cm, 5-μm particle size with a Rainin Spheri-5 RP_{18} precolumn 5 μm 30 × 4.6 mm, a Hewlett Packard 1050 Diode array detector, and a Bioanalytical Systems LC-4B amperometric electrochemical detector with a glassy carbon electrode. Detectors are set up in line, with the eluent passing first through the diode array detector. The delay between UV and EC detectors is approximately 2 sec.

A gradient is used consisting of a mixture of A [80/20 (v/v) methanol/

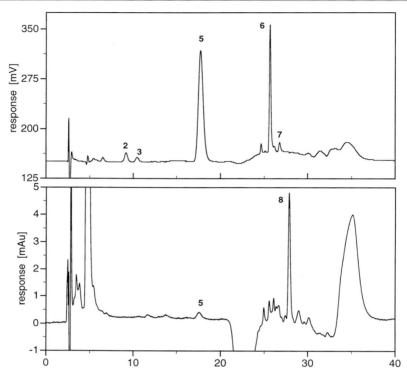

FIG. 4. Chromatogram of skin; peak identities are shown in Fig. 1 legend. *Top:* EC; *bottom:* UV 275 nm. From M. Podda *et al.*, *J. Lipid Res.* **37,** 893 (1996).

water and 0.2% (w/v) lithium perchlorate] and B [ethanol with 0.2% (w/v) lithium perchlorate] at a flow rate of 1 ml/min. The initial conditions are 39% A and 61% B. After 16 min the mobile phase is changed linearly over 2 min to 100% B; 100% B continues for 10 min, after which the system reverts linearly over 2 min to the initial conditions (39% A and 61% B). These changes in mobile phase perturb the baseline during both EC and UV detection at 21 to 24 min and at 33 to 37 min. Each run lasts 40 min. The diode array detector is set to 275 nm for the detection of ubiquinones; the electrochemical detector is in the oxidizing mode, potential 500 mV, full recorder scale at 50 nA. The output from the EC detector (Fig. 1B) demonstrates the separation of tocopherols, tocotrienols, and ubiquinols, which elute in the opposite order of their lipophilicity. The vitamin E forms elute as follows: γ-tocotrienol, α-tocotrienol, γ-tocopherol, and α-tocopherol. Subsequently, ubiquinol 9 and ubiquinol 10 are eluted as sharp peaks as ethanol forms an increasing proportion of the mobile phase. Ubiquinone 9 and ubiquinone 10 are eluted last and are detectable only

TABLE II
Lipophilic Antioxidants in Hairless Mouse Tissue[a]

Tissue	α-Tocopherol	γ-Tocopherol	α-Tocotrienol	γ-Tocotrienol	Ubiquinol 9	Ubiquinol 10	Ubiquinone 9	Ubiquinone 10
Brain	5.4 ± 0.1	0.01 ± 0.02	n.d.[b]	n.d.	1.6 ± 0.1	0.6 ± 0.1	10.2 ± 0.5	3.4 ± 0.5
Heart	24.2 ± 1.1	0.19 ± 0.05	0.08 ± 0.01	0.19 ± 0.05	18.5 ± 4.1	2.8 ± 0.7	244.6 ± 22.3	21.3 ± 7.5
Kidney	21.9 ± 0.6	0.35 ± 0.06	0.06 ± 0.04	0.15 ± 0.07	81.0 ± 28.5	10.9 ± 5.5	301.6 ± 123.8	30.8 ± 13.9
Liver	21.2 ± 2.9	0.29 ± 0.03	0.10 ± 0.04	0.19 ± 0.16	41.5 ± 15.5	1.7 ± 0.3	45.8 ± 17.8	n.d.
Skin	5.4 ± 1.6	0.04 ± 0.00	0.24 ± 0.20	0.76 ± 0.71	2.2 ± 0.3	0.4 ± 0.0	7.6 ± 1.9	n.d.
Skin and subcutis	5.7 ± 0.2	0.03 ± 0.01	0.14 ± 0.02	0.41 ± 0.10	2.6 ± 0.5	0.5 ± 0.1	8.1 ± 2.5	n.d.

[a] Data given in nmol/g tissue. From M. Podda et al., J. Lipid Res. **37**, 893 (1996).
[b] Not detectable.

TABLE III
PERCENTAGE DISTRIBUTION OF VITAMIN E HOMOLOGS IN HAIRLESS MOUSE TISSUES[a]

Tissue	α-Tocopherol	γ-Tocopherol	α-Tocotrienol	γ-Tocotrienol
Brain	99.8 ± 0.4	0.2 ± 0.4	—	—
Heart	98.1 ± 0.2	0.8 ± 0.2	0.3 ± 0.0	0.8 ± 0.2
Kidney	97.5 ± 0.6	1.6 ± 0.3	0.3 ± 0.2	0.6 ± 0.3
Liver	97.2 ± 1.5	1.4 ± 0.3	0.5 ± 0.3	0.9 ± 0.9
Skin	85.6 ± 7.2	0.6 ± 0.3	3.4 ± 1.5	10.4 ± 6.0
Skin and subcutis	90.8 ± 1.4	0.5 ± 0.2	2.2 ± 0.3	6.5 ± 1.3

[a] From M. Podda et al., J. Lipid Res. **37,** 893 (1996).

by the in-line UV-diode array detector (Fig. 2C). Note that the UV detector physically precedes the EC detector, so that peaks of compounds that are detected by both are detected 2 sec earlier by the UV detector.

Recovery and Reproducibility

Liver (100 mg) is homogenized as described earlier. The homogenate is divided into two equal aliquots, and 850 pmol of α- and γ-tocotrienols is added to one aliquot. Both samples are then extracted as described earlier. The percentage recovery is calculated by subtracting the amount of tocotrienols measured in liver from the amount found in liver with added internal standards and multiplying by 100. Recoveries of α-tocopherol, ubiquinol 9 and 10, and ubiquinone 9 and 10 from rat liver using the described extraction procedure have been reported previously by our laboratory to be >90%.[20] Using the present method, the recovery of added γ-tocopherol is 93%, α-tocotrienol 85%, and γ-tocotrienol 94%.[22]

The limit of electrochemical detection is 0.1 pmol for tocopherols and 0.3 pmol for ubiquinols. The detection limit of UV detection is 0.2 pmol for ubiquinone. This relatively low limit for UV detection is due to the very high extinction coefficient of ubiquinone.

The standard curves for ubiquinol 9, α-tocopherol, and γ-tocopherol quantitated from the output of the EC detector and for ubiquinone 9 from the UV detector are linear from 0.5 to 250 pmol for all lipophilic antioxidants analyzed.[22]

Representative chromatograms for liver (Fig. 2), heart (Fig. 3), and skin (Fig. 4) demonstrate the clear separation of these lipophilic antioxidants. After 40 min, no peaks are detectable at 206 nm, thus the sample elutes completely from the column and no postrun column washout is required before the next injection. In fact, a large series of samples (>100) can be run without postrun time or column cleaning.

[22] M. Podda, C. Weber, M. Traber, and L. Packer, *J. Lipid Res.* **37,** 893 (1996).

The concentration of lipophilic antioxidants is shown in Table II and the percentage distribution of the various vitamin E homologs is shown in Table III. Intriguingly, very different distributions of the various vitamin E forms in different organs are found, suggesting that these may be regulated independently in each tissue.

Conclusion

The method described herein allows the extraction and detection of the various vitamin E forms along with ubiquinol/ubiquinone in a single aliquot of tissue. The unexpected distribution of vitamin E forms in different tissues exemplifies that the selective determination of the major lipophilic antioxidants is important for understanding the relationship among diet, oxidative stress, and cellular regulatory processes.

Acknowledgments

Kenneth Tsang and Heike Beschmann provided excellent technical assistance. We gratefully acknowledge the efforts of Dr. Asat A. Qureshi, University of Wisconsin, who isolated tocotrienols for use as standards for this study. Funding was provided in part by NIH (CA47597), the Palm Oil Research Institute of Malaysia (PORIM), and the Marie Christine Held and Erika Hecker Foundations.

[31] High-Performance Liquid Chromatography–Coulometric Electrochemical Detection of Ubiquinol 10, Ubiquinone 10, Carotenoids, and Tocopherols in Neonatal Plasma

By BARBARA FINCKH, ANATOL KONTUSH, JENS COMMENTZ, CHRISTOPH HÜBNER, MARTIN BURDELSKI, and ALFRIED KOHLSCHÜTTER

Introduction

Oxidative stress is implicated as playing a major role in the pathogenesis of various disorders called "oxygen radical diseases of the premature" in prematurely born children.[1] Until recently the investigation of the balance between prooxidants and antioxidants in such premature babies was hampered by the small quantities of blood available for the necessary tests as

[1] J. L. Sullivan, *Am. J. Dis. Child.* **142**, 1341 (1988).

well as by the low concentrations of the lipophilic antioxidants.[2,3] In many investigations umbilical cord blood was thus used, which allowed a sufficient blood volume but entailed the disadvantage of excluding the possibility of measuring several samples over a longer time period. We therefore developed a micromethod to simultaneously determine the lipophilic antioxidants ubiquinol 10, ubiquinone 10, carotenoids, and tocopherols in plasma samples of neonates.[4] The described method, which is a modification of two other methods,[5,6] combines high-performance liquid chromatography (HPLC) with coulometric electrochemical detection and individual internal standardization. HPLC with coulometric electrochemical detection is characterized by total oxidation or reduction of the substance passing the electrode. Individual internal standardization is useful for volume control as well as for standardization of the oxidation of ubiquinol 10 to ubiquinone 10 during the preparation of the samples.[4] An overview of the presented technique to measure lipophilic antioxidants in neonatal plasma microsamples is given in Fig. 1.

Preparation of Standards and Internal Standards

Standards and internal standards (1–5 mg) are dissolved in 100 ml of ethanol (tocopherols from E. Merck, Darmstadt, Germany, ubiquinones from Sigma Chemical Co., Deisenhofen, Germany), hexane (β-carotene from Sigma Chemical Co., β-cryptoxanthin from Roth, Karlsruhe, Germany), or chloroform (ethyl β-apo-8′-carotenoate from Fluka, Neu-Ulm, Germany). Ubiquinol 9 and ubiquinol 10 are prepared by reducing the corresponding ubiquinones with sodium dithionite.[5] All standard solutions are stored sealed under argon at $-20°$. The concentration of each standard is measured photometrically in quartz cuvettes (1 cm) and is calculated using the following extinction coefficients: γ-tocotrienol, $E^{1\%} = 103.0$ at 298 nm[7]; γ-tocopherol, $E^{1\%} = 91.4$ at 298 nm[7]; α-tocopherol, $E^{1\%} = 75.8$ at 292 nm[7]; ubiquinone 7, $E^{1\%} = 221$ at 275 nm[8]; ubiquinone 9, $E^{1\%} = 185$ at 275 nm[8]; ubiquinone 10, $E^{1\%} = 165$ at 275 nm[8]; ethyl β-apo-8′-caroten-

---

[2] S. Sinha and M. Chiswick in "Vitamin E in Health and Disease" (L. Packer and J. Fuchs, eds.), p. 861. Dekker, New York, 1993.
[3] J. H. N. Lindemann, D. Van Zoeren-Grobben, J. Schrijver, A. J. Speek, B. J. H. M. Poorthuis, and H. M. Berger, *Pediatr. Res.* **26,** 20 (1989).
[4] B. Finckh, A. Kontush, J. Commentz, C. Hübner, M. Burdelski, and A. Kohlschütter, *Anal. Biochem.* **232,** 210 (1995).
[5] J. K. Lang, G. Kishorchandra, and L. Packer, *Anal. Biochem.* **157,** 106 (1986).
[6] G. Grossi, A. M. Bargossi, P. Fiorella, and S. Piazzi, *J. Chromatogr.* **593,** 217 (1992).
[7] M. Kofler, P. F. Sommer, H. R. Bolliger, B. Schmidli, and M. Vecchi, *Vitam. Horm.* **20,** 407 (1962).
[8] Y. Hatefi, *Adv. Enzymol.* **25,** 275 (1963).

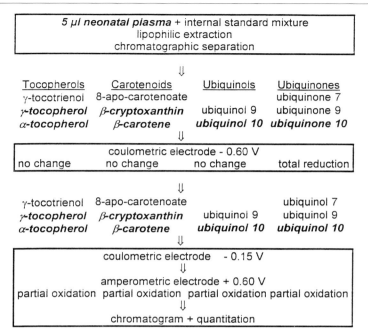

FIG. 1. Measurement of lipophilic antioxidants in neonatal plasma. Lipophilic antioxidants are extracted from a 5- to 10-μl plasma sample (bold: plasma antioxidants; other: internal standards) and the chromatographic separation is performed. By passing the first coulometric electrode the ubiquinones are reduced to the corresponding ubiquinols; the other substances remain unchanged. After passing the first coulometric electrode of the second cell, all compounds are partially oxidized by passing the second amperometric electrode. The signals of this electrode are transferred to the integrator and can be quantitated by calculating the area ratios between substances and internal standards.

oate, ε_{mol} = 115,000 at 445 nm[9]; β-cryptoxanthin, ε_{mol} = 136 at 451 nm[10]; and β-carotene, $E^{1\%}$ = 2592 at 453 nm.[11] The concentrations of ubiquinol 9 and ubiquinol 10 are calculated by measuring the difference in absorption before and after oxidation with 2 μl of 1 M potassium hydroxide: ubiquinol 9, $\Delta E^{1\%}$ = 158 at 275 nm[12] and ubiquinol 10, $\Delta E^{1\%}$ = 142 at 275 nm.[12] The purity of standards is checked by using available spectra for the standards

---

[9] S. Zeng, H. C. Furr, and J. A. Olson, *Am. J. Clin. Nutr.* **56**, 433 (1992).
[10] D. I. Thurnham, E. Smith, and P. S. Flora, *Clin. Chem.* **34**, 377 (1988).
[11] K. L. Simpson, S. C. S. Tsou, and C. O. Chichester, in "Methods of Vitamin Assay" (J. Augustin, B. P. Klein, D. Becker, and P. B. Venugopal, eds.), p. 185. Wiley, New York, 1985.
[12] A. Kröger, *Methods Enzymol.* **53**, 579 (1978).

and by monitoring the individual standards with our HPLC system using electrochemical detection as well as ultraviolet (UV) or visible light detection with the wavelength used for quantitation of the individual standard.

Plasma Sample Preparation and Chromatographic Separation

A neonatal plasma sample (5 or 10 μl) is transferred with a micropipette into an aluminum foil wrapped conical vial (Chromacol, London, Great Britain). An internal standard mixture in ethanol (50 μl; 61 pmol γ-tocotrienol, 4.2 pmol ubiquinone 7, 2.8 pmol ubiquinol 9, 0.1 pmol ubiquinone 9, 1.8 pmol ethyl β-apo-8′-carotenoate) and 2[6]-di-*tert*-butyl-*p*-cresol (50 μl, 111 mg/100 ml ethanol) are added. The vial is vortexed for 10 sec and again for 2 min after adding 500 μl hexane for lipid extraction. After phase separation following centrifugation (1000g, 5 min, 4°), 400 μl of the hexane phase is transferred to another conical vial, evaporated dry under a stream of argon, and redissolved in 40 μl of reagent alcohol:methanol (1:1, v/v; reagent alcohol: ethanol:2-propanol, 95:5, v/v). This sample is injected onto the HPLC system using an injection system with a 20-μl loop (Rheodyne, Cotati). The helium-degassed mobile phase consists of 13.4 mM lithium perchlorate in methanol:ethanol:2-propanol (88/24/10, v/v/v) and is delivered at a flow rate of 1.2 ml/min by a low pulse pump (Modell 2150, Pharmacia LKB, Germany), which is additionally connected to a pulse damper (ESA, Bedford, MA). It is also advisable to prime the electrodes by pumping the mobile phase at a low flow rate overnight before the day of measurement. The separation of the lipophilic antioxidants is performed at room temperature on a SuperPac Pep-S RP$_{C_2/C_{18}}$ column (250 × 4.0 mm I.D., 5-μm particle size, Pharmacia, Freiburg, Germany) preceded by a guard column (10 × 4.0 mm I.D., 5-μm particle size, Pharmacia, Freiburg, Germany).

Conditions of Electrochemical Coulometric Detection

After separation, the lipophilic antioxidants pass the detection system of a Coulochem Model 5100 A or 5200 electrochemical detector (ESA, Bedford, MA) one after the other; first the conditioning cell (Modell 5020; coulometric electrode: −0.60 V) and then the analytical cell (Modell 5011; first coulometric electrode: −0.15 V; second amperometric electrode: 0.60 V). The two cells, which are preceded by an in-line filter with a graphite filter element, work together in the reduction–oxidation mode. The current of the amperometric electrode in the analytical cell is monitored. Typical chromatograms of a standard mixture and of a neonatal plasma microsample such as derived through this method are shown in Fig. 2. A sufficient response at a voltage of 0.6 V has been shown in hydrodynamic voltammograms for tocopherols, carotenoids, ubiquinols, and ubiquinones.[4]

Quantitation of Lipophilic Antioxidants

The concentrations of lipophilic antioxidants in neonatal plasma are quantitated using the peak area ratios between sample peak and internal standard peak. γ-Tocotrienol is used as internal standard for γ-tocopherol and α-tocopherol. Ethyl β-apo-8′-carotenoate is used as an internal standard for β-cryptoxanthin and β-carotene. Ubiquinol 9 is used as internal standard for ubiquinol 10. Ubiquinone 9 or ubiquinone 7 is used as an internal standard for ubiquinone 10.

Precision and Detection Limits

The within-day precision using 5 or 10 μl plasma was between 3 and 14% (coefficient of variation) for all measured substances. Detection limits of the measured compounds varied between 21 and 60 fmol/20 μl injection volume, which correspond to 0.004 to 0.012 μmol/liter plasma (signal-to-noise ratio of 5).[4]

The detection limits for ubiquinol 9 (37 fmol/20 μl injection) and ubiquinone 9 (44 fmol/20 μl injection) as well as for ubiquinol 10 (51 fmol/20 μl injection) and ubiquinone 10 (60 fmol/20 μl injection) are very similar.[4] The internal standard ubiquinone 9 and the plasma-derived ubiquinone 10 are reduced to the corresponding ubiquinols by passing the first coulometric electrode (Fig. 1). The internal standard ubiquinol 9 and the plasma-derived ubiquinol 10 remain unchanged by passing the first coulometric electrode. Because the ubiquinones are monitored and quantitated in the analytical cell as having been transformed to the corresponding ubiquinols, it becomes clear that the detection limits of corresponding ubiquinols and ubiquinones have to be very similar; the slight variation reflects the difference in retention time.

Comments

Central to the development of the presently described method is the clinical necessity of working with tiny blood samples as obtained from premature babies. The work of other researchers is concentrated mainly on improving the basic methodology but not on scaling down the amount of sample necessary.

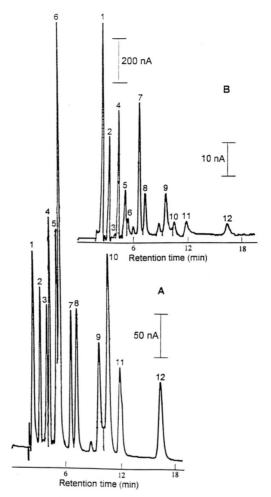

FIG. 2. (A) Chromatogram of a standard mixture; 10 pmol per standard was injected. Peak numbers and retention times are as follows (peak number; retention time): 2,6-di-*tert*-butyl-*p*-cresol (1; 2.1 min), γ-tocotrienol (2; 3.3 min), γ-tocopherol (3; 4.0), α-tocopherol (4; 4.3 min), ethyl β-apo-8′-carotenoate (5; 5.1 min), β-cryptoxanthin (6; 5.4 min), ubiquinone 7 (7; 6.7 min), ubiquinol 9 (8; 7.4 min), ubiquinol 10 (9, 9.7 min), β-carotene (10; 10.7 min), ubiquinone 9 (11; 12.0 min), and ubiquinone 10 (12; 16.5 min). (B) Chromatogram of a 10-μl plasma sample of a 6-day-old newborn [peak numbers as in (A)]. The concentrations measured were 1.11 μmol/liter γ-tocopherol, 19,96 μmol/liter α-tocopherol, 0.15 μmol/liter β-cryptoxanthin, 0.39 μmol/liter ubiquinol 10, and 0.02 μmol/liter ubiquinone 10. Chromatographic conditions: SuperPac Pep-S $RP_{C_2/C_{18}}$ column 250 × 4.0 mm I.D., 5-μm particle size; mobile phase, methanol/ethanol/2-propanol (88/24/10; v/v) containing 13 mM lithium perchlorate; flow rate 1.2 ml/min; applied potential, conditioning cell −0.60 V, detector 1, −0.15 V; detector 2, 0.60 V; gain, 10 × 20; integrator attenuation, 64 (standard mixture), 256 (plasma microsample 0–4.7 min), 16 (plasma microsample 4.7–18 min); chart speed, 0.25 cm/min. From B. Finckh, A. Kontush, J. Commentz, C. Hübner, M. Burdelski, and A. Kohlschütter, *Anal. Biochem.* **232**, 210 (1995), with permission.

The reduction of ubiquinones, for example, can also be performed by using a postseparation on-line reduction column instead of a coulometric electrode as has been shown.[13,14] Although this methodological approach is similar to ours, 10 to 20 times more plasma is necessary, no separation of γ-tocopherol and α-tocopherol is performed, and no internal standardization is used.

Internal standardization is also not used in the method of Sattler et al.[15] In this method the concentrations of lipophilic antioxidants in low density lipoproteins are measured using two amperometric electrochemical electrodes (-0.70 V; 0.60 V) in series. Disadvantages of the Motchnik et al.[16] method for measuring lipophilic plasma antioxidants also using the coulometric reduction–oxidation mode are the longer sample run times and the demand of 250 μl plasma.

A precolumn switching system[17] or an automatic valve for discarding column eluent within the first 2 min[18] is integrated into some of the HPLC systems to solve the problem of overloading the coulometric electrode with lipophilic plasma components, thus shortening the lifetime of the electrode considerably. We totally avoid this problem by using tiny plasma samples of neonates which, in addition, have low plasma lipid levels.[4]

The oxidation of ubiquinol 10 to ubiquinone 10 during sample preparation is minimized by the method of Yamamita and Yamamoto[14] and Lagendijk et al.[18] by injecting the hexane or propanol plasma lipid extracts directly onto the HPLC system, but this also demands much larger sample quantities. This problem was solved by using ubiquinol 9 and ubiquinone 9 as internal standards. HPLC with coulometric electrochemical detection in the oxidation–reduction–oxidation detection mode can also be used to simultaneously measure ubiquinol 10 and ubiquinone 10,[17,18] but only those compounds are measured that are undergoing reversible oxidation, also demanding greater quantities of plasma.

We overcame the difficulty of measuring low lipophilic antioxidant concentrations in small volumes of neonatal plasma by combining the advantage of on-line coulometric electrochemical detection with the advantage of internal standardization. The presented method can be applied in slightly modified form[4] to measure lipophilic antioxidants in plasma or low-

---

[13] H. Wakabayashi, S. Yamato, M. Nakajima, and K. Shimada, *Biol. Pharmacol. Bull.* **17**, 997 (1994).
[14] S. Yamahita and Y. Yamamoto, *Anal. Biochem.* **250**, 66 (1997).
[15] W. Sattler, D. Mohr, and R. Stocker, *Methods Enzymol.* **233**, 469 (1994).
[16] P. A. Motchnik, B. Frei, and N. B. Ames, *Methods Enzymol.* **234**, 269 (1994).
[17] P. O. Edlund, *J. Chromatogr.* **425**, 87 (1988).
[18] J. Lagendijk, J. B. Ubbink, and W. J. Hayward Vermaak, *J. Lipid Res.* **37**, 67 (1996).

density lipoproteins[19] of healthy adults,[20] vitamin E-deficient patients,[21] or hyperlipidemic patients[20] as has been shown.

Summary

A micromethod for the rapid simultaneous determination of several lipophilic antioxidants in plasma from newborn infants is presented. Because only 5 μl of plasma is required, the procedure lends itself for repetitive use in very immature infants at risk for developing so-called "oxygen radical diseases of the premature."[1] The method allows continuous monitoring of antioxidants in such patients and can easily be combined with monitoring other parameters of interest in this context. Reuse of blood samples taken routinely for the determination of hematocrit and bilirubin concentration is possible, reducing the blood volume required to be taken for the oxygen radical-related studies to virtually zero.

Acknowledgments

This study was supported in part by the parents' self-help group "Helft dem muskelkranken Kind," Hamburg, Germany. We thank Dr. J. Knispel, Hamburg, for his language advice.

[19] A. Kontush, C. Hübner, B. Finckh, A. Kohlschütter, and U. Beisiegel, *FEBS Lett.* **341**, 69 (1994).
[20] A. Kontush, A. Reich, K. Baum, T. Spranger, B. Finckh, A. Kohlschütter, and U. Beisiegel, *Atherosclerosis* **129**, 119 (1997).
[21] A. Kohlschütter, E. Mayatepek, B. Finckh, and C. Hübner, *J. Inher. Metab. Dis.* **20**, 581 (1997).

[32] Simultaneous Determination of Retinol, Tocopherols, Carotene, Lycopene, and Xanthophylls in Plasma by Means of Reversed-Phase High-Performance Liquid Chromatography

By Claude-Pierre Aebischer, Joseph Schierle, and Willy Schüep

Introduction

The role that is attributed to micronutrients, especially vitamins and carotenoids, in maintaining good health has received more attention. The prevention of certain cancers or cardiovascular diseases has been correlated to the food intake habit and particularly to the composition of nutrition. Although some controversial discussions on the impact of certain food ingredients or cooking styles are still ongoing more and more evidence is given that a nutrition rich in vegetables, poor in animal fats, and covers adequately the micro- and macronutrient requirements represents a good way of protection. Among the micronutrients, the vitamins, namely vitamin A, E, and C, and the carotenoids, such as α- and β-carotene, lycopene, or the xanthophylls such as cryptoxanthin, lutein, or zeaxanthin, belong to the class of substances of interest occurring in food. The way these compounds are being absorbed by the body is of great interest, and the concentrations of the individual components present in biological fluids or tissues play an important role in the interpretation of epidemiological studies. A prerequisite to performing such studies is the availability of adequate analytical methods, which have to be sensitive enough and simultaneously selective. In addition, efficient and inexpensive procedure should be on hand with a high degree of precision to meet the required constancy and reproducibility of the measurements over a long interval. The method described here was developed to assay a large number of plasma or serum samples. It has been tested over a long period of time and has been proven to be robust, efficient, and accurate. In view of the length of epidemiological studies being sometimes in the range of over a decade, special attention has been paid to the stability of the different analytes in the samples under storing conditions as well as to the aspects related to quality assurance.

Materials and Methods

Reagents and Solutions

If not otherwise stated, all reagents are of pro analysi grade, and all solutions are prepared fresh, stored at $-20°$, and protected from light. Water is nanopure grade (>18 MΩ, Millipore, Bedford, MA). α-Carotene, all-*trans*-β-carotene, and various *cis*-β-carotene isomers, all-*trans*-lycopene and various *cis*-lycopene isomers, β-cryptoxanthin, phytofluene, canthaxanthin, astaxanthin, lutein, zeaxanthin, retinol, α-, β-, γ-, and δ-tocopherol are from Hoffmann-La Roche Ltd (Basle, Switzerland). Absolute ethanol (Lichrosolv), methanol, *n*-hexane (Uvasol), 1,4-dioxane, dichloromethane, chloroform, 2-propanol, and 2,6-di-*tert*-butyl-4-methylphenol (BHT, puriss) are from E. Merck (Darmstadt, Germany). Tetrahydrofuran (THF) and ammonium acetate are from Fluka (Buchs, Switzerland), and acetonitrile (HPLC grade S) is from Rathbun Chemicals (Walkerburn, Scotland). *n*-Hexane is stabilized with BHT (350 mg/liter) and is stored at room temperature for 2 months.

Calibration Solutions

The carotenes are dissolved in small volumes of dichloromethane before being diluted with n-hexane to a final concentration of 0.5 µg/liter. The xanthophylls are dissolved in 1.5 ml chloroform and diluted up to 100 ml with 2-propanol (this stock solution is stable for 1 month at $-20°$) before being diluted to 0.5 µg/liter with n-hexane. The final concentration of dichloromethane or chloroform in the standard solution is lower than 0.1%. Retinol is dissolved directly in n-hexane, whereas tocopherols are dissolved in ethanol.

The concentration of the standard solution is measured photometrically at the appropriate wavelength (Table I) against n-hexane, n-hexane/BHT, or ethanol, respectively.

The standard solutions are treated as samples. Hence 400 µl water, 400 µl ethanol (200 µl for tocopherols), 200 µl standard solution, and 600 µl n-hexane/BHT (800 µl for tocopherols) are well mixed and processed as described later. The resulting solutions are used to calibrate the high-performance liquid chromatography (HPLC) system.

Material

All procedures involving organic solvents are done in glass tubes. The plastic stoppers, used during the plasma extraction, should be checked for interference. Shaking of the glass tubes is done on a horizontal mechanical shaker (Vetter AG, St. Leon, Switzerland), and Eppendorf tubes are shaken on a vortex (Bender Hobein, Zürich, Switzerland). To dry the sample extracts, a centrifugal evaporator (Savant Instruments, Farmingdale, NY) is used. A dispenser (Hamilton Microlab M, Bonaduz, Switzerland) is used to handle organic solvents and solutions, whereas aqueous solutions are dosed with Eppendorf pipettes and disposable tips (Eppendorf, Hamburg, Germany).

HPLC Equipment and Conditions

The HPLC system is assembled using a high-pressure pump (Kontron, T 414), an autosampler (Kontron, 360), a column oven (Kontron, Oven Controller 480), a photometric (GAT LCD 501, Stagroma, Wallisellen, Switzerland), and a fluorimetric (Perkin Elmer LS40, England) detector. Data are acquired via a chromatography server (Fissions Instruments, England) and analyzed by a VAX based multichannel data acquisition system (Multichrom, VG Data Systems Ltd., England). To avoid disintegration of the redissolved plasma extract, the autosampler is kept at approximately $25°$.

TABLE I
SPECIFIC COEFFICIENTS OF ABSORPTION USED TO ASSESS CONCENTRATION OF STANDARD SOLUTIONS[a]

Analyte	λ_{max} (nm)	$E^{(1\%/1cm)}$	Solvent
α-Carotene	446	2725[b]	n-Hexane[h]
all-*trans*-β-Carotene	450	2590[b]	n-Hexane[h]
9-*cis*-β-Carotene	445	2550[b]	n-Hexane[h]
13-*cis*-β-Carotene	443	2090[b]	n-Hexane[h]
15-*cis*-β-Carotene	447	1820[b]	n-Hexane[h]
all-*trans*-Lycopene	472	3450[c,d]	n-Hexane[h]
5-*cis*-Lycopene	470	3466[e]	n-Hexane[h]
7-*cis*-Lycopene	470	2901[e]	n-Hexane[h]
15-*cis*-Lycopene	470	2072[e]	n-Hexane[h]
β-Cryptoxanthine	449	2386[d]	Petroleum ether[i]
Lutein	450	2500[f]	n-Hexane[h]
Zeaxanthin	450	2450[f]	n-Hexane[h]
α-Tocopherol	297	84	n-Hexane
	292	75.8[f]	Ethanol
β-Tocopherol	296	89.4[g]	Ethanol
γ-Tocopherol	298	96.0	n-Hexane
	298	91.4[g]	Ethanol
δ-Tocopherol	298	87.3[g]	Ethanol
Retinol	325	1826	n-Hexane

[a] (1% solution on 1-cm path length) at corresponding wavelength.
[b] J. Schierle, W. Härdi, N. Faccin, I. Bühler, and W. Schüep, in "Carotenoids, Isolation and Analysis" (G. Britton, S. Liaaen-Jensen, and H. Pfander, eds.), Vol. 1A, p. 265. Birkhäuser, Basel, 1995.
[c] J. Schierle *et al.*, *Food Chem.* **59**, 459 (1997).
[d] G. Britton, in "Carotenoids, Spectroscopy" (G. Britton, S. Liaaen-Jensen, and H. Pfander, eds.), Vol. 1B, p. 13. Birkhäuser, Basel, 1995.
[e] U. Hengartner *et al.*, *Helv. Chim. Acta* **75**, 1848 (1992).
[f] S. Weber, in "Analytical Methods for Vitamins and Carotenoids in Feed" (H. E. Keller, ed.), p. 83. Roche, Basel.
[g] W. Schüep and R. Rettenmaier, *Methods Enzymol.* **234**, 294 (1994).
[h] Containing 2% (v/v) dichloromethane or toluene.
[i] Similar n-hexane.

The separation is done on a reversed-phase column (Primesphere C_{18}-HC 5 μm 110 Å, 250 by 4.6 mm, Phenomenex, Torrance, CA) at 30° with a mixture of acetonitrile (684), tetrahydrofuran (220), methanol (68), and a 1% (w/v) ammonium acetate soution (28) at a flow of 1.6 (ml/min). The resulting backpressure on the system is between 40 and 70 bar. The detectors are programmed as shown in Table II. The photometric detector acquires data simultaneously on two channels (flip-flop mechanism set to 0.05 sec), different only in the range scale. This allows for low and high

TABLE II
Wavelength Program of Fluorescence and Photometric Detector

Run time (sec)	Fluorescence	
	$\lambda_{excitation}$ (nm)	$\lambda_{emission}$ (nm)
0 to 240	330	470[a]
240 to 600	298	328[a]
600 to 1250	349	480[a]

	Absorbance
	$\lambda_{absorbance}$ (nm)
20 to 470	450[b]
470 to 780	472[b]
780 to 1250	450[b]

[a] Autozero after wavelength change is done automatically.
[b] Autozero is done during the first 20 sec.

concentration samples, optimal sensitivity and limits of quantification, and a substantial extension of the dynamic linear range of the analytical method.

Plasma and Serum Extraction

Plasma or serum samples are usually stored at $-20°$, thawed slowly (about 30 min), and mixed well on a vortex. If necessary, plasma can be spun down at $4°$ for 5 min at $12,000g$ in an Eppendorf centrifuge. Two hundred microliters of plasma or serum is transferred to a 4-ml glass tube (inner diameter 8 mm), diluted with 200 μl water, and deproteinized with 400 μl absolute ethanol. The suspension is mixed vigorously on a vortex for 30 sec. To extract lipophilic molecules, 800 μl of n-hexane is added. For 7 min, the biphasic system is well mixed on a mechanical shaker (not vortex) and is spun down for 10 min at $4°$ at $2000g$. Subsequently, 400 μl of the clear supernatant is transferred to an Eppendorf tube and dried on a Speed-Vac at room temperature and 40 mbar. The residue is then redissolved in 150 μl of a mixture of methanol (1) and 1,4-dioxane (1) and further diluted with 100 μl acetonitrile. One hundred microliters of the resulting clear solution is injected into the HPLC system.

Calibration

One point calibrations (through zero) are done using vitamin, carotene, and xanthophyll solutions within the dynamic linear range.

Quality Assurance Scheme

To assess daily and long-term laboratory performance, we prepare our own internal control plasma. Heparinized, freshly frozen plasma is received from the Red Cross center in Basle, Switzerland. Two liters of pooled human plasma is mixed with 80 ml of carotene-rich bovine plasma, resulting in a higher carotene level without lowering the concentration of the other analytes. Until use, this control plasma is stored in 1-ml portions in Eppendorf tubes as $-80°$. Control plasma is measured daily in double each 10 samples. The daily average and the measured ranges are monitored for each parameter on quality control charts. In addition, the analytical methods are monitored by the participation in external quality assurance programs (NIST, Gaithersburg, MD and St. Helier, Surrey, England).

Results and Discussion

General Aspects

The presented isocratic method has been used for more than 7 years in our laboratory (typical chromatograms are shown in Figs. 1 and 2). About 150,000 plasma or serum samples have been analyzed and the method continues to be very robust and reliable. To guarantee reproducible results, the method has been monitored with internal control plasma and also by the external quality assurance programs organized by the National Institute of Standards and Technology (NIST, Gaithersburg, MD) and the St. Heliers Hospital (Carshalton Surrey, UK). The use of control charts was helpful in recognizing trends and outliers.

Plasma and Serum Extraction

To extract the lipophilic analytes from the plasma, it is first diluted with water, then the proteins are precipitated with ethanol, and finally the extraction is done with n-hexane. All these steps have to be validated for the ratio of volumes, temperatures, and mixing times. The critical points shall be discussed.

An important step during the extraction procedure is the dilution of the plasma with water. Unfortunately with n-hexane, not all analytes showed optimal extraction behavior at the same dilution ratio (Fig. 3). However, all desired analytes had an optimum water to plasma ratio between 1 and 2. Retinol and α-tocopherol were best extracted at a ratio of 2, whereas carotenes and xanthophylls were better extracted at a ratio of 1. In this regard, the extraction procedure has to be adapted according to the needs

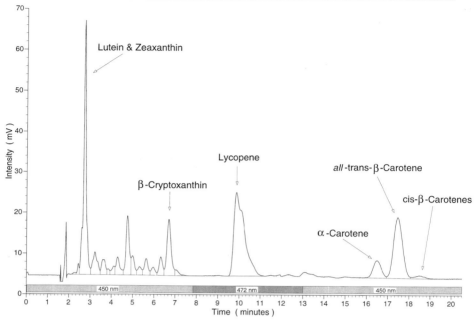

Fig. 1. Typical HPLC chromatogram of a human blood plasma analyzed with the described reversed-phase method (see text) showing visible detection of the carotenes and xanthophylls.

or followed by a second extraction step, which would increase the analysis time considerably.

The extraction procedure was then checked for its overall efficiency. The same plasma sample was extracted consecutively three times with n-hexane. In general, the first extraction was sufficient for all analytes, except retinol, which was extracted in average 93.5% (Table III). All analytes achieved an equilibrium state of distribution between the aqueous and organic phase within 3 min, giving the minimum shaking time for the extraction.

Additionally, it was recognized during extraction optimization that mixing the plasma/water/ethanol/n-hexane emulsion on a vortex was not sufficient to extract all the analytes quantitatively. Therefore the extraction was done on a horizontal mixing device in 4-ml tubes in a way that the remaining void volume was at least 2 ml.

Neither heparin nor ethylenediaminetetraacetic acid (EDTA), which are used for plasma preparation, showed any interferences with any of the analytes (data not shown).

In large-scale studies, the use of plasma versus serum is usually preferred

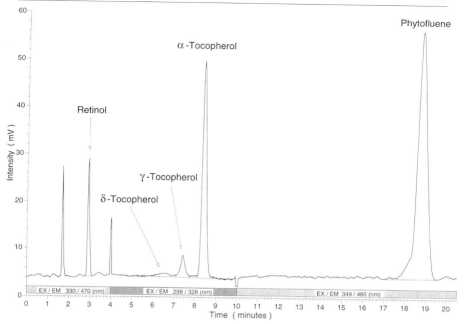

FIG. 2. Fluorescence detection for retinol, tocopherols, and phytofluene (which is not quantified routinely).

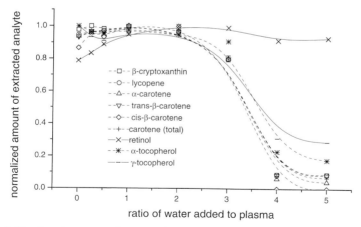

FIG. 3. For the given extraction procedure, the effect of water added to plasma, before protein precipitation with ethanol, is shown. An optimum water to plasma ratio for carotenes and xanthophylls lies closer to 1, whereas tocopherols and retinol are best extracted at a water to plasma dilution ratio of 2.

TABLE III
EXTRACTION EFFICIENCY OF PROCEDURE[a]

Analyte	Extraction 1		Extraction 2		Extraction 3	
	μg/liter	%	μg/liter	%	μg/liter	%
β-Cryptoxanthin	222	96.7	7.5	3.3	0	0
Lycopene (total)	346	97.2	10.1	2.8	0	0
α-Carotene	96.3	99.3	0.7	0.7	0	0
all-*trans*-β-Carotene	480	97.2	13.7	2.8	0	0
cis-β-Carotene	29	100	0	0	0	0
β-Carotene (total)	509	97.4	13.7	2.6	0	0
Retinol	497	93.5	34.8	6.5	0	0
α-Tocopherol[b]	10.46	95.7	0.47	4.3	0	0
γ-Tocopherol	400	100	0	0	0	0
δ-Tocopherol	50	100	0	0	0	0

[a] To assess the extraction efficiency of the procedure the same plasma sample has been extracted three times with *n*-hexane.
[b] Data in mg/liter.

as the reproducibility of its preparation at various laboratories is better. Still, some trials are done with serum samples, which of course we were supposed to analyze. Data have not yet been published, checking plasma versus serum extraction for vitamins, carotenes, and xanthophylls. Therefore, we processed plasma and serum from the same donor with the described method. Although no differences could be observed statistically for vitamins and carotenoids, the plasma levels for all analytes were slightly, but systematically, lower than serum levels ($-2.0 \pm 0.6\%$, Table IV). At a level of significance of 99%, no differences between plasma and serum extraction were shown. Because of the very low variance on the reversed-phase data set for vitamin E ($p = 0.008$) and total carotene ($p = 0.004$), slightly higher levels of significance were calculated. These findings may be disregarded as the mean values for both analytes differ in the same range and have the same trend as all other analytes.

Validation Parameters

Linearity, limit of detection, and recoveries of the described method have been published previously.[1] Some of the results are summarized in Table V. The intraday (identical materials, instruments, technicians) and interday repeatability (varying conditions such as materials, instruments,

[1] D. Hess, H. Keller, B. Oberlin, R. Bonfanti, and W. Schüep, *Int. J. Vit. Nutr. Res.* **61**, 232 (1991).

TABLE IV
EXTRACTION PROCEDURE FOR PLASMA AND SERUM SAMPLES FROM INDIVIDUAL DONOR

Analyte	Mean (μg/liter)		Variance[a]		n	F	p^b
	Plasma	Serum	Plasma	Serum			
	Reversed-phase HPLC (presented method)						
Retinol	731	745	179	179	6	3.281	0.100
α-Tocopherol	14.68[d]	15.07[d]	0.050	0.034	6	10.66	0.008
Carotene (total)[c]	446	460	63.1	27.8	6	13.88	0.004
Lycopene	572	583	84.8	54.7	6	5.856	0.036
β-Cryptoxanthin	200	203	6.7	15.8	6	2.967	0.116
	Normal-phase HPLC						
Retinol	698	706	42.9	32.9	4	3.810	0.099
α-Tocopherol	14.72[d]	15.01[d]	0.028	0.029	4	6.080	0.048
Carotene (total)[c]	551	566	30.91	139.6	4	4.932	0.068
Zeaxanthin	89.3	90.8	4.25	4.91	4	0.981	0.360
Lutein	336	344	26.3	33.7	4	4.544	0.077

[a] n, F, number of analysis, percentile of F distribution; p, level of significance where differences occur.
[b] Data have been analyzed with a one-side ANOVA (Origin, MICROCAL).
[c] Sum of α-carotene and β-carotene.
[d] mg/liter.

and technicians) have been determined with an internal control plasma over a period of almost 2 years and are expressed as the average of the standard deviations in percentages of multiple measurements (intraday), as the standard deviation in percentages of the average values between different days (interday) (see Table V).

The upper and lower limits of quantification can be estimated as the beginning and the end of the dynamic linear range. At these concentrations, the interval of confidence ($p = 0.95$) was found to be lower than 7% for all analytes. At the concentration levels quantified for internal control plasma, the corresponding intervals of confidence (at $p = 0.95$) for all analytes are shown in Table VI.

Peak Identification

For peak identification, the retention time of pure substances has been used. Additionally, all analytes, except β-cryptoxanthin, have been quantified on various normal-phase systems (results not shown). β-Tocopherol coelutes with γ-tocopherol but could not be detected in human plasma in a normal-phase system (data not shown).

TABLE V
VALIDATION PARAMETER FOR ISOCRATIC REVERSED-PHASE HPLC METHOD

Analyte	Limit of detection[a] (μg/liter)	Linear dynamic range[b] (μg/liter)	Repeatability[c] Intraday[d] (%)	Interday[e] (%)
Retinol	20	20–3,000	3.0	5.7
α-Tocopherol	20	20–20,000	2.0	1.9
γ-Tocopherol	20	20–10,000	5.4	2.8
α-Carotene	5	5–20,000	1.9	2.5
all-*trans*-β-Carotene	5	5–21,000	1.5	4.4
Lycopene	5	5–12,000	2.1	2.6
β-Cryptoxanthin	10	10–13,000	1.8	3.7

[a] x-fold signal of the baseline noise, where the standard deviation ($n-1$) is still smaller than 7% of the mean signal.
[b] The visible detector was run in a flip-flop mode, meaning that data were acquired simultaneously on two channels with two different ranges. This particular instrumental setup increased the dynamic linear range by a factor of 10.
[c] Determined at the levels of the internal control plasma given as the standard deviation in percentages of the average values.
[d] Identical materials, instruments, and technicians.
[e] Varying materials, instruments, and technicians.

α-Carotene is baseline resolved from all-*trans*-β-carotene, an important factor for plasma or serum analyses within human studies. The peak assigned as *cis*-β-carotene consists mainly of 13-*cis*- and 15-*cis*-β-carotene. 9-*cis*-β-Carotene underlies the all-*trans*-β-carotene peak, leading to a small overestimation of the all-*trans*-β-carotene concentration. However, in contrast to tissue samples, 9-*cis*-β-carotene in human serum is detectable only in very low concentrations.[2,3]

With the present HPLC system, lycopene isomers were only partially resolved and therefore quantified as the grand total related to all-*trans*-lycopene. We have analyzed our internal control plasma on a normal-phase system that resolves all-*trans*-lycopene from about a dozen partially identified cis isomers.[4] The lycopene in pooled human plasma consisted of up to 60% cis isomers (Table VI). Compared to the normal phase method, the reversed-phase method resulted in an overall concentration of lycopene

[2] W. Stahl, W. Schwarz, and H. Sies, *J. Nutr.* **123,** 847 (1993).
[3] J. von-Laar, W. Stahl, K. Bolsen, G. Goerz, and H. Sies, *J. Photochem. Photobiol. B* **33,** 157 (1996).
[4] J. Schierle, W. Bretzel, I. Bühler, N. Faccin, K. Steiner, and W. Schüep, *Food Chem.* **59,** 459 (1997).

TABLE VI
AVERAGE CONCENTRATIONS OF ANALYTES MONITORED IN INTERNAL CONTROL PLASMA OVER 2-YEAR PERIOD

Analyte	$\bar{x} \pm T^c$	n	s_x, P^d
α-Tocopherol	11.5 ± 0.13 mg/liter	66	0.22, 0.95
α-Tocopherol[a]	11.0 ± 0.05 mg/liter	67	0.18, 0.95
γ-Tocopherol	0.46 ± 0.023 mg/liter	67	0.04, 0.95
Retinol	475 ± 4.9 μg/liter	70	27.0, 0.95
Retinol[a]	465 ± 0.9 μg/liter	80	5.2, 0.95
trans-β-Carotene	471 ± 3.8 μg/liter	118	20.8, 0.95
cis-β-Carotene	31.4 ± 0.38 μg/liter	118	2.0, 0.95
α-Carotene	93.3 ± 0.4 μg/liter	67	2.3, 0.95
Total Carotene[a]	610 ± 1.3 μg/liter	80	7.3, 0.95
Total Lycopene	341 ± 1.6 μg/liter	72	9.0, 0.95
all-trans-Lycopene[a]	126.5 ± 0.75 μg/liter	72	4.6, 0.95
5-cis-Lycopene[a]	91.4 ± 0.57 μg/liter	72	3.5, 0.95
xz-cis-(9-, 13-, 15-, 7-, 5,5'-)Lycopene[a,b]	99.0 ± 0.98 μg/liter	72	6.0, 0.95
β-Cryptoxanthin	262 ± 2.0 μg/liter	118	9.8, 0.95

[a] Normal phase.
[b] Sum of baseline-resolved, but only partially identified cis isomers of lycopene.
[c] Average ± interval of confidence.
[d] Standard deviation ($n - 1$), percentile of student's distribution.

(sum of all isomers) that averaged 8% higher. Still, the present method reflected in an acceptable way the total lycopene concentration of the pooled human plasma. Specific absorption coefficients are still not known for all of the major separated and partially identified cis isomers (5-, 13-, 15-, 9-, 7- and 5,5'-di-cis-lycopene) (see Table I). Consequently, all-cis isomers have to be quantified with the specific absorption coefficient of all-trans-lycopene, which leads to an underestimation of the total lycopene concentration in plasma or serum.

Tocopherols and retinol have been detected by fluorescence, which enhances the selectivity and the sensitivity of the two analytes. The two xanthophylls lutein and zeaxanthin almost coelute with retinol, hence possible interferences were checked. Retinol has an emission maximum at 470 nm; lutein and zeaxanthin also absorb light at this wavelength. Only at concentrations of lutein or zeaxanthin higher than 1.8 mg/liter (corresponding to a plasma concentration of 7 mg/liter) could quenching effects be observed (decreasing emission by retinol at 470 nm).

In the time frame of 2 to 7 min, more polar molecules eluted from the reversed-phase column, mostly degradation and oxidation products, but also oxygen-containing xanthophylls. Some of these were run on the HPLC system described. Astaxanthin eluted at 2.4 min, lutein and zeaxanthin

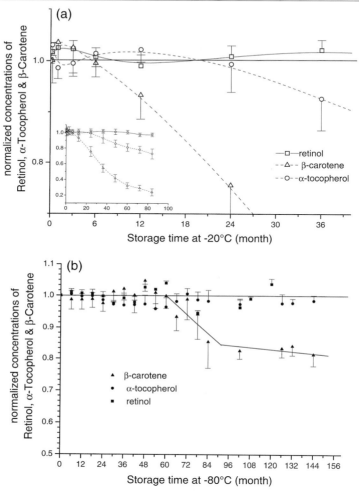

Fig. 4. Stability of retinol, α-tocopherol, and β-carotene at −20° (a) and −80° (b) in plasma samples.

coeluted at 2.7 min, canthaxanthin at 3.2 min, apo-8'-carotinoic acid ethyl ester at 5.7 min, and β-cryptoxanthin at 6.8 min. However, only lutein/zeaxanthin and β-cryptoxanthin were routinely quantified using the described reversed-phase method.

Long-Term Stability of Plasma

In order to use pooled human or animal plasma as internal control samples, stability of the monitored analytes has to be assessed over time. We

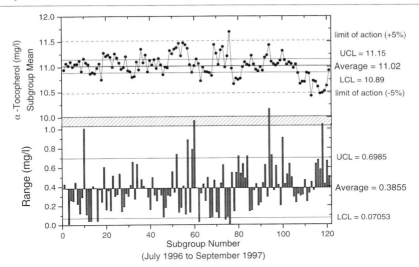

FIG. 5. Typical quality control chart for the described method generated with ORIGIN (Microcal). Data for α-tocopherol are shown. UCL/LCl, upper/lower confidence limit. For this analyte and laboratory the limit of action is set at ±5%.

therefore prepared plasma from 5 or 10 normal healthy human individuals immediately after blood taking and stored the samples at either −20 or −80°. Vitamin A, vitamin E, and β-carotene were measured at regular intervals over almost 12 years. During the first 5 years, the plasma samples were analyzed with normal-phase methods[5] and afterward with the described reversed-phase method. β-Carotene was only stable at −20° for 6 months, whereas at −80° it lasted for up to 5 years. The α-tocopherol concentration was unchanging at −20° for 2 years and at −80° for 10 years. Astonishingly, retinol showed no losses at all at both temperatures for up to 8 to 10 years (see Fig. 4).

Quality Assurance Concept

To control and monitor laboratory work, one can either buy and run certified reference materials or prepare one's own control plasma (which of course has no assigned values). In the latter case, the internal quality assurance concept data should be linked with an external quality assurance program. In the case of a high sample throughput laboratory, the second option is certainly the less expensive one. It is also necessary to determine that the control samples do not change at all during the periods used (see earlier).

[5] J. P. Vuilleunier, H. Keller, D. Gysel, and F. Hunziker, *Int. J. Vit. Nutr. Res.* **53**, 265 (1983).

For our laboratory, the second concept has turned out to be very efficient, inexpensive, and easy to use. Although the determined stability of the monitored analytes was much longer at $-80°$, the internal control plasma was used for a maximum of 2 years, before it was replaced by a new batch. The results were entered on a spreadsheet and presented as quality control charts (Fig. 5; ORIGIN Microcal, Northhampton, MA). The daily average and the range (minimum to maximum) are shown and have helped the laboratory staff to recognize trends or outliers during routine plasma analyses.

[33] Assessment of Prooxidant Activity of Vitamin E in Human Low-Density Lipoprotein and Plasma

By PAUL K. WITTING, DETLEF MOHR, and ROLAND STOCKER

Introduction

The oxidation of low-density lipoprotein (LDL) is implicated as an initiating event in atherogenesis,[1,2] the major cause of death in Western society. This has led to a veritable explosion of research into peroxidation of LDL lipids and its inhibition by antioxidants, particularly α-tocopherol (α-TOH), biologically the most active form of vitamin E[3] and the most abundant lipid-soluble antioxidant in human LDL.[4] α-Tocopherol can be a powerful inhibitor of lipid peroxidation.[3] For example, even small amounts of the vitamin strongly inhibit the autoxidation of polyunsaturated lipids (LH) in homogeneous solution. This is due to the ability of α-TOH to rapidly trap the chain-propagating lipid peroxyl radical (LOO·) [Eq.(1)] and the resulting α-tocopheroxyl radical (α-TO·) participating in a radical–radical termination reaction(s) [Eq. (2)], giving rise to nonradical products (NRP).

$$\text{LOO·} + \alpha\text{-TOH} \xrightarrow{k_H} \text{LOOH} + \alpha\text{-TO·} \quad (1)$$

$$\text{LOO·} + \alpha\text{-TO·} \rightarrow \text{NRP} \quad (2)$$

Despite scores of publications (reviewed in Refs. 5 and 6), confusion remains, however, as to whether α-TOH retards or promotes LDL lipid

---

[1] D. Steinberg, S. Parthasarathy, T. E. Carew, J. C. Khoo, and J. L. Witztum, *N. Engl. J. Med.* **320**, 915 (1989).
[2] J. A. Berliner and J. W. Heinecke, *Free Radic. Biol. Med.* **20**, 707 (1996).
[3] G. W. Burton and K. U. Ingold, *Acc. Chem. Res.* **19**, 194 (1986).
[4] H. Esterbauer, G. Jürgens, O. Quehenberger, and E. Koller, *J. Lipid Res.* **28**, 495 (1987).
[5] H. Esterbauer, J. Gebicki, H. Puhl, and G. Jürgens, *Free Radic. Biol. Med.* **13**, 341 (1992).

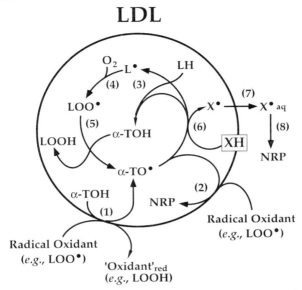

SCHEME 1. Tocopherol-mediated peroxidation (TMP) of LDL lipids and its inhibition by coantioxidants (XH). LDL lipid peroxidation initiated by one-electron oxidants can give rise to α-TO·, which in turn initiates and propagates TMP of LDL lipids. In the presence of XH or competing termination reactions involving radical–radical termination reactions [e.g., Eq. (2)], which give rise to nonradical products (NRP), the chain-carrying α-TO· is quenched and lipid peroxidation is inhibited.

peroxidation initiated by one-electron oxidants. This conflict is probably the consequence of both the physical properties of lipoproteins[7,8] and the ratio of oxidant:LDL[7,9] employed in the different studies.

Briefly, lipoproteins are emulsions of lipid droplets physically too small to harbor more than one radical at a time. In addition, α-TO·, thermodynamically the most stable lipid radical that can be formed, is too lipophilic to escape an oxidizing LDL particle within the time frame in which even the "unreactive" α-TO· can initiate and propagate tocopherol-mediated peroxidation (TMP) of LDL lipids [Scheme 1, reactions (3)–(5)]. In effect, and unlike the situation for an oxidizing LDL particle, for the termination of lipid peroxidation to occur [reaction (2) in Scheme 1], the frequency

---

[6] R. Stocker, *Curr. Opin. Lipidol.* **5**, 422 (1994).
[7] V. W. Bowry and R. Stocker, *J. Am. Chem. Soc.* **115**, 6029 (1993).
[8] K. U. Ingold, V. W. Bowry, R. Stocker, and C. Walling, *Proc. Natl. Acad. Sci. U.S.A.* **90**, 45 (1993).
[9] J. Neuzil, S. R. Thomas, and R. Stocker, *Free Radic. Biol. Med.* **22**, 57 (1997).

with which LDL encounters radicals (i.e., the radical flux) and the reactivity of these radicals (i.e., whether an "encounter" likely results in a reaction) together govern the fate of α-TO$\cdot$: under conditions of high fluxes of reactive aqueous radicals termination is favored, and vitamin E prevents lipid peroxidation in LDL. Conversely, under conditions of low fluxes of less reactive radicals, TMP [reactions (3)–(5), Scheme 1] is favored and vitamin E can promote lipid peroxidation in isolated LDL.

In addition to being the potential lipid peroxidation chain-transfer agent,[7] LDL vitamin E is also the most reactive redox moiety present at the lipid–water interface. By "scavenging" aqueous radicals, α-TOH in fact aids the entry of radicals into the lipoprotein particle. This phase-transfer activity can also result in an overall prooxidant activity of vitamin E, depending on the experimental conditions employed (see earlier). For example, a 10-fold increase in the concentration of α-TOH resulted in both a 6-fold increase in the proportion of aqueous peroxyl radicals scavenged by the vitamin and a 4.5-fold increase in the extent of lipid peroxidation when reconstituted plasma was exposed to the peroxyl radical generator 2,2'-azobis(2-amidopropane) dihydrochloride (AAPH).[9]

Here we introduce a series of simple tests that manipulate both the phase- and the chain-transfer activity of α-TOH in LDL and that can be used to evaluate whether the vitamin acts as a pro- or antioxidant under the given conditions. We emphasize that while these tests can provide useful information on the mechanism of action of α-TOH *in vitro*, they do not give information on the action of vitamin E *in vivo*, nor propose that α-TOH acts as a prooxidant *in vivo*. Rather, we suggest that LDL α-TOH effectively prevents the formation of lipid hydroperoxides (LOOH), as long as coantioxidant(s) (XH in Scheme 1) are available to eliminate α-TO$\cdot$ and the coantioxidant-derived radical diffuses from the particle and gives rise to NRP in the aqueous phase[10] [reactions (6)–(8), Scheme 1].

Methods and Applications

Reagents

Phosphate buffer (pH 7.4, 50 mM in phosphate) is prepared from nanopure water or deuterium oxide (D$_2$O), and the highest purity reagents available. Buffers are stored over Chelex 100 (100 mg/100 ml buffer, Bio-Rad, Richmond, CA) at 4° for at least 24 hr. This treatment effectively removes contaminating transition metals, as verified by the ascor-

---

[10] V. W. Bowry, D. Mohr, J. Cleary, and R. Stocker, *J. Biol. Chem.* **270,** 5756 (1995).

bate autoxidation method.[11] AAPH (molecular weight 271) and 2,2'-azobis(2,4-dimethylvaleronitrile) (AMVN; molecular weight 248) are from Polysciences (Warrington, PA) and are prepared as fresh reagents in phosphate buffer and ethanol, respectively. Tocopherols are from Henkel Corporation (La Grange, IL) (α-TOH) or Kodak (Rochester, NY) (γ-TOH, δ-TOH, and α-TOH acetate) and are prepared as 18 mM stock solutions in dimethyl sulfoxide (DMSO), using extinction coefficients of $\varepsilon_{294\ nm} \sim 3058$, $\varepsilon_{298\ nm} \sim 3866$, and $\varepsilon_{298\ nm} \sim 3672\ M^{-1}\ cm^{-1}$, for α-, γ-, and δ-TOH, respectively. Cholesteryl linoleate, soybean phosphatidylcholine, copper(II) sulfate, D$_2$O (100-ml bottles), and reduced glutathione (GSH) are from Sigma Chemicals (Sydney, Australia). Ebselen [2-phenyl-1,2-benziso-selenazol-3(2H)-one, a glutathione peroxidase mimic; molecular weight 274.2] is from Daiichi Pharmaceuticals (Tokyo, Japan) and prepared as a 10 mM ethanolic stock. Ebselen is also available from BIOMOL (Plymouth Meeting, PA). Authentic samples of hydroperoxides of cholesteryl linoleate (Ch18:2-OOH) and phosphatidylcholine (PC-OOH) are prepared from their purified lipids by oxidation with AMVN and stored as ethanolic stocks at $-20°$.[12] Lipoprotein-depleted plasma (LPDP) is obtained by gel filtration (PD-10 column, Pharmacia, Uppsala, Sweden) of the bottom fraction of plasma previously subjected to density ultracentrifugation using Procedure C in Ref. 12. All organic solvents employed were of HPLC grade quality (Merck).

Preparation of Lipoproteins

Native and Tocopherol-Enriched LDL. For enriched LDL, freshly obtained plasma (10 ml) is supplemented with 200 μl of stock solution of the various tocopherol analogs or dimethyl sulfoxide (DMSO) alone (control), and incubated under argon for 6 hr at 37°.[13] Low-density lipoprotein is then isolated from tocopherol-enriched and control plasma by 4-hr density-gradient ultracentrifugation at 15° using a TL100.4 rotor (Beckman, Palo Alto, CA) (Procedure B in Ref. 12). Excess KBr and remaining low molecular weight water-soluble antioxidants are removed by size exclusion chromatography (PD-10 column). The concentration of LDL is determined using the bicinchoninic acid assay,[14] with bovine serum albumin (Sigma Diagnostics) as a standard, and modifying the protocol described in the manufacturer's instructions by the addition of 2% (w/v) sodium dodecyl sulfate and

[11] G. R. Buettner, *Methods Enzymol.* **186,** 125 (1990).
[12] W. Sattler, D. Mohr, and R. Stocker, *Methods Enzymol.* **233,** 469 (1994).
[13] H. Esterbauer, M. Dieber-Rotheneder, G. Striegl, and G. Waeg, *Am. J. Clin. Nutr.* **53,** 314S (1991).
[14] P. K. Smith, R. J. Krohn, G. T. Hermanson, A. K. Mallia, F. H. Gartner, M. D. Provenzano, E. K. Fujimoto, N. M. Goeke, B. J. Olson, and D. C. Klenk, *Anal. Biochem.* **150,** 76 (1985).

assuming that LDL contains 1 mol apolipoprotein B–100/mol LDL particle.[15] Alternatively, LDL concentration is determined by measurement of cholesterol content by HPLC[12] and assuming 550 molecules of cholesterol per LDL particle.

α-TOH-Depleted LDL. α-TOH-depleted LDL is obtained by equilibrating freshly prepared, gel-filtered LDL ($\approx$1 mg protein/ml) at 37° for 5 min before the addition of 50 mM AAPH. Under these conditions of high peroxyl radical flux, LDL α-TOH is consumed rapidly with accumulation of only small amounts of lipid hydroperoxides and no measurable protein modification except some loss of thiol groups.[7] Oxidation is stopped by placing the LDL samples on ice. By removing aliquots at various times, LDL with different fractional α-TOH content is obtained (this is useful for the tests described later, although it will not be dealt with here any further). "Control" LDL, containing all of its endogenous α-TOH, is obtained by adding AAPH to LDL and placing the sample on ice for the same period of time. At this temperature, there is no substantial decomposition of AAPH, as has been verified by indistinguishable levels of α-TOH in freshly isolated and "control" LDL, and the absence of chemiluminescence detectable[12] hydroperoxides of cholesteryl esters (CE-OOH) in LDL following such incubation.[9]

Under these conditions, and at 37°, complete consumption of LDL α-TOH is achieved within 18–22 min of oxidation. Care should be taken that the oxidation does not proceed beyond α-TOH depletion, as this results in comparatively massive lipid peroxidation, degradation of some of the lipid hydroperoxides, and more substantial protein oxidation as judged by loss of tryptophan (fluorescence detection with excitation at 280 and emission at 335 nm). As LDL samples from different donors vary slightly, we recommend that one first oxidize an aliquot of the LDL sample to determine precisely the time required for $\approx$90–100% of α-TOH to be oxidized. This procedure of *in vitro* depletion of α-TOH can also be applied to lipoproteins other than LDL and, in principle, to any lipid emulsion, as long as high fluxes of peroxyl radicals are employed.

AAPH is removed from control and α-TOH-depleted LDL by two sequential gel filtration steps using PD-10 columns equilibrated with cold buffer. Two columns are required to remove all AAPH.[9] The small amounts of lipid hydroperoxides present in such AAPH-oxidized lipoproteins are reduced to the corresponding nonreactive alcohols by treatment of the LDL (0.3–0.5 mg protein/ml) with GSH (300 μM) and ebselen (10 μM) at 37° for 30 min. Control lipoproteins are also treated with GSH and ebselen. Conversion of hydroperoxides to the corresponding alcohols can

---

[15] R. E. Morton and T. A. Evans, *Anal. Biochem.* **204**, 332 (1992).

be verifed by HPLC with UV$_{234\ nm}$ detection,[16] whereas the absence of hydroperoxides is verified by HPLC with postcolumn chemiluminescence detection (see later). Finally, ebselen and GSH are removed by gel filtration of LDL through two sequential PD-10 columns as described earlier. This procedure allows the *in vitro* preparation of peroxide-free LDL (and other lipoproteins) with different α-TOH content without marked changes in the lipid compositon.

α-TOH-Replenished LDL. In vitro α-TOH-depleted, lipid hydroperoxide-free LDL (0.1–0.2 mg protein/ml) obtained as described earlier is combined with LPDP (2:1, v/v) and supplemented with α-TOH [dissolved in DMSO; final DMSO concentration ≤3% (v/v)] or α-TOH acetate [dissolved in ethanol; final ethanol concentration ≤3% (v/v)] at a final concentration five times that of the corresponding native, unoxidized LDL. Control, native LDL is treated similarly with DMSO only. Following incubation at 37° for 4 hr under unaerobic conditions, LDL is reisolated by ultracentrifugation as described earlier. The resulting LDL band locates at a position indistinguishable from that of native, untreated LDL, indicating that the *in vitro* α-TOH depletion and replenishment procedures do not alter the physical properties of the lipoprotein. The α-TOH-replenished LDL is subsequently gel filtered (one PD-10 column). The replenishment procedure results in LDL that contains one to two times the vitamin E level of the native lipoprotein and is free of detectable lipid hydroperoxides.

Deuterium-Labeled LDL. The phenolic proton of α-TOH readily exchanges with deuterium.[17] Thus, deuterium-labeled LDL (D$_2$O–LDL) can be prepared by percolating freshly isolated native LDL (2 ml of 0.5–1 mg protein/ml) through two PD-10 columns preequilibrated with 50 mM phosphate buffer prepared in D$_2$O.[18] PD-10 columns are equilibrated with 3 bed volumes of D$_2$O buffer immediately prior to use; D$_2$O–LDL obtained is used immediately for subsequent experiments. As D$_2$O is hydroscopic, small quantities of D$_2$O buffers are prepared ≤24 hr prior to use and using fresh D$_2$O.

Total lipid extracts of LDL are prepared by extracting 1 volume LDL (0.3–0.4 mg protein/ml) with 3 volumes chloroform and 1 volume methanol, centrifuging the mixture (1000 rpm, 5 min; Beckman GS-6R table-top centrifuge), and collecting the (bottom) chloroform layer.[19] The lipid phase is evaporated to dryness and the residue is redissolved in benzene to yield a homogeneous solution of total LDL lipids. Lipids may be redissolved at

[16] L. Kritharides, W. Jessup, J. Gifford, and R. T. Dean, *Anal. Biochem.* **213,** 79 (1993).
[17] K. U. Ingold and J. A. Howard, *Nature* **195,** 280 (1962).
[18] P. K. Witting, V. W. Bowry, and R. Stocker, *FEBS Lett.* **375,** 45 (1995).
[19] E. G. Blight and W. J. Dyer, *Can. J. Biochem. Physiol.* **37,** 911 (1959).

varying concentrations, e.g., to mimic bulk lipid oxidation. Aliquots of this extract (5 volumes) are underlayered with D_2O or H_2O (1 volume), respectively, prior to oxidation initiation with AMVN (0.5 mM) at 37° to assess the deuterium kinetic isotope effect (DKIE) on lipid peroxidation in homogeneous solution.

Assessment of Oxidation of Lipoproteins

Measurement of Lipid Hydroperoxides and Analyses of Lipid-Soluble Antioxidants. The extraction of LDL and subsequent analysis of the aqueous methanol phase for phosphatidylcholine hydroperoxides and the organic phase for cholesterol, cholesteryl esters, CE-OOH, and lipid-soluble antioxidants have been described previously in detail[12] using reversed-phase high-performance liquid chromatography (HPLC) with postcolumn chemiluminescence (for lipid hydroperoxides), electrochemical (for antioxidants in reduced and oxidized forms), and $UV_{210\ nm}$ detection (for cholesterol and cholesteryl esters).

We now routinely use a Supelcosil (Supelco, Bellefonte, PA) LC-18 (25 × 0.46 cm, 5-μm particle size) column equipped with a 2-cm guard column and eluted at 1 ml/min with ethanol/methanol/2-propanol/water (196.7:60:10:1, v/v/v/v) containing 0.8 g/liter $LiClO_4 \cdot H_2O$ (solvent A in Scheme 2). We use a linear arrangement of detectors, composed of two

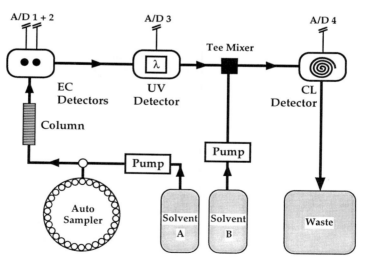

SCHEME 2. Schematic representation of the HPLC system employing serial electrochemical (EC), variable wavelength UV/VIS (UV), and chemiluminescence (CL) detection. Solvents A and B are as described in the text. A/D (1–4) represent data interface systems capable of receiving two independent channels of detector output.

amperometric electrochemical (LC-BAS 4C, Bioanalytical Systems), followed by a variable wavelength UV/VIS and a CLD-110 chemiluminescence detector (Tohoku Electronic Industrial Co., Miyagi, Japan) equipped with a 314-μl spiral flow cell heated to 50° (CL detector in Scheme 2). The electrochemical detectors are operated in dual mode with potentials of -450 mV (reductive mode, upstream) and 600 mV (oxidative mode, downstream) vs Ag|AgCl|KCl reference electrode. The potential at the reductive electrode is lower than the -700 mV described previously.[12] This decreases the size and extent of tailing for the solvent front in this mode, allowing a more accurate determination of ubiquinone 10; it also increases the "life" of the working electrode. For postcolumn chemiluminescence detection, we employ a second pump with an aqueous borate buffer at 1.5 ml/min (solvent B, Scheme 2). Solvent B is prepared by dissolving 76.3 g of sodium tetraborate decahydrate in 2 liters of H_2O (100 mM final concentration), adjusting the pH to 10.0 using sodium hydroxide (6.3 g per 2 liters), and diluting this solution with 2 liters of methanol (Merck) containing 708.8 mg isoluminol (6-amino-2,3-dihydro-1,4-phthalazinedione; Sigma; 1 mM final concentration) and 4 mg microperoxidase (Type MP-11; Sigma). This amount of microperoxidase is 34 times less than that used originally,[20] resulting in further cost savings for this expensive chemical. If greater sensitivity is required, the concentration of microperoxidase can be increased.

This HPLC system allows the simultaneous determination of α-TOH, ubiquinol 10, and ubiquinone 10, and estimation of lycopene and β-carotene (by electrochemical detection) (see Ref. 12), cholesterol, cholesteryl arachidonate, cholesteryl linoleate, and cholesteryl oleate (UV detection), and hydroperoxides of cholesteryl arachidonate and cholesteryl linoleate (chemiluminescence detection) within 25 min run time. Unlike coulometric analysis, amperometric electrochemical detection involves chemical modification to only a minute proportion of the analyte, and the upstream electrochemical detection does not significantly decrease the amounts of CE-OOH detected downstream by chemiluminescence (97% vs 100%, chemiluminescence response with electrochemical detector in "on" vs "off" mode, on injecting 10 pmol of Ch18:2-OOH). If such a small decrease in CE-OOH response is significant to the overall content of CE-OOH in the sample or if amperometric electrochemical detectors are not available, it is possible to substitute fluorescence (to analyze α-TOH, excitation 290 nm and emission 230 nm) for electrochemical detection. Under these conditions, α-TOH, however, is detected with lower sensitivity than electrochemical detection operated in oxidative mode. In addition, ubiquinol 10, ubiquinone 10, and estimations of lycopene and β-carotene are not obtained.

[20] Y. Yamamoto, M. H. Brodsky, J. C. Baker, and B. N. Ames, *Anal. Biochem.* **160**, 7 (1987).

Oxidized cholesteryl ester may also be determined by this assay using $UV_{234\ nm}$ rather than/or in combination with postcolumn chemiluminescence detection. For this, the variable wavelength UV detector is programmed to switch from 210 to 234 nm at 7.0 min and back to 210 nm at 12.0 min. Under these conditions, both cholesteryl ester hydroperoxides and hydroxides [CE-O(O)H] show similar retention times ($\approx$8.3–8.8 min, UV detection); the retention time for CE-OOH measured by chemiluminescence detection is $\approx$0.5 min delayed. Analysis of δ- and γ-TOH is performed using reversed-phase HPLC with electrochemical detection.[21]

Assays for Prooxidant Activity of Vitamin E in LDL or Other Lipid Emulsions

Lipoprotein particle size, α-TOH content, and radical flux all significantly influence lipid peroxidation kinetics.[7,8] Under conditions of identical and relatively low rates of initiation (R_i), the chain length of lipid peroxidation decreases in the order of very low-density lipoprotein > LDL > high-density lipoprotein,[7,22] i.e., decreasing ratios of lipoprotein particle volume to surface area. This order also reflects the decreasing ability to sustain a single radical per particle by virtue of a relatively higher incidence of termination of α-TO· at the surface of the particle under identical R_i.

Modulating Phase Transfer Activity of Vitamin E in Lipoprotein Emulsions Changes Lipid Oxidizability. Enrichment of lipoproteins with α-TOH modulates the chain-transfer activity of the vitamin. For example, under comparable and low R_i, cholesteryl esters and phosphatidylcholine in α-TOH-enriched LDL peroxidize faster than those in the corresponding native lipoprotein particle.[23] Carefully matched experiments with AAPH-initiated LDL lipid peroxidation show that enrichment of LDL with any of the members of vitamin E affords *faster* rates of lipid hydroperoxide formation (Fig. 1). However, while increasing the loading of α-TOH from its normal 8–10 to 45 molecules of α-TOH/per LDL particle accelerates the rate of lipid hydroperoxide accumulation by a factor of ~5, smaller rate increases were produced by similar enrichment of the LDL with γ-TOH or δ-TOH (factors of ~4 and 2, respectively). These results clearly demonstrate that with vitamin E in LDL we have a situation in which the usual paradigms of inhibited autoxidation are inverted: The most effective phenolic antioxidant for lipids in homogeneous solution—α-TOH— becomes the best prooxidant for LDL lipids. Conversely, other forms of

---

[21] G. A. Pascoe, C. T. Duda, and D. J. Reed, *J. Chromatogr.* **414**, 440 (1987).
[22] D. Mohr and R. Stocker, *Arterioscl. Thromb.* **14**, 1186 (1994).
[23] V. W. Bowry, K. U. Ingold, and R. Stocker, *Biochem. J.* **288**, 341 (1992).

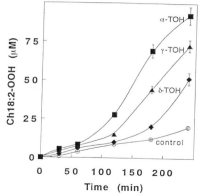

FIG. 1. α-TOH is the best vitamin E analog for promoting LDL lipid peroxidation. Purified LDL (0.4 mg protein/ml) isolated from native human plasma or plasma enriched *in vitro* with α-, γ-, or δ-TOH was oxidized with AAPH (4 mM) and Ch18:2-OOH accumulation monitored. Initial concentrations of α-, γ-, and δ-TOH for enriched and control LDL samples were 52, 51, 47, and 3.2 μM, respectively, whereas residual [α-TOH] for γ- and δ-TOH-enriched LDLs were 3.4 and 3.3 μM, respectively. From P. K. Witting, V. W. Bowry, and R. Stocker, *FEBS Lett.* **375,** 45 (1995) with permission.

vitamin E that show less effective peroxyl radical-scavenging ability[24] provide comparatively better protection for LDL lipids than α-TOH.

Inverse Deuterium Kinetic Isotope Effect (DKIE). The ability of α-TOH to transfer radicals from the aqueous compartment into LDL can be modified by exchanging the phenolic hydrogen with deuterium,[18] effectively attenuating the rate of formation of α-TO· by virtue of the lower LOO· scavenging ability of α-TOD vs α-TOH [Eq. (1)]. Figure 2A shows that this exchange results in an overall twofold decrease ($k_H/k_D \sim 2$ measured at 37°) in both the rates of accumulation of CE-OOH (e.g., inset shows the ratio of CE-OOH in D_2O vs H_2O) and the consumption of α-TOH in LDL exposed to low concentrations of AAPH, despite deuterium exchange reducing the radical scavenging ability of α-TOH.[17] Thus, if lipid peroxidation were to proceed in LDL via the "conventional" mechanism of vitamin E antioxidant activity, a reduction in radical-scavenging ability is predicted to enhance rather than diminish the rate of lipid peroxidation,[25] as shown in Fig. 2B for the AMVN-mediated oxidation of LDL lipid extract underlayered with H_2O or D_2O buffer.[18] In contrast, this *inverse* DKIE observed with intact LDL is fully consistent with the reduced phase-transfer activity of α-TOD vs α-TOH, decreasing TMP of LDL lipids.

[24] G. W. Burton, L. Hughes, and K. U. Ingold, *J. Am. Chem. Soc. Chem.* **105,** 5950 (1983).
[25] G. W. Burton and K. U. Ingold, *J. Am. Chem. Soc.* **103,** 6472 (1981).

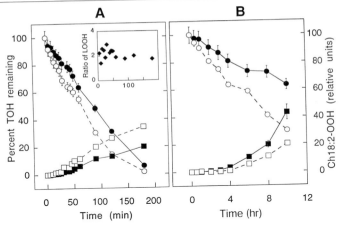

FIG. 2. α-TOH (circles) consumption and Ch18:2-OOH (squares) accumulation monitored for peroxyl radical-initiated peroxidation of LDL lipid emulsion and LDL lipid extract. (A) LDL (0.3–0.4 mg protein/ml) isolated from and suspended in deuterated or protonated buffer was oxidized with 4 mM AAPH in D_2O (filled symbols) or H_2O (open symbols). (B) Total LDL lipid extract in benzene was underlayered with D_2O (filled symbols) or H_2O (open symbols) and oxidized with 0.5 mM AMVN. Initial concentrations of α-TOH varied from 4.3–5.5 (A) to 52–53 (B) μM. Units for Ch18:2-OOH are 0.025 (A) and 1 (B) μM, respectively. (Inset) Ratio of Ch18:2-OOH formed in protonated over deuterated system. From P. K. Witting, V. W. Bowry, and R. Stocker, *FEBS Lett.* **375**, 45 (1995) with permission.

Using this test, it is important to control for potential differences in the rate of radical generation in H_2O vs D_2O buffer. For example, if enzymes were to be employed as sources of one-electron oxidants, exchange of hydrogen with deuterium could affect the catalytic action independent of phenolic proton exchange in α-TOH. In the case of AAPH and LDL, the observed inverse deuterium kinetic isotope effect is not due to an (unprecedented) solvent effect on the efficiency of peroxyl radical generation by AAPH. Thus, oxygen consumption for the peroxyl radical generator dissolved in either H_2O or D_2O buffer is not significantly different. In addition, the results are not restricted to aqueous peroxyl radicals as similar inverse deuterium kinetic isotope effects are observed with Cu^{2+} or metmyoglobin/H_2O_2 as the initiators of LDL lipid peroxidation.[18,26]

Dependency of LDL Lipid Oxidizability on α-TOH. A prooxidant activity of vitamin E is implied if the oxidizability of LDL lipids is *dependent* on the presence and/or is directly proportional to the initial levels of α-TOH. This can be assessed readily by comparing the oxidizability of

---

[26] P. K. Witting and R. Stocker, 1998, unpublished data.

native vs α-TOH-depleted and α-TOH-supplemented LDL under otherwise identical oxidizing conditions. Such studies have been carried out with LDL, HDL, and reconstituted human plasma, using a variety of one-electron oxidants (water- and lipid-soluble peroxyl radicals, Cu^{2+} ions, $\cdot$OH, myeloperoxidase/H_2O_2/Cl^-/tyrosine, and horseradish peroxidase/H_2O_2) or conditions that give rise to one-electron oxidants (Ham's F10 medium, macrophages, soybean lipoxygenase, SIN-1) to initiate lipid peroxidation.[9,27]

Most strikingly, the initiation of lipid peroxidation in (hydroperoxide-free) LDL by soybean lipoxygenase, Cu^{2+} ions,[9] or myeloperoxidase/H_2O_2/Cl^-/tyrosine[27] requires the presence of α-TOH. In the case of Cu^{2+} ions this is true even when Cu^{2+} : LDL ratios of up to 20 : 1 are used. Such findings strongly indicate an active participation of vitamin E in the promotion of lipoprotein lipid peroxidation. Perhaps more commonly, a prooxidant action of vitamin E is implied more indirectly from lower rates in lipid peroxidation in *in vitro* α-TOH-depleted LDL than the corresponding native samples. Such a finding can even apply to highly reactive $\cdot$OH.[9] Again, appropriate control experiments employing α-TOH- and/or α-TOH acetate-replenished LDL are important. If a diminished oxidizability of a lipid emulsion is truly the result of α-TOH depletion, then full replenishment with α-TOH but not α-TOH acetate is expected to restore lipid oxidizability. An example of this is shown in Fig. 3 for LDL undergoing oxidation initiated by aqueous peroxy radicals. Converse arguments and testing can be conducted with α-TOH-supplemented vs native lipoproteins (as described earlier).

Pro- vs Antioxidant Activity of LDL Vitamin E. As pointed out earlier, the extent of proxidant action of α-TOH in isolated, coantioxidant-free LDL exposed to a single radical oxidant is dependent of the frequency with which the lipoprotein encounters a radical "hit," i.e., the radical flux. This necessitates testing of a given oxidizing agent and lipid emulsion at different oxidant : target ratios. Under conditions where TMP prevails, the paradigms of kinetics of inhibited lipid peroxidation (i.e., that occurring in the presence of α-TOH) are no longer adhered to: lipid peroxidation *decreases* with an *increasing* rate of radical generation. As R_i increases, the incidence of termination reactions involving the chain carrying α-TO$\cdot$ also increases [Eq. (2)], thereby reducing the extent of lipid peroxidation and allowing vitamin E to increasingly appear to act as a "conventional" antioxidant. This feature is characterized by a "switching" of the activity of α-TOH from a pro- to an antioxidant as the radical flux increases. Examples

[27] P. K. Witting, J. M. Upston, and R. Stocker, in "Subcellular Biochemistry" (V. Kagan and P. J. Quinn, eds.), p. 345. Plenum Press, London, 1998.

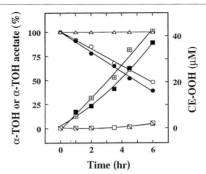

FIG. 3. Replenishment of *in vitro* vitamin E-depleted LDL with α-TOH but not α-TOH acetate restores its lipid peroxidizability to aqueous ROO·. Ascorbate- and peroxide-free LDL (0.15–0.2 mg protein/ml) was oxidized in the presence of 0.2 mM AAPH. Initial concentrations of α-TOH were 2.1 and 2.4 μM in native and α-TOH-replenished samples, whereas that of α-TOH-acetate was 2.1 μM. Symbols listed correspond to α-TOH analog or CEOOH for native (●, ■), α-TOH-depleted (none, □), α-TOH-repleted (○, ⊞), or α-TOH acetate-replenished (△, ◨) LDL, respectively. From J. Neuzil, S. R. Thomas, and R. Stocker, *Free Radic. Biol. Med.* **22,** 57 (1997) with permission.

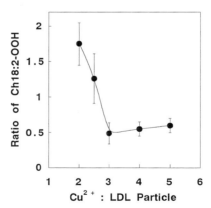

FIG. 4. Titration of Cu^{2+} into LDL dispersions suspended in deuterated or protonated buffer. Results show mean values ± SEM of two separate experiments carried out with different LDL (0.2–0.3 mg protein/ml) with addition of Cu^{2+} to give the final concentrations indicated. Ratios of cholesteryl linoleate hydroperoxide were determined for individual experiments by obtaining the relative ratio of Ch18:2-OOH produced in the protonated over deuterated systems throughout the 2 hr each oxidation experiment was monitored. From P. K. Witting, V. W. Bowry, and R. Stocker, *FEBS Lett.* **375,** 45 (1995) with permission.

of the use of this test are given by Neuzil et al.[9] for ROO· or ·OH and LDL and in Fig. 4 for Cu^{2+} ion induced LDL lipid peroxidation, where the "switching point" is achieved at Cu^{2+}:LDL ratios ~3.[18]

Conclusions

We have presented procedures for testing the prooxidant activity of α-TOH in lipoproteins and lipid emulsions. The unusual features observed are detected readily if the sensitive and accurate methods described for the analysis of lipid peroxidation and varying ratios of oxidant to target lipid emulsion are used. The methods outlined here allow investigations on the mechanism of *in vitro* lipoprotein lipid peroxidation and how this process is affected by various (co-)antioxidants surrounding or associated with lipoproteins. In addition, the methods should, in principle, allow investigation of a possible role of TMP in cellular lipid peroxidation and antioxidation, particularly once analytical tools comparable to those for lipoproteins are available for the assessment of cellular lipid peroxidation.

Acknowledgment

This work was supported by the Australian National Health and Medical Research Council (Grant 970998 to R.S.).

Section VI

Carotenoids and Retinoids

[34] Screening of Dietary Carotenoids and Carotenoid-Rich Fruit Extracts for Antioxidant Activities Applying 2,2'-Azinobis(3-ethylenebenzothiazoline-6-sulfonic acid Radical Cation Decolorization Assay

By Nicoletta Pellegrini, Roberta Re, Min Yang, and Catherine Rice-Evans

Introduction

A number of assays are available for screening the antioxidant activity of pure compounds, food constituents, and body fluids,[1–11] but few of these methods are applicable to lipophilic substances such as carotenoids and food extracts containing them. The ABTS [2,2'-azinobis(3-ethylbenzothiazoline-6-sulfonic acid)] radical cation has been used to screen the relative radical-scavenging abilities of flavonoids and phenolics through their properties as electron- or H-donating agents.[12–17] The structural characteristics of the stable green radical cation in the near-infrared region of the spectrum are shown in Fig. 1, with peaks at 630, 734, and 812 nm. On interaction with antioxidants the radical is reduced, suppressing the absorbance of the

---

[1] R. J. Delange and A. N. Glazer, *Anal Biochem.* **177**, 300 (1989).
[2] G. Cao, H. M. Alessio, and R. G. Cutler, *Free Radic. Biol. Med.* **14**, 303 (1993).
[3] G. Cao, C. P. Verdon, A. H. B. Wu, H. Wang, and R. L. Pryor, *Clin. Chem.* **41**, 1738 (1995).
[4] D. D. M. Wayner, G. W. Burton, K. U. Ingold, and S. Locke, *FEBS Lett.* **187**, 33 (1985).
[5] T. Metsa-Ketela, *in* "Bioluminescence and Chemiluminscence: Current Status" (P. Slanley and L. Kricka, eds.), p. 389. Wiley, Chichester, 1991.
[6] J. K. Votila, A. L. Kirkola, H. Rorarius, R. J. Tuihala, and T. Metsa-Ketela, *Free Radic. Biol. Med.* **16**, 5 (1994).
[7] E. Lissi, M. Salim-Hanna, C. Pascula, and M. D. Del Castillo, *Free Radic. Biol. Med.* **18**, 153 (1995).
[8] R. McKenna, F. J. Kezdy, and D. E. Epps, *Anal. Biochem.* **196**, 443 (1991).
[9] W. A. Pryor, J. A. Cornicelli, and L. J. Devall, *J. Organ. Chem.* **58**, 3521 (1993).
[10] B. G. Flecha, S. Llesuy, and A. Boveris, *Free Radic. Biol. Med.* **10**, 93 (1991).
[11] T. P. Whitehead, G. H. G. Thorpe, and S. R. J. Maxwell, *Anal. Chim. Acta* **266**, 265 (1992).
[12] C. A. Rice-Evans, N. J. Miller, and G. Paganga, *Free Radic. Biol. Med.* **20**, 933 (1996).
[13] N. J. Miller, C. A. Rice-Evans, M. J. Davies, V. Gopinathan, and A. Milner, *Clin. Sci.* **84**, 407 (1993).
[14] N. J. Miller, J. Sampson, L. Candeias, P. Bramley, and C. A. Rice-Evans, *FEBS Lett.* **384**, 240 (1996).
[15] N. J. Miller and C. A. Rice-Evans, *Redox Rep.* **2**, 161 (1996).
[16] N. J. Miller and C. A. Rice-Evans, *Free Radic. Res.* **26**, 195 (1997).
[17] C. A. Rice-Evans, N. J. Miller, G. P. Bolwell, P. Bramley, and J. Pridham, *Free Radic. Res.* **22**, 375 (1995).

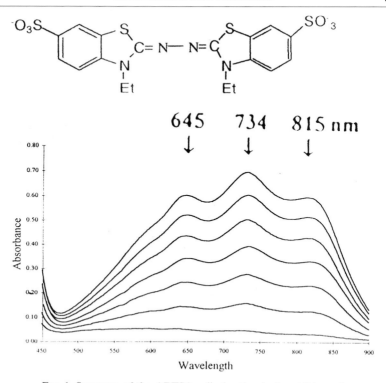

Fig. 1. Spectrum of the ABTS·+ radical cation in the visible region.

green radical cation to an extent and on a time scale dependent on the antioxidant activity or the reducing properties of the substance in question. This method is not confounded by other factors that contribute to the antioxidant activity in other model systems, such as metal chelation, partitioning abilities, or interference from other radicals involved in the chemistry of the system.

The basis of the method (Fig. 1) is the generation of a long-lived specific ABTS·+ radical cation chromophore and the relative abilities of antioxidants to quench the radical in relation to that of Trolox, a vitamin E analog. Thus the Trolox equivalent antioxidant activity is defined as the concentration of Trolox with equivalent antioxidant activity to a 1 mM concentration of the substance under investigation. The method is carried out as a decolorization assay, appropriate for lipophilic (and aqueous) systems in which the antioxidant is added to the preformed radical cation produced on the one-electron oxidation of ABTS and is applied to the monitoring of lipophilic antioxidants such as carotenoids and lipophilic extracts of nutritional components.

Methods

Materials

Trolox (Hoffman-La Roche) (6-hydroxy-2,5,7,8-tetramethylchroman-2-carboxylic acid, Aldrich Chemical Co., The Old Brickyard, Gillingham, Dorset SP8 4BR, UK) is used as the antioxidant standard. ABTS, diammonium salt, and potassium persulfate (dipotassium peroxodisulfate) are from Sigma-Aldrich (Fancy Road, Poole, Dorset BH12 4QH, UK). β-Carotene and lycopene are from Antioxidant Analysts, Suppliers and Consultants (Bitterne, Hampshire, SO18 6PP, UK). Ferulic acid and chlorogenic acid are from Extrasynthese (ZI Lyon Nord, BP 62 69730 Genay, France).

All the solvents are HPLC grade (Rathburn Chemicals Ltd., Caberston Road, Walkerburn, Peebleshire, Scotland EH43). Elgastat UHP double-distilled water (18 + Ω grade) is used to prepare ABTS and potassium persulfate stock solutions. Experiments are performed on the Hewlett-Packard spectrophotometer Model HP 8453 (Heathside Park Road, Cheadle Heath, Stockport Cheshire SK3 0RB, UK) fitted with peltier electronics. Tomatoes are from a major U.K. supermarket.

Preparation of Preformed ABTS Radical Cation

The total antioxidant activity of pure carotenoids and extracts is measured by the $ABTS^{\cdot+}$ radical cation decolorization assay involving preformed $ABTS^{\cdot+}$ radical cation. $ABTS^{\cdot+}$ is prepared by reacting ABTS with potassium persulfate according to the scheme in Fig. 2, and the mixture is allowed to stand in the dark at room temperature for 12–16 hr before use. Because ABTS and potassium persulfate react stochiometrically at a ratio of 1:0.5, this results in incomplete oxidation of the ABTS. Oxidation of the ABTS commences immediately, but the absorbance is not maximal and stable until ca. 6 hr has elapsed. The radical cation is stable in this form for more than 2 days in storage in the dark at room temperature. Prior to assay, the solution is diluted in ethanol to give an absorbance at 734 nm of 0.70 ± 0.02 in a 1-cm cuvette and is equilibrated to 30°, the temperature at which all assays are performed.

Trolox Reference Standard for Relative Antioxidant Activities

A 5 mM stock solution of Trolox is prepared in ethanol and subsequently diluted further in ethanol for introduction into the assay system at concentrations within the activity range of the assay (0 to 20 μM final concentration) for preparing a standard curve to which all data are referred (Fig. 3).

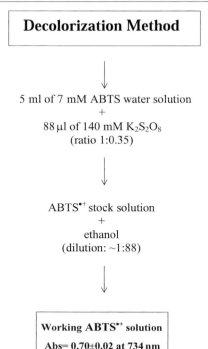

Fig. 2. Preparation of the ABTS·+ radical cation.

The ethanolic ABTS·+ (1 ml) solution is added to aliquots of Trolox standards (0 to 20 μM final concentration). The analyticals are vortex mixed for 30 sec and absorbance readings at 734 nm are taken 1 min immediately postaddition and then every minute for 30 min. Appropriate solvent blanks are run in each assay. Triplicate determinations are made at each dilution of the standard, and the percentage inhibition is calculated of the blank absorbance at 734 nm and then plotted as a function of Trolox concentration (Fig. 3).

Preparation of Antioxidant Standards

Lycopene and β-carotene are dissolved in dichloromethane, and concentrations are validated according to the extinction coefficient. Dilutions are prepared in the concentration range from 0.3 to 9 μM (final concentrations). Ferulic acid and chlorogenic acid are diluted in ethanol to a final concentration of 5 mM and then subsequently diluted for introduction into the assay

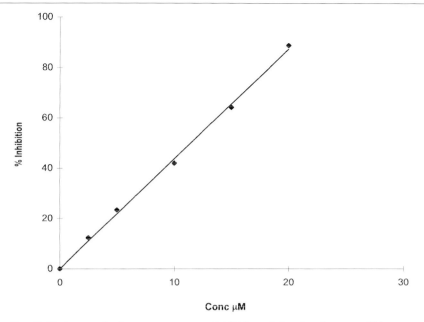

FIG. 3. Trolox standard curve: percentage inhibition of the absorbance at 734 nm as a function of concentration.

system to concentrations within the range of 2 to 10 μM (final concentration).

Extraction of Tomato Fruit

Tomatoes (169 g wet weight) are chopped into small pieces and placed in a round-bottom flask for freeze-drying. The resulting material (9.15 g dry weight) is then powdered under liquid nitrogen and divided into individual samples of 100 mg. Each 100-mg sample is extracted according to the scheme depicted in Fig. 4, with 10 ml dichloromethane and 10 ml water, vortex mixed for 2 min, and centrifuged at 1000 rpm for 15 min to enhance the separation of the two phases. The extraction process is repeated by adding a further 10-ml aliquot of dichloromethane to the water layer. The two extracts are combined and subjected to rotary evaporation at 30° to remove the organic solvent, and the extract is reconstituted in 1 ml of dichloromethane. The lipophilic extracts are now prepared for antioxidant activity measurement. The precipitated residue within the aqueous phase is treated with 20 ml methanol and the sample is refluxed for 30 min. The solution is filtered, and the filtrate rotary is evaporated and subsequently

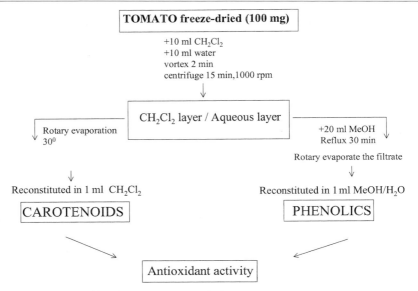

FIG. 4. Extraction of antioxidants from tomato fruit.

reconstituted in 1 ml aqueous methanol (50%). The samples are now ready for analysis.

Assay Protocol: Decolorization Assay

The procedure for the interaction of the pure antioxidant compounds with the $ABTS^{\cdot+}$ radical cation is as described for the Trolox standards and is measured as a function of time of interaction and concentration of antioxidant. The activity of antioxidants is estimated at a minimum of three different concentrations within the range of the dose–response curve, and the mean value is derived as the TEAC (Trolox equivalent antioxidant capacity) value. To determine the TEAC values of the pure antioxidants, the percentage inhibition of the absorbance at 734 nm is calculated, referred to the standard curve for the percentage inhibition for Trolox, and normalized to a 1 mM concentration. Each compound is assayed on a minimum of 3 separate days (i.e., at least 27 different determinations). The unit of antioxidant activity (the TEAC) is defined as the concentration (mmol/liter) of Trolox having the equivalent antioxidant activity to a 1 mM or 1-mg/ml solution of the substance under investigation or expressed as mmol/kg for foods.

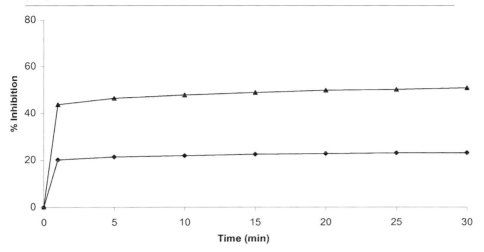

FIG. 5. Antioxidant activity of lycopene. The percentage inhibition (at 2 and 4 μM, final concentration) is plotted as a function of time.

Results

The interaction of pure carotenoids with $ABTS^{·+}$ is undertaken by the addition of 1 ml $ABTS^{·+}$ radical cation solution in ethanol (initial absorbance 0.70 ± 0.02), as described previously for the Trolox standard interactions, to solutions of carotenoids in dichloromethane (final concentrations 0.3 and 9 μM). After addition the solution is vortex mixed for 0.5 min, and the absorbance at 734 nm is read 1 min after mixing and every 5 min up to 30 min. The percentage inhibition is calculated for each concentration of each carotenoid and plotted as a function of time (Fig. 5). Data show that the reaction is essentially complete at 1–2 min and thus the time point of 2.5 min is selected for measuring the absorbance. Figure 6 illustrates the linear relationship between the percentage inhibition and the carotenoid concentration at 2.5 min for concentrations up to 8 μM for the pure compounds, with correlation coefficients of 0.995 and 0.983 for lycopene and β-carotene, respectively. Calculation of the antioxidant activity in terms of the TEAC gives values of 3.01 ± 0.13 mM for lycopene and 2.47 ± 0.03 mM for β-carotene. These are close to the previously reported values using an assay for the $ABTS^{·+}$ radical cation formed through manganese dioxide decolorization.[14]

Analysis of the lipophilic component of the tomato extract demonstrates that 2.5 min is the appropriate time point for monitoring the antioxidant activity (Fig. 7) and that the degree of suppression of the absorption of

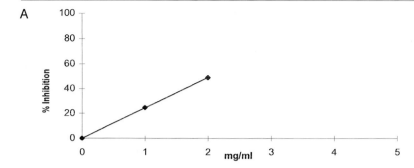
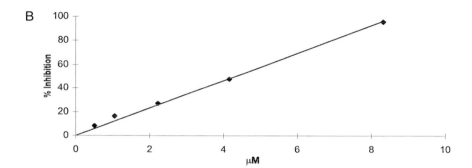
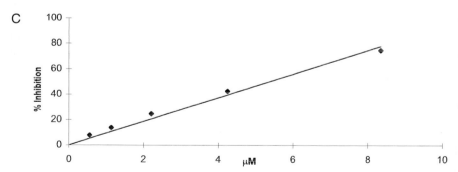

FIG. 6. Plot of the influence of concentration on the percentage inhibition of absorbance at 734 nm, at 2.5 min, for lipophilic tomato extract (A), lycopene (B), and β-carotene (C).

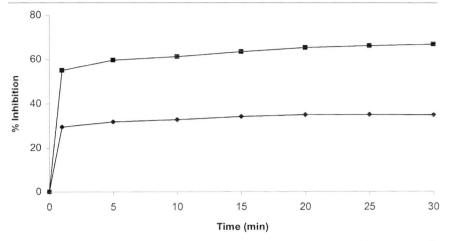

FIG. 7. Lipophilic tomato extract (1 and 2 mg dry weight in 1 ml of dichloromethane). The percentage inhibition of absorbance at 734 nm, plotted as a function of time, shows that the reaction is complete after 1 min.

the radical cation is directly proportional to the concentration of the extract. The antioxidant activity of the tomato extract relative to Trolox is 5.7 ± 0.2 mmol/kg dry weight (Table I).

There are many components in tomatoes with antioxidant properties in addition to the carotenoids. The aqueous methanolic extract was also examined for its antioxidant activity. Contributing antioxidants include

TABLE I
TOTAL ANTIOXIDANT ACTIVITY OF TOMATO EXTRACTS AND PURE COMPONENTS[a]

Compounds	TEAC (mM)	TAA (mmol/kg dry wt)
Tomato extracts		
Lipophilic		5.72 ± 0.21 (4)
Aqueous/methanol		19 ± 0.1 (4)
Pure compounds		
Lycopene	3.01 ± 0.13 (3)	
β-Carotene	2.47 ± 0.03 (3)	
Ferulic acid	1.85 ± 0.13 (3)	
Chlorogenic acid	1.07 ± 0.05 (3)	

[a] Data shown are mean values ± SD of three/four complete sets of experiments. Numbers in parentheses represent the number of experiments.

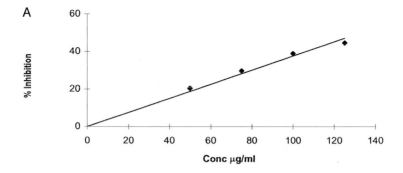
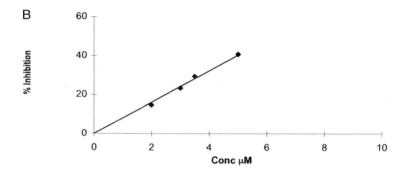
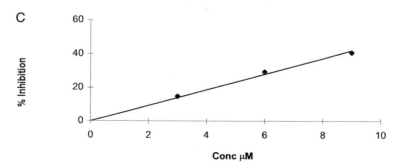

Fig. 8. Plot of the influence of concentration on the percentage inhibition of absorbance at 734 nm, at 2.5 min, for aqueous methanolic tomato extract (A) and standard phenolics, ferulic acid (B) and chlorogenic acid (C).

phenolics such as chlorogenic acid and conjugates of ferulic acid and of p-coumaric acid, as well as flavonoid glycosides of naringenin and rutin.[18] Figure 8A demonstrates the linearity of the concentration dependency of the aqueous methanolic tomato extract in scavenging the ABTS$^{\cdot+}$ radical cation in comparison to the standard phenolics, chlorogenic acid and ferulic acid, at the 2.5-min time point. Calculation of the TEAC for the standards and the antioxidant activity of the aqueous methanolic tomato extract (Table I) gives values of 1.85 mM for ferulic acid, 1.05 mM for chlorogenic acid, and 19 mmol/kg dry weight for the tomato extract.

Conclusions

Results demonstrate that the decolorization of the ABTS radical cation is an efficient, accurate assay for screening the antioxidant activities of lipophilic substances and food extracts. The inhibitory response of the radical cation is proportional to the antioxidant concentration, and the time point selected (2.5 min) for the analysis demonstrates that this is an ideal measuring point when the reaction is complete. Results for the carotenoids are of the same order as previously reported data using alternative systems for generating the ABTS$^{\cdot+}$ radical cation.[12,14]

It is of interest to note that, in this variety of tomatoes, the antioxidant activity of the aqueous methanolic extract, rich in flavonoids and phenolics, is three times that of the lipophilic extract to which the carotenoids contribute, but clearly the relative proportions, as well as the individual contributions, will vary for different varieties and different origins. It is essential to examine the relative bioavailability, absorption, and bioactivity of dietary carotenoids and flavonoids/phenolics to understand the importance of these diverse nutritional components in the maintenance of health and disease prevention.

Acknowledgments

We acknowledge the Ministry of Agriculture, Fisheries, and Food (UK) for financial support for this research (CR-E). NP acknowledges support from the European Union for a visiting fellowship.

---

[18] L. Bourne and C. A. Rice-Evans, *Methods Enzymol.* **299**, 89 (1998).

[35] Matrix-Assisted Laser Desorption Ionization–Postsource Decay-Mass Spectrometry

By Thomas Wingerath, Dieter Kirsch, Raimund Kaufmann, Wilhelm Stahl, and Helmut Sies

Introduction

Analytical protocols for the identification of carotenoids in biological samples often rely on comparison of high-performance liquid chromatography (HPLC) retention times and UV/visible spectra with standard compounds. Carotenoids are strong absorbers in the wavelength range of 400–500 nm, and most carotenoids can be identified by their characteristic absoption maxima. However, the spectral shifts of such maxima induced by esterification are usually small, e.g., about 2 nm, and are nonspecific. It is important to note that identification based on HPLC retention times is equivocal.[1,2] Characterization can be improved by using mass spectrometry (MS), which provides molecular weight and characteristic fragment ions for structural elucidation.[1-11]

Several ionization methods have been reported for the mass spectrometric analysis of carotenoids, including electron impact (EI),[6-8] chemical ionization (CI),[4] fast atom bombardment (FAB),[1-3,5] electrospray ionization (ESI),[9,10] and atmospheric pressure chemical ionization (APCI).[10,11] Although mass spectrometry with EI and CI can be used to identify carotenoids, the thermal lability of carotenoids renders these ionization techniques difficult. FAB-MS offers some improvement in this regard, but sensitivity is limited. The combination of liquid chromatography (LC) and

---

[1] H. H. Schmitz, R. B. van Breemen, and S. J. Schwartz, *Methods Enzymol.* **213,** 322 (1992).
[2] R. B. van Breemen, H. H. Schmitz, and S. J. Schwartz, *Anal. Chem.* **65,** 965 (1993).
[3] S. Caccamese and D. Garozzo, *Org. Mass Spectrom.* **25,** 137 (1990).
[4] W. R. Lusby, F. Khachik, G. R. Beecher, and J. Lau, *Methods Enzymol.* **213,** 111 (1992).
[5] R. B. van Breemen, H. H. Schmitz, and S. J. Schwartz, *J. Agric. Food Chem.* **43,** 384 (1995).
[6] C. R. Enzell and S. Back, in "Carotenoids" (G. Britton, S. Liaaen-Jensen, and H. Pfander, eds.), Vol. 1B, p. 261. Birkhäuser, Basel, 1995.
[7] H. Budzikiewicz, H. Brezinka, and B. Johannes, *Monatsh. Chem.* **101,** 579 (1970).
[8] M. E. Rose, in "Carotenoid Chemistry and Biochemistry" (G. Britton and T. W. Goodwin, eds.), p. 167. Pergamon, Oxford, 1982.
[9] R. B. van Breemen, *Anal. Chem.* **67,** 2004 (1995).
[10] R. B. van Breemen, *Anal. Chem.* **68,** 299 (1996).
[11] R. B. van Breemen, C.-R. Huang, Y. Tan, L. C. Sander, and A. B. Schilling, *J. Mass Spectrom.* **31,** 975 (1996).

ESI-MS or APCI-MS provides a tool with sufficient sensitivity and selectivity for the quantitation of carotenoids in complex mixtures.[9–11]

This article describes the use of a highly sensitive technique, matrix-assisted laser desorption ionization (MALDI)-MS,[12] for the molecular weight determination of carotenoids. We present MALDI-postsource decay (PSD)[13] fragment ion analysis performed on model carotenoids and the novel use of MALDI-MS and MALDI-PSD-MS in combination with HPLC separation and UV/visible absorption spectroscopy to characterize carotenoids and their fatty acid esters present in biological samples.

Experimental

Standards and Samples. Reference samples of α- and β-carotene, lycopene, β-cryptoxanthin, lutein, zeaxanthin, canthaxanthin, and astaxanthin are gifts from Dr. J. Bausch, Hoffmann-La Roche (Basel, Switzerland). Straight, long-chain carotenol mono- and bis-fatty acid esters are prepared by partial synthesis from the parent carotenols and fatty acyl chlorides (Merck, Germany).[14] Synthesis of the monohydroxyxanthophyll 2',3'-anhydrolutein is carried out as described by Khachik *et al.*[15] Purity of the standards was verified by HPLC analysis. Tangerine juice concentrate (*Citrus reticulata*) (Krings Fruchtsaft AG, Mönchengladbach, Germany) is used as a biological sample for carotenoid analysis.

Extraction Procedures. Tangerine juice concentrate (0.5 g) is diluted in 100 ml of water (10% sodium chloride; w/w) and extracted five times with 50 ml of diethyl ether. The combined organic layers are washed several times with 50 ml of water and dried over sodium sulfate. The extract is filtered, and the solvent is removed on a rotary evaporator at 28°. The residue is dissolved in 5 ml of dichloromethane/*n*-hexane (1/1) and stored at $-20°$ until analysis. For HPLC analysis, an appropriate volume is evaporated under a stream of nitrogen and redissolved in HPLC solvent A. All steps are performed under diminished light.

An ethereal solution of tangerine juice concentrate extract is saponified with saturated methanolic potassium hydroxide for 30 min under nitrogen in the dark at room temperature. The mixture is diluted with diethyl ether and washed several times with water (10% sodium chloride). The organic

[12] M. Karas and F. Hillenkamp, *Anal. Chem.* **60,** 2299 (1988).
[13] R. Kaufmann, B. Spengler, and F. Lützenkirchen, *Rapid Commun. Mass Spectrom.* **7,** 902 (1993).
[14] T. Wingerath, W. Stahl, D. Kirsch, R. Kaufmann, and H. Sies, *J. Agric. Food Chem.* **44,** 2006 (1996).
[15] F. Khachik, G. R. Beecher, M. B. Goli, W. R. Lusby, and J. C. Jr. Smith, *Anal. Chem.* **64,** 2111 (1992).

layer is dried over sodium sulfate, filtered, and evaporated to dryness, and the residue is dissolved in the appropriate solvent for chromatographic analysis.

Approximately 200 mg of green lettuce (*Lactuca sativa* L.) leaves is cut into small pieces and homogenized in an Ultra Turrax (IKA, Kriens, Switzerland) together with 1 ml of water. The extraction and saponification of carotenoids are performed as described for tangerine juice concentrate.

HPLC Equipment. High-performance liquid chromatography is carried out with a Merck/Hitachi Model 655 A-12 ternary solvent delivery system equipped with a Merck/Hitachi Model L-4200 UV/visible detector and a Merck/Hitachi L-5000 LC controller (Merck, Darmstadt, Germany). For peak identification, we use a Model 168 diode array detector (Beckman, Munich, Germany). Absorption spectra of carotenoids are recorded between 230 and 600 nm. Analytical separations are performed on a 5-μm Suplex pKb 100 column (250 mm length $\times$ 4.6 mm I.D.) (Supelco, Bellefonte, PA) with a 20-mm guard column. For semipreparative separations a 10-μm C_{18} reversed-phase column is used (Lichrospher) (Merck, Germany).

HPLC Procedures Used for Carotenoid Isolation. Eluent A consists of methanol/acetonitrile/dichloromethane/*n*-hexane (10/85/2.5/2.5, v/v/v/v) and is applied isocratically for 5 min; from 5 to 40 min a linear gradient is applied to methanol/acetonitrile/dichloromethane/*n*-hexane (10/45/22.5/22.5, v/v/v/v). The final composition is held for another 10 min. The flow rate is 0.7 ml/min.

Eluent B consists of a mixture of methanol/acetonitrile/dichloromethane/*n*-hexane (20/40/20/20, v/v/v/v) and is employed for semipreparative HPLC to purify carotenol fatty acid esters prepared by partial synthesis. The flow rate is 4 ml/min. Carotenoids are detected at 450 nm.

MALDI-MS and MALDI-PSD-MS Equipment. Investigations are carried out with two time-of-flight (TOF) MS instruments built in-house; technical details have been described elsewhere.[16,17] Some main features are summarized as follows: Both instruments use a nitrogen laser (VSL 337 ND Laser Sciences, Inc., Cambridge, MA) for desorption. In the linear TOF MS, the formed ions have been accelerated up to a kinetic energy of 20 keV. The field-free drift path has a length of 110 cm; ion detection is performed by means of a dual microchannel plate with its front face biased to ground.

The second instrument is a reflectron-type instrument employing a gridded two-stage reflectron at the end of a 204-cm first field-free drift

---

[16] B. Spengler, D. Kirsch, and R. Kaufmann, *J. Phys. Chem.* **96,** 9678 (1992).

[17] R. Kaufmann, T. Wingerath, D. Kirsch, W. Stahl, and H. Sies, *Anal. Biochem.* **238,** 117 (1996).

region. Its second field-free drift path (reflectron to detector) extends over 175 cm. For delayed extraction (DE),[18-20] a two-stage acceleration ion source with pulsed high voltages is applied. The formed ions are accelerated up to kinetic energies between 10 and 15 keV. For ion detection a dual microchannel plate is employed. Typical values of mass resolution obtained under DE are 2500–4000 M/dM (full width at half-maximum). Thus, MALDI-PSD spectra shown in this work are recorded under DE conditions. A precursor ion selector allows to select ions from a mixture of about 50 M/dM. It is a mandatory device for PSD fragment ion mass analysis in nonhomogeneous samples. PSD fragment ion spectra are recorded sequentially over 12–15 mass windows by incremental reduction of the reflectron voltages. In each spectral window, 30–100 single-shot spectra are summed depending on sample conditions. Signal averaging and processing are carried out using a PC with Ulisses software (version 7.31, Chips at Work, Germany). Compilation of a set of PSD mass windows is performed by a subroutine contained in the "MALDI-PSD Peptide Sequencer" software (version 6.0, copyright by Frank Lützenkirchen, ILM, Germany).

MALDI specimens are prepared by dissolving the analyte in about 20 μl acetone. To 10 μl of this solution, 10 μl acetone saturated with 2,5-dihydroxybenzoic acid (Fluka, Germany) is added. A 5- to 10-μl aliquot (sample load of about 100 pmol) of this mixture is pipetted on the surface of a slightly preheated sample holder, on which the solvent evaporates within a few seconds. Transfer of the sample into the instrument is performed as quickly as possible.

Results and Discussion

Molecular Ion Species and Sensitivity. MALDI ionization produces abundant molecular ions of both nonpolar carotenes and polar oxocarotenoids. Although protonated molecules are typically produced during MALDI ionization,[21] most of the carotenoids and fatty acid esters investigated so far produced mixed populations of radical ($M^{·+}$) and protonated ($[M + H]^+$) molecular ions.[17] For carotenes the signal intensities of the radical molecular ions are dominant. The presence of a keto and/or a hydroxyl group leads to the appearance of protonated $[M + H]^+$ ions, which were occasionally even more abundant than the $M^{·+}$. These results are comparable with observations in FAB-MS.[3,6]

[18] R. S. Brown and J. J. Lennon, *Anal. Chem.* **67**, 3990 (1995).
[19] M. L. Vestal, P. Juhasz, and S. A. Martin, *Rapid Commun. Mass Spectrom.* **9**, 1044 (1995).
[20] R. Kaufmann, P. Chaurand, D. Kirsch, and B. Spengler, *Rapid Commun. Mass Spectrom,* **10**, 1199 (1996).
[21] R. Kaufmann, *J. Biotechnol.* **1174**, 1 (1995).

MALDI-PSD fragment ion mass analysis was employed to obtain structurally significant fragmentation for the identification of carotenoids.[14,17] Typical sample loads for PSD fragment ion analysis were in the range of about 100 pmol but only a few femtomoles were consumed during a full session of PSD fragment ion mass analysis.

MALDI-PSD Fragment Ion Mass Analysis of Model Carotenoids. MALDI-PSD spectra of β-carotene, α-carotene, and lycopene (molecular ions at m/z 536) are shown in Fig. 1. These structural isomers are differentiated readily by characteristic fragment ions in PSD spectra.

The most abundant fragment ion in the mass spectrum of β-carotene (top) is observed at m/z 444.2 and corresponds to the elimination of toluene from the precursor ion ($[M-92]^{\cdot+}$). The end groups of β-carotene are known to be almost stable to mass spectrometric conditions and do not undergo specific fragmentation.[3,4,6] The positions of elimination of polyene chain toluene are given in Fig. 2A.[22]

In contrast to β-carotene, the double bond in the terminal ring of α-carotene (α-ionone ring) is not conjugated to the polyene chain. The unconjugated 4,5-double of α-carotene makes the characteristic losses of 56 and 123 mass units from the precusor ion favorable as given in Fig. 2B. $[M-56]^{\cdot+}$ at m/z 480.8 results from retro-Diels–Alder (RDA) fragmentation of the α-ionone ring and is highly diagnostic for this end group; $[M-123]^+$ at m/z 413.1 is due to loss of the ring at the doubly allylic position.[8] In Fig. 1 (middle) the MALDI-PSD spectrum of α-carotene is shown. Like β-carotene the most abundant fragment ion in the PSD spectrum corresponds to the elimination of toluene ($[M-92]^{\cdot+}$) at m/z 444.4. Additionally, a combination of elimination of toluene and retro-Diels–Alder fragmentation ($[M-92-56]^{\cdot+}$) at m/z 388.7 is detectable.

Lycopene (Fig. 1, bottom) is an acyclic carotenoid that forms $[M-92]^{\cdot+}$ (m/z 444.4) ions by elimination of toluene from the polyene chain as also described for α- and β-carotene. Lycopene can be characterized by allylic cleavage of the terminal isoprene group ($[M-69]^+$) to form a diagnostic fragment ion at m/z 467.3 (see also Fig. 2C). The abundance of this ion under the conditions of MALDI occasionally exceeds the intensity of the $[M-92]^{\cdot+}$ signal. Additional prominent signals occur at m/z 398.2 ($[M-69-69]^{\cdot+}$) and m/z 375.5 ($[M-69-92]^+$). This pattern is qualitatively consistent with B/E linked-scan spectra of lycopene.[2,3,8] However, in the MALDI-PSD spectrum the signals at m/z 398.2 and 375.5 are more intense (up to 50% precursor ion intensity) than those in B/E linked-scan spectra.

As described for α- and β-carotene the structural isomers of zeaxanthin and lutein (molecular ions at m/z 568) are differentiated readily by their

---

[22] B. Johannes, H. Brezinka, and H. Budzikiewicz, *Org. Mass Spectrom.* **9**, 1095 (1974).

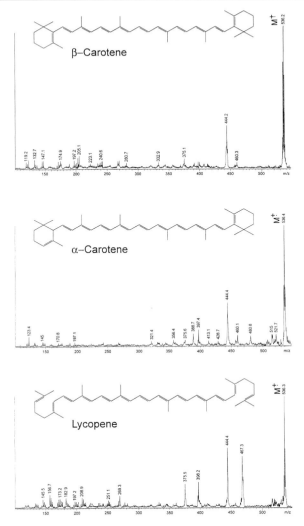

FIG. 1. MALDI-PSD spectra of β-carotene, α-carotene, and lycopene.

PSD spectra (see Fig. 3). Zeaxanthin and lutein are analogous to β- and α-carotene regarding their terminal end groups but differ from carotenoids in that both terminal rings are hydroxylated. The PSD spectrum of zeaxanthin (Fig. 3, top) looks very simple, and the fragment ions correspond to the well-known elimination of toluene ($[M-92]^{·+}$) at m/z 476.1, and loss of

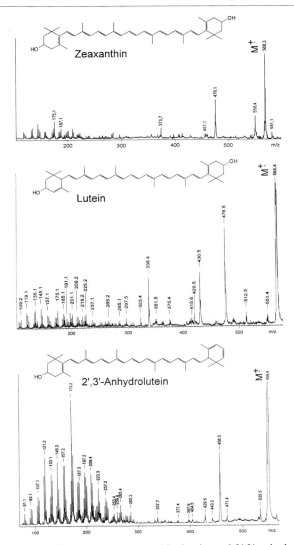

FIG. 3. MALDI-PSD spectra of zeaxanthin, lutein, and 2′,3′-anhydrolutein.

FIG. 2. Fragmentation scheme of carotenoids and carotenoid esters in MALDI-PSD: (A) elimination of toluene from the polyene chain [M-92], (B) end group-specific diagnostic fragments of α-ionone rings, (C) end group-specific diagnostic fragments of acyclic carotenoids, (D) proposed elimination mechanism of the hydroxylated α-ionone ring, and (E) elimination of polyene chain [M-80] species from carotenoid epoxides.

water ([M-18]$^{\cdot+}$) at m/z 550.4. No specific fragmentation of the terminal rings is observed.

Characteristic cleavages of lutein (Fig. 3, middle) occur in analogy to α-carotene such as loss of toluene ([M-92]$^{\cdot+}$) at m/z 476.5, retro-Diels–Alder reaction ([M-56]$^{\cdot+}$) at m/z 512.5, and a combination of both loss of toluene and retro-Diels–Alder fragmentation ([M-92-56]$^{\cdot+}$) at m/z 420.5. Additional signals correspond to the loss of a hydroxyl group ([M-17]$^+$) at m/z 551.4 or loss of water at m/z 550.4. The rather prominent fragment ion at m/z 338.4 in the PSD spectrum of lutein was not observed in conventional MS and is the result of consecutive losses of toluene and the terminal ring ([M-92-138]$^{\cdot+}$).

The elimination of the terminal ring occurs in different ways. In addition to the strong signal at m/z 430.5 ([M-138]$^{\cdot+}$) there is a comparably weak signal at m/z 429.4 ([M-139]$^+$), which indicates allylic cleavage of the ring, as described for α-carotene. A possible elimination mechanism of the hydroxylated terminal ring is shown in Fig. 2D. This mechanism assumes a hydrogen shift between the ring and the polyene chain followed by loss of the end group. There is no indication that the intense peak at m/z 430.5 is formed by the loss from the protonated molecular ion of lutein because the isotopic pattern shows no difference to the theoretical intensities of the radical molecular ion. This observation confirms the postulated mechanism, as the other pathway leads to more stable transition states, expressed as higher signal intensities.

The fragment ion spectrum of the monohydroxyxanthophyll 2',3'-anhydrolutein is shown in Fig. 3 (bottom). This compound contains two double bonds within one of its terminal rings, which are not in conjugation with the chain double bonds. Because of this structure cleavage of the ring is detected in the PSD spectrum at m/z 429.5 ([M-121]$^+$). Retention of the charge by the terminal ring gives rise to the fragment ion at m/z 121.2 ([M-429.5]$^+$). Other fragment ions, such as loss of a methyl group ([M-15]$^+$) at m/z 535.5, elimination of toluene ([M-92]$^{\cdot+}$) at m/z 458.5, and a combination of both loss of toluene and a methyl group ([M-92-15]$^+$) at m/z 443.2, and toluene and the terminal ring ([M-92-121]$^+$) at m/z 337.7 are observed. Additionally, a wealth of prominent signals are observed below m/z 300. Polyene bond cleavages and retention of the charge at either the polyene chain or the terminal ring favor these fragmentation pattern. Cleavage of the 10'/9' double bond gives rise to the most abundant fragment ion at m/z 173.2 ([M-377.4]$^+$) and the corresponding ion at m/z 377.4 ([M-173.2]$^+$).

Abundant [M-80 (C_5H_8)]$^{\cdot+}$ fragments are observed in all spectra of 5,6- and 5,8-epoxides investigated and are an important diagnostic feature for this group of molecules (see Fig. 2E).[6,17,22] The presence of one or two epoxide functions in the 5,6- and/or 5,8-position favors the characteristic cleavage and rearrangement in the polyene chain. Thus, numerous diagnostic fragments have been recorded.[17] The most abundant signal is usually the m/z (204 + R) signal (R = H or OH at position 3). The MALDI-PSD spectrum of violaxanthin (molecular ion at m/z 600.4) is shown in Fig. 4 (top). Structurally significant fragment ions are formed by the loss of the $[M-80]^{·+}$ species at m/z 520.1, loss of toluene ($[M-92]^{·+}$) at m/z 507.9, loss of toluene and a hydroxyl group ($[M-92-17]^+$) at m/z 490.9, and the consecutive losses of toluene and a C_5H_8 fragment ($[M-92-80]^{·+}$) at m/z 428.2. In addition, a rather prominent fragment ion is observed at m/z 221.3 (204 +

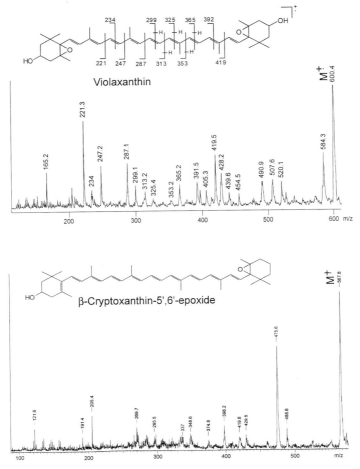

FIG. 4. MALDI-PSD spectra of violaxanthin and β-cryptoxanthin 5',6'-epoxide.

OH). Polyene bond cleavages that give rise to specific signals in the PSD spectrum of violaxanthin are also indicated in Fig. 4.

Characteristic fragment ions observed in the PSD spectrum of a β-cryptoxanthin epoxide (molecular ion at m/z 567.8) (Fig. 4, bottom) are [M-80]$^{\cdot+}$ (m/z 488.6), [M-92]$^{\cdot+}$ (m/z 475.6), [M-139]$^{\cdot+}$ (m/z 429.5), [M-80-139]$^{+}$ (m/z 348.6), [M-92-139]$^{+}$ (m/z 337), and (204 + H) at m/z 205.4. The signals at m/z 205.4 and 429.5, and the respective secondary fragment ions at m/z 348.6 and 337, suggest the localization of the epoxide and the hydroxyl function at different β-ionone rings (for structure, see Fig. 2E), thus the structure of β-cryptoxanthin 5′,6′-epoxide is postulated.

The presence of carbonyl groups as in canthaxanthin or astaxanthin also facilitated polyene fragmention. The fragmentation patterns, however, are qualitatively and quantitatively different from those of the epoxide species.[17] The type of fragment that appears to be diagnostic for this group results from the cleavage of the 8/7 (8′/7′) bond(s). Cleavage of the 10/11 (10′/11′) bond(s) furnished fragment ions at m/z 203 in the PSD spectrum of canthaxanthin, respectively, at m/z 219 in the spectrum of astaxanthin.[17]

Fatty acyl mono- and bisesters of oxocarotenoids, present in fruits[14,23] and vegetables,[24,25] are particularly challenging compounds for mass spectrometric analysis and therefore mass spectrometric studies on carotenoid fatty acid esters are scarce.[23,24] Molecular ions are rather unstable under EI and CI (methane and isobutane as reagent gas) conditions and reach only 0–10% relative intensity.[4,24,26] Better results have been obtained by chemical ionization with ammonia.[24]

MALDI-PSD spectra of carotenoid fatty acid esters are characterized by their simplicity. The vast majority of fragment ions belong to only four groups of characteristic ions; three of them are formed by the loss of either one or both fatty acids: [M-R$_1$COOH]$^{\cdot+}$, [M-R$_2$COOH]$^{\cdot+}$, and [M-R$_1$COOH-R$_2$COOH]$^{\cdot+}$. The dominating reaction is obviously an analogous McLafferty rearrangement where the charge remains with the conjugated π-electron system. In contrast to EI or CI ionization,[4,6,26] MALDI-PSD does not furnish ionized fatty acids. Depending on the nature of the xanthophyll, all three species of fragments can be accompanied by either loss of toluene or a C_5H_8 fragment. Finally, there are also abundant [M-92]$^{\cdot+}$ and/or [M-80]$^{\cdot+}$ fragments without loss of a fatty acid.[17]

---

[23] F. Khachik, G. R. Beecher, and W. R. Lusby, *J. Agric. Food Chem.* **37,** 1465 (1989).
[24] F. Khachik and G. R. Beecher, *J. Agric. Food Chem.* **36,** 929 (1988).
[25] F. Khachik, G. R. Beecher, and W. R. Lusby, *J. Agric. Food Chem.* **36,** 938 (1988).
[26] P. Barry, R. P. Evershed, A. Young, M. C. Prescott, and G. Britton, *Phytochemistry* **31,** 3163 (1992).

Similar to the parent carotenoid, the PSD spectrum of zeaxanthin monomyristate (Fig. 5, top) (molecular ion at m/z 778.3) shows a simple fragmentation pattern such as loss of toluene ([M-92]$^{\cdot+}$) at m/z 686.6, loss of myristic acid ([M-228]$^{\cdot+}$) at m/z 550.2, and a combination of both eliminations ([M-92-228]$^{\cdot+}$) at m/z 458.2.

In the same way the PSD spectrum of zeaxanthin laurate myristate (Fig. 5, middle) (molecular ion at m/z 960.5) exhibits abundant fragment ions from loss of toluene ([M-92]$^{\cdot+}$) at m/z 868.2, loss of lauric acid ([M-200]$^{\cdot+}$) at m/z 760, loss of myristic acid ([M-228]$^{\cdot+}$) at m/z 732, and consecutive losses of toluene and lauric acid ([M-92-200]$^{\cdot+}$) at m/z 668.3, toluene and myristic acid ([M-92-228]$^{\cdot+}$) at m/z 640.5, and toluene and both fatty acids ([M-92-200-228]$^{\cdot+}$) at m/z 440.2.

Characteristic fragment ions in the PSD spectrum of lutein myristate palmitate (Fig. 5, bottom) (molecular ion at m/z 1016.6) resulted from loss of toluene ([M-92]$^{\cdot+}$) at m/z 924.2, loss of myristic acid ([M-228]$^{\cdot+}$) at m/z 788.5, loss of palmitic acid ([M-256]$^{\cdot+}$) at m/z 761.2, loss of a combination of toluene and myristic acid ([M-92-228]$^{\cdot+}$) at m/z 696.4, respectively, palmitic acid ([M-92-256]$^{\cdot+}$) at m/z 668.9, loss of both fatty acids ([M-228-256]$^{\cdot+}$) at m/z 533, and loss of toluene and both fatty acids ([M-92-228-256]$^{\cdot+}$) at m/z 440.5.

Additionally, fragment ions that help to identify the parent carotenoid are observed in the lower mass range. Elimination of the respective fatty acid from the α-ionone ring furnished a ring structure analogous to 2′,3′-anhydrolutein (for structure, see Fig. 3, bottom). Loss of the ring at the doubly allylic position appears in the PSD spectrum of lutein myristate palmitate at m/z 411.3 [M-fatty acids-terminal ring]$^+$ and m/z 319.4 [M-fatty acids-toluene-terminal ring]$^+$. Retention of the charge by the terminal ring gives rise to the signal at m/z 121 (see also PSD spectrum of 2′,3′-anhydrolutein in Fig. 3).

Thus, MALDI-PSD analysis can overcome many of the difficulties encountered in the conventional mass spectometric analysis of carotenoids and carotenol fatty acid esters and can be useful for structural determination or confirmation of this group of compounds isolated from natural sources such as plant or animal tissues.

Carotenoids in Tangerine Juice Concentrate. Figure 6 shows the HPLC chromatogram of carotenoids present in tangerine juice concentrate extracts before (A) and after saponification (B). Oxocarotenoids (xanthophylls), carotenes, carotenol mono-, and bis-fatty acid esters were identified. Table I lists all major carotenoids in tangerine juice concentrate extract separated by HPLC in the order of elution.[14] The tentative identification of these compounds is based on the comparison of HPLC retention and

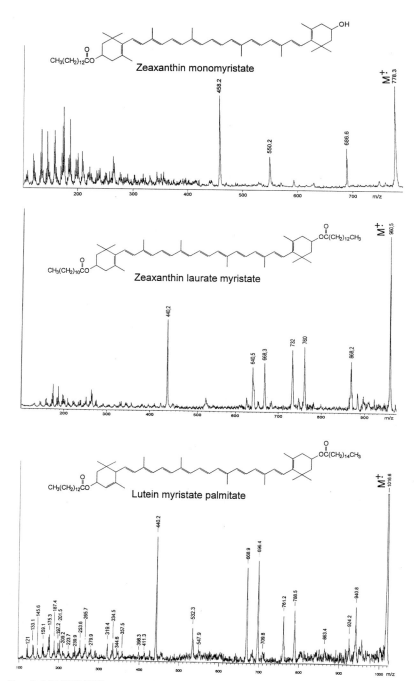

FIG. 5. MALDI-PSD spectra of zeaxanthin monomyristate, zeaxanthin laurate myristate, and lutein myristate palmitate.

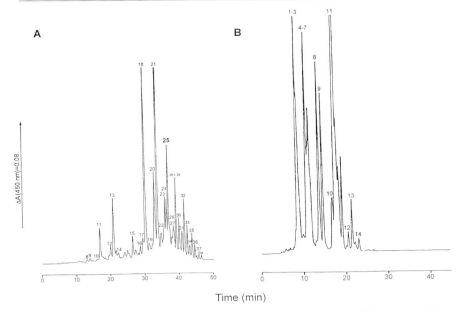

FIG. 6. HPLC chromatogram of unsaponified (A) and saponified (B) tangerine juice concentrate extracts with eluent A monitored at 450 nm. For peak identification see Table I. Modified from T. Wingerath et al., J. Agric. Food Chem. **44,** 2006 (1996).

UV/visible spectra with synthetic references. In most cases further structural analysis was performed by means of MALDI and/or MALDI-PSD mass spectrometry (see Table I).

PSD analysis of peak 25 (Fig. 7) yielded a leading precursor ion signal at m/z 790.5. Characteristic fragment ions at m/z 698.9 ($[M-92]^{·+}$), 535.5 ($[M-256]^{·+}$), and 442.3 ($[M-92-256]^{·+}$) identify the fatty acid as palmitic acid, confirming the structure of β-cryptoxanthin palmitate. In line with the results obtained on analysis of synthetic carotenoids with terminal β-ionone rings, the parent carotenoid β-cryptoxanthin does not undergo a specific fragmentation.

Carotenoids in Green Lettuce. The HPLC chromatogram of green lettuce extract, monitored at 450 nm, is shown in Fig. 8 (top). Carotenoids identified include neoxanthin, violaxanthin, lutein 5,6-epoxide, antheraxanthin, lactucaxanthin, lutein, zeaxanthin, and β-carotene. These carotenoids were tentatively identified based on HPLC retention times and UV/visible spectra. The MALDI-PSD spectrum and structure of lactucaxanthin are shown in Fig. 8 (bottom). The presence of a hydroxyl group and two α-ionone rings

TABLE I
PEAK IDENTIFICATION OF CAROTENOIDS OF TANGERINE JUICE CONCENTRATE EXTRACT SEPARATED BY HPLC (ELUENT A)

Chemical class	Peak No.	Carotenoid	γ_{max}^{a} (nm)	UV/visible absorption comparison with synthetic sample ($r > 0.990$)[b]	MALDI M^{+} (m/z)[c]	PSD analysis
Xanthophylls	1	all-*trans*-Violaxanthin	439	0.999	600	+[d]
	2	all-*trans*-Luteoxanthin	420	—[e]	600	+
	3	all-*trans*-Neoxanthin	434	—	600	+
	4	all-*trans*-Taraxanthin	445	0.998	584	—
	5	all-*trans*-Antheraxanthin	445	0.995	584	+
	6	all-*trans*-Mutatoxanthin	427	0.998	584	+
	7	all-*trans*-β-Cryptoxanthin 5,6,5',6'-diepoxide	442	—	584	—
	8	all-*trans*-Lutein	445	0.995	568	+
	9	all-*trans*-Zeaxanthin	451	0.997	568	+
	10	all-*trans*-α-Cryptoxanthin	446	—	—	+
	11	all-*trans*-β-Cryptoxanthin	451	0.996	552	+
Hydrocarbon Carotenoids	12	all-*trans*-α-Carotene	442	0.993	536	+
	13	all-*trans*-β-Carotene	450	0.995	536	+
	14	13-*cis*-β-Carotene	445	—	536	—
Carotenol Mono-fatty acid esters	15	β-Cryptoxanthin caprate	451	0.999	706	+
	16	Lutein monomyristate	444	0.999	778	—
	17	α-Cryptoxanthin laurate	445	—	734	—
	18	β-Cryptoxanthin laurate	449	0.999	734	+
	19	Zeaxanthin monopalmitate	451	0.999	804	+
	20	β-Cryptoxanthin palmitoleate	452	—	788	+
	21	β-Cryptoxanthin myristate	449	0.999	762	+
	23	β-Cryptoxanthin oleate	452	0.999	816	+
	25	β-Cryptoxanthin palmitate	450	0.999	790	+

Carotenol	22	Zeaxanthin dicaprate	449	—	876	—
Bis-fatty acid esters	24	Mutatoxanthin laurate myristate	427	—	977	—
	26	Mutatoxanthin laurate oleate	426	—	1031	—
	27	Lutein caprate laurate	445	—	905	—
	28	Mutatoxanthin dimyristate	429	—	1005	—
	29	Antheraxanthin dimyristate	445	—	1005	—
	30	Zeaxanthin laurate myristate	450	0.996	961	+
	31	Antheraxanthin myristate oleate	446	—	1059	+
	32	Antheraxanthin myristate palmitate	446	0.999	1033	+
	33	Zeaxanthin dimyristate	450	0.999	989	—
	34	Taraxanthin palmitate stearate	445	—	1089	—
	35	Antheraxanthin dipalmitate	445	0.993	1061	+
	36	Zeaxanthin myristate palmitate	452	0.992	—	—
	37	Lutein dipalmitate	448	0.994	—	—
	38	Zeaxanthin dipalmitate	451	0.998	—	—

[a] Absorption maxima in eluent A. Data modified from T. Wingerath, W. Stahl, D. Kirsch, R. Kaufmann, and H. Sies, *J. Agric. Food Chem.* **44**, 2006 (1996).
[b] Correlation coefficient comparing the UV/visible spectra of reference compounds with those of fruit juice carotenoids. Spectra were obtained by photodiode array spectroscopy. Calculation was achieved with "gold software" (Beckman, Munich).
[c] Mass assignments represent average masses of the compounds rounded to the next integer.
[d] Structural elucidation based on PSD analysis.
[e] Not measured.

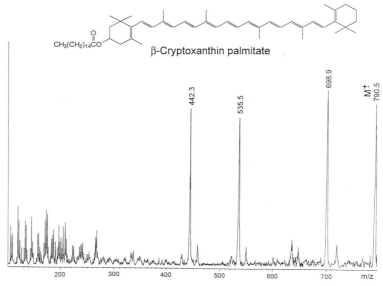

FIG. 7. MALDI-PSD spectrum of peak 25 in tangerine juice concentrate extract. Peak 25 was confirmed to be β-cryptoxanthin palmitate. Modified from T. Wingerath et al., J. Agric. Food Chem. **44**, 2006 (1996).

in the molecule structure are indicated by characteristic fragment ions, including the loss of water ([M-18]$^{\cdot+}$) at m/z 550.5, retro-Diels–Alder reaction ([M-56]$^{\cdot+}$) at m/z 512.6, retro-Diels–Alder reaction and loss of a hydroxyl group ([M-56-17]$^+$) at m/z 495, respectively, a combination of loss of toluene and retro-Diels–Alder fragmentation ([M-92-56]$^{\cdot+}$) at m/z 419.5, and the consecutive losses of two retro-Diels–Alder fragments combined with loss of a hydroxyl group ([M-56-56-17]$^+$) at m/z 439.1. In analogy to lutein, loss of the terminal ring ([M-138]$^{\cdot+}$) at m/z 429.8 is observed. Fragment ions at m/z 374.6 and 292.5 correspond to a combination of loss of the terminal ring and retro-Diels–Alder fragmentation ([M-138-56]$^{\cdot+}$), or consecutive losses of both terminal rings ([M-138-138]$^{\cdot+}$), respectively.

Conclusion

Molecular ions of nonpolar carotenes, polar oxocarotenoids, and carotenoid fatty acid esters are observed in MALDI. In contrast to other ionization techniques, PSD spectra of carotenoid compounds are simple

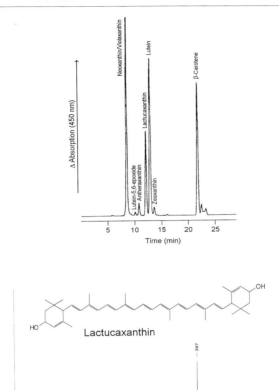

FIG. 8. HPLC chromatogram of carotenoids present in green lettuce extract with eluent A monitored at 450 nm (top) and MALDI-PSD spectrum of lactucaxanthin (bottom).

and correlate well with structural features. Although many additional carotenoid fragment ions are detected under the conditions of MALDI, some of the observed fragment ions are identical to those reported for conventional mass spectrometry, such as $[M-92]^{\cdot+}$ for nearly all carotenoids, $[M-56]^{\cdot+}$ for an α-ionone ring (e.g., α-carotene, lutein), $[M-138]^{\cdot+}$ for a hydroxylated α-ionone ring (e.g., lutein, lactucaxanthin), $[M-69]^{+}$

for acylic carotenoids (e.g., lycopene), [M-80]$^{\cdot+}$ for epoxidized carotenoids (e.g., violaxanthin), and [M-fatty acid(s)]$^{\cdot+}$ for carotenoid esters (e.g., zeaxanthin laurate myristate). This feature makes MALDI-PSD well suited for the identification of unknown carotenoids isolated from natural sources.

[36] Carotenoid Photobleaching

By ALAN MORTENSEN and LEIF H. SKIBSTED

Introduction

Carotenoids are strongly colored naturally occurring pigments found in plants and animals, although only plants and algae can synthesize carotenoids. They usually vary in color from yellow to red, but may turn blue-green when associated with proteins as found in the shell of lobsters. Carotenoids serve many functions in plants and animals. One of the most spectacular in animals is the color they provide to the plumage of birds. Carotenoids also have many other important biological functions in animals and humans,[1] one of the most important being the provitamin A activity of some carotenoids, whereas the role as photoprotector in the yellow spot of the eye[2] and the role as antioxidant are less certain. The function of carotenoids in plants is closely associated with photosynthesis. Carotenoids serve two main functions in the photosynthetic apparatus of plants: (i) as an auxiliary light-harvesting pigment and (ii) as quencher of excited states of other molecules. Carotenoids are therefore found closely associated with chlorophyll in the photosystem II reaction centers and the light-harvesting complexes where they may quench triplet chlorophyll and singlet oxygen. Because carotenoids are exposed to strong light in the photosynthetic apparatus, they face the risk of being photobleached. Carotenoids have thus been found to be bleached irreversibly in photosystem II when the usual electron transfer pathways are blocked, as may happen at high light intensities.[3,4]

With their extended conjugated electron systems, carotenoids are easily oxidized, and photobleaching therefore usually implies oxidation of the

---

[1] N. I. Krinsky, *Pure Appl. Chem.* **66**, 1003 (1994).
[2] W. Schalch, I. Emerit, and B. Chance (eds.), in Free Radicals and Aging, p. 280. Birkhäuser Verlag, Basel, 1992.
[3] J. De Las Rivas, A. Telfer, and J. Barber, *Biochim. Biophys. Acta* **1142**, 155 (1993).
[4] A. Telfer, J. De Las Rivas, and J. Barber, *Biochim. Biophys. Acta* **1060**, 106 (1991).

carotenoid. Carotenoids may be photobleached by two different mechanisms: (i) reactions between the excited state carotenoid formed by absorption of light and other compounds or (ii) formation of excited states (singlet or triplet) or radicals of other molecules by photolysis and subsequent reaction of these reactive species with carotenoids. Both of these processes may lead to degradation of the carotenoid, although physical processes, e.g., quenching of chlorophyll and singlet oxygen by energy transfer, which do not lead to degradation, may also occur.

The possibility of photobleaching of carotenoids is important in a number of biological systems. Food and beverages may lose their appeal due to loss of carotenoids and hence color. The decrease in provitamin A activity should also be considered. The photosynthesis of plants in certain geographic regions may be impaired due to photodamage to the photosynthetic apparatus, an issue becoming increasingly important due to the loss of ozone in the upper atmosphere. Assays for antioxidative capacity have also been based on light-induced reactions of carotenoids.[5] Understanding of the mechanisms behind photobleaching of carotenoids is thus an important area that has received some attention.

Experimental Methods

Two different methods have been developed in order to study the various aspects of photobleaching of carotenoids: (i) static methods based on steady-state photolysis to study the quantum yield of photodegradation and the product distribution and (ii) dynamic methods based on flash photolysis techniques to study the mechanisms and intermediates of photobleaching. A combination of both techniques may provide a more detailed picture of the photobleaching of carotenoids than either technique alone.

Steady-State Photolysis

By employing steady-state photolysis it is possible to determine the quantum yields of photodegradation of carotenoids and study the product distribution obtained by photolysis. The quantum yield of photodegradation is defined as the fraction of molecules degraded in relation to how many photons are absorbed and quantifies the light sensitivity of a given compound. The photodegradation quantum yield is usually below 1 but may be higher than unity in the case of a (free radical) chain reaction. The product distribution provides information about the reactions taking place on photolysis.

[5] A. Mortensen and L. H. Skibsted, *Free Radic. Res.* **27**, 229 (1997).

Unsensitized Photolysis. Different experimental methods have been developed and are available. Photobleaching of carotenoids may hence be studied in both aqueous (homogeneous and heterogeneous) and nonaqueous systems, and under varying experimental conditions, i.e., temperature and oxygen partial pressure.

AQUEOUS SOLUTIONS. The pyridinium salt of crocetin, the water-soluble carotenoid from saffron, is dissolved in 0.01 M NaOH (concentration of crocetin: 2×10^{-5} M). Photolysis solutions are prepared by diluting this stock solution with 0.01 M NaOH aqueous solutions to yield a total concentration of crocetin of 4×10^{-6} M and with varying concentrations of glycerol or sucrose in order to study the influence of water activity.[6]

Twenty-five milligrams of β-carotene or lutein[7] or astaxanthin[8] and 1.0 ml Tween 20 are dissolved in chloroform to 25 ml. Of this stock solution, 0.5 ml is evaporated to dryness, and the residue is dissolved in 0.1 M citrate buffer, pH 5.5, in water or heavy water. The concentration of carotenoid in these photolysis solutions with carotenoid solubilized is approximately 1×10^{-5} M.

NONAQUEOUS SOLUTIONS. Astaxanthin, canthaxanthin, or β-carotene[8,9] are dissolved directly in chloroform, acetone, vegetable oil, ethanol, or toluene. The stock solutions are diluted to 1×10^{-5} M or 6.5×10^{-6} M (toluene) carotenoid.

PHOTOLYSIS. The photolysis solutions are irradiated with monochromatic light in either a 1- or a 2-cm quartz cell.[6–9] Monochromatic light is provided by a high-pressure mercury lamp (313, 334, 366, 405, and 436 nm), a low-pressure argon–mercury lamp (254 nm),[7–9] or a xenon lamp equipped with a monochromator (Fig. 1).[6] In the case of mercury lamps, the desired wavelength is chosen by the use of appropriate interference or short-wave filters. The photolysis cells are thermostatted so that the influence of temperature on the quantum yield of photodegradation could be determined, and the solutions are stirred during photolysis by means of magnetic stirring. The solutions are either in contact with air or purged by mixtures of nitrogen and oxygen (oxygen content 3, 50, or 100%) to determine the influence of oxygen partial pressure.

QUANTUM YIELDS. The photodegradation quantum yield is defined as the number of molecules degraded divided by the number of molecules excited by light in a given experiment. Usually, the photodegradation quan-

---

[6] K. Jørgensen, M. R. Olsen, and L. H. Skibsted, *Z. Lebensm. Unters. Forsch.* **195,** 555 (1992).
[7] K. Jørgensen and L. H. Skibsted, *Z. Lebensm. Unters. Forsch.* **190,** 306 (1990).
[8] A. G. Christophersen, H. Jun, K. Jørgensen, and L. H. Skibsted, *Z. Lebensm. Unters. Forsch.* **192,** 439 (1991).
[9] B. R. Nielsen, A. Mortensen, K. Jørgensen, and L. H. Skibsted, *J. Agric. Food Chem.* **44,** 2106 (1996).

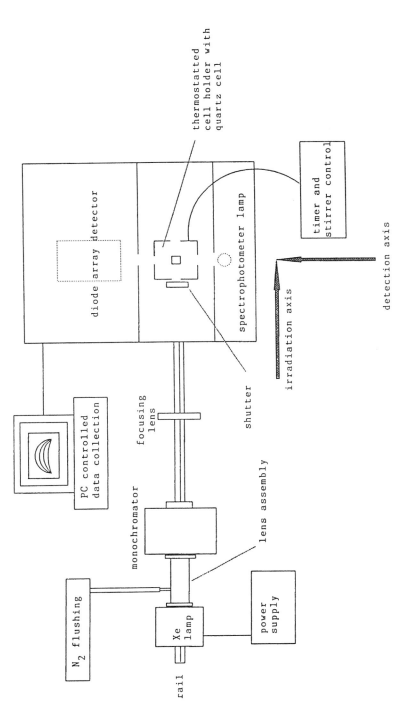

FIG. 1. Diagram of a photolysis system with a xenon lamp for photolysis, a monochromator for the selection of excitation wavelength, and a spectrophotometric diode array detection. [From K. Jørgensen, M. R. Olsen, and L. H. Skibsted. Z. Lebensm. Unters. Forsch. **195**, 555 (1992); reproduced with permission from Springer-Verlag.]

tum yield increases with decreasing wavelength and also depends on solvent and oxygen partial pressure.

The extent of photodegradation may conveniently be followed by UV–VIS spectrophotometry measuring the amount of carotenoid left unreacted as a function of photolysis time, i.e., the total irradiation dose. The combination of a xenon lamp with monochromator and a photodiode spectrophotometer provides a very convenient experimental setup, which may be computerized (Fig. 1). It is important to make sure that the degradation products do not absorb in the same region as the monitored wavelength(s) or to apply corrections as the observed quantum yield will be lower than the true quantum yield of photodegradation. The photodegradation quantum yield Φ may be calculated from

$$\Phi = \frac{A(0) - A(t_i)}{Q(t_i)A(0)}$$

where A is the absorbance at time 0 or t_i and Q is the total number of photons absorbed from time 0 to time t_i. Q is given by

$$Q(t_i) = \frac{I}{n}(1 - 10^{-A}) \sum_i (t_i - t_{i-1})$$

where I is the intensity (number of photons per time interval), n is the number of carotenoid molecules ($= cVN_A$), A is the absorbance at the irradiation wavelength, and $t_i - t_{i-1}$ is a small but finite time interval. A plot of the fraction of residual absorption versus Q is hence linear

$$\frac{A(t_i)}{A(0)} = 1 - \Phi Q(t_i)$$

This linearity typically holds until around one-third of the carotenoid is degraded.[7] Light intensities should be determined by chemical actinometry, and several methods are available for different wavelength regions such as ferrioxalate and Reinecke salt.[6–9]

In addition to photodegradation, carotenoids may also suffer thermal degradation during photolysis experiments. Thermal degradation was insignificant compared to photobleaching when irradiating at 313, 334, or 366 nm. However, thermal degradation was important when irradiating at 405 and 436 nm because longer irradiation times were needed due to a lower photodegradation quantum yield at these wavelengths (see later).[9] The photodegradation quantum yield depends only weakly on temperature, the dependence being different at different wavelengths,[8] and low temperature photolysis may be preferred in order to minimize thermal degradation.

Photolysis of β-carotene in chloroform results in a decrease in absorption above 400 nm and an increase between 300 and 400 nm due to formation of photolysis products (Fig. 2). The photodegradation quantum yield is dependent on a number of physical parameters such as irradiation wavelength, solvent, oxygen partial pressure, and, to a lesser degree, temperature. A shorter irradiation wavelength, higher solvent polarity, and higher oxygen partial pressure all give a higher quantum yield. The photodegradation quantum yield of canthaxanthin is thus roughly an order of magnitude higher in chloroform and acetone than in toluene and vegetable oil.[8,9] The photodegradation quantum yield is strongly dependent on irradiation wavelength. As Fig. 3 shows, the photodegradation quantum yield decreases by three orders of magnitude in going from 313 to 436 nm. The influence of oxygen partial pressure on the photodegradation quantum yield is also significant (Fig. 4). The results presented in Fig. 4 show that photodegradation of carotenoids in toluene proceeds via two different pathways: one oxygen dependent and the other oxygen independent. The photodegradation quantum yield of lutein and β-carotene solubilized in water was proportional to the square root of the oxygen partial pressure and no oxygen-independent photodegradation was observed.[7] This square-root dependence is indicative of a complex chain reaction taking place.[7]

Sensitized Photolysis. Whereas photolysis of carotenoids in the absence of sensitizers is a measure of the reactions of excited state carotenoid, sensitized photolysis is a measure of the reaction between ground state carotenoid and other excited state molecules. In sensitized photolysis a triplet state of a sensitizer is formed on light absorption. This triplet state

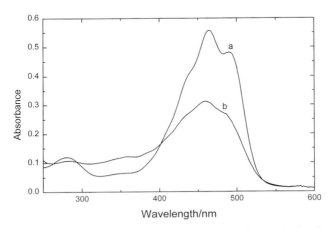

FIG. 2. Absorption spectra of 5 μM β-carotene in chloroform. Unphotolyzed (a) and photolyzed (b) for 3 hr at 510 nm.

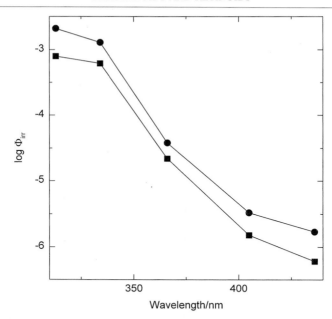

FIG. 3. Photodegradation quantum yield of β-carotene (●) and canthaxanthin (■) in toluene as a function of irradiation wavelength. [From B. R. Nielsen, A. Mortensen, K. Jørgensen, and L. H. Skibsted, *J. Agric. Food Chem.* **44**, 2106 (1996), reproduced with permission from The American Chemical Society.]

sensitizer may either react directly with the carotenoid or react with oxygen to produce singlet oxygen, which may go on to react with the carotenoid. A number of sensitizers have been used to study the photobleaching of carotenoids: toluidine blue,[10] 1-nitronaphthalene,[11] methylene blue,[12] and rose bengal.[7] Other photosensitizers, such as the food color erythrosine, may also be used. Sensitized photolysis has been performed in aqueous solution[7,11] with the carotenoid solubilized in micelles or in homogeneous methanol,[11] ethanol,[10] or benzene[12] solutions.

AQUEOUS SOLUTION. Preparation of the photolysis solutions is identical to the procedure used in unsensitized photolysis[7] (see earlier) except for the addition of rose bengal ($1.1 \times 10^{-5} \ M^{-1}$).

NONAQUEOUS SOLUTIONS. The carotenoid and the sensitizer are dissolved directly in the solvent.[10–12]

[10] P. Mathis and A. Vermeglio, *Photochem. Photobiol.* **15**, 157 (1972).
[11] J. H. Tinkler, S. M. Tavender, A. W. Parker, D. J. McGarvey, L. Mulroy, and T. G. Truscott, *J. Am. Chem. Soc.* **118**, 1756 (1996).
[12] J. F. Rabek and D. Lala, *J. Polym. Sci. Polym. Lett. Ed.* **18**, 427 (1980).

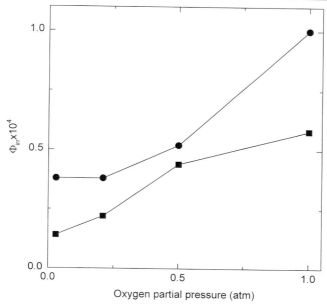

Fig. 4. Photodegradation quantum yield of β-carotene (●) and canthaxanthin (■) in toluene as a function of oxygen partial pressure at 366 nm. [From B. R. Nielsen, A. Mortensen, K. Jørgensen, and L. H. Skibsted, *J. Agric. Food Chem.* **44**, 2106 (1996), reproduced with permission from The American Chemical Society.]

PHOTOLYSIS. Sensitized photolysis has been performed with a number of light sources: mercury lamp,[7] tungsten lamp,[12] and pulsed lasers[10,11] (see later). Triplet toluidine blue and 1-nitronaphthalene bleach carotenoids by electron transfer to generate a carotenoid radical cation. The mechanism in the case of methylene blue-sensitized photobleaching is less certain but may involve a methylene blue radical or singlet oxygen (or both).[12] Rose bengal-sensitized photobleaching of water-solubilized carotenoids proceeds via both an oxygen-dependent and an oxygen-independent pathway.[7] In heavy water the photodegradation quantum yield was two to four times higher than in water, indicating that singlet oxygen was involved in the oxygen-dependent pathway. In the absence of oxygen triplet carotenoid was believed to be formed as an intermediate in photobleaching.[7] However, another possibility is the formation of carotenoid radical cation formed by electron transfer from ground-state carotenoid to triplet-state rose bengal, a reaction pathway shown to apply to sensitized bleaching by toluidine blue and 1-nitronaphthalene.[10,11]

Flash Photolysis

Flash photolysis allows the formation and decay of unstable intermediates formed during photolysis to be followed in real time, and today mainly flash methods based on pulsed lasers are used. Laser flash photolysis may provide not only the reaction mechanisms but also the rates of independent reaction steps. Just as in the case of steady-state photolysis sensitized[10,11] (see earlier) and unsensitized photolysis can be performed with laser flash photolysis.

Solutions. Stock solutions ($10^{-4}\ M^{-1}$) of carotenoids in either chloroform or carbon tetrachloride are stored cold and in the dark.[13,14] Stock solutions are used within a week during which no appreciable degradation of the carotenoid has taken place (as monitored by UV/VIS absorption spectroscopy). Stock solutions are diluted to the required concentration of carotenoid (5–30 μM).

Photolysis. The samples are thermostatted at 20°. The samples may be deoxygenated prior to laser photolysis by three freeze-pump-thaw cycles. Laser flash photolysis of the carotenoids is performed in standard 1 × 1-cm fluorescence quartz cells. The samples are usually photolyzed at 355 nm (with an intensity of approximately 60 mJ per pulse) although excitation light of 532 nm can be used as well. Optical filters are used in order to minimize the extent of photodegradation induced by the monitoring light.[13,14]

Photobleaching of carotenoids in chloroform is a two-step process.[13,14] The first step is very fast (complete within the time span of the laser pulse, i.e., 10 nsec) and would require a better time resolution, i.e., pico- or femtosecond, to be studied in detail. This first fast step is followed by a slower (pseudo) first-order bleaching, which is complete in around 1 msec (Fig. 5A). The second order rate constant of slow bleaching is 10^8–$10^9\ M^{-1}\ s^{-1}$; the carotenes lycopene and β-carotene are bleached faster than the hydroxyxanthophylls zeaxanthin and lutein, whereas the ketoxanthophylls canthaxanthin, astaxanthin, and β-apo-8'-carotenal are bleached at the slowest rate.[13] Concomitant with the bleaching of carotenoid is the formation of near-infrared absorbing intermediates (Fig. 5B). One species is formed instantaneously, corresponding to the instantaneous bleaching. This species shows maximum absorption in the region 835–960 nm, depending on the carotenoid (Fig. 5B). The initially formed transient species decays to another species absorbing in the region 855–1040 nm (Fig. 5B). This species has been identified as a carotenoid radical cation.

[13] A. Mortensen and L. H. Skibsted, *Free Radic. Res.* **26**, 549 (1997).
[14] A. Mortensen and L. H. Skibsted, *Free Radic. Res.* **25**, 355 (1996).

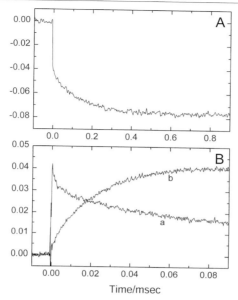

FIG. 5. Transient absorption time traces obtained by laser flash photolysis at 355 nm of lycopene in chloroform. (A) 10 μM lycopene monitored at 524 nm and (B) 30 μM lycopene monitored at (a) 820 or (b) 1100 nm.

$$\text{Car} + \text{CHCl}_3 \xrightarrow{h\nu} \text{Intermediate} \rightarrow \text{Car}^{\cdot+} + \text{CHCl}_2^{\cdot} + \text{Cl}^-$$

The intermediate has not been identified unambiguously, but it could possibly be a carotenoid radical cation/chloroform radical anion ion pair.[13,14] The slow bleaching (Fig. 5A) is due to reaction between the carotenoid and the chloroform radical $\text{CHCl}_2^{\cdot+}$ formed by the instantaneous bleaching of carotenoid. Photobleaching is independent of whether oxygen is present or not.[13,14] The carotenoid radical cation decays by second-order kinetics:

$$2\text{Car}^{\cdot+} \rightarrow \text{Product(s)}$$

The decay, however, is not by disproportionation to regenerate the carotenoid and a two-electron oxidized carotenoid. The degradation products from photolysis absorb in the region 300–400 nm (Fig. 2), which could indicate formation of *cis* isomers and shorter polyenes as stable end products.

Photobleaching of carotenoids in carbon tetrachloride shows an identical pattern as photobleaching in chloroform, i.e., a two-step bleaching. However, only a very minute amount of carotenoid radical cation is observed.[13] This could be due to either a much shorter lifetime of the carotenoid radical cation in carbon tetrachloride than in chloroform or that the radical cation is not formed as an intermediate during photolysis.

The radical cation has been shown to be an intermediate in both sensitized[10,11] and unsensitized[13,14] photobleaching of carotenoids. However, the nature of the stable end products has not been established.

Degradation Products

Photolysis of carotenoids in apolar solvents without a sensitizer does not to a significant extent lead to bleaching, and the study of degradation products has thus mainly been performed in the presence of sensitizers. A number of different sensitizers have been used: rose bengal,[15] toluidine blue,[16] chlorophyll,[17] and hypericin.[18]

A number of light sources, sensitizers, solvents, and column materials have been employed. Only one example is given here.

Solution. Two milliliters of a 0.005% solution of β-carotene was mixed with 4 ml of 4×10^{-5} M toluidine blue O in methanol.[16]

Photolysis. The solutions were irradiated with light from a continuous wave helium–neodymium (cw He–Ne) laser at 632.8 nm (25 mW at the sample). The samples were either irradiated in the presence of air or a stream of nitrogen was passed over the sample during photolysis.[16]

Isolation of Photodegradation Products. After irradiation, methanol and toluidine blue O were removed by washing with water. The hexane phase was dried over anhydrous sodium sulfate.[16] Chromatographic isolation was performed by thin-layer chromatography (TLC) or on a small column consisting of a mixture of MgO and Hyflo Super Cel (1:2); in the case of TLC, 13% calcium sulfate was added. TLCs were developed with hexane/acetone (96/4) and the columns by hexane followed by hexane with 2 and 5% acetone. Identification of the compounds was by UV/VIS spectroscopy.[16]

Photodegradation Products. Oxygen is important for the photodegradation of carotenoids. β-Carotene was thus not affected by irradiation in the presence of toluidine blue O in the absence of oxygen,[16] and sensitization by hypericin only led to the formation of *cis* isomers in the absence of oxygen.[18] In the presence of oxygen a number of oxygenated compounds were formed in addition to *cis* isomers. Thus, compounds with hydroxy,

---

[15] S. Isoe, S. B. Hyeon, and T. Sakan, *Tetrahed. Lett.* 279 (1969).
[16] K. Hasegawa, J. D. MacMillan, W. A. Maxwell, and C. O. Chichester, *Photochem. Photobiol.* **9**, 165 (1969).
[17] K. Tsukida, S.-C. Chô, and M. Yokota, *Chem. Pharm. Bull.* **17**, 1755 (1969).
[18] G. R. Seely and T. H. Meyer, *Photochem. Photobiol.* **13**, 27 (1971).

epoxy, and keto groups have been isolated.[15-19] Acetylenic and allenic structures are also formed as a result of sensitized photolysis.[17] Photobleaching under mild conditions seems to retain the basic carotenoid carbon skeleton, whereas exhaustive photobleaching leads to fragmentation. Thus, dihydroactinidiolide, β-ionone, and 6-hydroxy-2,2,6-trimethylcyclohexanone have all been isolated and identified after exhaustive photobleaching of β-carotene in the presence or absence of rose bengal.[15,19]

Application to Food Systems

In addition to the model systems mentioned in this article, photobleaching of carotenoids has also been studied in systems more relevant to the natural occurrence of carotenoids, e.g., foods and beverages. In addition to imparting color, carotenoids serve two purposes in food: (i) they may act as photoprotecting agents just as in photosynthetic systems, i.e., quenching of reactive excited states, and (ii) as primary antioxidants by quenching free radicals, including free radicals generated by photolysis. During these actions, the carotenoids may of course be degraded. For experimental details, see the relevant papers.

In tomato pulp the extent of lycopene degradation was higher when the pulp was stored under light than when it was stored in the dark,[20] and in a vegetable juice α- and β-carotene were found to be photobleached faster than lycopene.[21] The photodegradation of all three carotenoids followed first-order kinetics with half-lives ranging from 2 to 11 days (approximately 2500 lux). Lutein, lycopene, and β-carotene were found to increase the amount of peroxides in the triglyceride fraction of rapeseed oil exposed to fluorescent light (10000 lux for a couple of days) compared to rapeseed oil without carotenoids under the same conditions.[22,23] The carotenoids themselves were bleached by the fluorescent light. These three carotenoids hence acted as prooxidants. However, in the presence of γ-tocopherol the prooxidant effect of the carotenoids was inhibited and their bleaching was retarded. In fact, lutein and β-carotene in combination with γ-tocopherol were found to be better at inhibiting peroxide formation than γ-tocopherol alone.[22,23] A number of polyphenolic antioxidants extracted from green tea were also found to inhibit bleaching of β-carotene in beverages and margarine exposed to UV light (254 nm for a couple of days).[24]

[19] S. Isoe, S. B. Hyeon, S. Katsumura, and T. Sakan, *Tetrahed. Lett.*, 2517 (1972).
[20] S. K. Sharma and M. Le Maguer, *Food Res. Int.* **29**, 309 (1996).
[21] C. A. Pesek and J. J. Warthesen, *J. Food Sci.* **52**, 744 (1987).
[22] K. Haila and M. Heinonen, *Lebensm. Wiss. Technol.* **27**, 573 (1994).
[23] K. M. Haila, S. M. Lievonen, and M. I. Heinonen, *J. Agric. Food Chem.* **44**, 2096 (1997).
[24] L. Unten, M. Koketsu, and M. Kim, *J. Agric. Food Chem.* **45**, 2009 (1997).

Radical Efficiency Assay

The assays developed for studying the photobleaching of carotenoids may be employed to examine the reactions between various antioxidants. The interaction among lycopene and α-, β-, γ-, or δ-tocopherol has thus been studied by laser flash photolysis.[5]

Photolysis. The experimental details are as described earlier for laser flash photolysis[13,14] except for the presence of 1 mM tocopherol in a 10 μM solution of lycopene in chloroform.

Photolysis of lycopene in chloroform results in the formation of lycopene radical cations and chloroform radical anions (see earlier). The chloroform radicals react with either lycopene or tocopherol, resulting in the formation of tocopheroxyl radicals. By following the formation and decay of the transient species it was demonstrated that α-tocopherol could reduce the lycopene radical cation whereas lycopene could reduce the δ-tocopheroxyl radical. The lycopene radical cation and the β- and γ-tocopheroxyl radicals were in an equilibrium.[5] An antioxidant hierarchy involving lycopene and the four tocopherol homologues can thus be established by this use of laser flash photolysis of a carotenoid: α-tocopherol $>$ β-tocopherol $\sim$ γ-tocopherol $\sim$ lycopene $>$ δ-tocopherol.

Concluding Remarks

As described earlier, photobleaching of carotenoids depends on a number of parameters: solvent, oxygen partial pressure, sensitizers, and of course light intensity and spectral distribution. Although the results are not unanimous as to whether oxygen is really needed in photobleaching, oxygen certainly plays an important role. In addition, polar solvents have been found to increase the extent of photobleaching of carotenoids, as have sensitizers. The photodegradation quantum yield is strongly wavelength dependent, with a higher yield at shorter wavelengths, and an exponential relationship between light energy and quantum yield has been suggested.[7]

One important finding emerging from the studies of carotenoid photobleaching is the importance of the carotenoid radical cation as an intermediate both in sensitized and in unsensitized photobleaching.[10,11,13,14] This species has not been observed during steady-state photolysis due to its rather short lifetime, which requires flash photolysis techniques for detection. However, this species may prove to be important even under conditions where triplet-state carotenoid or singlet oxygen has been suggested to be involved in photobleaching. The distribution of photodegradation products, however, illustrates the complexity of photobleaching of carotenoids, and this subject is far from fully understood.

Photobleaching of carotenoids may also be used to study the interaction between carotenoids and other antioxidants directly by laser flash photolysis[5] or indirectly by steady-state photolysis,[22–24] which eventually may lead to antioxidant hierarchy.

Acknowledgment

This work was supported by the FØTEK program through LMC—Center for Advance Food Studies.

[37] Interactions between Vitamin A and Vitamin E in Liposomes and in Biological Contexts

By Maria A. Livrea and Luisa Tesoriere

Introduction

Vitamin A reacts effectively with peroxyl radicals in chemical systems[1–3] and may behave as a lipid antioxidant of biological relevance.[4,5] Injection of very low doses has been shown to protect membranes from various rat tissues against oxidative stress induced *in vitro*[4] and *in vivo*.[4,5] In addition, low-density lipoprotein (LDL) isolated from human plasma 8 hr after a single oral administration of vitamin A exhibits a markedly enhanced resistance to Cu^{2+}-dependent oxidation as a result of a very small increase of retinol and retinyl esters in LDL.[6] Although oxygen partial pressure, retinol concentration, and radical fluxes can contribute to enhance or minimize the antioxidant activity of retinol,[3] the extent of the effects observed in the *in vivo* and *ex vivo* studies suggests that other factors may enhance the antioxidant efficacy of vitamin A in biological contexts.

---

[1] L. Tesoriere, M. Ciaccio, A. Bongiorno, A. Riccio, A. M. Pintaudi, and M. A. Livrea, *Arch. Biochem. Biophys.* **307,** 217 (1993).
[2] L. Tesoriere, A. Bongiorno, A. M. Pintaudi, R. D'Anna, D. D'Arpa, and M. A. Livrea, *Arch. Biochem. Biophys.* **326,** 57 (1996).
[3] L. Tesoriere, D. D'Arpa, R. Re, and M. A. Livrea, *Arch. Biochem. Biophys.* **343,** 13 (1997).
[4] M. Ciaccio, M. Valenza, L. Tesoriere, A. Bongiorno, R. Albiero, and M. A. Livrea, *Arch. Biochem. Biophys.* **302,** 103 (1993).
[5] L. Tesoriere, M. Ciaccio, M. Valenza, A. Bongiorno, E. Maresi, R. Albiero, and M. A. Livrea, *J. Pharmacol. Exp. Ther.* **269,** 430 (1994).
[6] M. A. Livrea, L. Tesoriere, A. Bongiorno, A. M. Pintaudi, M. Ciaccio, and A. Riccio, *Free Radic. Biol. Med.* **18,** 401 (1995).

Interactions and recycling are very common mechanisms in the action of antioxidants. These phenomena investigated in solution,[7,8] micelles,[9] membranes,[10] and lipoproteins[11,12] strongly suggest that in biological contexts the radical scavenging antioxidants may act not individually, but rather cooperatively or even synergistically with each other. Interactions between vitamin A and vitamin E, the major lipid antioxidant of membranes and other lipid systems,[13] have been investigated. Studies carried out by incorporating a variable proportion of all-*trans*-retinol and α-tocopherol in soybean phosphatidylcholine liposomes and the antioxidant effectiveness of all-*trans*-retinol in retinal membranes, whether deprived or not of endogenous α-tocopherol, as well as the consumption kinetics of the antioxidants, are reported.

Chemicals and Equipment

Soybean phosphatidylcholine (PC), all-*trans*-retinol, α-tocopherol, butylated hydroxytoluene (BHT), and thiobarbituric acid (TBA) are from Sigma Chemical Company (St. Louis, MO); 2,2'-azobis(2-amidinopropane) hydrochloride (AAPH) is from Polyscience, Inc. (Warrington, PA); and Chelex 100 ion-exchange resin is from Bio-Rad Laboratories (Munchen, Germany). Phosphate-buffered paline (PBS: 0.9% NaCl in 5 mM phosphate buffer, pH 7.4) used throughout these studies is chromatographed over Chelex 100, and suitable plastic labware is used to minimize the effects of adventitious metals. High-performance liquid chromatography (HPLC) analyses are carried out with a Gilson system (Middleton, WI) consisting of a Rheodyne injector, a Model 305 pump, and a Model 118 multiwavelength detector. Chromatograms are recorded by a HP 3395 integrator (Hewlett Packard, Palo Alto, CA). All operations are carried out under red light to avoid possible photooxidation of fatty acids and to preserve light-sensitive all-*trans*-retinol and α-tocopherol. Spectrophotometric analyses are performed with a Beckman DU 640 (Palo Alto, CA) spectrophotometer.

Liposomal Oxidation

The liposome suspension is prepared by adding, in this order, a chloroform solution of soybean phosphatidylcholine and variable amounts of all-

[7] J. E. Packer, T. F. Slater, and R. L. Willson, *Nature* **278**, 737 (1979).
[8] E. Niki, T. Saito, A. Kawakami, and Y. Kamiya, *J. Biol. Chem.* **259**, 4177 (1984).
[9] R. Coates Barclay, S. J. Locke, and J. M. MacNeil, *Can. J. Chem.* **61**, 1288 (1983).
[10] A. Costantinescu, D. Han, and L. Packer, *J. Biol. Chem.* **268**, 10906 (1993).
[11] V. W. Bowry and R. Stocker, *J. Am. Chem. Soc.* **115**, 6029 (1993).
[12] V. E. Kagan, E. Serbinova, T. Forte, G. Scita, and L. Packer, *J. Lipid Res.* **33**, 385 (1992).
[13] G. Burton and K. U. Ingold, *J. Am. Chem. Soc.* **103**, 6472 (1981).

trans-retinol (0.5 to 20 μM) and/or α-tocopherol (5 μM) as ethanol solutions, in a round-bottom tube kept in an ice bath. Solvent is evaporated to dryness, after each addition, under a nitrogen stream, and the thin film obtained is mixed with PBS to reach a 10 mM final lipid concentration and is vortexed for 10 min. The resulting multilamellar dispersion is then transferred into an Avestin Liposofast (Avestin, Inc., Ottawa, Canada) small volume extrusion device provided with a polycarbonate membrane of 100 nm pore size designed to obtain a homogeneous population of large unilamellar liposomes.

To obtain a constant rate of chain initiation, which is essential to kinetic studies, azo initiators are commonly used. Oxidation is carried out in a water bath at 37°, under air, in the presence of 2 mM AAPH, added to the suspensions in a small PBS volume. Aliquots of liposomes (20 μl) are taken at 10-min intervals and dissolved in 50 volumes of absolute ethanol, and spectra are then recorded in the range of 200 to 300 nm. Conjugated diene hydroperoxide production is evaluated by the increase in absorbance at 234 nm, using a molar absorption coefficient of 27,000.[14]

Retinal Membrane Oxidation

Retinas from freshly excised bovine eyes are homogenized in PBS (1.5 ml/retina). The homogenate is centrifuged at 700g for 10 min, and the supernatant is precipitated at 110,000g for 60 min to obtain the 110,000g postnuclear pellet, which is referred to as retinal membranes. The preparation is stored at $-80°$ and is utilized within 48 hr to minimize possible autoxidation of lipid components. To clear membranes of endogenous vitamin E, UV irradiation of membrane suspensions (2 mg protein ml^{-1} is carried out by exposure to a solar light simulator (LOT Oriel, Italy, wavelength range 295–400 nm) for 10 min in an ice bath at a 30-cm distance. This short exposure does not cause formation of detectable amounts of TBA-reactive substances (TBARS).[15]

Homogenization of membranes is carried out in PBS, to reach a protein concentration of 2 mg ml^{-1}, and incubation is performed at 37° in the presence of 10 mM AAPH, under air. all-*trans*-Retinol (0.75 nmol mg protein^{-1}) is added to the suspension as an ethanol solution, and the mixture is allowed to stand at room temperature for 10 min. Final ethanol concentration is 0.2%, which does not affect the peroxidation assay. Under these conditions, 80% of the added retinol is incorporated, as monitored by HPLC

[14] W. A. Pryor and L. Castle, *Methods Enzymol.* **105,** 293 (1984).
[15] E. Pelle, D. Maes, G. A. Padulo, E.-K. Kim, and W. P. Smith, *Arch. Biochim. Biophys.* **283,** 234 (1990).

analysis carried out on the resedimented membranes. At suitable time intervals, lipid peroxidation accumulation products are evaluated as TBARS from 1.0-ml portions of the incubation mixture and quantitated spectrophotometrically as malondialdehyde (MDA), using a molar absorption coefficient of 156,000.[16] BHT (0.03%) is added to the TBA reagent to prevent artifactual lipid peroxidation during the assay procedure.

The induction times caused by endogenous antioxidants, or generated after incorporation of all-*trans*-retinol into the membranes, are calculated from the intercept with the abscissa of the extension of the linear portion of the peroxidation curve relevant to the propagation phase.

Retinol and α-Tocopherol Analysis

The consumption of all-*trans*-retinol and α-tocopherol during peroxidation of either liposomal or membrane suspensions is assayed by extracting aliquots of the relevant incubation mixtures, followed by HPLC quantitation. all-*trans*-Retinol and α-tocopherol are extracted from 1 ml of either suspension by mixing with 2 volumes of absolute ethanol, followed by two successive extractions with 6 and 2 volumes of petroleum ether. The yield of this procedure is 85% and data are corrected accordingly. The organic extracts are gathered, dried under nitrogen, resuspended in several microliters of suitable solvent, and injected on top of a Supelco (Bellefonte, PA) Supelcosil LC-18 HPLC column (25 × 0.46 cm). Analysis is carried out by eluting with 2% water in methanol at 1.5 ml min^{-1}. Detection of all-*trans*-retinol and α-tocopherol is at a wavelength of 320 and 290 nm, respectively. Under the conditions described, all-*trans*-retinol elutes after 4 min and α-tocopherol after 12 min. An automatic wavelength change after 9 min allows the detection of both compounds in the same sample, when necessary. Quantitation is by reference to a standard curve constructed with 1 to 100 ng of either all-*trans*-retinol or α-tocopherol and by relating the amount of the compound injected to the peak area.

The endogenous amounts of all-*trans*-retinol and α-tocopherol in retinal membranes are measured by extracting samples (2 mg protein) with organic solvent followed by HPLC analysis, as described earlier.

Synergistic Antioxidant Activity between all-*trans*-Retinol and α-Tocopherol in Unilamellar Soybean Phosphatidylcholine Liposomes

The inhibition rate (R_{inh}) and inhibition periods (τ) measured in the presence of either the individual antioxidants or a variety of combinations

---

[16] J. A. Buege and S. D. Aust, *Methods Enzymol.* **LII,** 302 (1978).

of all-*trans*-retinol and α-tocopherol are reported in Table I. Whereas α-tocopherol at 5 μM causes a 82% decrease of the propagation rate, all-*trans*-retinol alone is poorly effective to contrast the production of conjugated dienes during the AAPH-induced oxidation of PC liposomes. It has been observed that concentrations as low as 1.0–10.0 μM do not bring about either a measurable inhibition period or a decrease of the propagation rate.[2] A substantial decrease of the propagation rate is evident at a retinol concentration as high as 20 μM (Table I). The length of the inhibition periods observed for α-tocopherol alone is extended markedly when α-tocopherol and all-*trans*-retinol are assayed in combination (Table I). Because all-*trans*-retinol does not bring about any inhibition time when used alone ($\tau_A = 0$), kinetic data provide evidence of synergistic interactions. The amount of synergism, as expressed by $\tau_{(E+A)} - (\tau_E + \tau_A)$, is a function of the molar ratio of the two antioxidants. An almost linear variation is observed for a 0.1 to 1.0 molar ratio of all-*trans*-retinol/α-tocopherol (column 11, Table I). In the same experiments, the propagation rate throughout the inhibition period, $R_{inh(E+A)}$, is considerably lower than that observed for α-tocopherol alone and decreases as the concentration of all-*trans*-retinol increases. Because the rate of chain initiation in these assays is controlled and constant, the decrease of R_{inh} indicates that species other than, or in addition to, α-tocopherol are involved in the scavenging of lipoperoxyl radicals. The kinetics of consumption of α-tocopherol and all-*trans*-retinol in the course of lipid peroxidation may help gain some insight into the mechanism of the synergistic interactions. When all-*trans*-retinol and α-tocopherol are combined (Fig. 1), the rate of the temporal disappearance of *both* antioxidants is delayed significantly, with respect to assays where they act separately, which suggests a reciprocal protection. As shown in Fig. 1, the inhibition period determined by the combination of the antioxidants does not coincide with maintenance of either of the antioxidants. Therefore, regeneration mechanisms cannot be postulated to account for the synergistic interactions.

Reciprocal Protective Effects of all-*trans*-Retinol and α-Tocopherol during Membrane Lipid Peroxidation

Retinal membranes that are endowed with high amounts of vitamin E[17,18] are a good model to investigate interactions between all-*trans*-retinol and α-tocopherol in that they offer the advantage of a sizable pool of α-

[17] C. D. Farnsworth and E. A. Dratz, *Biochim. Biophys. Acta* **443**, 556 (1976).
[18] R. J. Stephens, D. S. Negi, S. M. Short, F. J. G. M. van Kuijk, E. A. Dratz, and D. W. Thomas, *Exp. Eye Res.* **47**, 237 (1988).

TABLE I

INHIBITING EFFECT OF α-TOCOPHEROL (E) AND ALL-*trans*-RETINOL (A)[a]

PC^b (mM)	R_P^b (M^{-8} sec^{-1})	E^b (M^{-6})	A^b (M^{-6})	$R_{inh(E)}^b$ (M^{-8} sec^{-1})	$R_{inh(A)}^b$ (M^{-8} sec^{-1})	$R_{inh(E+A)}^b$ (M^{-8} sec^{-1})	τ_E (sec)	τ_A (sec)	$\tau_{(E+A)}$ (sec)	$\tau_{(E+A)} - (\tau_E + \tau_A)$ (sec)
10.0	26.0	5.0	20.0	4.68	15.2		1980	0		
		5.0	0.5			3.81			2130	150
		5.0	1.0			3.36			2270	290
		5.0	2.5			3.12			2760	780
		5.0	5.0			2.60			3300	1320
		5.0	10.0			1.90			3480	1500

[a] Combinations of E + A on the peroxidation rate (R_P) of soybean phosphatidylcholine (PC) liposomes initiated by 2.0 mM AAPH in PBS at pH 7.4. Amounts are given in moles for ease of comparison, as PC, E, and A are concentrated in liposomes, whereas AAPH is in the aqueous phase. An average molecular weight of 900 is considered for phosphatidylcholine.[32] Kinetic data are obtained from a computer-assisted analysis (Table Curve 2D, Jandel, CA) of the experimental peroxidation curves. R_P is measured as the amount of lipid hydroperoxides formed per second in the absence of any antioxidant or after the inhibition period (τ). $R_{inh(E)}$ represents the lipoperoxyl radical propagation rate in the presence of all-*trans*-retinol. The inhibition rate in the presence of α-tocopherol, $R_{inh(E)}$, and in the presence of α-tocopherol and all-*trans*-retinol, $R_{inh(E+A)}$, is calculated by the coordinates of the intercept of the tangents to the parts of the curve representing the inhibition and propagation phases. Values in each series are means of at least three experiments with SD ≤ 6%.

[b]

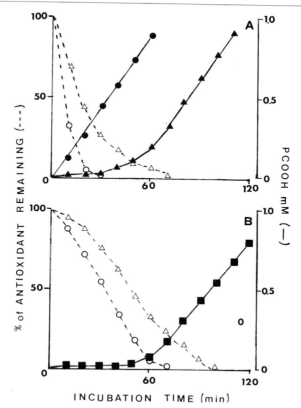

Fig. 1. Time course of the consumption of α-tocopherol (△) and all-*trans*-retinol (○) incorporated separately (A) or in combination (B) in 10 m*M* soybean PC liposomes during AAPH-induced peroxidation (closed symbols). Liposomes contain 5 μ*M* α-tocopherol and/ or 5 μ*M* all-*trans*-retinol. The uninhibited reaction curve was equivalent to the curve in the presence of all-*trans*-retinol (●). Each point represents the mean of four experiments carried out with different preparations. SD of PC-OOH values was within 10%, and of α-tocopherol and all-*trans*-retinol values was within 15%. [Modified from L. Tesoriere *et al.*, *Arch. Biochem. Biophys.* **326**, 57 (1996).]

tocopherol in its natural localization. It has been reported that bovine retinal membranes may vary largely in their vitamin E content, according to the season and diet of the animals.[19] At the same time, the amount of free retinol may vary considerably according to the bleaching degree of

---

[19] E. R. Berman, *in* "Biochemistry of the Eye," p. 309. Plenum Press, New York, 1991.

rhodopsin.[20,21] In the experiments that are summarized later, the mean amount of endogenous α-tocopherol, measured in 10 different membrane preparations, was 0.73 ± 0.08 nmol mg protein^{-1} and of all-*trans*-retinol was 0.1 ± 0.012 nmol mg protein^{-1}.

The incorporation into native retinal membranes of an amount of all-*trans*-retinol comparable with that of endogenous α-tocopherol causes a significant increase of the resistance to peroxidative stress induced by AAPH. However, membranes that have been deprived of the endogenous vitamin E by UV light exposure do not exert a significant resistance to peroxidation, even after incorporation of all-*trans*-retinol (Fig. 2A). These findings provide evidence that the effects of the antioxidants are cooperative rather than additive, i.e., the observed inhibition period is beyond the sum of the inhibition periods of the individual compounds. Finally, all-*trans*-retinol appears more rapidly consumed when incorporated in membranes deprived of α-tocopherol than in native ones (Fig. 2B), and the addition of all-*trans*-retinol to the membranes slows down the consumption of endogenous α-tocopherol (Fig. 2C).

Concluding Remarks

Cooperative interactions between antioxidants are very common mechanisms in the biological defense against peroxidation, particularly when α-tocopherol is involved. Synergism with ascorbic acid[10–12,22–24] and coenzyme Q_{10}[11,25] is due to "regeneration" of α-tocopherol from its tocopheroxyl radical. Data demonstrating some synergism between α-tocopherol and β-carotene in microsomal membranes[26] showed that α-tocopherol may be "consumed" to avoid autoxidation of β-carotene and reduce its prooxidant effects. In contrast, other investigations in homogeneous solution, while confirming synergism between β-carotene and α-tocopherol,[27] show that electron transfer occurs from carotene to tocopheroxyl radical. Similar findings reported in studies carried out with β-carotene and phenolic compounds[28] suggest that β-carotene may protect α-tocopherol.

[20] W. F. Zimmerman, *Vis. Res.* **14**, 795 (1974).
[21] A. Bongiorno, L. Tesoriere, M. A. Livrea, and L. Pandolfo, *Vis. Res.* **31**, 1099 (1991).
[22] T. Doba, G. W. Burton, and K. Ingold, *Biochim. Biophys. Acta* **835**, 298 (1985).
[23] C. T. Ho and A. C. Chan, *FEBS Lett.* **306**, 269 (1992).
[24] A. C. Chan, K. Tran, T. Raynor, P. R. Ganz, and C. K. Chow, *J. Biol. Chem.* **266**, 17290 (1991).
[25] K. U. Ingold, V. W. Bowry, R. Stocker, and C. Walling, *Proc. Natl. Acad. Sci. U.S.A.* **90**, 45 (1993).
[26] P. Palozza and N. I. Krinsky, *Arch. Biochem. Biophys.* **297**, 184 (1992).
[27] F. Bohm, R. Edge, E. J. Land, D. J. McGarvey, and T. G. Truscott, *J. Am. Chem. Soc.* **119**, 621 (1997).
[28] A. Mortensen and L. H. Skibsted, *Free Radic. Res.* **25**, 515 (1996).

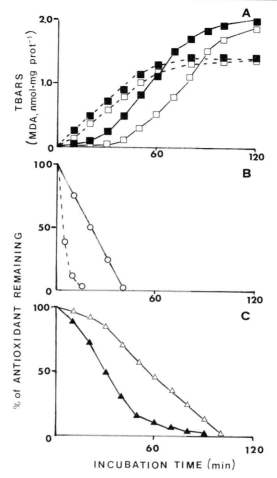

FIG. 2. Time course of lipid peroxidation (A) and consumption of all-*trans*-retinol (B) and α-tocopherol (C) in native retinal membranes (solid line) or in UV-treated membranes (dotted line) in the absence (closed symbols) or in the presence (open symbols) of added all-*trans*-retinol. Each point is the mean of six experiments carried out with different membrane preparations. SD of TBARS values was within 10% and of α-tocopherol and all-*trans*-retinol values was within 15%. [Modified from L. Tesoriere *et al.*, *Biochem. Mol. Biol. Int.* **37,** 1 (1995).]

In the experiments reported here, synergistic effects between all-*trans*-retinol and α-tocopherol are evident in chemical bilayers, as well as in natural membranes. In addition, when all-*trans*-retinol and α-tocopherol are allowed to act in combination, consumption of both antioxidants during the lipid peroxidation is delayed, in comparison to the assays where they act separately. Despite evidence for reciprocal protective effects, the mechanisms underlying these interactions remain puzzling in some respects. Consumption kinetics do not show regeneration mechanisms for any of the antioxidants. Because vitamin A easily undergoes autoxidation, an antioxidant protection of all-*trans*-retinol by α-tocopherol[29–31] could lead to the enhancement of antioxidant activity of all-*trans*-retinol. Nevertheless, if some α-tocopherol serve to prevent autoxidation of all-*trans*-retinol, it would also be expected to be consumed more rapidly in the presence of all-*trans*-retinol excess, which is in contrast with the experimental evidence. In the light of the complex antioxidant mechanism of retinol,[3] a possibility is that the antioxidant protection by α-tocopherol results in a much more effective activity of all-*trans*-retinol so that it can strongly compete with α-tocopherol for the scavenging of peroxyl radicals. This could finally preserve the membrane lipid and α-tocopherol itself.

[29] L. A. Mejia, in "Vitamin A Deficiency and Its Control" (J. C. Bauerfeind, ed.), p. 69. Academic Press, Orlando, Florida, 1986.
[30] W. G. Robison Jr., T. Kuwabara, and J. G. Bieri, *Invest. Ophthalmol. Vis. Sci.* **18,** 683 (1979).
[31] W. G. Robison Jr., T. Kuwabara, and J. G. Bieri, *Invest. Ophthalmol. Vis. Sci.* **19,** 1030 (1980).
[32] Y. Yamamoto, E. Niki, Y. Kamiya, and H. Shimasaki, *Biochim. Biophys. Acta* **795,** 332 (1984).

[38] On-line Solid-Phase Extraction and Isocratic Separation of Retinoic Acid Isomers in Microbore Column Switching System

By THOMAS E. GUNDERSEN and RUNE BLOMHOFF

Introduction

Vitamin A (retinol) is involved in the regulation of proliferation and differentiation of many cell types during fetal development as well as for most cell types throughout life. The discovery of several nuclear receptors for retinol metabolites has provided an understanding of the underlying mechanism of action of vitamin A. The three retinoic acid receptors (RARα, β, and γ) and the three retinoid X receptors (RXRα, β, and γ) belong to the steroid/thyroid receptor superfamily of ligand-dependent transcription factors. Physiological ligands for all the RARs appear to be all-*trans*-retinoic acid, 9-*cis*-retinoic acid, 3,4-didehydroretinoic acid, and 4-oxoretinoic acid. The RXRs are more selective and are only activated by 9-*cis*-retinoic acid.

It has been shown that RARα and RXRβ mRNA are widely distributed in most tissues and cells, whereas the expression of RARβ, RARγ, RXRα, and RXRγ mRNA is more specific. The nuclear retinoid receptor binds to retinoic acid responsive elements as RXR–RAR heterodimers or as RXR–RXR homodimers. The RXRs also bind to various other response elements as a heterodimer partner for a number of other nuclear receptors such as the vitamin D receptor and the thyroid hormone receptor.[1]

Very little is known, however, about the concentration of the active ligands, the retinoic acid metabolites, in various cells and tissues. It is, e.g., not known whether the ligand is induced by certain stimuli or whether it is present in most cells at a sufficient concentration to work as a "cofactor" for the receptors when they are expressed. This loss of knowledge is mainly due to the lack of methods with high enough sensitivity to quantitate the physiological levels of the active retinoids. We have previously published a paper dealing with on-line solid phase extraction and isocratic separation of retinoic acid isomers.[2] This setup has been improved as our experience has grown. Speed, sensitivity, robustness, and recovery have been optimized as a result of these changes. For a detailed discussion of the technique, the reader should refer to Gundersen et al.[2]

Standards

Retinoids are a class of compounds highly sensitive to elevated temperatures, light, and oxygen. For this reason, standards should be stored cold ($-80°$) under argon and protected from white light at all times. The use of yellow or red light is preferred when handling both standards and samples. Standards with known concentrations should be made by weighing and measuring the optical density at the absorbance maximum (λ_{max}) of an ethanol or methanol solution. When the molar extension coefficient (ε) is known, the exact concentration can be calculated from Lambert–Beers law (Table I). For further details and additional retinoids, the reader is referred to Ref. 3. The all-*trans*, 13-*cis* and 9-*cis* isomers of retinoic acid are available commercially. The availability of other standards such as 4-oxo, 4-hydroxy, or the glucuronides of retinoic acid is very often dependent on the generosity of institutions like Hoffman-LaRoche Inc. Disdier et al.[4] found an alcoholic solution of retinoic acid to isomerize or degrade to less than 50% of the

---

[1] P. Kastner, P. Leid, and P. Chambon, *in* "Vitamin A in Health and Disease" (R. Blomhoff, ed.). Dekker, New York, 1994.
[2] T. E. Gundersen, E. Lundanes, and R. Blomhoff, *J. Chromatogr. B* **691,** 43 (1997).
[3] H. C. Furr, A. B. Barua, and J. A. Olson (eds.), "The Retinoids: Biology, Chemistry, and Medicine," 2nd ed., p. 179. Raven Press, New York, 1994.
[4] B. Disdier, H. Bun, J. Catalin, and A. Durand, *J. Chromatogr. B* **683,** 143 (1996).

TABLE I
MOLAR EXTENSION COEFFICIENT (ε) AND ABSORBANCE MAXIMUM (λ_{max}), IN ETHANOL, FOR SOME SELECTED RETINOIDS[a]

Retinoid	λ_{max}	ε
all-*trans*-Retinol	325	52,770
13-*cis*-Retinol	328	48,305
11-*cis*-Retinol	319	34,890
9-*cis*-Retinol	323	42,300
all-*trans*-Retinal	383	42,880
13-*cis*-Retinal	375	35,500
11-*cis*-Retinal	380	24,935
9-*cis*-Retinal	373	36,100
all-*trans*-Retinoic acid	350	45,300
13-*cis*-Retinoic acid	354	39,750
11-*cis*-Retinoic acid	342	27,780
9-*cis*-Retinoic acid	345	36,900
9,13-Di-*cis*-retinoic acid	346	34,500
all-*trans*-Retinoic acid methyl ester	354	44,340
13-*cis*-Retinoic acid methyl ester	359	38,310
all-*trans*-Retinylpalmitate	325	49,260
all-*trans*-3,4-Didehydroretinol	350	41,320
13-*cis*-3,4-Didehydroretinol	352	39,080
9-*cis*-3,4-Didehydroretinol	348	32,460
all-*trans*-3,4-Didehydroretinoic acid	370	41,750
13-*cis*-3,4-Didehydroretinoic acid	372	38,740
9-*cis*-3,4-Didehydroretinoic acid	369	36,950
all-*trans*-4-Oxoretinoic acid	360	58,220
13-*cis*-4-Oxoretinoic acid	361	39,000
Anhydroretinol	371	97,820
all-*trans*-TMMP-retinol	325	49,800
all-*trans*-TMMP-retinoic acid	361	41,400
N-4-(Hydroxyphenyl)-all-*trans*-retinamide (in methanol)	362	47,900

[a] Data from H. C. Furr, A. B. Barua, and J. A. Olson (eds.), "The Retinoids: Biology, Chemistry, and Medicine," 2nd ed. Raven Press, New York, 1994.

initial concentration in 1 hr under natural light. After 24 hr, less than 10% was left. When the same alcoholic solution was stored under yellow light, no isomerization or degradation was observed.[4]

Wyss[5] investigated the stability of retinoids in plasma and found that all-*trans*-retinoic acid, 13-*cis*-retinoic acid, and their 4-oxo metabolites are stable for only 3 months when stored at $-20°$. This was extended to 9 months when stored at $-80°$.[5] There is no convincing support that addition of an antioxidant[6] or purging with an inert gas[7] prior to storage or extraction will affect the result. The introduction of an antioxidant may, however, interfere with the analyte of interest.

Sample Extraction

The extraction of retinoic acid from biological fluids such as serum or plasma is hampered by the fact that retinoic acid, in contrast to most other retinoids that are insoluble in water, is rather water soluble.[8] Retinoic acid will therefore not be extracted efficiently when water-immiscible solvents are used. Water-rich samples can be dried with anhydrous sodium sulfate or by lyophilization, but such procedures can be laborious and introduce additional uncertainties. Extraction with ethyl acetate or hexane after deprotonation with an alcohol and acidification with acetic or hydrochloric acid are used in most methods. Strong acids or bases should, however, be avoided due to the risk of hydrolyzing the glucuronides of retinoic acid. Because endogenous retinoic acid is present at low nanomolar concentrations in serum or plasma, several milliliters of sample is therefore often required.[3,9] For further details, the reader should refer to Refs. 3 and 9.

Treating serum or plasma with 2–3 volumes of a water-miscible solvent such as acetonitrile or methanol will precipitate the proteins, liberate retinoic acid, and generate a single phase with quantitative recoveries of retinoic acid.[4] Reducing the volume of this supernatant by evaporation with an inert gas is, however, not convenient because of the high water content. A direct injection of the supernatant is, however, also possible. The maximum volume that can be injected without affecting the peak shapes depends significantly on the column inner diameter and the water content of the supernatant in combination with the mobile phase. Substances with large capacity factors (k') are affected to a lesser degree. The volume necessary

---

[5] R. Wyss, *J. Chromatogr. B* **671**, 381 (1995).
[6] A. B. Barua, H. C. Furr, D. Janick-Buckner, and J. A. Olson, *Food Chem.* **46**, 419 (1993).
[7] N. E. Craft, E. D. Brown, and J. C. Smith, *Clin. Chem.* **34**, 44 (1988).
[8] E. Z. Szuts and F. I. Harosi, *Arch. Biochem. Biophys.* **287**, 297 (1991).
[9] C. A. Frolic and J. A. Olson, *in* "The Retinoids" (M. B. Sporn, A. B. Roberts, and D. S. Goodman, eds.), Vol. 1. Academic Press, New York, 1984.

to apply depends on the levels of analyte, along with the sensitivity of the detection technique. If direct injection is not feasible, the supernatant can be concentrated with solid-phase extraction (SPE). This technique can be performed off-line or on-line. On-line SPE has the advantage of providing full light protection during the entire analytical procedure, losses are kept to a minimum, and automation is easily accomplished. The instrumentation for on-line SPE is normally quite complex and expensive.

On-line SPE can be performed with an advanced automated sample processor (AASP) or by column switching. The AASP utilizes prepacked disposable SPE cartridges and is far more expensive compared to the use of the short HPLC columns used in column switching techniques. Such short columns can be dry packed easily with an inexpensive bulk packing material.

In our laboratory, we have used a 2.1-mm inner diameter, 10-mm long PEEK column (Jour Research, Onsala, Sweden) in a column switching system. The 5-μm PAT frit is removed and replaced when the column is repacked. An uncapped C_{18} material with a large particle size and pore diameter is used (Bondapak C_{18}, 37–53 μm, 300 Å, Waters, Milford, MA). The totally inert frit material, in combination with a large particle size, allows for hundreds of samples to be injected without any increase in backpressure. When using such an on-line SPE system, recoveries for retinoic acid isomers were in the range of 97–100%.

Therefore, optimized for high recoveries, column switching techniques are ideal for retinoic acid analysis, providing full protection from light and minimal exposure to high concentrations of organic solvents during sample cleanup and concentration. Operation, however, can be more complex and requires somewhat higher operator skill.

Column Diameter and Detection

The traditional column diameter in high-performance liquid chromatography (HPLC) is 3.9–4.6 mm, although microbore columns with 2.1 and 1.0 mm diameter have been available for years. There are several interesting aspects of micro-LC. A 1-mm I.D. column will consume 1/20 of solvent compared to a 4.6-mm column. The cost is reduced and the benefit for the environment is obvious. For a given injection volume the peak volume is proportional to the square of the column inner diameter. Thus for a 1-mm column, one can theoretically expect to obtain a 20-fold increase in sensitivity compared to a 4.6-mm column. This is, however, seldom the case as the maximum injection volume is also reduced by a factor of 20.

The reduced peak volume can give rise to peak broadening in the detector cell. Flow cells with a smaller volume may be applied, but the benefit is highly dependent on the design of the cell. Keeping the length

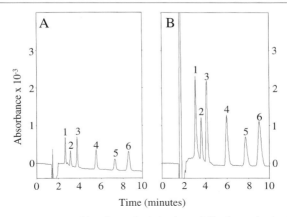

FIG. 1. Chromatograms resulting from the injection of 10 μl standard solution of 4-oxo-13-*cis*-retinoic acid (1), all-*trans*-4-oxoretinoic acid (2), all-*trans*-TMMP-retinol (IS), 13-*cis*-retinoic acid (4), 9-*cis*-retinoic acid (5), and all-*trans*-retinoic acid (6) on the column switching system, with a 250 × 4.6-mm (A) and a 250 × 2.1-mm (B) inner diameter pkb-100 column. The flow of P3/M2 consisting of acetonitrile : 1-butanol : methanol : 2% ammonium acetate : glacial acetic acid [69 : 2 : 10 : 16 : 3, (v/v)] was 1.8 (A) and 0.4 (B) ml/min.

of the light path constant must necessarily reduce the inner diameter of the cell. Increased turbulence in the cell, caused by the reduced inner diameter, can give greater variations in lost incident light, which in turn cause an increased noise. The benefit of small volume detector cells and narrow columns is more pronounced with short retention times.[10]

To realize the theoretical increase in sensitivity of micro-LC, on-line SPE with column switching can be utilized. The actual volume injected onto the analytical column (AC) is the sum of the internal volume of the SPE column and the tubing to the connected switching valve. The liquid contained here is transferred to the AC when the six-port valve is changed. This volume is constant and not dependent on the volume applied to the SPE column in the first step. An additional band focusing effect can be achieved on top of the analytical column if the composition of this liquid is of weak eluting strength compared to the mobile phase. The chromatograms in Fig. 1 illustrate the increase in peak height obtained by using a 2.1-mm compared to a 4.6-mm inner diameter analytical column.

Instrumentation, Materials, and Reagents

The instrumentation needed for the fully automated three-pump column switching system is rather complex, but, when established, this system is

[10] P. Cucera, J. Chromatogr. Library Vol. 28. Elsevier, Amsterdam, The Netherlands.

very flexible and can be applied to automated analyses of a wide range of drugs from most sample types with only a minor alteration in sample preparation. The pumps in micro-LC must provide a flow of about 10% of normal flow rates and maintain high-pressure specifications. In addition, the detector cell volume can be critical, depending on the column length and the retention of the substance. The inner diameter of the tubing must be smaller, and dead volumes must be kept to a minimum.[10] A schematic illustration of the micro-LC column switching system in the authors' laboratory is shown in Fig. 2. Of the three pumps, only P3 must meet the high specifications needed for micro-LC.

For on-line degassing of the mobile phases, a Shimadzu DGU-3A is utilized. The three HPLC pumps (P1, P2, and P3) needed are of the type Shimadzu LC-10AD or AT. The autoinjector, a Shimadzu SIL-10 A with a 2-ml titanium loop (Rheodyne, Cotati, CA), is equipped with a Peltier sample cooler. The set of valves needed for the column switching system are one PEEK Rheodyne injector 7125 and one Shimadzu FCV-12AH with a six-port automatically Rheodyne valve (V1). For recovery tests and manual HPLC analyses, a manual injection valve is placed in front of the analytical column.

The analytical column (AC) is a 250 × 2.1-mm I.D. Suplex pkb-100 from Supelco Inc. (Bellefonte, PA). The precolumn (PC), a 10 × 2.1-mm I.D. PEEK column with 5-μm PAT frits (Jour Research), is dry packed with Bondapak dimethyl-octadecylsilyl bulk packing material, 37–53 μm (Waters, Millipore Corp. Milford, MA). A 0.5-μm PEEK in-line filter (F) (Upchurch Scientific, Oak Harbor, WA) is located in front of the analytical column. A 20 × 4.6-mm C_{18} Pelliguard guard column from Supelco is placed in front of the T-piece to stop trace organic compounds from M2 to be concentrated on the PC. A restrictor is needed to balance the pressure on

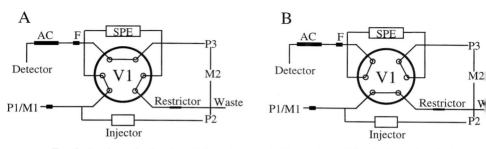

FIG. 2. A schematic drawing of the column switching system. (A) Initial position during concentration and cleanup of the sample. (B) System in the back-flush mode, resulting in transfer of the concentrate to the analytical column.

both sides of the switching valve. Similar pressure will spare the analytical column for sudden pressure changes when the valve is turned. Mobile-phase M2 is shared by P2 and P3. Initially, these two phases are identical, coming from the very same bottle. The addition of water (M1) through the T-piece (Upchurch Scientific) generates M3, having a strongly reduced elution strength, the same pH, but lower ionic strength. The close matching of the two liquid phases and the pressure on both sides of the SPE cartridge ensure a minimal solvent front when the valve is turned.

Mobile-phase M1 consists of doubly distilled deionized water and M2 of acetonitrile : 1-butanol : methanol : 2% ammonium acetate : glacial acetic acid [69 : 2 : 10 : 16 : 3 (v/v)]. The flow of M2 delivered by P3 is 0.4 ml/min. M1/P1 and M2/P2 are mixed in a ratio of 0.5 to 2.2 ml/min, resulting in a final water content of 84.5% in M3. M4 is generated when the flow of P1 is set to 0.5 ml/min and P2 to 2.2 ml/min. This mixing ratio generates a phase with 38.7% water. A 10% ammonium acetate solution is made up by dissolving 100 g in 1000 ml distilled water and filtering it through a 0.22-mm filter HA type (Millipore). This solution is diluted further to give a 2% solution. Detection is performed using a Shimadzu SPD-M10A diode array detector with a titanium flow cell and a Shimadzu SPD-10AV dual-channel UV detector connected in series. An interface and communication module (Shimadzu CBM-10) is needed for control of the system from the HPLC manager CLASS LC-10 (version 1.6). The entire system is controlled from a personal computer with the HPLC software based on MS Windows 95.

Absolute ethanol is obtained from A/S Vinmonopolet (Oslo, Norway). Acetonitrile, methanol, acetic acid, 1-butanol, and ammonium acetate (all p.a.) are from E. Merck (Darmstadt, Germany). Water is deionized and glass distilled. all-*trans*-Retinol, 13-*cis*-retinol, all-*trans*-retinoic acid, 13-*cis*-retinoic acid, and 9-*cis*-retinoic acid are supplied by Sigma (St. Louis, MO). Both 4-oxo-all-*trans*-retinoic acid and 4-oxo-13-*cis*-retinoic acid are provided by F. Hoffman-La Roche (Basel, Switzerland). Helium and argon, both of grade 4.8, are obtained from Hydro Gas A/S (Oslo, Norway). Human serum, obtained from a blood bank (Ullevål Hospital, Oslo, Norway), and all standards are kept under argon at $-20°$ and are protected from light at all times.

Analytical Procedure

A 500-μl serum sample is transferred to a 1.5-ml amber vial, diluted with 1000 μl of acetonitrile containing the internal standard, flushed with argon, sealed with an 11-mm Teflon/alumina crimp cap, mixed, centrifuged, and placed in the cooled sample rack (5°). A 1000-μl aliquot of the supernatant is injected on the precolumn while the start signal to the integrator is delayed for 5 min. M2 delivered by P2 carries the sample toward the precolumn (SPE in Fig. 2) with a flow of 0.5 ml/min. On-line dilution of M2 with M1 delivered at a flow of 2.2 ml/min by P1 is carried out through a T-piece located prior to a 1-m long 1-mm I.D. steel tubing. After 5 min, the position of the six-port switching valve V1 (Fig. 2) is changed and the start of the integration is executed. At this time, the flow of M1 is reduced to 0.5 ml/min and M2 is increased to 2.2 ml/min creating M4. Thus, the flow lines carrying the sample are purged with M4. After 1 min, when the extracted compounds have been transferred to the analytical column by M2 delivered by P3, valve V1 is switched to its original position and M4 also cleans the precolumn. The six-port switching valve is then, after another 5 min, set to the initial position again. Shortly before the next injection, the flow of pump 1 and pump 2 is reset to 0.5 and 2.2 ml/min, respectively, conditioning the flow lines and precolumn with M3. In Fig. 3, the peak areas of all-*trans*-retinoic acid from spiked human serum are plotted against the injection volume. Data demonstrate that the recovery of retinoic acid on the SPE column under the given conditions is not dependent on the injected volume.

Detection of the retinoids is carried out at 360 nm with a dual-channel UV detector. The diode array detector provides qualitative information together with peak purity information. For injections directly onto the analytical column, a manual injector (Rheodyne 9125) placed in front of the AC is used.

Chromatographic Separation

An isocratic separation of retinoic acid isomers in a reversed-phase system is most likely to be achieved on a column with a high degree of residual silanol activity giving rise to strong hydrogen bindings. The

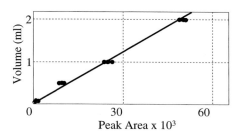

FIG. 3. Supernatants (50, 100, 500, 1000, and 2000 μl, $N = 3$) spiked with all-*trans*-retinoic acid were injected onto the SPE column. The peak areas were integrated and plotted against the injected volume.

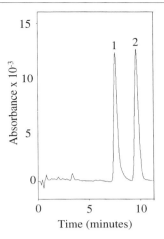

FIG. 4. Chromatogram of all-*trans*-retinoic acid (2) and the corresponding methyl ester (1) under identical conditions.

geometrical orientation between the carboxylic acid moiety and the six-carbon ring structure seems to be crucial for the separation. This is supported by the fact that the corresponding retinol isomers are not separated at all under the same conditions. The methyl ester of retinoic acid will normally give a higher retention than the free acid, but under the described conditions the methyl ester elutes several minutes earlier, as illustrated in Fig. 4, indicating that loss of the proton donor ability is the reason for the reduced retention. To obtain a similar separation on an end-capped column, gradient elution will usually be needed. An isocratic separation is often preferred due to a shorter run time, simpler instrumentation and operation, and lesser baseline fluctuations in combination with some detectors (e.g., electrochemical detection).

The degree of secondary interactions with the residual silanol groups is strongly affected by the amount of methanol in the mobile phase. Methanol will effectively break the hydrogen bonding between the acid moiety in retinoic acid and the residual silanol groups on the stationary phase. For this reason, acetonitrile and water are preferred as the main components whereas methanol is used for selectivity adjustments. The observation that acetonitrile is the solvent of choice with non end-capped columns, whereas methanol is to be used with fully end-capped columns, has also been made by others.[11] To suppress the ionization of retinoic acid, an acetate buffer

[11] R. W. Curley, D. L. Carson, and R. N. Ryzewski, *J. Chromatogr.* **370**, 188 (1986).

is used. The pK_a of retinoic acid is between 6 and 8, which is dependent on the medium. A pH of 3 is therefore sufficient to ensure a total protonization. The strong contribution from secondary interactions will, however, hamper the use of Acitretin as an internal standard. This commonly used internal standard will coelute with 13-*cis*-retinoic acid. As a substitute, the retinol analog (TMMP-retinol) is used. Figure 5 shows the chromatogram resulting from the analysis of a human serum sample. An aliquot of 1000 μl supernatant was injected corresponding to 333 μl serum.

Assay Performance

The described automated assay for retinoic acid measurement is found to be linear in the region of 100 fmol–3 nmol with a coefficient of variation (r^2) better than 0.998 for all five retinoids. As was illustrated in Fig. 3, the recovery is independent on the volume injected (50–2000 μl). The volume of supernatant applied can therefore be adjusted according to the concentration of retinoic acid. The limit of quantification is in the range of 80–200 fmol for the five retinoids. Recovery, tested at 0.2, 2, and 30 pmol injected, is in between 97–100% for all five retinoids and the internal standard. The within-day repeatability tested at the LOQ level is in the range of 2.8–3.3% ($N = 6$). The between-day reproducibility is in the range of 4.7–5.6% ($N = 6$).

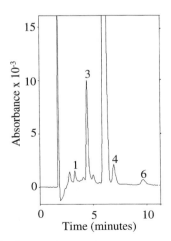

FIG. 5. Chromatogram showing endogenous concentrations of retinoic acid in a 333-μl human serum sample: (1) 13-*cis*-4-oxoretinoic acid (0.9 ng/ml), (4) 13-*cis*-retinoic acid (2.1 ng/ml), and (6) all-*trans*-retinoic acid (1.2 ng/ml). (3) TMMP-retinol is used as internal standard.

Summary

On-line solid-phase extraction coupled with micro-HPLC by column switching is an ideal technique for the analysis of retinoic acid in serum or plasma. The advantages are mainly contributed to an automated sample workup and low detection limits. On-line processing of the sample ensures minimal losses and full light protection during the entire procedure. Critical steps such as evaporation, extraction, and multiple transfers are avoided. Furthermore, the precision of highly automated methods is generally better than manual methods.

We have successfully coupled a 2.1-mm I.D. analytical column with a 2.1-mm extraction column. This setup allows for large amounts of supernatant to be injected onto precolumns for concentration and cleanup. By means of column switching, this concentrate is transferred to the microcolumn with a highly reduced volume. The reduced diameter of the analytical column and the on-line solid-phase extraction allow for the fully automated quantification of as little as 100 fmol all-*trans*-retinoic acid in human serum. The detection limits obtained with these column switching techniques can compete with LC-MS.

This new micro-HPLC method will be useful for the quantitation of endogenous retinoic acid metabolites, which are present at very low concentrations in biological material. Furthermore, more sensitive methods might also lead to the discovery of hitherto unknown retinoic acid metabolites. The combination of on-line SPE and micro-HPLC has, to our knowledge, not been used previously for retinoic acid analysis. The development of isocratic separation methods for retinoic acid isomers made this possible.

[39] Purification and Characterization of Cellular Carotenoid-Binding Protein from Mammalian Liver

By M. R. LAKSHMAN and MANJUNATH N. RAO

Introduction

Epidemiological studies have shown that an increase in the intake of fruits and vegetables rich in carotenoids leads to a decreased risk in heart disease[1] and certain types of cancer of the lung, colon, and cervix.[2] High

---

[1] D. Steinberg and workshop participants, *Circulation* **85**, 2338 (1992).
[2] R. G. Ziegler, A. F. Subar, N. E. Craft, G. Ursin, B. H. Peterson, and B. I. Graubard, *Cancer Res.* **52**, 2060S (1992).

dose of β-carotene is clinically approved as a treatment for patients with erythropoietic protoporphyria and is now being evaluated as a cancer preventive agent. Xin et al.[3] have shown that β-carotene also has properties to modulate gene expression. Working with C3H/10T1/2 cells they have shown that carotenoids upregulate gap junctional communication and correlate with inhibition of neoplastic transformation. Humans, ferrets, horses, and certain breeds of cattle are known to absorb and store β-carotene in various organs. It is now well known that absorbed β-carotene is transported by low-density lipoproteins.[4,5] However, the exact mechanism of tissue absorption of intact β-carotene, its storage, and transport into the liver is not clearly understood.

Purification and characterization of a mammalian carotenoid-binding protein have not been reported so far. However, there are a number of reports regarding the isolation and purification of carotenoproteins from other lower forms of plants and animals. In cyanobacteria, carotenoproteins that are either water soluble or detergent soluble have been reported. Water-soluble carotenoproteins have been isolated from three genera of cyanobacteria,[6] and detergent-soluble carotenoproteins have been isolated from cytoplasmic membranes of *Synechocystis* sp. PCC6714[7] and *Synechococcus* strain PCC7942.[8] The carotenoid-binding protein from *Synechococcus* sp. (*Anacystis nidulans* R2) has been cloned (*cpbA*) and sequenced.[9] Zagalsky et al.[10] have reported a carotenoprotein from lobster carapace called α-crustacyanin that binds astaxanthin. These workers purified the protein as crystals, sequenced the amino acids, and found that it is only 25% homologous to the mammalian retinol-binding protein. A lutein-binding protein with a molecular size of 36 kDa was isolated from the midgut region of the silkworm *Bombyx mori*.[11] Perhaps the difficulty involved in characterizing a carotenoprotein complex from a mammalian tissue may have been due to the inability of tissue processing and also a specific technique to isolate the protein from the tissue. This article discusses a unique and a novel method of immobilizing β-carotene to a column matrix that helped us in purifying to homogeneity a cellular carotenoid-binding protein (CCBP) from ferret liver.

[3] Li-Xin Zhang, R. V. Cooney, and J. S. Bertram, *Cancer Res.* **52,** 5707 (1992).
[4] N. I. Krinsky, D. G. Cornwell, and J. L. Oncley, *Arch. Biochem. Biophys.* **73,** 233 (1958).
[5] S. Ando and M. Hatano, *Oncorhyncus Keta J. Lipid Res.* **29,** 1264 (1988).
[6] T. K. Holt and D. W. Krogmann, *Biochim. Biophys. Acta* **637,** 408 (1981).
[7] G. S. Bullerjahn and L. A. Sherman, *J. Bacteriol.* **167,** 396 (1986).
[8] K. Masamoto, H. C. Reithman, and L. A. Sherman, *Plant Physiol.* **84,** 633 (1987).
[9] K. J. Reddy, K. Masamoto, M. M. Sherman, and L. A. Sherman, *J. Bacteriol.* **171,** 3486 (1989).
[10] P. F. Zagalsky, E. E. Eliopoulos, and J. B. C. Findlay, *Biochem. J.* **274,** 79 (1991).
[11] Z. E. Jouni and M. A. Wells, *J. Biol. Chem.* **271,** 14722 (1996).

Purification and Assay Methods

Animals and Diet

Male ferrets (*Mustela putoris furo*) (body weight ca. 600 g) are from Marshall Farms, North Rose, New York. After 1 week of quarantine, they are maintained on a high protein ferret diet (Purina, St. Louis, MO) for 1 week. This diet has 40% protein, 13% fat, and 22% carbohydrate. This diet is fortified with 2 g β-carotene in the form of beadlets [Hoffman-La Roche, Nutley, NJ: β-carotene 10% (w/w)] and 10 g taurocholate per kg, and the animals are fed *ad libitum* for a period of 4 weeks. The animals are euthanatized by aortic exsanguination under pentobarbital anesthesia (50 mg/kg, IP) and the livers are saved for CCBP isolation.

Chemicals

All chemicals, solvents, and reagents are of analytical or ultrapure grade.

Carotenoid Protein Complex Isolation

The procedure is essentially according to that described previously for rat liver.[12] Briefly, each liver is homogenized using a Polytron homogenizer (Brinkman Instruments, Westbury, NY) with 10 volumes of the homogenizing buffer [50 mM morpholinoethanesulfonic acid (MES), 1 mM EDTA, 20% glycerol, 0.2% *n*-octyl-D-glucopyranoside, 5 mM CHAPS, 0.5% Triton X-100, 50 µg/ml butylated hydroxytoluene (BHT) and 1 µg/ml of each of the following protease inhibitors: phenylmethylsulfonyl fluoride (PMSF), aprotinin, and leupeptin]. Unless otherwise specified, all procedures are carried out under F40 Gold fluorescent light at 4°. Following ultracentrifugation at 100,000g, the supernatant fraction is subjected to 0–50% ammonium sulfate (AS) fractionation. The ammonium sulfate precipitate can be stored at −80° for nearly 6 months, and the active CCBP can be purified. The AS fraction is redissolved in elution buffer (50 mM Tris–HCl and 50 mM ammonium bicarbonate buffer, pH 7.2, containing 0.5% Triton X-100 and 1 µg/ml of each of the following: PMSF, aprotinin, and leupeptin). After dialysis against the elution buffer, this fraction is incubated with β-[^{14}C]carotene (2 × 10^6 dpm) for 1 hr at 25°.

DEAE-Sephacel Column Chromatography

The dialyzed β-[^{14}C]carotene labeled AS fraction from the previous step is fractionated further on a DEAE-Sephacel column (1.5 × 25-cm

---

[12] C. Okoh, I. Mychkovsky, and M. R. Lakshman, *J. Nutr. Biochem.* **4**, 569 (1993).

bed; Sigma Chemical Co.) equilibrated with elution buffer. The column is initially washed with 2 bed volumes of the elution buffer containing 10 mM NaCl and then with 2 bed volumes of the same buffer containing 100 mM NaCl. Finally, the carotenoid–protein complex is eluted from the column as a yellow band with the elution buffer containing 350 mM NaCl at a flow rate of 0.4 ml per minute. The eluted fractions are monitored for their absorption at 280 and 465 nm in a Shimadzu UV/VIS spectrophotometer (Shimadzu UV-160, Columbia, MD). Fractions 9–19 corresponding to the peak of [^{14}C] radioactivity are pooled, concentrated by the use of a Speed-Vac Concentrator (Forma Scientific, Inc., Marietta, OH), and stored at 4° for further studies.

Sephadex G-75 Column Chromatography

The concentrated fraction containing CCBP from the previous step is subjected to gel-filtration chromatography on a Sephadex G-75 column (1 × 25 cm) equilibrated with the elution buffer. The eluted fractions are monitored for their absorption at 280 and 465 nm. Fractions 20–25 corresponding to the major peak of ^{14}C radioactivity are combined, concentrated by the use of a Speed-Vac concentrator, and stored at 4° for further studies.

Release of Apoprotein Component from Complex

The yellow-colored concentrated CCBP fraction obtained from the Sephadex G-75 column chromatography step is treated with an equal volume of cold acetone (−20°), shaken gently, and left aside at −20° for 20–30 min followed by centrifuging at −20° at 1300g for 20 min. The organic supernatant fraction is removed while the apoprotein pellet is dissolved in the original volume of TAB buffer (50 mM Tris–HCl and 50 mM ammonium bicarbonate buffer, pH 7.2). The apoprotein is reprecipitated with an equal volume of cold acetone as before and the pellet is again redissolved in the original volume of TAB buffer. Its protein content is estimated.[13]

Affinity Chromatography

Because CCBP shows high affinity for β-carotene, we decided to explore whether it was possible to make an affinity column with immobilized β-carotene as the affinity ligand. However, β-carotene is a long chain hydrocarbon with no functional groups, which makes its immobilization difficult, if not impossible. Therefore, we took a novel approach of using Pharmalink

---

[13] M. M. Bradford, *Anal. Biochem.* **72,** 248 (1976).

Immobilization Kit (Pierce, Rockford, IL), which is based on the principles of the Mannich reaction to immobilize the ligand. The Mannich reaction consists of the condensation of formaldehyde or any other aldehyde with ammonia, primary or secondary amines, and a ligand molecule having possibly an active hydrogen atom. The Pharmalink Gel included in the kit is immobilized diaminodipropylamine, which can be used as a source of the primary amine. The 5,6-ethylenic bond on the β-ionone ring, especially with the activator methyl group at C-5 position, can sufficiently activate the hydrogen atom at C-4 or C-4' to participate in this reaction as shown in Fig. 1. However, the ligand binding to the matrix can also occur at other positions as β-carotene has a number of conjugated double bonds with adjacent methyl groups to activate a hydrogen atom for participation in this reaction.

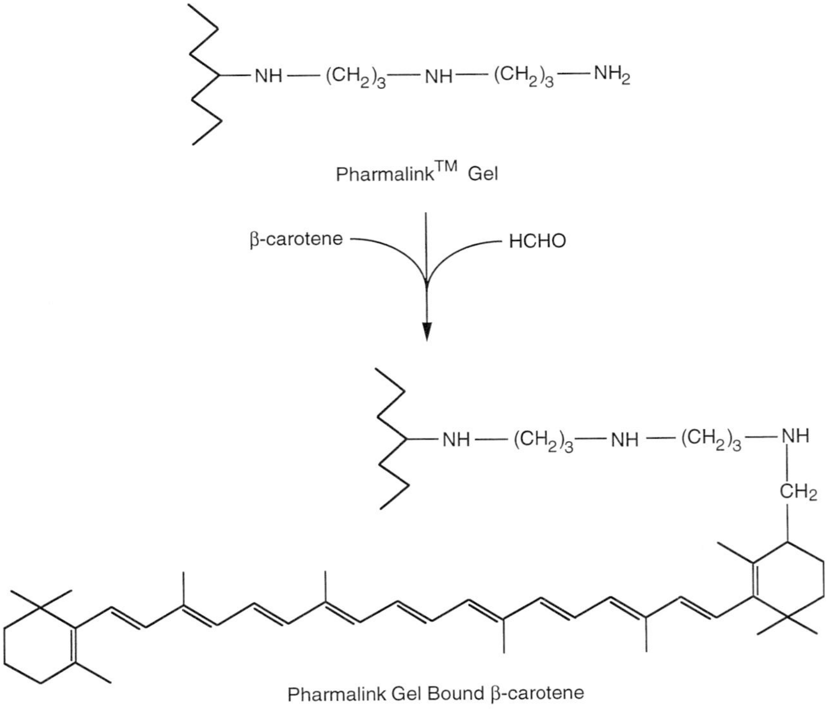

FIG. 1. Chemistry of β-carotene affinity gel. β-Carotene in the presence of formaldehyde reacts with the immobilized diaminodipropylamine through an active hydrogen at C-4 of the β-ionone ring, thus becoming immobilized onto the agarose matrix. See text for further explanation.

Preparation of Pharmalink-Immobilized β-Carotene

All operations are carried out in the dark or, where necessary, under F40 gold fluorescent light to minimize the oxidation of β-carotene. The following procedure is essentially according to manufacturer's specifications using their kit, which contains all the coupling reagents. Briefly, the storage solution in the Pharmalink column (2 ml prepacked column) is drained completely, and the gel is equilibrated with the coupling buffer [2 ml buffer plus 2 ml dimethyl sulfoxide (DMSO)]. The coupling buffer is also drained as before. Then, 10 mg of purified β-carotene dissolved in 2 ml DMSO containing BHT (50 μg/ml) is mixed with the Pharmalink gel followed by the addition of Pharmalink coupling reagent in the reaction bottle initially at 37° for 1 hr followed by incubation at 4° for 24 hr with gentle mixing end over end. The column matrix is washed thoroughly with 30 ml of the TAB buffer containing 50% ethanol until no more unbound β-carotene is eluted from the column (evidenced by absorption spectrum). Ninety percent of the added β-carotene is bound to the matrix. The ligand binding to the affinity matrix increases to 100% when only 0.2 mg of β-carotene is used for binding. The washed column with immobilized β-carotene is ready for affinity chromatography after equilibration with 5 volumes of TAB buffer. Figure 1 shows the proposed structure of immobilized β-carotene using the Pharmalink gel, although other sites of attachment of the chromophore may be possible, as indicated earlier.

Bound β-carotene leaches out easily during the washes if there is even a slight contamination of Triton X-100 in the eluting buffers. Thus all procedures have to be performed in the absence of Triton X-100. Apart from the solvents stated earlier, we tested other solvents such as acetone and tetrahydrofuran (THF) to solubilize β-carotene and use it for binding. Although the solubility of β-carotene in THF is severalfold higher than in DMSO, its binding is very poor using THF as the solvent. The column is generally stable for 1–2 weeks when stored at 4° in the dark. However, it cannot be stored indefinitely because of the labile nature of β-carotene. In our hands, we also found TAB buffer to be the best buffer system to carry out these chromatographic procedures.

Affinity Chromatographic Purification of CCBP

The CCBP apoprotein fraction, isolated by cold acetone treatment of the complex fraction from Sephadex G-75 column chromatography step (~3 mg protein), is subjected to affinity chromatography on the Pharmalink Immobilized β-carotene (PIC) column prepared earlier. After applying the protein, the column is washed initially with 20 ml of the TAB buffer to remove the unbound protein (evidenced by the decrease in absorbance at

280 nm). The bound protein is then eluted from the affinity column with 20 ml of the TAB buffer containing 250 mM NaCl and collected as 1-ml fractions. Aliquots of fractions showing protein peaks are tested for their homogeneity by sodium dodecyl sulfate–polyacrylamide gel electrophoresis (SDS–PAGE) on a 10% polyacrylamide gel (Bio-Rad, Richmond, CA). Fractions showing a single homogeneous band are pooled, concentrated in a Speed-Vac concentrator, dialyzed against TAB buffer, and stored at 4°. This purified protein is tested for its binding affinity to various carotenoid ligands, as described later.

The purified CCBP appears to be quite a stable protein. The protein exhibits its binding properties to β-carotene even after 4 weeks of storage at 4°. Storage at $-20°$ in TAB buffer for a month also does not appear to alter the binding properties significantly. The binding capacity tends to increase slightly (10–20%) at both pH 6.0 and pH 8.0. However, drastic changes in the pH of the binding buffer (below pH 6.0 and above pH 8.0) tend to lower or even abolish the binding capacity of the protein. Binding capacity is also affected by the salt concentration in the buffer. High salt concentrations such as 300–500 mM NaCl surprisingly appear to reduce the binding ability of the protein by as much as 25%, although it is well known that high salt concentrations favor hydrophobic binding reactions.

Binding Assay

The standard binding assay (unless specified otherwise) is as follows: 30 μg of purified CCBP or a nonspecific protein such as bovine serum albumin (BSA) or the 43-kDa protein from ferret liver in 950 μl TAB buffer is incubated with 8 nmol of β-carotene in 50 μl acetone at 37° for 60 min. This is followed by thorough extraction of each reaction mixture with 8 ml light petroleum five times to remove the unbound carotenoid. The absorption spectrum of each resulting aqueous reaction mixture is determined. In addition to β-carotene, the binding of the following ligands to CCBP is tested under standard binding conditions: β-carotene, cryptoxanthin, zeaxanthin, lycopene, astaxanthin, and retinol. We found that solvents other than acetone such as DMSO or THF or even ethanol do not yield reliable binding data in the previous binding assay. This is probably due to the differences in the dielectric constants of these solvents, which may alter the protein structure, even though the protein may not be denatured. We also found that the inclusion of Triton X-100 in the binding reaction also leads to erroneous results.

Competitive Binding-Assay of CCBP with Alternate Ligands

The competitive binding assay is very similar to the standard binding assay described earlier except for the following details: (i) β-[^{14}C]carotene (specific activity 168,000 dpm/nmol) is used instead of unlabeled β-carotene. (ii) The incubation with labeled β-carotene is carried out both in the absence and in the presence of 20-fold excess of the following unlabeled ligands: α-carotene, β-carotene, cryptoxanthin, astaxanthin, lycopene, and retinol. At the end of the incubation period, 100 μl of the reaction mixture is loaded onto a Sephadex G-25 column matrix (biospin disposable column with an Eppendorf collection tube: 0.65 × 3-cm packed dimensions, Bio-Rad) preequilibrated with TAB buffer. The column is centrifuged at 4° at 1100 g for 5 min, and the ^{14}C radioactivity in the eluted fraction is determined using a Beckman LS-6500 liquid scintillation spectrometer, which shows a ^{14}C counting efficiency of 95%. This procedure results in the quantitative recovery of CCBP bound β-[^{14}C]carotene in the Eppendorf collection tube, whereas the unbound β-[^{14}C]carotene is completely trapped on the column. Control experiments with (i) labeled β-carotene only and (ii) labeled β-carotene incubated with a nonspecific protein (BSA) show negligible recovery of the label in the eluate.

Precautions

The use of biospin column chromatography for this assay is very helpful in obtaining reliable and fast results. Use of Sephadex gel above the G-50 grade does not give reliable results, as the gel tends to collapse during centrifugation. Also, the volume of sample that can be applied to the column cannot exceed 100 μl. Here again use of Triton X-100 or any other detergent should be totally avoided.

Binding Assay for Scatchard Analysis

This method is identical to the competitive binding assay described earlier except that the CCBP (62 nM) is incubated with increasing concentrations of β-[^{14}C]carotene (62.5–3200 nM) both in the presence and in the absence of 20-fold excess of unlabeled β-carotene and the amount of β-[^{14}C]carotene bound is determined after the spin column procedure. The specific binding (total minus nonspecific) is subjected to Scatchard analysis using the LIGAND computer program.

Gel Electrophoresis

All fractions are tested by 10% SDS–PAGE or 6% native-PAGE essentially as described by Lemmli and Favre[14] and are stained with Silver Stain Plus (Bio-Rad) or Coomassie blue stain.

[14] U. K. Lemmli and Favre, M. *J. Mol. Biol.* **80,** 575 (1973).

Labeling of Apo-CCBP with ^{125}I

Purified apo-CCBP (50 µg) is labeled with ^{125}I using the Bolton–Hunter reagent essentially as described.[15] After extensive dialysis the specific activity of the ^{125}I-labeled apo-CCBP is 2×10^5 cpm/g of which 95% was trichloroacetic acid (15%, w/v) precipitable.

Major Findings

A 67-kDa protein has been purified to homogeneity from ferret liver that showed a high degree of specificity to β-carotene. The purification steps involved ion-exchange, gel-filtration, and affinity chromatography, which are described next.

DEAE-Sephacel Chromatography

Ion-exchange chromatography of the crude AS fraction of the 100,000 g fraction of ferret liver homogenate yielded a yellow fraction that was eluted with the elution buffer containing 0.35 M NaCl showing the characteristic β-carotene absorption spectrum (data not shown). It was also found that 78% of the ^{14}C radioactivity applied to the column was associated with this protein complex. SDS–PAGE on 10% gel of this fraction revealed the existence of four major bands and several minor bands when stained with Coomassie blue. The peak fraction from the DEAE-Sephacel column had an absorbance of 0.156 at 280 nm and 0.284 at 465 nm (A_{465}/A_{280} ratio of 1.82).

Sephadex G-75 Chromatography

The peak labeled β-carotene complex fraction isolated from the DEAE-Sephacel chromatography step was subjected to Sephadex G-75 gel filtration chromatography. Several early fractions were eluted exhibiting minor radioactive peaks, but none of them had β-carotene spectrum (data not shown). However, fractions 20–25 showed a major radioactive peak along with a characteristic β-carotene spectrum. Fractions 20–25 accounted for 74% of the ^{14}C radioactivity applied to the column. The peak fractions 21 and 22 had ^{14}C/protein ratios of 6×10^5 and 6.1×10^5 dpm/280 nm absorbance unit, respectively. The peak fraction had an absorbance of 0.24 at 280 nm and 0.46 at 465 nm (A_{465}/A_{280} ratio 1.92). SDS–PAGE of this fraction showed a major band of 67 kDa and several minor bands of 50 kDa (data not shown). It was clear that the β-carotene-binding protein was

---

[15] A. E. Bolton, W. M. Hunter, *Biochem. J.* **133**, 529 (1973).

still not homogeneous. A 43-kDa protein band was extracted with TAB buffer and saved for binding assay as a nonspecific ferret liver protein. To purify further, the apoprotein fraction was isolated from pooled fractions 20–25 by removing the chromophore with cold acetone as described in the assay methods section.

Affinity Chromatography on Immobilized β-Carotene Column

The apoprotein fraction, isolated from the complex after Sephadex-G75 chromatography, when subjected to affinity chromatography on the PIC column (Fig. 1), yielded a fraction that was eluted with TAB buffer (no detergent) containing 250 mM NaCl. SDS–PAGE of an aliquot of this fraction on 10% gel showed a single homogeneous band with a molecular mass of 67 kDa (Fig. 2). Significantly, no detergent-containing buffer was used to elute the apoprotein from the affinity column. Thus the apoprotein is totally water soluble. This affinity-purified protein fraction was used for all subsequent analyses. The elution profile of bovine serum albumin (BSA), a nonspecific protein, was tested to demonstrate the specificity of this affinity column. It was found that BSA was eluted completely in the void volume when chromatographed on the affinity column under identical conditions. This experiment proves beyond doubt the specificity of the affinity column to a specific carotenoid-binding protein.

Table I shows the summary of purification of CCBP. It must be pointed out that because the crude liver homogenate failed to show any high affinity binding with labeled β-carotene, the true fold purification of the homogeneous CCBP should be much higher than that reported. Thus, taking the crude AS fraction to have a relative binding of 1 arbitrary unit, CCBP was purified 15-fold at the DEAE-Sephacel chromatography step, 30-fold at

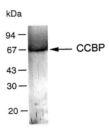

FIG. 2. SDS–PAGE of CCBP purified from ferret liver. Twenty micrograms of CCBP taken after affinity column purification, in 20 μl of TAB buffer, was mixed with glycerol/BCP buffer such that final concentration was 10% glycerol. It was then loaded onto a 10% SDS–polyacrylamide gel and electrophoresed as described in the text. Later, the gel was stained using the Silver Stain Plus kit from Bio-Rad to visualize the protein band. CCBP protein moved parallel to the 67-kDa marker.

TABLE I
PURIFICATION OF CCBP[a]

Fraction	Protein (μg)	β-carotene (bound pmol)	Specific binding (pmoles/μg protein)	Purification (-fold)
AS fraction	3500	77.8	0.022	1
DEAE-Sephacel	500	166.8	0.33	15
Sephadex G-75	245	162.3	0.66	30
Affinity	50	667	13.35	607

[a] The binding assay was carried out with the indicated amounts of each fraction using 8 nmol β-[^{14}C]carotene (168,000 dpm/nmol) in the absence (total binding) and presence of 20-fold excess nonradioactive β-carotene (nonspecific binding) under standard conditions, and the amount specifically bound (total minus nonspecific binding) is expressed in pmoles β-carotene. The fold purification was calculated based on the specific binding of each fraction. Each value is the average of three independent determinations.

the Sephadex G-75 chromatography step, and finally 607-fold after the affinity chromatography step. The final yield of the purified protein was approximately 500 μg from 5 g of the liver.

The purified protein showed a high affinity binding with β-carotene with its characteristic absorption spectrum (Fig. 3) consisting of a shoulder peak at 460 nm and two prominent peaks at 482 nm and 516 nm apart from a protein peak at 280 nm. There was a 32-nm bathochromic shift of

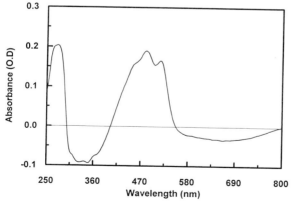

FIG. 3. Absorption spectrum of the β-carotene bound to CCBP. Purified CCBP was incubated with 8 nmol of β-carotene under standard conditions and subjected to gel filtration on a Sephadex G-25 spin column as described in the text. The absorption spectrum of the β-carotene bound complex was monitored in a Shimadzu UV-VIS spectrophotometer between 250 and 800 nm. Note the protein absorption peak at 280 nm and a shift in the λ_{max} of the complex to 482 nm with the appearance of a third peak at 516 nm.

its λ_{max} compared to that of β-carotene in light petroleum. The bound chromophore could be extracted from the CCBP complex with light petroleum only after it was treated with an equal volume of acetone. In contrast, the nonspecific proteins, BSA and the ferret liver 43-kDa protein, showed low-affinity binding to β-carotene, as evidenced by the lack of the characteristic β-carotene absorption spectrum in the standard binding assay. This was because the weakly bound β-carotene was extracted completely with light petroleum, even without denaturation with acetone.

Figure 4 shows the Scatchard plot of the specific binding of β-carotene to CCBP as a function of increasing concentration of β-carotene. The nonspecific binding ranged from 7 to 13% of the total binding at the ligand concentrations tested. The analysis of the specific binding data by the LIGAND program showed two classes of binding sites with an apparent

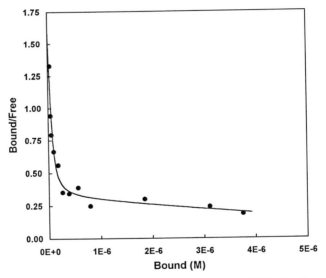

FIG. 4. Scatchard analysis of specific binding of β-carotene CCBP. Duplicate samples of CCBP (62 nM) were incubated with the indicated increasing concentration of β-[^{14}C]carotene (62.5–3200 nM) in TAB buffer at 37° for 90 min, both in the presence and in the absence of 20-fold excess of unlabeled β-carotene. At the end of 90 min, each reaction mixture was subjected to spun column purification of holo-CCBP using Sephadex G-25 equilibrated in TAB buffer. The filtrate containing the bound β-[^{14}C]carotene was analyzed for radioactivity in a Beckman LS-6500 scintillation spectrometer. Analysis of specific binding data by the LIGAND program showed two classes of binding sites with an apparent K_d of 56 × 10^{-9} M for the high-affinity site and an apparent K_d of 32 × 10^{-6} M for the low-affinity site. The nonspecific binding ranged from 7 to 13% of the total binding at the ligand concentrations tested.

K_d of 56×10^{-9} M for the high-affinity site and an apparent K_d of 32×10^{-6} M for the low-affinity site. The B_{max} for β-carotene binding to the high-affinity site was 1.16 mol/mol. It can be seen in Table I that when the binding assay was carried out on a large scale, 13.35 pmol of β-carotene was specifically bound per microgram of purified CCBP. This amounts to 0.89 mol of β-carotene bound per mole of CCBP, a value comparable to that obtained from the Scatchard analysis. Thus, it is reasonable to conclude that CCBP binds β-carotene mole per mole at the high-affinity site. In contrast, the calculated B_{max} of 145 mol/mol by the LIGAND program for the low-affinity site may not have physiological relevance because of its very high K_d of 32 μM.

The purified ^{125}I-labeled apo-CCBP was complexed with β-carotene under standard conditions, and both the holo- and the apo-CCBP were subjected to native PAGE on a 6% polyacrylamide gel followed by autoradiography. Figure 5 shows that both holo-CCBP (lane 1) and apo-CCBP (lane 2) moved as homogeneous bands, although apo-CCBP moved faster than the holo-CCBP. Because the mobility of proteins in native PAGE is

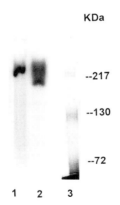

FIG. 5. Autoradiogram of native PAGE profile of purified apo-CCBP and holo-CCBP. Purified ^{125}I-labeled apo-CCBP (10 μg; specific activity 2×10^5 cpm/μg) in 100 μl of TAB buffer was incubated with 2.7 nmol of β-carotene for 1 hr at 37° and was subjected to gel filtration on a Sephadex G-25 spin column. ^{125}I-Labeled holo-CCBP was isolated in the eluate. Both holo- and apo-CCBP were mixed with glycerol/BCP-loading dye and electrophoresed on a 6% native polyacrylamide gel as described in the text. Later, the gel was dried and exposed to autoradiographic film for 3 hr. A single slow-moving radioactive band can be seen clearly in lane 1 for holo-CCBP, whereas a faster moving band can be seen in lane 2 for apo-CCBP. However, no molecular size can be assigned to these bands based on their mobility on native PAGE. Bio-Rad kaleidoscope prestained molecular weight markers (Bio-Rad Laboratories, CA) consisting of myosin (217 kDa), β-galactosidase (130 kDa), BSA (72 kDa), carbonate dehydratase (42 kDa), soybean trypsin inhibitor (31 kDa), and lyzozyme (18 kDa) were run on lane 3. The 217-, 130-, and 72-kDa markers can be seen as diffused bands.

TABLE II
COMPETITION BY VARIOUS LIGANDS FOR CCBP
BINDING TO β-[^{14}C]Carotene[a]

Ligand	Inhibition (%)
β-Carotene	100
α-Carotene	94
Cryptoxanthin	84
Astaxanthin	5
Lycopene	+13
Retinol	0

[a] β-[^{14}C]Carotene binding to purified CCBP was determined in the absence and presence of 20-fold excess of the indicated ligands as described in the text, and the results are expressed as percentage inhibition (or percentage activation marked by the plus sign), taking the inhibition by 20-fold excess of β-carotene to be 100%.

by their net charge, native PAGE is not a reliable method to assess the molecular size of any protein.[16] Significantly, ^{125}I-labeled apo-CCBP moved as a single sharp band with a molecular size of 67 kDa on a 10% SDS–PAGE gel (data not shown), thus confirming our finding of the mobility of unlabeled apo-CCBP (Fig. 2).

Among the alternate ligands tested for binding with CCBP, only α-carotene and cryptoxanthin showed any binding as evidenced by their corresponding absorption spectra (data not shown). In contrast, β-ionone ring-substituted carotenoids such as zeaxanthin or astaxanthin or a carotenoid without an intact β-ionone ring such as lycopene or a shortened molecule with one intact β-ionone ring such as retinol showed no binding as evidenced by the lack of their characteristic absorption spectra.

Competitive binding of β-[^{14}C]carotene by 20-fold excess of each alternative ligand was determined and the results are shown in Table II. Each value is the average of triplicate experiments. It is obvious that whereas α-carotene and cryptoxanthin inhibited labeled β-carotene binding by 94 and 84%, respectively, none of the other ligands tested showed any competition. Thus, CCBP showed a high degree of specificity toward carotenoids with at least one unsubstituted β-ionone ring but not toward other carotenoids or retinol.

[16] T. B. Nielsen and J. A. Reynolds, *Methods Enzymol.* **48**, 3 (1978).

Other Carotenoproteins

It is significant to point out that in contrast to a molecular size of 67 kDa for the mammalian CCBP found in this study, carotenoproteins from various lower organisms vary widely between 18 and 350 kDa in their sizes. Thus, α-crustacyanin, an astaxanthin-binding protein from carapace of the lobster, is a 350-kDa protein,[17] whereas a lutein-binding protein isolated from the midgut of silkworm B. mori is a 36 kDa protein.[11] Bacterial and plant carotenoproteins are 35- and 18-kDa proteins, respectively.[7,18] However, mammalian retinoid-binding proteins are in the 15-kDa range.[19,20]

Amino acid sequences for some of the retinoids and carotenoproteins have been reported in the literature. However, except for the fact that they all belong to the family of lipid-binding proteins, there is no other similarity. Each lipid-binding protein appears to be an independent protein. The CCBP also appears to be a unique protein with a large molecular size. Keen et al.[17] have shown that there is only a 25% similarity in the amino acid sequence between retinol-binding protein and crustacyanin. No consensus binding sequences between these two proteins have been found or reported. They also report that there is the formation of a cavity in the three-dimensional structure of crustacyanin into which the lipid sits in and forms a hydrogen bond with a threonine or tyrosine in the calyx of the protein. Work is currently underway to deduce the amino acid sequence of CCBP from its cDNA sequence. We speculate that a similar interaction occurs between the CCBP and β-carotene.

The CCBP from the mammalian source in the present study exhibits three peaks at 460, 482, and 516 nm (Fig. 3) with a 32-nm bathochromic shift in its λ_{max} compared to the absorption spectrum of β-carotene in light petroleum. Interestingly, carotenoid-binding proteins from *Mangifera indica*[21] and from cyanobacterium[7] also had a similar absorption spectra with λ_{max} at 498 and 476 nm, respectively. However, crustacyanin, the carotenoprotein from lobster carapace, showed a 160-nm bathochromic shift in its λ_{max} compared to the absorption spectrum of the parent carotenoid, astaxanthin.[10]

---

[17] J. N. Keen, I. Caceres, E. E. Eliopoulos, P. F. Zagalsky and J. B. C. Findlay, *Eur. J. Biochem.* **202**, 31 (1991).
[18] J. R. Zhou, E. T. Gugger, and J. W. Erdman, Jr., *J. Agr. Food. Chem.* **42**, 2386 (1994).
[19] D. E. Ong, J. E. Crow, F. Chytil, *J. Biol. Chem.* **257**, 13385 (1982).
[20] F. Chytil and D. E. Ong, in "The Retinoids" (M. B. Sporn, A. B. Roberts, and D. S. Goodman, eds.), p. 90. Academic Press, New York, 1984.
[21] C. Subbarayan and H. R. Cama, *Indian. J. Biochem.* **3**, 225 (1966).

Possible Physiological Roles of CCBP

The physiological role(s) of CCBP remains to be defined. In view of the possible protective roles of carotenoids against cancer, heart disease, and erythropoietic porphyria, a potential role for a specific binding protein may become central in the mechanism of action of carotenoids. Thus, CCBP may play a major role in the storage, transport, and targeting of β-carotene in mammalian systems. It may also act as the natural substrate for many of the metabolic reactions of β-carotene. By virtue of forming a stable high-affinity complex, it may protect the carotenoid from degradation. As a result, carotenoids bound to CCBP may be better antioxidants compared to free carotenoids and thus protect the system from oxidative damage.

Summary

A cellular carotenoid-binding protein was purified to homogeneity from β-carotene-fed ferret liver utilizing the following steps: ammonium sulfate precipitation, ion exchange, gel filtration, and affinity chromatography. The final purification was 607-fold. β-[^{14}C]Carotene copurified with the binding protein throughout the purification procedures. SDS–PAGE of the purified protein showed a single band with an apparent molecular mass of 67 kDa. Scatchard analysis of the specific binding of the purified protein to β-carotene showed two classes of binding sites; a high-affinity site with an apparent K_d of 56×10^{-9} M and a low-affinity site with a K_d of 32×10^{-6} M. The B_{max} for β-carotene binding to the high-affinity site was 1 mol/mol whereas that for the low-affinity site was 145 mol/mol. The absorption spectrum of the complex showed a 32-nm bathochromic shift in λ_{max} with minor peaks at 460 and 516 nm. Except for α-carotene and cryptoxanthin, none of the model carotenoids or retinol competed with β-carotene binding to the protein. Thus, a specific carotenoid-binding protein of 67 kDa size has been characterized in mammalian liver with a high degree of specificity for binding only carotenoids with at least one unsubstituted β-ionone ring.

Acknowledgments

This work was supported by NCI Grant CA39999. We gratefully acknowledge the generous gift of β-carotene beadlets and β-[^{14}C]carotene from Hemmige N. Bhagavan, Hoffmann-La Roche, Nutley, NJ. A part of this work was presented at a minisymposium on "Carotenoids" held during the FASEB annual meeting in April 1996 in Washington, D.C.

[40] Analysis of Zeaxanthin Distribution within Individual Human Retinas

By John T. Landrum, Richard A. Bone, Linda L. Moore, and Christina M. Gomez

Introduction

The macular pigment is highly organized both compositionally and spatially within the retina of the human eye. Studies have shown that the retina contains the two isomeric xanthophylls, lutein and zeaxanthin, with the greatest concentration at the center of the macula and diminishing with eccentricity.[1–4] Within the central macula, zeaxanthin is the dominant component, reaching proportions as great as 75% of the total, whereas in the peripheral retina, lutein predominates, usually being 67% or greater.[2] Figure 1 illustrates the relative proportion of lutein versus eccentricity from the fovea and variation in the macular pigment concentration across the same region. The concentration of the macular pigment is on the order of 1 mM within the central 10° (3 mm) of the retina. Such a high concentration of carotenoid within a tissue is exceptional. Typical carotenoid concentrations in other human tissues are of the order of 2 μM, three orders of magnitude less that that within the central macula.[5] Data show the macular pigment increases when dietary supplements increase serum xanthophyll levels.[6]

The retina is aerobic and highly active metabolically while simultaneously being illuminated by bright visible light.[7] The function for the macular pigment is not proven. It has been speculated that it may serve to protect the retina from photooxidation by attenuating the intensity of blue

---

[1] R. A. Bone and J. T. Landrum, *Methods Enzymol.* **213**, 360 (1992).

[2] R. A. Bone, J. T. Landrum, L. Fernandez, and S. L. Tarsis, *Invest. Ophthalmol. Vis. Sci.* **29**, 843 (1988).

[3] R. Bone, J. T. Landrum, L. M. Friedes, C. M. Gomez, M. D. Kilburn, E. Menendez, I. Vidal, and W. Wang, *Exp. Eye Res.* **64**, 211 (1997).

[4] G. J. Handelman, E. A. Dratz, C. C. Reay, and F. J. G. M. Van Kujik, *Invest. Ophthalmol. Vis. Sci.* **29**, 850 (1988).

[5] H. H. Schmitz, C. L. Poor, E. T. Gugger and J. W. Erdman, Jr., *Methods Enzymol.* **214**, 102.

[6] J. T. Landrum, R. A. Bone, H. Joa, M. D. Kilburn, L. L. Moore, and K. E. Sprague, *Exp. Eye Res.* **65**, 57 (1997).

[7] J. Ahmed, R. D. Brown, R. Dunn, Jr., *Invest. Ophthalmol. Vis. Sci.* **34**, 516 (1993).

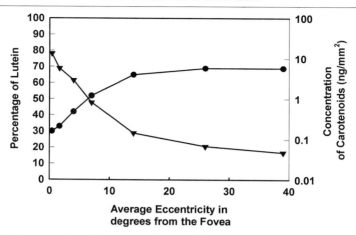

Fig. 1. Concentration of the macular pigment (▼) decreases through two orders of magnitude with eccentricity away from the inner macula. The proportion of lutein (●) is seen to be maximal (67%) in the peripheral retina, declining in percentage relative to zeaxanthin and reaching a minimum (31%) in the inner macula.

light reaching posterior structures.[8,9] Illuminated, aerobic tissues in plants that are rich in carotenoids are protected from the fatal effects of singlet oxygen.[10] In humans, β-carotene and canthaxanthin have been proven to be protective against singlet oxygen generated in the skin of individuals suffering from erythropoietic porphyria.[11] Singlet oxygen, $^1O_2$, and excited state triplets, ^3S, capable of generating $^1O_2$ are quenched readily by carotenoids.[12] Significant data have accumulated showing that photic damage by blue light is a significant problem in the mammalian retina.[13–16] Direct evidence that the macular pigment protects the human retina from photic

[8] D. M. Snodderly, *Am. J. Clin. Nutr.* **62**(Suppl.), 1448S (1995).
[9] W. Schalch, in "Free Radicals and Aging" (E. Emerit and B. Chance, eds.), p. 280. Birkhauser Verlag, Basel, Switzerland, 1992.
[10] N. I. Krinsky, *Photophysiology* **3**, 123 (1968).
[11] M. M. Mathews-Roth, *Methods Enzymol.* **213**, 479 (1992).
[12] P. Di Mascio, A. R. Sundquist, T. P. A. Devasagaya, and H. Sies, *Methods Enzymol.* **213**, 429 (1992).
[13] W. T. Ham, Jr. and W. A. Mueller, in "Laser Application in Medicine and Biology" (M. L. Wolbarsht, ed.), p. 191, Plenum Press, New York, 1989.
[14] W. T. Ham, Jr., H. A. Mueller, J. J. Ruffolo, J. E. Millen, S. F. Cleary, R. K. Guerry, and D. Guerry, *Curr. Eye Res.* **3**, 165 (1984).
[15] J. D. Gottsch, S. Pou, L. A. Bynoa, and G. M. Rosen, *Invest. Ophthalmol. Vis. Sci.* **31**, 1674 (1990).
[16] T. G. M. F. Gorgels and D. van Norren, *Invest. Ophthalmol. Vis. Sci.* **36**, 851 (1995).

injury exists in the observations of Haegerstrom-Portnoy,[17] Jaffe and Wood,[18] and Weiter et al.[19] showing that photic damage to the retina is diminished in the pigmented macular region relative to the less pigmented periphery. Additionally, circumstantial evidence points to a correlation between macular pigment density and a reduction in the risk for age-related macular degeneration (AMD).[20,21]

A photoprotective function for lutein and zeaxanthin in the retina could involve one or more different mechanisms. A passive photoprotection mechanism functions via absorption of blue light in the 400- to 490-nm absorption window of the carotenoids. Typically, about 60% of the blue light, $\lambda = 450$ nm, reaching the inner retinal layers is absorbed by the macular pigment. An active photoprotection mechanism involving the macular xanthophylls would require quenching of triplet sensitizers preventing generation of singlet oxygen and/or direct deactivation of singlet oxygen. Some debate exists about the significance of this second function,[22] particularly with respect to AMD, which manifests itself in physical changes observed primarily in the outer retinal layers, including Bruch's membrane, the retinal pigment epithelium, and the outer segments.[23] This argument is credible in that the greatest concentration of macular pigment is physically separated from these structures and the region of highest aerobic metabolism by a distance of nearly 50 μm.[7,8,24,25] Clearly, given the lifetimes of excited state triplets (ca. 10^{-9} sec) and singlet oxygen (10^{-3}–10^{-6} sec),[26,27] diffusion of these species is limited to much smaller distances. While singlet oxygen has been shown to diffuse distances of as great as 500 Å,[26] this is only ~0.1% the distance from the outer segments to the receptor axons where the highest concentration of carotenoid is observed. Nevertheless,

---

[17] G. Haegerstrom-Portnoy, *J. Opt. Soc. Am. A* **5**, 2140, (1988).
[18] G. J. Jaffe and I. S. Wood, *Arch. Ophthalmol.* **106**, 445 (1988).
[19] J. J. Weiter, F. C. Delori, and C. K. Dorey, *Am. J. Ophthalmol.* **106**, 286 (1988).
[20] J. M. Seddon, U. A. Ajani, R. D. Sperduto, R. Hiller, N. Blair, T. C. Burton, M. D. Farber, E. S. Gragoudas, J. Haller, D. T. Hiller, L. A. Yannuzzi, and W. Willet, *J. Am. Med. Assoc.* **272**, 1413 (1994).
[21] Eye Disease Case-Control Study Group, *Arch. Ophthalmol.* **110**, 104 (1993).
[22] D. V. Crabtree and A. J. Adler, *Med. Hypoth.* **48**, 183 (1997).
[23] A. C. Bird, in "Age Related Macular Degeneration: Principles and Practice" (G. R. Hampton and P. T. Nelsen, eds.), p. 63. Raven Press, New York, 1992.
[24] D. M. Snodderly, P. K. Brown, F. C. Delori, and J. D. Auran, *Invest. Ophthalmol. Vis. Sci.* **25**, 660 (1984).
[25] D. M. Snodderly, J. D. Auran, and F. C. Delori, *Invest. Ophthalmol. Vis. Sci.* **25**, 674 (1984).
[26] D. Bellus, in "Singlet Oxygen" (R. Ranby and J. F. Rabek, eds.), p. 61. Wiley, New York, 1978.
[27] K. Gollnick, in "Singlet Oxygen" (R. Ranby and J. F. Rabek, eds.), p. 111. Wiley, New York, 1978.

Foote et al.[28] have shown that β-carotene is effective at quenching singlet oxygen at concentrations lower than 1×10^{-4} M. Di Mascio et al.[29] have shown that lutein and zeaxanthin have second order rate constants for reaction with singlet oxygen that are comparable to that of β-carotene. β-Carotene and lutein were found to reach their maximal quenching efficiency in solution at around 5×10^{-5} M. While it remains unproven, a $\sim 10^{-4}$ M xanthophyll concentration possibly exists within the outer retinal structures of the central retina. Indeed, spectrophotometric evidence supports the presence of carotenoids within the outer retinal structures in macaques,[8] and van Kujik et al.[30] have reported detection of lutein and zeaxanthin in rod outer segments.

An antioxidant function for the macular pigment would be expected to result in oxidation of lutein and/or zeaxanthin as a side reaction. Bleaching of carotenoids is observed to occur for in vitro systems.[31] Khachik et al.[32,34] have identified oxidative metabolites of lutein and zeaxanthin in both human serum and human retinas. These include the monoketo hydroxy-carotenoids, 3-hydroxy-β,ε-carotene-3'-one and 3-hydroxy-ε,ε-carotene-3'-one, and the diketocarotenoid, ε,ε-carotene-3,3'-dione. Additionally, the nondietary dihydroxycarotenoids epilutein ($3R, 3'S, 6'R$)-3,3'-dihydroxy-β,ε-carotene and ε,ε-carotene-3,3'-diol are found in human retina and serum.[32,33] These carotenoids originate metabolically through a sequence of oxidation–reduction steps involving the monoketo and diketo carotenoids.[35,36] Significant quantities of the 9Z- and 13Z-lutein and the 9Z- and 13Z-zeaxanthin are also observed.[32,33] The presence of these Z isomers supports the argument that in vivo oxidation of the carotenoid to the cation radical may be occurring.[37] The cation radical has a low barrier to rotation about the double bonds of the polyene chain, and reduction back to the carotenoid accounts for the build-up of the Z isomers.[37] The presence of these metabo-

[28] C. S. Foote, R. W. Denny, L. Weaver, Y. Chang, and J. Peters, *Ann. N. Y. Acad. Sci.* **171,** 139.
[29] P. Di Mascio, S. Kaiser, and H. Sies, *Arch. Biochem. Biophys.* **274,** 532 (1989).
[30] F. J. G. M. Van Kujik, W. G. Seims, and O. Sommerburg, *Invest. Ophthalmol. Vis. Sci.* **39,** S1030 (1997).
[31] M. Tsuchiya, G. Scita, H.-J. Freisleben, V. E. Kagan, and L. Packer, *Methods Enzymol.* **213,** 460 (1992).
[32] F. Khachik, C. J. Spangler, J. C. Smith, Jr., L. M. Canfield, A. Steck, and H. Pfander, *Anal. Chem.* **69,** 1873 (1997).
[33] F. Khachik, G. R. Beecher, and J. C. Smith, Jr., *J. Cell. Biochem.* **22,** 236 (1995).
[34] F. Khachik, P. S. Bernstein, and D. Garland, *Invest. Ophthalmol. Vis. Sci.* **38,** 1802 (1997).
[35] K. Schiedt, in "Carotenoids: Chemistry and Biology" (N. I. Krinsky, ed.), p. 247. Plenum Press, New York, 1990.
[36] K. Schiedt, S. Bischof, and E. Glinz, *Pure Appl. Chem.* **63,** 89 (1991).
[37] G. Gao, C. C. Wei, A. S. Jeevarajan, and L. D. Kispert, *J. Phys. Chem.* **100,** 5362, (1996).

lites in both the serum and the retina is strong support for the broader antioxidant function of the hydroxy carotenoids. These data leave open the question of whether the metabolites found in the retina are absorbed from the serum or if they are produced in the retina. The putative antioxidant function of the macular pigment remains ambiguous based on these data.

Goralczyk et al.[38] have studied the uptake of canthaxanthin in the retina of the cynomolgus monkey. They found that the reduction metabolites, 4-hydroxyechinenone and β,β-carotene-4,4'-diol (isozeaxanthin), account for as much as 40% of the canthaxanthin in the monkey retina. These data would appear to support the hypothesis that a functional metabolical pathway for the reduction of ketocarotenoids exists in the primate retina (the serum was analyzed only for canthaxanthin).

We have previously shown that zeaxanthin found in the retina is present as the three stereoisomers, $3R,3'R$-β,β-carotene-3,3'-diol, $3S,3'S$-β,β-carotene-3,3'-diol, and $3R,3'S$-β,β-carotene-3,3'-diol.[1,3,39] Only the RR isomer is present in the human diet in appreciable quantities. The two other isomers must originate from the metabolism of zeaxanthin or lutein. Oxidation of either lutein or RR-zeaxanthin will produce 3-hydroxy-β,ε-carotene-3'-one.[35,36] Subsequent nonstereospecific reduction of this keto carotenoid would produce a mixture of R,R- and R,S-zeaxanthin. Given that lutein contains an allylic hydroxyl group that has a significantly lower oxidation potential, it would be expected that it would be oxidized preferentially relative to zeaxanthin. This may account for the relative decrease in the proportion of lutein in the inner macula where the highest carotenoid concentrations are observed.[1] Schiedt et al.[35,36] have shown that a similar series of conversions occur in the retinas of avian species.

We report here a method by which sections of individual eyes may be analyzed for the content of the three stereoisomers of zeaxanthin, thereby determining their distribution across the retina. Using this method we have been able to analyze total quantities of individual isomers as small as 0.3 ng obtained from the extraction of retinal tissue.[3] An example of the application of this method to human serum is included that illustrates the significant differences in stereoisomer composition between the serum and the retina.

---

[38] R. Goralczyk, S. Buser, J. Bausch, W. Bee, U. Zühlke, and F. M. Barker, *Invest. Ophthalmol. Vis. Sci.* **38**, 741 (1997).

[39] R. A. Bone, J. T. Landrum, G. W. Hime, A. Cains, and J. Zamor, *Invest. Ophthalmol. Vis. Sci.* **34**, 2033 (1993).

Serum and Tissue Handling

Serum samples are collected in 5-ml Vacutainer serum separator tubes by venipuncture, and after allowing 30 min for coagulation the samples are centrifuged for 10 min and the serum removed by pipette. To a 200-μl aliquot of serum, 20 μl of an internal standard, containing 90 ng of lutein monohexyl ether,[40] is added as an ethanol solution. The serum proteins are precipitated by the addition of 2 ml of ethanol/water (50:50). The carotenoids are extracted by three successive additions of 2-ml aliquots of hexane, with homogenization on a vortex mixer for 1 min followed by centrifugation for 5 min after which the hexane layer is removed by pipette and dried into a 1.5-ml siliconized polyethylene sample tube under a stream of dry N_2. The resulting serum carotenoid extract is separated by reversed-phase high-performance liquid chromatography (HPLC) in a manner identical to that described for the carotenoid obtained from the retina *vide infra*.

Human donor eyes were obtained from the National Disease Research Interchange (Philadelphia, PA) where they were enucleated and fixed in formaldehyde within approximately 6 hr of death. Eyes were stored at 4° prior to analysis. Dissection of the retinas is carried out in a 0.9% saline solution and care is taken to minimize the exposure of the tissues to bright light. The retinas are draped on a 2.5-cm Lucite sphere that is raised from the solution, and the retina is sectioned by aligning three concentric trephines of 3, 11, and 21 mm diameter with the fovea.[1] The resulting tissue sections are a disk containing the yellow spot, 7.1 mm^2, and two concentric annuli having areas of 93 and 343 mm^2, respectively.

Extraction of Macular Carotenoids

Each tissue sample is individually extracted in a 5-ml tissue homogenizer using 2-ml of ethanol/water (1:1) to which an internal standard, 9.9 ng (500 μl) lutein monopropyl ether,[40] is added prior to the extraction process. The resulting homogenate is transferred to a large culture tube, rinsing the tissue homogenizer with three 2-ml aliquots of ethanol/water and two 5-ml aliquots of hexane, which are added to the culture tube. The homogenate and hexane are agitated on a vortex mixer for 1 min and centrifuged to separate the resulting emulsion. The hexane is transferred to a pear-shaped flask and dried under a stream of N_2.

Separation of Lutein and Zeaxanthin

Lutein and zeaxanthin present in the extract are quantified and separated by reversed-phase HPLC. The system employs a 250 × 2-mm C_{18}

---

[40] S. Liaaen-Jensen and S. Hertzberg, *Acta Chem. Scand.* **20**, 1703 (1966).

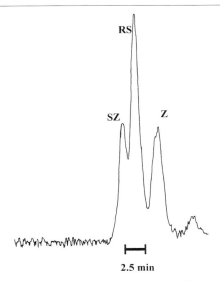

FIG. 2. A normal-phase HPLC chromatogram of a racemic mixture of zeaxanthin stereoisomers shows (3S, 3'S)-zeaxanthin (SZ), (3R, 3'S)-zeaxanthin (RS), and (3R, 3'R)-zeaxanthin (Z), eluting at 23.22, 24.72, and 27.33 min, respectively. The ratio of the integrals of the peaks is 0.83:2.0:1.0 (SZ:RS:Z), near the theoretical ratio, 1:2:1.

column packed with 3 μm Ultracarb ODS (Phenomenex, Torrance, CA). The mobile phase is 90% acetonitrile and 10% methanol to which 0.1% (v/v) of triethylamine is added to inhibit the degradation of carotenoids during elution. The flow rate is 0.2 ml/min. A UV/VIS detector is used to monitor the elution at 451 nm. The zeaxanthin peak, which contains all three stereoisomers (when present), is collected in a siliconized polyethylene microcentrifuge tube and carefully dried under a stream of N_2, concentrating the sample in the bottom of the tube for further derivatization and analysis. The mass of zeaxanthin is determined by comparison of the area of the internal standard and that of the zeaxanthin.

Preparation of Zeaxanthin Dicarbamate Diastereomers

Zeaxanthin fractions collected from the reversed-phase HPLC are transferred into a glove box containing a dry N_2 atmosphere in order to carry out the derivatization procedure, which was modified from Rüttiman et al.[41] The zeaxanthin is dissolved in 20 μl of anhydrous pyridine/benzene

---

[41] A. Rüttimann, K. Schiedt, and M. Vecci, *J. High Res. Chrom. Commun.* **6**, 612 (1983).

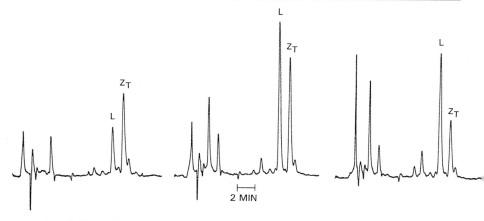

FIG. 3. HPLC chromatograms obtained with a reversed-phase column of macular pigment extracts from three differents sections of a single human retina. (a) inner: disk centered on the fovea obtained with a 3-mm trephine, area 7.1 mm^2. (b) Medial: annulus obtained with 3- and 11-mm trephines, area 93 mm^2. (c) Outer: annulus obtained with 11- and 21-mm trephines, area 343 mm^2. L, lutein; Z_T, combined zeaxanthin stereoisomers. The chromatograms have been truncated and do not show the internal standard.

(50/50 v/v, Aldrich, Milwaukee, WI). To the resulting solution, 1–2 μl of (S)-$(+)$-1-(1-naphthyl)ethyl isocyanate is added and the mix is capped, covered with aluminum foil to exclude light, and allowed to react for a period of ~48 hr.

Diastereomeric dicarbamate derivatives are prepared for chromatographic separation by the addition of 5 ml of hexane to the reaction mixture

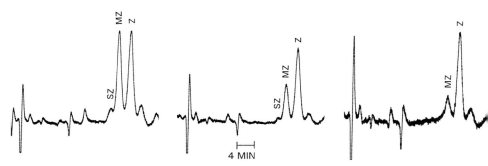

FIG. 4. HPLC chromatograms obtained with a normal-phase column of dicarbamate derivatives of zeaxanthin stereoisomers. These were obtained from the three different sections, as defined in Fig. 3, of a single human retina. In order of elution, the stereoisomers are $(3S, 3'S)$-zeaxanthin (SZ), $(3R, 3'S)$-zeaxanthin (RS), and $(3R, 3'R)$-zeaxanthin (Z).

TABLE I
MEAN CONCENTRATIONS OF CAROTENOIDS (pmol/mm^2) IN SECTIONS TAKEN FROM 16 NORMAL RETINAS

Section	Lutein	Zeaxanthin (all isomers)	RR-Zeaxanthin	RS-Zeaxanthin	SS-Zeaxanthin
Inner (7.1 mm^2)	2.4 ± 1.5	3.4 ± 2.3	1.7 ± 1.2	1.4 ± 0.88	0.22 ± 0.22
Medial (93 mm^2)	0.22 ± 0.25	0.14 ± 0.14	0.094 ± 0.095	0.037 ± 0.035	0.0076 ± 0.0081
Outer (343 mm^2)	0.065 ± 0.064	0.028 ± 0.027	0.020 ± 0.020	0.0061 ± 0.0065	0.0013 ± 0.0016

and extraction against an equal volume of water. The hexane is then removed by drying under a stream of N$_2$ gas, taking care to concentrate the dicarbamate product into the tip of a microcentrifuge tube during the drying process.

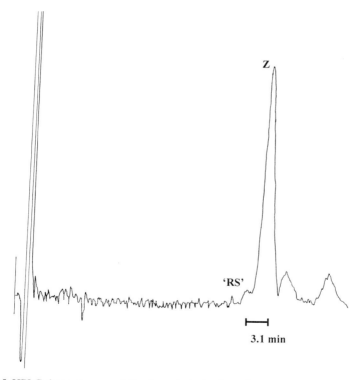

FIG. 5. HPLC chromatogram of dicarbamate derivatives of serum zeaxanthin on a normal-phase column showing the presence of a major component, identified by coinjection as (3R, 3'R)-zeaxanthin (Z), (see Fig. 6). The peak preceding the RR isomer (Z) may be the RS isomer ('RS') and represents a maximum of 6% of the total zeaxanthin in human serum.

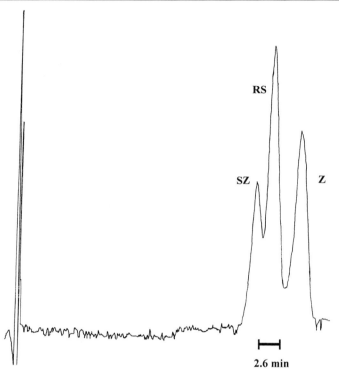

FIG. 6. HPLC chromatogram of the principal dicarbamate zeaxanthin diastereomer isolated from serum and coinjected with the racemic mixture. Peak enhancement of the (3R, 3'R)-zeaxanthin (Z) peak is observed. The ratio of the peak areas is 0.95 : 2.0 : 1.63 ($SZ:RS:Z$).

The resulting dicarbamate derivatives are dissolved in 20 μl of the mobile phase and analyzed by HPLC on a 250 × 2-mm normal-phase column packed with 5 μm Prodigy silica (Phenomenex). The mobile phase is 88% hexane and 12% isopropyl acetate at a flow rate of 0.2 ml/min. Detection is at 451 nm. No internal standard is utilized in the analysis as only relative proportions of the three stereoisomers are needed to determine the composition; the total mass of zeaxanthin is known from the chromatographic separation on the reversed-phase column. It is essential for purposes of identification that the dicarbamate derivatives of the zeaxanthin stereoisomers be compared to authentic stereoisomers prepared from rhodoxanthin[1] by the modification of the method of Maoka et al.[42] Zeaxanthin dicarbamate diastereomers can be collected separately for coinjection with

[42] T. Maoka, A. Arai, M. Shimizu, and T. Matsuno, *Comp. Biochem. Physiol. B* **83**, 121 (1986).

the racemic mixture. Peak enhancement during cochromatography provides the most reliable confirmation of isomer identity.[3,39]

Discussion

The elution order of the three authentic zeaxanthin dicarbamate diastereomers in a racemic mixture prepared from rhodoxanthin,[1,39] by a modification of the method reported by Maoka et al.[42] is illustrated in Fig. 2. The identity of the three isomeric zeaxanthin peaks has been established as described previously.[3,39] A set of sample reversed-phase chromatograms obtained from the inner, medial, and outer retinal sections of a single eye is shown in Fig. 3. The subsequent normal-phase separation of the derivatized macular pigment zeaxanthin isomers is presented in Fig. 4. Table I gives the mean concentrations of the carotenoids in these sections as determined from the analysis of 16 normal eyes.

An analysis of serum zeaxanthin reveals that the serum contains dominantly RR-zeaxanthin (Fig. 5). The principal peak was collected and combined with a sample of the racemic mixture. Its identity was confirmed through the observed enhancement of the all-E-RR-zeaxanthin dicarbamate peak (Fig. 6). The identity of the peak labeled RS remains to be firmly established. Coinjection with the racemic mixture has not been carried out, rather its identity was inferred from its relative chromatographic position and spectrum. The peaks that elute after the all-E-RR-zeaxanthin isomer may possibly be Z isomers. The concentration of zeaxanthin in the serum sample was determined to be 0.39 ± 0.02 μg/ml, and 94% of the carotenoid was confirmed to be the RR isomer, placing an upper limit of 6% on the abundance of any RS-zeaxanthin present in the serum. The serum zeaxanthin concentration in this sample was elevated abnormally (as was lutein) as the donor was on a diet rich in these two carotenoids. In the serum, the percentage of the component assumed to be the RS-zeaxanthin isomer (6%) is remarkable compared to the central macula, where RS represents $\sim$50% of the total zeaxanthin. These observations support the hypothesis that lutein and/or zeaxanthin undergoes oxidation in the retina followed by nonstereospecific reduction to regenerate the observed suite of stereoisomers. The presence and distribution of these stereoisomers appear to be consistent with, and support, a hypothesis of antioxidant function for the macular carotenoids.

Acknowledgments

Donor eyes were provided by the National Disease Research Interchange. Partial support was provided by NIH Grant GM08205.

Author Index

Numbers in parentheses are footnote reference numbers and indicate that an author's work is referred to although the name is not cited in the text.

A

Abramson, S. B., 258
Abul-Hajj, Y. J., 92
Adler, A. J., 459
Aebischer, C.-P., 348
Aejmelaeus, R. T., 6, 7, 9(16, 17), 11(14–16), 12(14–16), 14(16)
Afanas'ev, I. B., 92
Aguini, N., 276
Aherne, G. W., 277
Ahmed, J., 457, 459(7)
Ahonen, J.-P., 6(21), 7, 8(21), 9(21), 10(21), 11(21)
Ajani, U. A., 459
Akerboom, T. P. M., 271
Akoh, C. C., 319(25), 320, 323(25), 325(25)
Alagna, C., 171
Alaoui-Jamali, M., 258
Albert-Piela, M., 300
Albiero, R., 421
Alessio, H. M., 15, 20(6), 50, 51(6), 379
Alexander, D. W., 318(13), 319, 320(13), 329(13)
Al-Hashim, H., 92, 207, 230(8), 231(8)
Alho, H., 3, 6, 6(18–21), 7, 8(21), 9(16, 17, 21), 10(21), 11(14–16, 21), 12(14–16, 18–20), 14(16)
Allemann, L., 319(22), 320, 323(22)
Almazan, F., 193
Almeida, L., 96
Alomer, Y., 230
Amachi, T., 201
Amano, I., 107, 112(1)
Ameer, B., 93(31), 95
Ames, B. N., 3, 4(4), 8, 36, 83, 84, 86, 88, 88(1, 17), 331, 369
Ames, N. B., 347
Amselem, S., 297
Anderson, M. E., 258, 277, 282
Anderson, M. T., 248, 254(15)

Ang, C., 319(20), 320, 326(20)
Antoine, M., 277
Antonacci, D., 171
Antonelli, A., 151
Anzai, K., 28, 31, 32, 33
Appelkvist, E. L., 330
Arai, A., 466, 467(42)
Archier, P., 114, 123, 137(10)
Arichi, H., 184
Arichi, S., 184
Arner, E. S., 239
Arrick, B. A., 258
Arts, I. C. W., 202
Aruoma, O. I., 92, 207, 209, 230(8), 231(8)
Asensi, M., 267, 268(5), 270(5), 271(5), 274
Assen, N. A., 201
Assini, E., 50, 51(3)
Attaway, J. A., 120
Auran, J. D., 459
Aust, S. D., 424
Aviram, M., 201
Azzini, E., 15

B

Baba, K., 184
Baba, S., 92
Babson, J. R., 274
Back, S., 390, 393(6), 394(6), 398(6), 400(6)
Bae, H. D., 179
Baeuerle, P. A., 247
Bagnell, R., 294
Bailey, A., 23
Baker, E. R., 287
Baker, H., 287, 293
Baker, J. C., 369
Baker, P. F., 309, 310(3), 311(3), 315(3)
Balentine, D. A., 93(34), 95
Banhegyi, G., 83

Banks, M. A., 318(12), 319, 320(12)
Barber, J., 408
Barclay, L. R. C., 10(32), 14, 15, 20(3)
Barenholz, Y., 293, 294, 296(5), 297, 299(5)
Bargossi, A. M., 342
Barker, F. M., 461
Barlow-Walden, L., 14
Barna, J., 120
Barnes, P., 319(21), 320, 323(21), 325(21), 326(21)
Barnett, J., 35, 193
Barr, R., 331
Barron, D., 208
Barry, P., 400
Bartholomew, J. C., 84, 88(17)
Barua, A. B., 431, 432, 433, 433(3)
Baruchel, S., 258
Bates, C. J., 23
Batist, G., 258
Baum, K., 348
Bausch, J., 461
Beare-Rogers, J. L., 318(11), 319, 320(11), 329(11)
Beatty, P. W., 274
Beck, I., 92
Becker, K., 174
Beckman, J. S., 211, 212(34), 228(31, 34)
Beckman, K. B., 83
Bee, W., 461
Beecher, C. W. W., 185
Beecher, G. R., 73, 390, 391, 394(4), 400, 400(4), 460
Behrens, W. A., 318(11), 319, 320(11), 329(11)
Beisiegel, U., 35, 37, 43, 43(23), 44, 44(24), 45, 45(23, 26), 46, 46(24), 47, 48(23), 348
Bell, A., 208
Bell, G. D., 91
Bell, G. P., 231
Bellus, D., 459
Ben-Naim, M., 193, 199
Bentley, G. R., 318(9), 319, 320(9), 329(9)
Benzie, I. F. F., 15, 16, 16(2), 19, 20, 20(2), 21, 22, 23, 23(10), 24, 24(1, 2, 10), 25, 26, 27(2, 10), 50, 51(8)
Berends, H., 86
Berger, H. M., 342
Bergsten, P., 73, 77
Berlin, E., 318(12), 319, 320(12)
Berliner, J. A., 362
Berman, E. R., 427

Bernier, J. L., 92, 230
Bernini, W., 184
Bernstein, P. S., 460
Berry, E. M., 193, 199
Bertelli, A., 184
Bertram, J. S., 442
Bertrand, A., 122, 123(6), 137(6)
Bessis, R., 186, 187
Beutler, E., 270
Beyer, R. E., 331
Bezard, J., 185
Bieri, J., 8, 319(28), 320, 426(30), 430
Bird, A. C., 459
Birlouez-Aragon, I., 73
Bischof, S., 460, 461(36)
Blair, N., 459
Blais, L., 320, 324(34)
Blanchflower, W. J., 319(17, 29), 320
Blight, E. G., 367
Block, G., 207
Bloin, J., 166
Blomhoff, R., 430, 431
Blond, J. P., 185
Bloor, S. J., 113
Blumberg, J. B., 318(6), 319, 320(6)
Bockkisch, M., 84
Bode, A. M., 77, 78, 80, 81, 81(7), 85
Bohm, F., 428, 430(27)
Bolliger, H. R., 334, 342
Bolsen, K., 358
Bolton, A. E., 444, 449(13)
Bolwell, G. P., 91, 96, 97, 97(39), 105, 208, 210, 231(11), 233(28), 379
Bondjers, G., 6
Bone, R. A., 457, 461, 461(1, 3), 462(1), 467(1, 3, 39)
Bonfanti, R., 356
Bongiorno, A., 421, 425(2), 428, 430
Bonomo, R. P., 92
Bors, W., 207, 230
Bouma, J., 73
Bourgeois, C., 331, 333(14)
Bourne, L., 91, 93(47), 96, 97, 105
Bourzeix, M., 122, 136, 137(1, 2)
Boveris, A., 211, 230, 379
Bowry, V. W., 35, 48, 331, 363, 364, 364(7), 366(7), 367, 370, 370(7, 8), 371, 371(18), 372, 372(18), 374, 375(18), 422, 428, 428(11)
Boyd, M. R., 294
Bradford, M. M., 442, 455(11)

Braesen, J. H., 43, 44, 44(24), 46(24), 47
Braganza, J. M., 15, 51, 52(11), 60(11)
Bramley, P., 379, 385(14), 389(14)
Braun, A. M., 92
Braun, L., 83
Breen, P. J., 114, 119(6)
Breene, W. M., 319(27), 320, 323(27)
Bretzel, W., 358
Brezinka, H., 390, 394, 398(22)
Brigelius, R., 271
Britton, G., 351, 400
Brodie, A. E., 274
Brodskii, V., 92
Brodsky, M. H., 369
Brooks, G., 274
Brossaud, F., 182
Brown, E. D., 433
Brown, J., 92
Brown, J. E., 207, 230(10), 231(10)
Brown, P. K., 459
Brown, R. D., 457, 459(7)
Brown, R. S., 393
Brown, S., 91, 231
Brugirard, A., 175
Bruno, A., 94
Bruseghini, L., 275
Bryden, W. L., 318(14), 319, 320(14), 329(14)
BuAbbas, A., 208
Buckpitt, A. R., 243
Budin, J. T., 319(27), 320, 323(27)
Budzikiewicz, H., 390, 394, 398(22)
Bueding, E., 302, 306(9)
Buege, J. A., 424
Buettner, G. R., 365
Buhl, R., 258
Bühler, I., 351, 358
Bullerjahn, G. S., 442
Bun, H., 431, 433(4)
Bunting, J. R., 259
Burdelski, M., 331, 341, 342, 345(4), 346, 347(4)
Burger, R., 331
Burger, W. C., 330
Burkhardt, R., 176
Burr, J. A., 309, 310(5), 311(5), 315, 315(5), 316(5)
Burton, G. W., 4, 6(8), 8(8), 9(8), 10(8, 32), 14, 15, 20(3), 50, 309, 311, 317, 318, 319(4), 332, 362, 371, 379, 422, 428
Burton, N. K., 277
Burton, T. C., 459

Buser, S., 461
Bush, K. M., 211
Butler, J., 209
Butler, J. D., 73, 77
Buttriss, J. L., 318, 319(3), 323(3)
Bynoa, L. A., 458

C

Caccamese, S., 390, 393(3), 394(3)
Caceras, I., 455
Caceres, I., 454
Cai, L., 185
Cains, A., 461, 467(39)
Cama, H. R., 455
Campos, R., 319(32), 320
Candeias, L., 379, 385(14), 389(14)
Canfield, L. M., 460
Cantilena, L. R., 66, 72(9)
Cantin, A. M., 258
Cao, G., 15, 20(6), 50, 51(6, 7), 52, 57, 57(12), 58, 60(17), 379
Carew, T. E., 35, 48(1), 192, 362
Carreras, M. C., 211
Carruba, E., 171
Carson, D. L., 439
Carulli, N., 193, 199
Cassity, N. A., 318(13), 319, 320(13), 329(13)
Castelluccio, C., 96, 97(39), 210, 231, 233(28)
Castle, L., 423
Catalin, J., 431, 433(4)
Catignani, G., 8, 319(28), 320
Catteau, J. P., 92, 230
Cattell, D. J., 180
Catz, S. D., 211
Celeste, M., 157
Cerda, V., 157
Chambon, P., 431
Chan, A. C., 428
Chan, W. W., 331
Chance, B., 408
Chang, B. G., 182
Chang, R. L., 208
Chang, Y., 460
Chase, G. W., Jr., 319(25), 320, 323(25), 325(25)
Chathou, R. E., 259
Chaudière, J., 276, 277
Chaurand, P., 393

Cheeseman, K. H., 192
Cheminat, A., 181
Chen, H. Y., 208
Chen, J., 211, 212(34), 228(34)
Chen, L., 93(34), 95
Chen, S. C., 331
Chen, X., 258
Chen, Y., 91
Chen, Y.-L., 202
Cheynier, V., 114, 178, 181, 182
Chichester, C. O., 343, 418, 419(16)
Chimi, H., 233
Chiswick, M., 342
Chiu, D., 191
Chô, S.-C., 418, 419(17)
Chow, C. K., 319, 326(16), 428
Chow, S., 248
Christen, M., 300
Christophersen, A. G., 410, 412(8), 413(8)
Chrystal, R. G., 258
Chuang, J., 14
Chung, Y. K., 318(10), 319, 320(10), 329(10)
Chytil, F., 455
Ciaccio, M., 421
Cicocalteu, V., 153, 155(5), 164(5)
Cillard, J., 92, 233
Cillard, P., 92, 233
Cillers, J. J. L., 153
Cilliers, J. J. L., 173
Cladera, A., 157
Cladera-Fortaza, A., 159
Clancy, R. M., 258
Clark, R. W., 330
Clark, W. G., 92
Cleary, J., 364
Cleary, S. F., 458
Clemens, M. R., 191
Cliffe, S., 171
Clifford, M. N., 208
Coates Barclay, R., 422
Cobb, C. E., 83
Coen, S., 114, 123, 137(10)
Coffee, M., 185
Coggon, P., 107
Coign, M. J., 186
Commentz, J., 331, 341, 342, 345(4), 346, 347(4)
Comstock, G. W., 86
Conry-Cantilena, C., 66, 72(9)
Constable, A., 208

Constantinescu, A., 239
Cook, J. A., 248
Cooke, J. R., 65
Cooney, A. H., 208
Cooney, R. V., 442
Copeland, E., 208
Cornicelli, J. A., 379
Corry, P. M., 265
Corsaro, C., 92
Cort, W. M., 319(18), 320, 323(18)
Costantinescu, A., 422, 428(10)
Cotelle, N., 92, 230
Cotgreave, Y. A., 275
Coudray, C., 23
Coussio, J. D., 230
Cova, D., 93(51), 105
Crabtree, D. V., 459
Craft, N. E., 433, 441
Crane, F. L., 330, 331
Creasy, L. L., 185, 187
Crit, B., 27
Cross, C. E., 228
Croux, S., 92
Crow, J. E., 455
Crow, J. P., 211, 212(34), 228(34)
Cruickshank, A., 195, 201(33)
Csala, M., 83
Csallany, A. S., 318, 319, 319(2), 320(2), 326(16)
Cucera, P., 435, 436(10)
Cuenat, C., 122
Cuenat, P., 186, 187
Cunniff, P., 155
Cunningham, L., 80, 85
Curley, R. W., 439
Cutler, R. G., 15, 20(6), 50, 51(6), 379
Czochanska, Z., 178

D

Dabach, Y., 193, 199
Dabbagh, A. J., 36, 48(12)
Dabbagh, Y. A., 91
D'Agostino, S., 167, 171
Dahle, L. K., 199
Daimandis, E. P., 123
Dallner, G., 330, 331
D'Anna, R., 421, 425(2)

Darley-Usmar, V. M., 234
D'Arpa, D., 421, 425(2), 430(3)
Das, N. P., 93(32), 95
Davies, M. J., 13, 50, 59(4), 379
Davis, P. A., 196, 197(40), 198, 198(41), 199(41)
Davis, T. P., 76
Dean, R. T., 36, 333, 367
DeAngelis, B., 287
De Angelis, L., 93(51), 105
Debetto, P., 277
Dekker, M., 202
Delange, R. J., 379
De Las Rivas, J., 408
de la Torre, C., 123
de la Torre-Boronat, M. C., 114, 186, 187(11), 188(19)
del Castillo, M. D., 10(27), 12, 13(27), 15, 379
Delori, F. C., 459
Demacker, P. N. M., 201
Denis, M. P., 185
Denis, W., 154, 164(8)
Denny, R. W., 460
Derache, R., 175
Derdelinckx, G., 179
Deresinski, S. C., 248, 254(15)
de Rijke, Y. B., 201
DeRitter, E., 86
De Rosa, S. C., 248, 254(15)
Deutsch, J. C., 84
Devall, L. J., 379
Devary, Y., 330
Devasagaya, T. P. A., 458
de Vries, J. H. M., 93(50), 105, 186
de Whalley, C. V., 91, 231
Dey, G. R., 294
Dhariwal, K. R., 65, 66, 66(1), 72(9), 73(1), 74(7), 77, 81
Diamandis, E. P., 124, 128, 138, 139, 141, 150(2), 184, 185(4), 187
Di Bilio, A. J., 92
DiCola, D., 270
DiDonato, J. A., 330
Dieber-Rotheneder, M., 192, 193(21), 365
Dieffenbacher, A., 319(19), 320, 323(19), 325(19), 326(19)
Diez, C., 176
Dilengite, M. A., 193, 199
Di Mascio, P., 458, 460
Diplock, A. T., 318, 319(3), 323(3)

Dirr, A., 331
Disdier, B., 431, 433(4)
Djahansouzi, S., 43, 44, 44(24), 46(24), 47
Do, K. L., 84, 88(17)
Doba, T., 428
Doco, T., 182
Dollard, C. A., 164
Donko, E., 165
Dorey, C. K., 459
Dorian, R., 248, 259
Dormandy, T. L., 4, 191, 192
Dorozhko, I., 92
Dowben, R. M., 259
Dragsted, L. O., 93(27), 94
Draper, H. H., 319, 326(16)
Dratz, E. A., 425, 457
Drevon, C. A., 193, 199
Duda, C. T., 309, 310(11), 370
Dufour, J. H., 166
Dunn, R., Jr., 457, 459(7)
Durand, A., 431, 433(4)
Dyer, W. J., 367

E

Edamatsu, R., 29, 30
Eder, R., 120, 186
Edge, R., 428, 430(27)
Edlund, P. O., 347
Einsele, H., 191
Einsenberg, S., 193
Eisenrich, J. P., 228
Eitenmiller, R. R., 319(20, 25), 320, 323(25), 325(25), 326(20)
Ela, S. W., 248, 254(15)
Eliopoulos, E. E., 442, 454, 455
Elliot, G. E. P., 92, 207
Ellis, W. W., 274
Ellman, G., 8
Elson, C. E., 330
Emerit, I., 408
Engelhardt, U. H., 151, 206
England, L., 8, 36, 83
Enssle, B., 163
Enzell, C. R., 390, 393(6), 394(6), 398(6), 400(6)
Epps, D. E., 379
Erben-Russ, M., 230

Ercal, N., 258, 259, 260, 260(8), 262(8)
Erdman, J. W., Jr., 454, 457
Erhola, M., 6, 7, 12(18, 19)
Ernster, L., 331
Esau, P., 153, 174(6)
Escribano-Bailon, M. T., 114, 180, 181
Estela, J. M., 157
Esteras, A., 275
Esterbauer, H., 36, 37, 37(17), 38(20), 39(17), 41(20), 44(20), 48(17, 20), 190, 191, 192, 193, 193(1, 21), 194(4), 362, 365
Estrela-Ripoll, J. M., 159
Evans, T. A., 366
Evershed, R. P., 400
Every, D., 84
Eye Disease Case-Control Study Group, 459
Eyer, P., 268

F

Faccin, N., 351, 358
Fahey, R. C., 248, 259, 277
Fainaru, M., 37, 38(21), 39(21), 41(21), 43(21), 47(21)
Fantozzi, P., 176
Farber, M. D., 459
Farnsworth, C. D., 425
Farnsworth, N. R., 185
Farr, A. L., 264, 293
Farré, R., 23
Farrell, P. M., 319(31), 320
Fausto, M., 51
Faviet, A., 23
Favre, M., 443, 448(12)
Fawer, M. S., 171
Federici, G., 270
Feingold, K. R., 35
Fernandez, L., 457
Ferraro, G. E., 230
Ferrero, J. A., 274
Ferro-Luzzi, A., 15, 23, 50, 51(3)
Finckh, B., 37, 43, 43(23), 44, 44(24), 45, 45(23), 46(24), 48(23), 331, 341, 342, 345(4), 346, 347(4), 348
Findlay, J. B. C., 442, 454, 455
Fingerhut, R., 45
Finskh, B., 35
Fiorella, P., 342
Flecha, B. G., 379

Flora, P. S., 343
Floreani, M., 277
Folch, J., 301, 303(8)
Folin, O., 153, 154, 155(5), 164(5, 8)
Fong, H. H. S., 185
Fong, L. G., 35
Foo, L. Y., 178, 179
Food, L. Y., 180
Foote, C. S., 460
Forte, T., 422, 428(12)
Foster, D. O., 317
Fouchard, R. C., 318, 320, 324(34)
Fougeat, S., 277
Frackelton, A. R., 259
Fraga, C. G., 230
Frank, H., 191
Frank, O., 293
Frankel, E. N., 123, 138, 184, 185(2), 190, 191, 192, 193, 195, 196, 197(38–40), 198, 198(38, 39, 41), 199, 199(29, 41), 200
Frazer, D. R., 318(14), 319, 320(14), 329(14)
Freeman, B. A., 211
Fregoni, M., 184
Frei, B., 8, 27, 35, 36, 48(12), 83, 86, 191, 194(5), 201(5), 331, 347
Freisleben, H.-J., 460
Fridovich, I., 29
Friedes, L. M., 457, 461(3), 467(3)
Friedlander, M., 193
Frigola, A., 23
Fritsche, K. L., 318(13), 319, 320(13), 329(13)
Frolic, C. A., 433
Fry, S. C., 97
Fuchs, J., 309
Fuhr, U., 93(30), 94
Fuhrman, B., 201
Fujimoto, E. K., 365
Fujimoto, K., 192
Fujioka, M., 92
Fujiwara, K., 239
Fukui, S., 196, 230
Fuleki, T., 137
Furr, H. C., 343, 431, 432, 433, 433(3)
Furtuta, T., 92
Furukawa, T., 274

G

Gaag, M. V. D., 93(50), 105
Galensa, R., 107

Galleman, D., 268
Galletti, G. C., 151
Ganz, P. R., 428
Gao, G., 460
Gárate, M., 319(32), 320
García de la Asuncíon, J., 275
Gardana, C., 93(29), 94
Garland, D., 460
Garozzo, D., 390, 393(3), 394(3)
Garrido, A., 319(32), 320
Garrido, G. L., 176
Gartner, F. H., 365
Garzo, T., 83
Gascó, E., 274
Gattuso, A. M., 177
Gaydou, E., 92, 230
Gaziano, J. M., 35
Gebicki, J., 36, 191, 194(4), 362
Gérard-Monnier, D., 277
Gercken, G., 44, 45(26)
German, B., 123
German, J. B., 138, 193, 195, 196, 197(40), 198, 198(41), 199(29, 41), 200
Gerrish, C., 96, 97(30), 210
Gey, K. F., 27
Ghiselli, A., 15, 50, 51(3)
Giannessi, D., 184
Giavarini, F., 93(51), 105
Gierschner, K., 163
Gifford, J., 367
Ginsburg, I., 294, 296(5), 299(5)
Giovannini, L., 184
Gitler, C., 247
Glazer, A. N., 50, 51(2), 379
Glinz, E., 460, 461(36)
Gobbo, S., 93(34), 95
Godber, J. S., 319(26), 320
Godin, D. V., 258
Goeke, N. M., 365
Goerz, G., 358
Gogia, R., 78, 82(9)
Gohil, K., 8, 274, 332, 333(20), 336(20), 340(20)
Goldbach, M. H., 163
Goldberg, D. M., 122, 123, 124, 128, 138, 139, 141, 150(2), 184, 185(4), 187
Goldstein, B. D., 191
Goldstein, J. L., 163
Goli, M. B., 391
Gollnick, K., 459
Gomez, C. M., 457, 461(3), 467(3)

Goode, H. F., 13
Gopinathan, V., 13, 50, 59(4), 379
Goralczyk, R., 461
Goramaru, T., 92
Gordon, D. A., 330
Gorgels, T. G. M. F., 458
Gorham, J., 185
Goto, T., 107, 108, 112(1)
Gottardi, F., 174, 176(43)
Gottsch, J. D., 458
Gotz, M. E., 331
Grabber, J. H., 97
Graf, E., 97, 234
Gragoudas, E. S., 459
Grant, R., 123
Grasse, B. J., 193
Graubard, B. I., 441
Graumlich, J., 66, 72(9)
Gregson Dubs, J., 248, 254(15)
Griffith, O. W., 259, 277
Griffiths, L. A., 93(33), 95
Grossi, G., 342
Gryglewski, R. J., 92
Guerry, D., 458
Guerry, R. K., 458
Gugger, E. T., 454, 457
Guilland, J., 73
Gundersen, T. E., 430, 431
Guo, C., 57
Gupta, A. K., 265
Gutcher, G. R., 319(31), 320
Gutteridge, J. M. C., 3, 4, 13(2), 16, 192
Guttierez-Fernandez, Y., 180
Guyot, S., 178
Gyorffy, E., 83
Gysel, D., 361

H

Hachenberg, H., 197
Hackett, A. M., 92, 93(33), 95
Haddad, I. Y., 211
Haegerstrom-Portnoy, G., 459
Haest, C. W. M., 192
Hagen, D. F., 309
Hagen, T. M., 3, 4(4), 83, 84, 88, 88(1, 17)
Hahn, S. E., 138, 184, 185(4)
Haila, K. M., 419, 421(22, 23)
Haller, J., 459

Halliwell, B., 3, 13(2), 16, 27, 35, 36, 36(2), 92, 97(10, 11), 192, 207, 209, 211, 228, 228(39, 40), 230(8), 231(8), 234
Ham, A. J. L., 309, 313, 315, 316
Ham, W. T., Jr., 458
Hamiltonmiller, J. M. T., 208
Hammond, T. C., 260
Han, C., 208
Han, D., 239, 240, 246, 246(12), 247, 254, 254(7), 301, 303(7), 330, 422, 428(10)
Hanasaki, Y., 196, 230
Handelman, G. J., 239, 240, 246, 246(12), 457
Hanlon, M. C., 13
Hansen, S., 258
Hara, Y., 91, 107, 208
Haramaki, N., 239
Harats, D., 193, 199
Harborne, J. B., 207
Härdi, W., 351
Harosi, F. I., 433
Hartzell, W. O., 66, 77
Hasegawa, K., 418, 419(16)
Haslam, E., 178
Hatam, L. J., 309
Hatefi, Y., 330, 334, 342
Hatfield, D., 97
Hatina, G., 318, 319(1), 320(1), 323(1), 325(1)
Hattori, Y., 310
Haug, M., 163
Havsteen, B., 92, 94(12)
Hayashi, T., 309, 310(8)
Hayward Vermaak, W. J., 347
Hector, B. J., 186
Hedley, D. W., 248
Hegwood, C. P., 186
Heinecke, J. W., 362
Heinonen, M. I., 419, 421(22, 23)
Heller, W., 207
Helzlsouer, K. J., 86
Hemingway, R. W., 179
Hendelman, G. J., 301, 303(7)
Hénichart, J. P., 92, 230
Heredia, N., 122, 137(2)
Hermanson, G. T., 365
Herrmann, K., 107, 202
Hertzberg, S., 462
Hervonen, A., 6, 7, 9(16), 11(15, 16), 12(15, 16), 14(16)
Herzenberg, L. A., 248, 254(15)
Heseker, H., 23

Hess, D., 356
Hiai, H., 3
Hidalgo Arellano, I., 114
Hider, R. C., 92, 207, 230(8, 10), 231(8, 10)
Higgs, D. J., 73
Higgs, H. E., 73
Hill, E. G., 199
Hillenkamp, F., 391
Hiller, D. T., 459
Hiller, R., 459
Hime, G. W., 461, 467(39)
Hiramatsu, M., 29
Hirayama, O., 4
Hisanobu, Y., 109
Ho, C. T., 428
Hocquaux, M., 92
Hoefler, A. C., 107
Hoffman, S. C., 86
Hogarty, C. J., 319(20), 320, 326(20)
Hogberg, J., 84
Hogg, N., 234
Hollander, G., 193, 199
Hollman, P. C. H., 93(50), 105, 186, 202
Holm, P., 6, 7, 9(16), 11(16), 12(16), 14(16)
Holman, R. T., 199
Hølmer, G., 318(11), 319, 320(11), 329(11)
Holmgren, A., 239
Holt, T. K., 442
Hood, R. L., 333
Hoogenboom, J. J. L., 319(23), 320, 323(23), 326(23)
Hopia, A. I., 196
Horie, H., 107, 112(1)
Hornbrook, K. R., 83, 87(12)
Hostettmann, K., 150, 151(11)
Hoult, J. R., 91, 231
Howard, J. A., 367
Howdle, P. D., 13
Howell, S. K., 309
Howlet, B., 165
Høy, C.-E., 318(11), 319, 320(11), 329(11)
Hrstich, L. N., 182
Hu, M.-L., 191, 196, 197(38), 198(38)
Hu, P., 211
Huang, C.-R., 390, 391(11)
Huang, M.-T., 208
Huang, S.-H., 196
Huang, S.-W., 195, 196
Hübner, C., 35, 331, 341, 342, 345(4), 346, 347(4), 348

Huflejt, M., 83
Hughes, L., 317, 371
Huie, R. E., 210
Hukumoto, K., 4
Hulan, H. W., 320, 321(33), 329(33)
Hunter, W. M., 444, 449(13)
Hunziker, F., 361
Husain, S. R., 92
Hutner, S. H., 287
Hyeon, S. B., 418, 419, 419(15)

I

Ibrahim, A. R. S., 92
Ibrahim, R., 208
Igarashi, O., 318(7), 319, 320(7), 326(7), 329(7)
Iiyama, K., 97
Ikawa, M., 164
Ikeda, I., 323
Imanari, T., 73
Imasato, Y., 323
Ina, K., 107
Inama, S., 174, 176(45)
Indovina, M. C., 177
Indyk, H. E., 318(8), 319, 320(8), 326(8)
Ingham, J. L., 185
Ingold, K. U., 4, 6(8), 8(8), 9(8), 10(8, 32), 14, 15, 20(3), 50, 309, 311, 317, 318, 319(4), 332, 362, 363, 367, 370, 370(8), 371, 379, 422, 428
Inoue, M., 83
Ioannides, G., 208
Ioffe, B., 197
Ischiropoulos, H., 211, 212(34), 228(31, 34)
Isoe, S., 418, 419, 419(15)
Itaka, Y., 309
Itakura, H., 201
Iype, S. N., 248

J

Jaffe, G. J., 459
Jaffe, H. A., 258
Jagendorf, A., 165
Jahansson, M., 259
Janero, D. R., 192
Janetzky, B., 331
Jang, J. H., 91
Jang, M., 185
Janick-Buckner, D., 433
Janin, E., 173
Jankovic, I., 231
Jeandet, P., 186, 187
Jeevarajan, A. S., 460
Jennings, A. C., 171
Jerina, D. M., 208
Jerumanis, J., 179
Jessup, W., 91, 231, 367
Jirousek, L., 300
Joa, H., 457
Johannes, B., 390, 394, 398(22)
Johnson, J. A., 248
Johnson, J. B., 86
Johnson, J. V., 93(31), 95
Jones, A. D., 228
Jongen, W. M. F., 202
Jørgensen, K., 410, 411, 412(6-9), 413(7-9), 414, 414(7), 415, 415(7), 420(7)
Joseph, J., 234
Josimovic, L., 231
Jouni, Z. E., 442
Jovanovic, S. V., 207, 230(7), 231, 231(7)
Juan, I.-M., 202
Juelich, E., 164
Juhasz, P., 393
Jun, H., 410, 412(8), 413(8)
Jürgen, G., 190, 191, 193(1), 194(4)
Jürgens, G., 36, 362
Justesen, U., 93(27), 94

K

Kaasgaard, S. G., 318(11), 319, 320(11), 329(11)
Kabayashi, M. S., 254
Kabuto, H., 30
Kacprowski, M., 167
Kada, T., 107
Kagan, V. E., 301, 330, 422, 428(12), 460
Kaiser, S., 460
Kalef, E., 247
Kalen, A., 330
Kall, M., 93(27), 94
Kallury, K. M. R., 318, 320, 324(34)

Kalyanaraman, B., 234
Kamiya, Y., 422, 430
Kamp, D., 192
Kan, M., 4
Kandaswami, C., 202
Kanetoshi, A., 309, 310(8)
Kanner, J., 123, 193, 199(29), 200
Karas, M., 391
Karchesy, J. J., 179, 180
Karin, M., 330
Karten, B., 37, 43, 43(23), 44, 44(24), 45(23, 26), 46(24), 48(23)
Karumanchiri, A., 123, 124, 128, 138, 139, 141, 187
Kastner, P., 431
Katan, M. B., 93(50), 105, 186, 201, 202
Katoh, S., 4
Katsumura, S., 419
Katzhendler, Y., 293, 294, 296(5), 299(5)
Kaufmann, N. A., 193
Kaufmann, R., 390, 391, 392, 393, 393(17), 394(14, 17), 398(17), 399(17), 400(14), 401(14), 405, 406
Kaugmann, R., 393
Kaukinen, U., 6, 7, 9(16), 11(16), 12(16), 14(16)
Kaur, H., 211
Kawakami, A., 422
Kayden, H. J., 309
Kaysen, K. L., 309, 310(3), 311(3), 315(3)
Kearney, J. F., 191, 194(5), 201(5)
Keen, J. N., 454, 455
Kehrer, J. P., 309, 310(12)
Kehrl, J., 77
Keine, A., 151
Keller, H., 356, 361
Kellokumpu-Lehtinen, P., 6(18, 19), 7, 12(18, 19)
Kelly, D. R., 211
Kelner, M. J., 294
Kerojoki, O., 318, 319(5), 323(5)
Kezdy, F. J., 379
Khachik, F., 390, 391, 394(4), 400, 400(4), 460
Khan, S., 301
Khanna, S., 239
Khodr, H., 92, 207, 230(8, 10), 231(8, 10)
Khokhar, S., 202
Khoo, J. C., 35, 48(1), 192, 193, 362
Kilburn, M. D., 457, 461(3), 467(3)
Kim, E.-K., 423

Kim, K. Y., 35
Kim, M., 331, 419, 421(24)
Kimura, Y., 184
King, J., 66, 72(9)
Kinghorn, A., 185
Kinsella, J. E., 123, 184, 185(2), 193, 199(29), 200
Kiovistoinen, P., 318, 319(5), 323(5)
Kirkola, A. L., 379
Kirsch, D., 390, 391, 392, 393, 393(17), 394(14, 17), 398(17), 399(17), 400(14), 401(14)
Kirsch, K., 405, 406
Kishorchandra, G., 342
Kishore, K., 294
Kiso, M., 108
Kispert, L. D., 460
Kissinger, P. T., 23
Klenk, D. C., 365
Koda, H., 201
Kofler, M., 334, 342
Kohen, R., 293, 294, 296(5), 299(5)
Kohlschütter, A., 35, 37, 43, 43(23), 44, 44(24), 45, 45(23), 46(24), 48(23), 331, 341, 342, 345(4), 346, 347(4), 348
Kohno, M., 28, 29, 31, 32, 33
Koivistoinen, P., 318(15), 319, 323(15), 326(15)
Koketsu, M., 419, 421(24)
Kolhouse, J. F., 84
Koller, E., 190, 193(1), 362
Komatsu, Y., 109
Kondo, K., 201
Kondo, S., 208
Kontush, A., 35, 37, 43, 43(23), 44, 44(24), 45, 45(23, 26), 46, 46(24), 47, 48(23), 331, 341, 342, 345(4), 346, 347(4), 348
Kooy, N. W., 211
Korhammer, S., 186, 187(12)
Korshunova, G. A., 208
Koshiishi, I., 73
Kosower, E. M., 247, 248, 267
Kosower, N. S., 247, 267
Koster, A. S., 267
Kostyuk, V. A., 92
Koyama, K., 83
Kozwa, M., 184
Kramer, C., 172, 173(37)
Kramer, J. K. G., 318, 320, 321(33), 324(34), 329(33)
Kramling, T. E., 171, 172(36), 173(36)

Krebs, H. A., 276
Krinsky, N. I., 408, 428, 458
Kritharides, L., 367
Kröger, A., 343
Krogmann, D. W., 442
Krohn, R. J., 365
Kühnau, J., 195, 202
Kuhr, S., 206
Kummert, A. L., 93(30), 94
Kurata, H., 201
Kurtyka, A. M., 323
Kuwabara, T., 426(30), 430
Kuypers, F. A., 191

L

Lagendijk, J., 331, 347
Laippala, P., 6, 7, 9(16, 17), 11(14, 16), 12(14, 16), 14(16)
Lakritz, J., 243
Lakshman, M. R., 441, 442, 455(10)
Lala, D., 414, 415(12)
Lam, T. B. T., 97
Lamuela-Raventos, R. M., 113, 114, 119(6), 123, 136, 152, 173, 184, 185, 186, 186(8), 187, 187(11), 188(8, 19)
Land, E. J., 428, 430(27)
Landrum, J. T., 457, 461, 461(1, 3), 462(1), 467(1, 3, 39)
Lang, J. K., 8, 332, 333(20), 336(20), 340(20), 342
LaNotte, E., 171
Laranjinha, J., 96
Larson, E., 165
Latvala, M., 6(21), 7, 8(21), 9(21), 10(21), 11(21)
Lau, J., 390, 394(4), 400(4)
Lavy, A., 201
Lazano, Y., 182
Lazarev, A., 66, 72(9)
Lea, A. G. H., 179, 180
Leake, D., 91, 231
LeClair, I. O., 36
Lee, M.-J., 93(34), 95
Lee, Y. J., 265
Lees, M., 301, 303(8)
Legler, G., 164
Lehmann, J., 319(30), 320

Lehner, M., 267
Lehr, R. E., 208
Lehtimäki, T., 6(20), 7, 12(20)
Leibholz, J., 318(14), 319, 320(14), 329(14)
Leibovitz, B. E., 191
Leid, P., 431
Leinonen, J., 3, 6(20), 7, 12(20)
Le Maguer, M., 419
Lemmli, U. K., 443, 448(12)
Lennon, J. J., 393
Lepine, A. J., 318(10), 319, 320(10), 329(10)
Le Roux, E., 178, 182
Lester, R. L., 330
Leszcznska-Piziak, J., 258
Levartovsky, D., 258
Levine, M., 65, 66, 66(1), 72(9), 73, 73(1, 2), 74, 74(7), 77, 81, 83, 84(9)
Levy, E. J., 282
Lewin, G., 6
Lewis, B., 35
Liaaen-Jensen, S., 462
Liang, Y.-C., 202
Lichtenberg, D., 37, 38(21), 39(21), 41(21), 43(21), 47(21)
Liebler, D. C., 309, 310(3;5), 311(3, 5), 313, 314, 315, 315(3, 5), 316
Lievonen, S. M., 419, 421(23)
Lilley, T. H., 178
Lin, J.-K., 202
Lin, Y.-L., 202
Lindemann, J. H. N., 342
Lindsay, D. A., 317
Lissi, E., 10(27), 12, 13(27), 15, 379
Liu, S. C., 192
Liu, X., 265
Livrea, M. A., 421, 425(2), 428, 430, 430(3)
Llesuy, S., 379
Llorca, L., 166
Lloret, A., 267
Locke, S. J., 4, 6(8), 8(8), 9(8), 10(8, 32), 14, 15, 20(3), 50, 379, 422
Loft, S., 83, 86
Lönnrot, K., 6(21), 7, 8(21), 9(21), 10(21), 11(21)
Lopez, T. R., 83, 87(12)
Loria, P., 193, 199
Lowry, O. H., 264, 293
Lubin, B., 191
Lui, J., 30
Luisada-Opper, A., 293

Lundanes, E., 431
Lund-Katz, S., 193, 199
Lusby, W. R., 390, 391, 394(4), 400, 400(4)
Lützenkirchen, F., 391
Lykkesfeldt, J., 83, 86

M

Ma, Y.-S., 36
Macauley, J. B., 318(6), 319, 320(6)
Mackay, E. M., 92
MacMillan, J. D., 418, 419(16)
MacNeil, J. M., 422
MaCord, J. M., 29
Maderia, V., 96
Madigan, D., 180
Maeda, Y., 107
Maelandsmo, G., 193, 199
Maes, D., 423
Magee, J. B., 186
Mahan, D. C., 318(10), 319, 320(10), 329(10)
Maiani, G., 15, 23, 50, 51(3)
Maier, G., 171
Makino, K., 29
Makkar, H. P. S., 174
Mallia, A. K., 365
Malterud, K. E., 91
Mandl, J., 83
Mangiapane, H., 91, 231
Maoka, T., 466, 467(42)
Marcheselli, F., 51
Maresi, E., 421
Margheri, G., 174, 176(43–46)
Margolis, S. A., 76
Marjanovic, B., 207, 230(7), 231(7)
Markham, K. R., 94, 113
Martin, H. L., 319(30), 320
Martin, S. A., 393
Martino, V. S., 230
Marx, F., 84
Masamoto, K., 442, 455(7)
Masoud, A., 302, 306(9)
Mastrangeli, A., 258
Masui, T., 107
Masumizu, T., 29
Matalon, S., 211
Mathews-Roth, M. M., 458
Mathieson, L., 91

Mathis, P., 414, 415(10), 416(10), 418(10), 420(10)
Matsuda, R., 109
Matsumoto, A., 201
Matsumoto, S., 309
Matsuno, T., 466, 467(42)
Matsuo, M., 309, 310
Matsuo, N., 208
Matsuzaki, T., 107
Matthews, R. H., 259, 260, 260(8), 262(8)
Mattiri, F., 139
Mattivi, F., 186, 187(12)
Mattson, F. H., 193
Maume, B. F., 186, 187
Maurer, R., 319(22), 320, 323(22)
Maurette, M. T., 92
Mauri, P. L., 93(29), 94
Maurizio, M., 51
Maxwell, S., 6, 15, 20(4), 50, 195, 201(33), 379
Maxwell, W. A., 418, 419(16)
May, J. M., 36, 83
Mayatepek, E., 348
McClure, D., 318(12), 319, 320(12)
McCord, J. D., 113
McDonagh, E. M., 191
McGarvey, D. J., 414, 415(11), 416(11), 418(11), 420(11), 428, 430(27)
McGuire, S. O., 318(13), 319, 320(13), 329(13)
McKenna, R., 379
McMurray, C. H., 319(17, 29), 320
McMurrough, I., 180
McMurtrey, D., 122
McMurtrey, K. D., 187
Mefford, I. N., 66
Mehta, R. G., 185
Meister, A., 258, 261, 282
Mejia, L. A., 430
Melchiorri, D., 14
Melikian, N., 96
Melnychuk, D., 258
Melnyk, R. A., 320, 324(34)
Menendez, E., 457, 461(3), 467(3)
Mengelers, M. J. B., 93(50), 105, 186
Mentges-Hettcamp, M., 164
Metsä-Ketelä, T., 6, 6(18, 19, 21), 7, 7(13), 8(13, 21), 9(16, 17, 21), 10(13, 21), 11(14–16, 21), 12(14–16, 18, 19), 14(16), 379
Meunier, P., 186
Meydani, M., 318(6), 319, 320(6)
Meydani, S. N., 318(6), 319, 320(6)

Meyer, A. S., 200
Meyer, T. H., 418, 419(18)
Michel, C., 207, 230
Middleton, E., 202
Migliori, M., 184
Milbradt, R., 330
Milikian, N., 210, 233(28)
Millán, A., 275
Millen, J. E., 458
Miller, E., 193
Miller, N. J., 4, 13, 15, 20(5), 50, 59(4, 9), 91, 195, 201(34), 202, 207, 208, 231, 231(11), 379, 385(14), 389(12, 14)
Milner, A., 13, 50, 59(4), 379
Minakami, S., 331
Miniati, E., 175
Minn, J., 122, 187
Mirzoeva, O. K., 208
Mitchell, J. B., 248
Mitjavila, S., 175
Mitsuta, K., 29
Miura, S., 91, 208, 231(11)
Mizuta, Y., 29
Mogyoros, M., 247
Mohr, D., 35, 347, 362, 364, 365, 366(12), 368(12), 369(12), 370
Moldeus, P., 37, 38(22), 39(22), 43(22), 47(22), 49(22), 84
Molnar, G., 6(21), 7, 8(21), 9(21), 10(21), 11(21)
Moncada, S., 234
Montedoro, G., 175, 176
Monties, B., 173
Montreau, F. R., 166
Moon, R. C., 185
Moore, L. L., 457
Mori, A., 28, 29, 30
Mori, M., 208
Mortensen, A., 408, 409, 410, 412(9), 413(9), 414, 415, 416, 417(13, 14), 418(13, 14), 420(5, 13, 14), 421(5), 428, 430(28)
Morton, R. E., 366
Motchnik, P. A., 86, 347
Motokawa, Y., 239
Mounts, T. L., 319(24), 320, 325(24)
Moutounet, M., 114, 167, 181, 182
Moxon, R. E. D., 65
Muckel, C., 271
Mueller, W. A., 458
Mukhtar, H., 208

Mulcahy, R. T., 248
Muller, J. M., 247
Müller-Mulot, W., 319(22), 320, 323(22)
Müller-Platz, C. M., 164
Mullholland, C. W., 10
Mulroy, L., 414, 415(11), 416(11), 418(11), 420(11)
Murphy, M. E., 309, 310(12)
Mychkovsky, I., 442, 455(10)
Myers, D. S., 13
Myers, T. E., 164

N

Nagashgima, H., 108
Nagel, C. W., 122, 135
Nagy, S., 120
Naik, D. B., 294
Nakajima, M., 347
Nakamura, M., 309, 310(8)
Nakamura, Y., 107
Namiki, M., 208
Nathan, C. F., 258
Neal, R. A., 294
Neff, W. E., 192
Negi, D. S., 425
Negre-Salvayre, A., 230
Nenseter, M. S., 91, 193, 199
Neuzil, J., 363, 364(9), 366(9), 373(9), 375(9)
Newman, R. H., 178
Newmark, H. L., 208
Newton, G. L., 248, 259, 277
Ng, E., 123, 124, 139, 141, 187
Nielsen, B. R., 410, 412(9), 413(9), 414, 415
Nielsen, J. B., 83
Nielsen, L. B., 35
Nielsen, S. E., 93(27), 94
Nielsen, T. B., 454
Nieminen, M., 6(18, 19), 7, 12(18, 19)
Nieto, S., 319(32), 320
Niki, E., 309, 422, 430
Nilsson, J., 37, 38(22), 39(22), 43(22), 47(22), 49(22)
Nobel, Y., 275
Noda, Y., 28, 31, 32, 33
Noorthy, P. N., 294
Nordberg, J., 239
Nordestgaard, B. G., 35

Norkus, E. P., 86
Norling, B., 330
Norman, Y., 193
Nursten, H. E., 180

O

Oberlin, B., 356
Ochi, H., 6(20), 7, 12(20)
Oesterhelt, G., 319(22), 320, 323(22)
Ogasawara, T., 4
Ogawa, S., 196, 230
Ohmori, S., 29
Ojala, A., 6(18), 7, 12(18)
Okada, K., 6(20), 7, 12(20)
Okamoto, K., 3
Okamura-Ikeda, K., 239
O'Keefe, D. O., 248
Okoh, C., 442, 455(10)
Okuda, H., 184
Oliveros, E., 92
Olsen, M. R., 410, 411, 412(6, 7)
Olson, B. J., 365
Olson, J. A., 343, 431, 432, 433, 433(3)
Ong, D. E., 455
Op den Kamp, J. A. F., 191
Orrenius, S., 84, 275
Orthofer, R., 113, 152
Osawa, T., 91, 208
Oszmaianski, J., 122, 137(1)
Oszmianski, J., 136
Oztezcan, S., 260

P

Pacakova, V., 66
Pace-Asciak, C. R., 184, 185(4)
Pachla, L. A., 23
Packer, J. E., 422
Packer, L., 8, 28, 31, 32, 33, 83, 239, 240, 246, 246(12), 247, 249, 254, 254(7), 274, 288, 300, 301, 301(4), 303(7), 309, 330, 331, 332, 333(20), 334, 335, 336(20), 338, 339, 340, 340(20), 342, 422, 428(10, 12), 460
Padmaja, S., 210
Padulo, G. A., 423
Paganga, G., 91, 92, 93(48), 96, 103, 195, 201(34), 202, 207, 208, 210, 230(8), 231(8, 11), 233(28), 379, 389(12)
Palatini, P., 277
Palek, J., 192
Palladini, G., 93(51), 105
Pallardo, F. V., 267, 274, 275
Palozza, P., 428
Pan, X.-M., 35
Pandolfo, L., 428
Pannala, A. S., 92, 97(10, 11), 207, 211, 220, 222, 223, 225, 226, 227, 228(40), 229, 230, 234
Papaconstantinou, E., 160
Pargament, G. A., 211
Park, J. B., 66, 72(9)
Park, J.-Y., 84, 88(17)
Parker, A. W., 414, 415(11), 416(11), 418(11), 420(11)
Parker, R. A., 330
Parks, E. J., 193, 196, 198, 198(41), 199(29, 41), 200
Parks, J. G., 138
Parthasarathy, S., 35, 48(1), 192, 193, 362
Pascoe, G. A., 309, 310(11), 370
Pascual, C., 10(27), 12, 13(27), 15
Pascula, C., 379
Pasquini, N., 175
Pastena, B., 171
Pasternack, A., 6(20), 7, 12(20)
Pataki, G., 211
Patton, S., 192, 194(14)
Pearce, B. C., 330
Pearson, A. B., 318(9), 319, 320(9), 329(9)
Pearson, D. A., 200
Peleg, H., 138, 200
Pelle, E., 423
Pellegrini, N., 379
Peltola, J., 6(21), 7, 8(21), 9(21), 10(21), 11(21)
Pennington, J. A. T., 309
Perego, R., 93(51), 105
Perez-Ilzarbe, J., 181
Peri, C., 175, 176(49)
Perry, T. L., 258
Pesek, C. A., 419
Peters, J., 460
Peters, R. C., 318(12), 319, 320(12)
Petersen, M. A., 86
Peterson, B. H., 441
Peterson, D. M., 330

Petrone, M., 277
Peyron, 186
Pezet, R., 122, 186, 187
Pezzuto, J. M., 185
Pfander, H., 460
Pfeilsticker, K., 84
Pflieger, G., 164
Philips, L., 315
Phillips, M. C., 193, 199
Phiniotis, E., 165
Phul, H., 36
Piazzi, S., 342
Pick, U., 239
Pieri, C., 51
Pietilä, T., 6(21), 7, 8(21), 9(21), 10(21), 11(21)
Pietrzik, K., 23
Pietta, P. G., 93(29), 94
Piironen, V., 318, 318(15), 319, 319(5), 323(5, 15), 326(15)
Pinchuk, I., 37, 38(21), 39(21), 41(21), 43(21), 47(21)
Pintaudi, A. M., 421, 425(2)
Pirrone, L., 177
Pirttilä, T., 6, 11(15), 12(15)
Piskula, M., 230
Plá, R., 275
Plass, G., 192
Plopper, C. G., 243
Pobanz, K., 187
Pocklington, W. D., 319(19), 320, 323(19), 325(19), 326(19)
Podda, M., 330, 331, 334, 335, 338, 339, 340
Poderosso, J. J., 211
Poeggler, B., 14
Polidori, E., 176
Pompei, C., 175, 176(49)
Pont, V., 122, 187
Poor, C. L., 457
Poorthuis, B. J. H. M., 342
Pope, M. T., 160
Popov, I. N., 6
Porter, J. L., 178
Porter, L. J., 178, 179, 182
Potapovitch, A. I., 92
Potter, D. W., 274
Pou, S., 458
Poulsen, H. E., 83, 86
Poussa, T., 6(19), 7, 12(19)
Powell, R., 301
Prescott, M. C., 400

Price, S. F., 113, 114, 119(6), 135
Pridham, J., 96, 210, 233(28), 379
Priemé, H., 83
Prieur, C., 181, 182
Prior, R. L., 50, 51(7), 52, 57, 57(12), 58, 60(17)
Probanz, K., 122
Provenzano, M. D., 365
Pryor, R. L., 379
Pryor, W. A., 228, 379, 423
Puhl, H., 36, 37, 37(17), 38(20), 39(17), 41(20), 44(20), 48(17, 20), 191, 192, 193(21), 194(4), 362
Pushkareva, M. A., 208
Puskas, F., 83
Putnam, D. H., 319(27), 320, 323(27)

Q

Qju, J. H., 191
Qu, Z. C., 83
Quehenberger, O., 362
Quenhenberger, O., 190, 193(1)
Quideau, S., 97
Qureshi, A. A., 330

R

Rabek, J. F., 414, 415(12)
Rabin, R., 248
Rabl, H., 192, 193(21)
Radi, R., 211
Rahmani, M., 233
Raju, P. A., 248
Ralph, J., 97
Ramis-Ramos, G., 159
Rammell, C. G., 318(9), 319, 319(23), 320, 320(9), 323(23), 326(23), 329(9)
Ramos, T., 122, 136, 137(1)
Randall, R. J., 264, 293
Rankin, S. M., 91, 231
Rantalaiho, V., 6(20), 7, 12(20)
Rao, M. N., 441
Rapp, J. H., 35
Rausch, W. D., 331
Rava, A., 94
Raynor, T., 428
Raynor, W. J., 319(31), 320

Razaq, R., 92, 97(11), 234
Re, R., 379, 421, 430, 430(3)
Reaven, P., 193
Reay, C. C., 457
Recchioni, R., 51
Reddy, K. J., 442, 455(7)
Redegeld, F. A. M., 267
Reed, D. J., 274, 309, 310(11), 370
Regnström, J., 37, 38(22), 39(22), 43(22), 47(22), 49(22)
Reich, A., 43, 44, 44(24), 46(24), 348
Reichmann, H., 331
Reiter, R. J., 14
Reithman, H. C., 442
Remmer, H., 191
Rendall, N., 105
Reniero, F., 186, 187(12)
Rettenbaier, R., 351
Reynolds, D. L., 23
Reynolds, J. A., 454
Rhodes, M. E., 120
Ricardo da Silva, J. M., 122, 137, 137(2), 181
Riccio, A., 421
Rice-Evans, C., 4, 13, 15, 20(5), 50, 59(4, 9), 91, 92, 93(47, 48), 96, 97, 97(10, 11, 30), 103, 105, 195, 201(34), 202, 207, 208, 210, 211, 228(40), 230(8, 10), 231, 231(8, 10, 11), 233(28), 234, 379, 385(14), 389(12, 14)
Richardson, N., 13
Richer, S. P., 78, 82(9)
Richoz, J., 208
Ridnour, L. A., 258
Riederer, P., 331
Rigaud, J., 114, 178, 181, 182
Rikans, L. E., 83, 87(12)
Risch, B., 107
Ristau, O., 269
Ritter, G., 171
Rivas-Gonzalo, J. C., 180
Robak, J., 92
Robison, W. G., Jr., 426(30), 430
Roederer, M., 248, 254(15)
Roelofsen, B., 191
Roggero, J.-P., 114, 123, 137(10)
Rohrer, G., 319(22), 320, 323(22)
Romero-Pérez, A. I., 114, 123, 186, 187(11), 188(19)
Rongliang, Z., 230
Rorarius, H., 379

Rose, M. E., 390, 394(8)
Rose, R. C., 77, 78, 80, 81, 82(9), 83, 85
Rosebrough, N. J., 264, 293
Rosec, J.-P., 122, 137(2)
Rosen, G. M., 458
Rosette, C., 330
Rossi, J. A., Jr., 153, 154(2), 155, 156(9), 157, 161(9)
Rotheneder, M., 37, 38(20), 41(20), 44(20), 48(20)
Rotrosen, D., 73
Rouseff, R. L., 93(31), 95
Roy, S., 239, 240, 247, 249, 254, 254(7)
Royall, J. A., 211
Rubin, S. H., 86
Ruffolo, J. J., 458
Rumsey, S. C., 65, 73(2)
Rupec, M. R. A., 247
Rustan, A. C., 193, 199
Rüttimann, A., 462
Ryzewski, R. N., 439

S

Sacchetta, P., 270
Saigo, H., 109
Saito, T., 422
Sakan, T., 418, 419, 419(15)
Salagoity-Auguste, M.-H., 122, 123(6), 137(6)
Salah, N., 91, 208, 231(11)
Salgues, M., 165
Salim-Hanna, M., 10(27), 12, 13(27), 15, 379
Salminen, K., 318, 318(15), 319, 319(5), 323(5, 15), 326(15)
Salter, A., 231
Saltini, C., 258
Salvayre, R., 230
Sampson, J., 96, 210, 233(28), 379, 385(14), 389(14)
Sander, D. N., 208
Sander, L. C., 390, 391(11)
Sano, M., 91, 208
Santa-Maria, G., 176
Santos-Buelga, C., 180
Saran, M., 207, 230
Sarni-Manchado, P., 182
Sashwati, R., 288
Sastre, J., 267, 274, 275

AUTHOR INDEX 485

Sattler, W., 347, 365, 366(12), 368(12), 369(12)
Sayer, J. M., 208
Sbachi, M., 187
Sbaghi, M., 186
Scalbert, A., 154, 173
Scalia, M., 92
Schafer, Z., 37, 38(21), 39(21), 41(21), 43(21), 47(21)
Schalch, W., 408, 458
Schaper, T. D., 164
Schaur, R. J., 193
Schecter, R. L., 258
Schell, D. A., 77, 81(7)
Schiavon, M., 175
Schiedt, K., 460, 461(35, 36), 462
Schierle, J., 348, 351, 358
Schilling, A. B., 390, 391(11)
Schippling, S., 46
Schlegel, J., 275
Schlotten, G., 167
Schmidli, B., 334, 342
Schmidt, A. P., 197
Schmidt, K., 319(22), 320, 323(22)
Schmitz, H. H., 390, 394(2), 457
Schneeman, B. O., 196, 198, 198(41), 199(41)
Schneider, V., 172
Schnitzer, E., 37, 38(21), 39(21), 41(21), 43(21), 47(21)
Schoffa, G., 269
Schofield, D., 15, 51, 52(11), 60(11)
Schou, A., 93(27), 94
Schrijver, J., 342
Schüep, W., 348, 351, 356, 358
Schultz, T. P., 122, 187
Schwartz, H. J., 390, 394(2)
Schwartz, R., 193, 199
Schwartz, S. J., 390
Schwarz, K., 196
Schwarz, W., 358
Scita, G., 422, 428(12), 460
Scott, B. C., 92, 207, 209, 230(8), 231(8)
Seddon, J. M., 459
Seely, G. R., 418, 419(18)
Seims, W. G., 460
Sen, C. K., 239, 247, 249, 254, 254(7), 256, 288, 300, 301(4)
Sen, K., 250
Serafini, M., 15, 50, 51(3)
Serbinova, E., 301, 330, 422, 428(12)
Serry, M. M., 91

Sewerynec, E., 14
Seybert, D. W., 13
Shapiro, A. C., 318(6), 319, 320(6)
Sharma, S. K., 419
Sharp, R. J., 4
Sheppard, A. J., 309
Sherman, L. A., 442, 455(7)
Sherman, M. M., 442, 455(7)
Shigenaga, M. K., 3, 4(4), 83, 88, 88(1)
Shimada, K., 347
Shimasaki, H., 430
Shimizu, M., 466, 467(42)
Shimmei, M., 28, 31, 32, 33
Shimoi, K., 107
Shin, T.-S., 319(26), 320
Shirahama, H., 309, 310(8)
Shivji, G. M., 208
Shoji, K., 208
Short, S. M., 425
Shvedova, A., 301
Sichel, G., 92
Siegel, F. L., 248
Siemann, E. H., 187
Sies, H., 267, 271, 294, 330, 358, 390, 391, 392, 393(17), 394(14, 17), 398(17), 399(17), 400(14), 401(14), 405, 406, 458, 460
Simic, M. G., 207, 230(7), 231(7)
Simpson, K. L., 343
Singh, S., 92, 97(10), 207, 211, 228(40), 234
Singleton, V. L., 113, 135, 137, 152, 153, 154(2), 155, 156, 156(9), 157, 161, 161(9), 162, 163(16), 164, 165, 165(11), 166, 166(11), 169(11), 171, 172, 172(36), 173, 173(36–38), 174(6)
Sinha, S., 342
Skibsted, L. H., 408, 409, 410, 411, 412(6–9), 413(7–9), 414, 414(7), 415, 415(7), 416, 417(13, 14), 418(13, 14), 420(5, 7, 13, 14), 421(5), 428, 430(28)
Slater, A., 91, 275
Slater, T. F., 192, 194(15), 422
Slinkard, K., 156, 165(11), 166, 166(11), 169(11)
Sloane-Stanley, G. H., 301, 303(8)
Sloots, L. M., 201
Slowing, K. V., 185
Smith, E., 343
Smith, J. C., 433
Smith, J. C., Jr., 391, 460
Smith, P. K., 365

Smith, W. E., 309
Smith, W. P., 423
Snodderly, D. M., 458, 459
Sofic, E., 52, 57, 57(12)
Solakivi, T., 6(17), 7, 9(17)
Soleas, G. J., 122, 123, 124, 128, 138, 139, 141, 150(2), 184, 185(4), 187
Solfrizzo, M., 208
Somers, T. C., 167
Sommer, P. F., 334, 342
Sommerburg, O., 460
Sorrell, M. F., 293
Souquet, J. M., 114, 178, 181, 182
Soyland, E., 193, 199
Spangler, C. J., 460
Speek, A. J., 342
Spengler, B., 391, 392, 393
Sperduto, R. D., 459
Spitz, D. R., 258, 259, 260, 260(8), 262(8)
Sprague, K. E., 457
Spranger, T., 43, 44, 44(24), 45, 46(24)
Springer, T., 348
Squadrito, G. L., 228
Srivastava, S. K., 270
Staal, F. J. T., 248
Stadler, R. H., 208
Stahl, W., 358, 390, 391, 392, 393(17), 394(14, 17), 398(17), 399(17), 400(14), 401(14), 405, 406
Stalik, K., 66
Stamler, J. S., 258
Stanley, K. K., 35
Stanley, W. C., 274
Staprans, I., 35
Starka, J., 300
Steck, A., 460
Steenken, S., 207, 230(7), 231(7)
Stefan, C., 275
Stein, O., 193, 199
Stein, Y., 193, 199
Steinberg, D., 35, 48(1), 192, 193, 362, 441
Steiner, K., 358
Stephens, R. J., 425
Stocker, R., 35, 36, 37(18), 48, 48(18), 331, 333, 347, 362, 362(6), 363, 364, 364(7, 9), 365, 366(7, 9, 12), 367, 368(12), 369(12), 370, 370(7, 8), 371, 371(18), 372, 372(18), 373, 373(9), 374, 375(9, 18), 422, 428, 428(11)
Stocks, J., 4, 191, 192

Stone, B. A., 97
Stone, W. L., 36
Stoscheck, C. M., 164
Strain, J. J., 10, 15, 16, 16(2), 19, 20, 20(2), 21, 22, 23, 23(10), 24(1, 2, 10), 25, 26, 27(2, 10), 50, 51(8)
Striegl, G., 37, 38(20), 41(20), 44(20), 48(20), 365
Ström, K., 37, 38(22), 39(22), 43(22), 47(22), 49(22)
Suarna, C., 36, 333
Subar, A. F., 441
Subbarayan, C., 455
Sud'ina, G. F., 208
Suematsu, S., 109
Sugano, M., 208, 323
Sullivan, A. R., 172, 173(37)
Sullivan, J. L., 341, 348(1)
Sumbutya, N. V., 208
Sun, Y., 93(34), 95
Sund, R. B., 91
Sundquist, A. R., 458
Surico, G., 208
Suzuki, T., 107
Suzuki, Y., 107
Swain, T., 163
Swanson, C., 301
Syvaoja, E.-L., 318, 318(15), 319, 319(5), 323(5, 15), 326(15)
Szaro, R. P., 259
Szuts, E. Z., 433

T

Tagaki, M., 4
Takata, K., 171
Takatsuki, K., 83
Takayanagi, R., 331
Takeshige, K., 331
Tamashita, K., 208
Tamura, M., 309, 310(8)
Tan, Y., 390, 391(11)
Tanahashi, H., 201
Tappel, A. L., 191, 196, 197(38, 39), 198(38, 39)
Tarsis, S. L., 457
Tate, S. S., 261
Tavender, S. M., 414, 415(11), 416(11), 418(11), 420(11)

Tavernier, J., 175
Taylor, G., 105
Taylor, P., 319(21), 320, 323(21), 325(21), 326(21)
Tedesco, C. J. G., 191
Teisedre, P. L., 138
Teissedre, P. L., 123, 199, 200
Telfer, A., 408
Tel-Or, E., 83
Terada, S., 107
Terao, J., 230
Tesoriere, L., 421, 425(2), 427, 428, 429, 430, 430(3)
Tessier, F., 73
Thomas, C. F., 185
Thomas, D. W., 425
Thomas, S. R., 363, 364(9), 366(9), 373(9), 374, 375(9)
Thompson, J. N., 318, 319(1), 320(1), 323(1), 325(1)
Thompson, M., 92
Thomson, A. D., 293
Thomson, J., 91, 231
Thorpe, G. H. G., 6, 15, 20(4), 50, 195, 201(33), 379
Thurnham, D. I., 23, 343
Tietze, F., 259, 268, 277
Tijburg, L., 91, 208, 231, 231(11)
Timberlake, C. F., 179
Tinkler, J. H., 414, 415(11), 416(11), 418(11), 420(11)
Tirosh, O., 293, 294, 296(5), 299(5)
Tjani, C., 73
Tolliver, T. J., 319(28), 320
Tomas, C., 157
Tomas-Mas, C., 159
Tomita, I., 91, 107, 208
Tomita, T., 91, 208
Tomita, Y., 208
Tomlinson, G., 123, 138
Tonon, D., 174, 176(43–45)
Torina, G., 171
Tosic, M., 207, 230(7), 231(7)
Tournaire, C., 92
Toyokuni, S., 3, 6(20), 7, 12(20)
Traber, K., 300, 301(4)
Traber, M. G., 330, 331, 334, 335, 338, 339, 340
Tran, K., 428
Trela, B. C., 187
Tribble, D., 48

Tripodi, A., 193, 199
Tritschler, H., 239, 240, 246, 301
Trollat, P. J., 186
Troly, M., 230
Trousdale, E., 165
Truscott, T. G., 414, 415(11), 416(11), 418(11), 420(11), 428, 430(27)
Tsang, E., 123, 187
Tsopu, S. C. S., 343
Tsuchiya, M., 460
Tsukida, K., 418, 419(17)
Tuihala, R. J., 379

U

Ubbink, J. B., 331, 347
Ublacker, G. A., 248
Udeani, G. O., 185
Ueda, T., 318(7), 319, 320(7), 326(7), 329(7)
Unten, L., 419, 421(24)
Upston, J. M., 373
Urano, S., 310
Ursin, G., 441

V

Valenza, M., 421
Valenzuela, A., 319(32), 320
Valladao, M., 114, 119(6)
van Bennekom, W. P., 267
van Breeman, R. B., 390, 394(2)
van Breemen, R. B., 390, 391(9–11)
Van den Berg, J. J. M., 191
Van den Dobbelsteen, D. J., 275
van der Berg, H., 23
Vanderslice, J. T., 73
Van der Vliet, A., 228
van Kujik, F. J. G. M., 425, 457, 460
van Leeuwen, S. D., 186
van Norren, D., 458
van Schaik, F., 23
van Trijp, J. M. P., 93(50), 105
Van Zoeren-Grobben, D., 342
Varga, N., 208
Varo, P., 318, 318(15), 319, 319(5), 323(5, 15), 326(15)
Varolomeev, S. D., 208

Varvaro, L., 208
Vatassery, G. T., 309
Vecchi, M., 334, 342, 462
Venema, D. P., 202
Verdon, C. P., 50, 51(7), 379
Vermaak, W. J., 331
Vermeglio, A., 414, 415(10), 416(10), 418(10), 420(10)
Versini, G., 174, 176(46)
Vestal, M. L., 393
Vettore, L., 191
Vicente, T. S., 319(18), 320, 323(18)
Vidal, I., 457, 461(3), 467(3)
Vierira, O., 96
Viguie, C., 274
Villa, A., 319(32), 320
Villa, D., 177
Viña, J., 267, 274, 274(1), 275, 276
Vinson, J. A., 91
Vitenberg, A. G., 197
von-Laar, J., 358
Votila, J. K., 379
Vrhovesk, U., 186
Vuilleunier, J. P., 361

W

Waeg, G., 36, 37(17), 39(17), 48(17), 192, 193(21), 365
Wakabayashi, H., 347
Walker, B. E., 13
Walker, R., 208
Waller, H. D., 191
Wallet, J. C., 92, 230
Walling, C., 363, 370(8), 428
Wang, H., 50, 51(7), 57, 379
Wang, T., 258
Wang, W., 457, 461(3), 467(3)
Wang, Y., 65, 66, 66(1), 72(9), 73(1), 74, 77, 83, 84(9)
Wang, Y. H., 318(14), 319, 320(14), 329(14)
Wang, Y. M., 309
Warner, C. G., 319(24), 320, 325(24)
Warthesen, J. J., 419
Washko, P. W., 65, 66, 66(1), 72(9), 73, 73(1), 74, 74(7), 77, 81, 83, 84(9)
Watanabe, J., 91, 208
Waterhouse, A. L., 113, 114, 119(6), 123, 128, 136, 138, 139, 141, 184, 185, 185(2), 186, 186(8), 187, 187(11), 188(8, 19), 193, 199, 200
Watson, B. T., 114, 119(6), 135
Wayner, D. D. M., 4, 6(8), 8(8), 9(8), 10(8, 32), 14, 15, 20(3), 50, 379
Waysk, E. H., 319(18), 320, 323(18)
Weaver, L., 460
Webb, A., 311, 317, 318, 319(4), 332
Weber, C., 330, 331, 334, 335, 338, 339, 340
Weber, S., 351
Webster, N. R., 13
Wehr, C. M., 84, 88(17)
Wei, C. C., 460
Weihrauch, J. L., 309
Weinmuller, M., 331
Weintraub, R. A., 93(31), 95
Weiter, J. J., 459
Welch, K. J., 294
Welch, R. W., 65, 66, 66(1), 72(9), 73, 73(1), 77
Wells, F. B., 258
Wells, M. A., 442
Weltman, J. K., 259
Wendelin, S., 120, 186
Wermeille, M., 93(33), 95
West, C., 270
Westerlund, D., 259
White, D. A., 91, 231
Whitehead, T. P., 6, 15, 20(4), 50, 379
Whiteman, M., 211, 228(39)
Whitesell, R. R., 83
Widmer, C., 330
Wightman, J. D., 135
Wiklund, O., 6
Wildenradt, H. L., 152
Willet, W., 459
Williams, B. D., 319(18), 320, 323(18)
Williams, C. R., 92, 207
Williams, V., 179
Willson, R. L., 422
Wilson, M. T., 234
Wingerath, T., 390, 391, 392, 393(17), 394(14, 17), 398(17), 399(17), 400(14), 401(14), 403, 405, 406
Winters, R. A., 258, 259, 260(8), 262(8)
Wirta, O., 6(20), 7, 12(20)
Witt, E. H., 240, 301
Witting, P. K., 362, 367, 371, 371(18), 372, 372(18), 373, 374, 375(18)
Witztum, J. L., 35, 48(1), 192, 193, 362

Wojtaszk, P., 97
Wolfender, J. L., 150, 151(11)
Wood, A. W., 208
Wood, I. S., 459
Woolemberg, A., 269
Wootton, R., 35
Wright, J. J., 330
Wrolstad, R. E., 135
Wu, A. H. B., 50, 51(7), 379
Wu, C. W., 259
Wu, F. Y. H., 259
Wulf, L. W., 122, 135
Wyss, R., 433

X

Xu, M. W., 86
Xu, R., 196, 198, 198(41), 199(41)

Y

Yadan, J.-C., 276, 277
Yagi, H., 208
Yamada, K., 208
Yamahita, S., 347
Yamamoto, Y., 347, 369, 430
Yamanoi, S., 310
Yamato, S., 347
Yan, D. G., 91
Yan, J., 124, 128, 139, 141, 187
Yan, L.-J., 331
Yanada, J., 208
Yang, C. S., 93(34), 95
Yang, M., 379
Yang, Z.-Y., 93(34), 95
Yannuzzi, L. A., 459

Yao, Q., 230
Yarbrough, L. R., 259
Ye, Y., 211
Yegudin, J., 258
Yen, G. C., 208
Yi, O.-S., 200
Yodoi, J., 3
Yokota, M., 418, 419(17)
Yokotsuka, K., 135
Yong, J., 230
Yoshida, Y., 107, 108, 112(1)
Yost, R. A., 93(31), 95
Young, A., 400
Youting, C., 230
Yowe, D. L., 84, 88(17)
Yu, R., 77

Z

Zagalsky, P. F., 442, 454, 455
Zamor, J., 461, 467(39)
Zaspel, B. J., 318, 319(2), 320(2)
Zaya, J., 165
Zeng, S., 343
Zhang, L.-X., 442
Zheng, R. L., 230
Zhingjian, J., 230
Zhou, J. R., 454
Zhou, M., 91
Zhou, Y. C., 230
Zhu, L., 211, 228(31)
Ziegler, R. G., 441
Zielinska, E., 208
Ziemelis, G., 167
Zimmerlin, A., 97
Zimmerman, W. F., 428
Zollner, H., 193
Zühlke, U., 461
Zukowski, J., 259, 260(8), 262(8)

Subject Index

A

AAPH, *see* 2,2′-Azobis[2-amidinopropane] hydrochloride
ABTS, *see* 2,2′-Azinobis(3-ethylbenzothiazoline-6-sulfonic acid)
ADT, *see* Anethole dithiolthione
Age-related macular degeneration, protection by macular pigments, 459
AMD, *see* Age-related macular degeneration
Amidothionophosphates
　antioxidant activity mechanism, 294–295
　peroxyl radical reactivity, 299
　reactivity
　　hydrogen peroxide decomposition, 296
　　lipid hydroperoxide reduction, 297–300
　　sodium hypochlorite, 296–297
　synthesis
　　2-hydroxyethylamidoethyl thionophosphate, 295
　　N,N',N''-tripropylamidothionophosphate, 296
Anethole dithiolthione
　effects on nuclear factor-κB, 300–301
　high-performance liquid chromatography with electrochemical detection
　　cell culture, 302
　　chromatography conditions, 302–303
　　extraction, 301–306
　　instrumentation, 301
Anthocyanins, *see* Phenolic antioxidants
Ascorbic acid, *see also* Ferric reducing/antioxidant power and ascorbic acid concentration assay
　high-performance liquid chromatography with electrochemical detection
　　amperometric detection, 66
　　chromatography conditions, 68, 70–72
　　coulometric detection, 66–67, 72
　　hydrodynamic voltammetry plot, 79, 82
　　instrumentation, 67–68, 78–79
　　materials, 68, 78
　　passivation, 72
　　sample preparation, 67, 72–73, 79–82
　　standards, 72
　hydroxyl and superoxide anion radical scavenging, 28, 32, 34
　radiolabeled compound detection, 75–76
　recycling from dehydroascorbic acid measurement from various systems, 83–84
　rat hepatocyte assay
　　aging effects on results, 87–88
　　hepatocyte isolation, 84–86
　　high-performance liquid chromatography with electrochemical detection, 86–87
　　incubation conditions, 86
　　materials, 84
　　pH, 85–86
　stability in solution, 65, 79–80
2,2′-Azinobis(3-ethylbenzothiazoline-6-sulfonic acid)
　radical cation spectrum, 379–380
　Trolox equivalent antioxidant capacity assay
　　decolorization assay, 384, 389
　　extraction and analysis of tomato, 383–385, 387, 389
　　materials, 381
　　preparation of preformed radical cation, 381
　　principle, 380
　　standards, preparation, 381–383
2,2′-Azobis[2-amidinopropane]hydrochloride
　decomposition and antioxidant consumption, 4
　plasma oxidizability assay, 38–40

C

Caffeine, high-performance liquid chromatography analysis in tea with catechins

chromatography conditions, 109, 111
comparison of green teas, 112–113
instrumentation, 108–109
quantitative analysis, 111–112
sample extraction, 109
standards, 109
Carotenoids, *see also* Cellular carotenoid-binding protein; Lycopene; Vitamin A; Zeaxanthin
functions, 408, 441–442
high-performance liquid chromatography, 390, 392
lutein
  distribution in retina, 457
  oxidative metabolites, 460–461
  photooxidation protection of retina, 457–460
matrix-assisted laser desorption ionization–mass spectrometry
  extraction, 391–392
  fragment ion mass analysis, 394–395, 398–401, 406–408
  green lettuce analysis, 403, 406
  instrumentation, 392–393
  molecular ion species and sensitivity, 393–394
  overview of ionization techniques, 390–391
  postsource decay, 391, 393
  standards and samples, 391
  tangerine juice concentrate analysis, 401, 403
photobleaching
  application to food systems, 419
  degradation products, 418–419
  flash photolysis, 416–418
  mechanisms, 409
  oxygen role, 420
  radical efficiency assay, 420
  steady-state photolysis
    quantum yield, 409–410, 412
    sensitized photolysis, 413–415
    unsensitized photolysis, 410, 412–413
Trolox equivalent antioxidant capacity assay
  decolorization assay, 384, 389
  extraction and analysis of tomato, 383–385, 387, 389

materials, 381
preparation of preformed 2,2′-azino-bis(3-ethylbenzothiazoline-6-sulfonic acid) radical cation, 381
principle, 380
standards, preparation, 381–383
Catechins, *see also* Phenolic antioxidants
absorption and excretion in humans, 95
antioxidant activity, 207–208
high-performance liquid chromatography
  chromatography conditions, 103–105
  food sample preparation, 100, 103
  retention times, 105–106
  standards, 100
  tea analysis with caffeine
    chromatography conditions, 109, 111, 203–204, 206
    comparison of green teas, 112–113, 203
    extraction, 204
    instrumentation, 108–109
    precision, 204, 206
    quantitative analysis, 111–112
    sample extraction, 109
    stability of samples, 206
    standards, 109, 204
  urine sample
    dietary manipulation, 97, 100, 106
    preparation, 103
inhibition of tyrosine nitration by peroxynitrite, 215–216, 231–232
tea types and structures, 107–108, 202–203
CCBP, *see* Cellular carotenoid-binding protein
Cellular carotenoid-binding protein
absorption spectra studies of binding, 451–452, 455–456
binding assays, 447–448, 452–454, 456
functions, 456
gel electrophoresis, 448, 453–454
iodine-125 labeling, 449
ligand specificity, 454–455
purification from ferret liver
  animals and diet, 443
  anion-exchange chromatography, 443–444, 449
  apoprotein component release from complex, 444

β-carotene affinity chromatography
  chromatography conditions, 446–447, 450
  column preparation, 444–446
  gel filtration, 444, 449–450
  homogenization, 443
  purification table, 450–451
  species comparison, 442, 455
Chemiluminescence
  free radical measurement in lipoproteins, 4–6
  total peroxyl radical-trapping potential assay enhancement
    accuracy assessment, 9
    contributions of individual components, 10–14
    effects on results
      age, 11–12
      antioxidant supplementation of diet, 10
      disease states, 12, 14
      fasting, 9–10
      gender, 11–12
      metabolic rate, 14
      pH of assay buffer, 13
      smoking, 10
    low-density lipoprotein assay, 8–9
    plasma assay, 7–8
    principle, 6–7
    stoichiometric peroxyl radical-scavenging factors, 10
    total antioxidant reactivity, 12–13
Coenzyme Q
  functions, 330–331
  simultaneous determination with vitamin E
    animal maintenance, 334–335
    chromatography conditions, 337–338, 340, 344, 347
    electrochemical detection, 333–334, 342, 344–345
    extraction, 331–332
    materials, 334
    plasma sample preparation, 344
    recovery and reproducibility, 340–341, 345
    saponification, 332–333
    standards, 336–337, 342–344, 347
    storage of samples, 332

Coumaric acid, peroxynitrite interactions
  m-coumaric acid, 223, 234
  o-coumaric acid, 222–223, 233–234
  p-coumaric acid, 219–222, 233

D

Dehydroascorbic acid
  high-performance liquid chromatography assays
    coulometric electrochemical detection following reduction
      hydrodynamic voltammetry plot, 79, 82
      instrumentation, 67–68, 78–79
      materials, 68, 78
      reduction reaction, 75–76, 81
      sample preparation, 67, 72–73, 79–82
      standards, 74–75
    derivatization, 73–74
    radiolabeled compound detection, 75–76
  recycling to ascorbic acid, see Ascorbic acid
  stability in solution, 65, 79–80
DHLA, see Dihydrolipoic acid
Dihydrolipoic acid, high-performance liquid chromatography with electrochemical detection
  cell culture, 243
  chromatography, 241
  coulometric detection, principle, 240–241
  current–voltage response curve, 242
  electrodes, 240
  extraction, 243, 245–246
  instrumentation, 241
  standards, 242–243
Dithiolthione, see Anethole dithiolthione

E

Ebselen, antioxidant activity, 294
Electrochemical detection, see High-performance liquid chromatography
Electron spin resonance
  hydroxyl radical scavenging assay

ascorbic acid contribution, 28, 32, 34
electron spin resonance measurements, 29–30
reaction conditions, 30
sample preparation, 29
standards, 30
superoxide anion radical scavenging assay
ascorbic acid contribution, 28, 32, 34
calibration, 32
electron spin resonance measurements, 29–30
reaction conditions, 28–31
sample preparation, 29
ESR, *see* Electron spin resonance

F

Ferric reducing/antioxidant power assay
advantages in total antioxidant power determination, 15–16
automated assay, 16, 18
comparison with oxygen radical absorbance capacity assay, 58–60
manual assay, 18–19
plasma assay, 22–23
precision and sensitivity, 21
principle, 16–17
reaction characteristics of pure antioxidants, 19–21
reagent preparation, 17
sample preparation, 18
standards and controls, 17–18
Ferric reducing/antioxidant power and ascorbic acid concentration assay
calculations, 25–27
plasma assay, 26–27
principle, 23–25
standards, 24–25
Ferulic acid, *see also* Phenolic antioxidants
antioxidant activity, 96–97
α_1-antiprotease protection against hypochlorous acid inactivation, 209
biosynthesis, 208
high-performance liquid chromatography
chromatography conditions, 103–105
food sample preparation, 100, 103
retention times, 105–106
standards, 100
urine sample
dietary manipulation, 97, 100, 106
preparation, 103
peroxynitrite interactions, 223, 228, 234
Flavonoid antioxidants, *see also* Phenolic antioxidants
absorption and excretion in humans, 92–95
high-performance liquid chromatography
chromatography conditions, 103–105
food sample preparation, 100, 103
retention times, 105–106
standards, 100
urine sample
dietary manipulation, 97, 100, 106
preparation, 103
Flow cytometry, thiol assay
advantages, 247, 257
cell preparation, 248–249
data collection, 251–252
differential assessment of thiols, 249–251, 258
monobromobimane modification of thiols, 248–249, 251–253, 257–258
monochlorobimane detection of glutathione, 248
Folin–Ciocalteu reagent, total phenol analysis
advantages, 177–178
chemistry of reaction, 160–161
cinchonine precipitations, 175–176
comparison with other techniques, 169–171
flavonoid–nonflavonoid separations, 170–173
flow automation, 156–158
Folin–Denis reagent comparison, 154–155
incubation conditions and color development, 155–156
interference
ascorbic acid, 166–167
nucleic acids, 164
oxidation, 164
proteins, 164–165
sugars, 165–166
sulfite, 167–169
sulfur dioxide, 167–168

molar absorptivity of specific phenols, 161–163
phloroglucinol addition, 173
preparation of reagent, 155
sample preparation, 159
standards and blanks, 158–159
tannin separations, 174–175
values for various wines, 176–177
volume of reaction, 156
FRAP assay, see Ferric reducing/antioxidant power assay
FRASC assay, see Ferric reducing/antioxidant power and ascorbic acid concentration assay

G

Gas chromatography headspace assay, volatile lipid oxidation products
  diet effects on lipid peroxidation, 198–200
  instrumentation, 197
  overview, 196–197
  phenolic antioxidant effects, *in vitro* assays, 200
  sample preparation, 197–198
  wine phenolics, evaluation, 199–201
Gas chromatography–mass spectrometry
  phenolic antioxidants in wine
    costs, 151
    extraction and characterization, 142–143
    gas chromatography conditions, 141
    instrumentation, 139, 141–142
    linearity, 148
    mass spectrometry conditions, 142
    materials, 139
    peak identification, 143–145
    precision, 149–150
    recovery, 148–149
    resolution, 145, 147–148
    sample preparation, 139
    sensitivity, 149–150
    standards, 139–140
  α-tocopherol
    calibration curve preparation, 311, 317–318
    extraction and derivatization, 311–312
    instrumentation, 312
    oxidation product analysis from rat liver, 314–317
    selected ion monitoring, 313–314
GC–MS, see Gas chromatography–mass spectrometry
Ginko biloba extract, flavonoid antioxidants, absorption and excretion in humans, 94
Glutathione
  depletion in disease, 258
  flow cytometry assay
    advantages, 247, 257
    cell preparation, 248–249
    data collection, 251–252
    differential assessment of thiols, 249–251, 258
    monobromobimane modification of thiols, 248–249, 251–253, 257–258
    monochlorobimane detection of glutathione, 248
  high-performance liquid chromatography assay with electrochemical detection
    cell culture, 243
    chromatography, 241
    coulometric detection, principle, 240–241
    current–voltage response curve, 242
    electrodes, 240
    extraction, 243, 245–246
    instrumentation, 241
    overview, 81–82
    standards, 242–243
  high-performance liquid chromatography of N-(1-pyrenyl)maleimide derivatives
    derivatization reaction, 259, 261–264
    glutathione disulfide quantitation, 259–260, 262
    materials, 260
    metabolic modulation studies, 265–266
    sample preparation, 264
    standard curves, 261–262
  lipoic acid effect on cellular levels, 253–254
  recycling, 247
7-trifluoromethyl-4-chloro-N-methylquinolium colorimetric assay
  absorbance properties, 277, 279–281
  incubation conditions, 284–286

instrumentation, 283
interferences, 281–283
reaction mechanism, 280, 286
reagents, 283
reproducibility, 281
sample preparation, 283–284
sensitivity, 281, 286
Glutathione disulfide/glutathione ratio
apoptosis correlation, 275–276
glutathione determination
blood sample preparation, 271
glutathione S-transferase assay, 271–273
protein precipitation, 268, 271
tissue sample preparation, 271–272
glutathione disulfide determination
accuracy requirements, 276
blood sample preparation, 268–269
N-ethylmaleimide trapping of glutathione, 267–270
high-performance liquid chromatography, 268, 270–271
tissue sample preparation, 269
mitochondrial DNA oxidation, correlation, 274–275
oxidative stress indicator, 267, 274
validation of assay, 273–274

H

High-performance liquid chromatography
Anethole dithiolthione, electrochemical detection
cell culture, 302
chromatography conditions, 302–303
extraction, 301–306
instrumentation, 301
ascorbic acid assay with electrochemical detection
amperometric detection, 66
chromatography conditions, 68, 70–72
coulometric detection, 66–67, 72
hydrodynamic voltammetry plot, 79, 82
instrumentation, 67–68, 78–79
materials, 68, 78
passivation, 72
recycling assay, 86–87

sample preparation, 67, 72–73, 79–82
standards, 72
carotenoids, 390, 392
dehydroascorbic acid assays
coulometric electrochemical detection following reduction
hydrodynamic voltammetry plot, 79, 82
instrumentation, 67–68, 78–79
materials, 68, 78
reduction reaction, 75–76, 81
sample preparation, 67, 72–73, 79–82
standards, 74–75
derivatization, 73–74
radiolabeled compound detection, 75–76
glutathione
electrochemical detection, 81–82
N-(1-pyrenyl)maleimide derivatives
derivatization reaction, 259, 261–264
glutathione disulfide quantitation, 259–260, 262
materials, 260
metabolic modulation studies, 265–266
sample preparation, 264
standard curves, 261–262
nitrated hydroxycinnamic acids, 214, 219, 222–223, 228
phenolic antioxidants
catechin analysis in tea with caffeine
chromatography conditions, 109, 111, 203–204, 206
comparison of green teas, 112–113, 203
extraction, 204
instrumentation, 108–109
precision, 204, 206
quantitative analysis, 111–112
sample extraction, 109
stability of samples, 206
standards, 109, 204
chromatography conditions, 103–105
food sample preparation, 100, 103
retention times, 105–106
standards, 100
urine sample
dietary manipulation, 97, 100, 106
preparation, 103

SUBJECT INDEX 497

wine analysis
  absorbance characterization of peaks, 119–120, 132–133
  anthocyanins, 119–120, 135–136
  catechins, 119
  difficulty of separations, 113–114
  flavonols, 121
  instrumentation, 124, 131
  linearity, 125, 127–129, 132
  low pH gradients, 114, 116–117, 124–125, 131
  pH shift gradient, 114–116
  polystyrene reversed-phase chromatography, 117
  precision, 129, 132
  procyanidins, 137
  recovery, 129, 132
  resolution, 125, 132
  sample preparation, 123
  sensitivity, 122–123, 130, 132
  standards, 118–119, 123–124, 130–131, 133, 135
retinoic acid isomer separation
  chromatography conditions and instrumentation, 434–437, 441
  data analysis, 437–438
  extraction, 433–435, 441
  resolution of peaks, 438–440
  sensitivity and reproducibility, 440–441
  standards, 431, 433
simultaneous determination of vitamin A, vitamin E, lycopene, and xanthophylls with reversed-phase high-performance liquid chromatography
  calibration, 352
  extraction from plasma and serum, 352–354, 356
  instrumentation and chromatography conditions, 350–352
  peak identification, 357–360
  quality assurance, 353, 356–357, 361–362
  reagents and solutions, 349
  stability of samples, 360–361
  standards, 350
tannins, size separation
  degree of polymerization, analysis, 182, 184
  gradient, 182

  peak identification, 180–181
  sample preparation, 180–181
thiol and disulfide analysis with electrochemical detection
  cell culture, 243
  chromatography, 241
  coulometric detection, principle, 240–241
  current–voltage response curve, 242
  electrodes, 240
  extraction, 243, 245–246
  instrumentation, 241
  standards, 242–243
vitamin E
  diol column chromatography, 320, 323–325
  fluorescence detection, 325
  materials, 321
  milk replacer diet effects in piglet tissues, 328–329
  oils and diets, extraction, 322
  separation of components, 323–324
  simultaneous determination with coenzyme Q
    animal maintenance, 334–335
    chromatography conditions, 337–338, 340, 344, 347
    electrochemical detection, 333–334, 342, 344–345
    extraction, 331–332
    materials, 334
    plasma sample preparation, 344
    recovery and reproducibility, 340–341, 345
    saponification, 332–333
    standards, 336–337, 342–344, 347
    storage of samples, 332
  standards, 322–323
  tissue pulverization and extraction, 321–322, 329
  triacylglycerol influence on quantitation, 325–329
HPLC, *see* High-performance liquid chromatography
Hydroxycinnamic acids, *see also* Phenolic antioxidants; *specific compounds*
  antioxidant activity, 96–97
  biosynthesis, 208
  high-performance liquid chromatography

chromatography conditions, 103–105
food sample preparation, 100, 103
retention times, 105–106
standards, 100
urine sample
dietary manipulation, 97, 100, 106
preparation, 103
inhibition of tyrosine nitration by peroxynitrite, 216, 219, 232–233
peroxynitrite interactions
m-coumaric acid, 223, 234
o-coumaric acid, 222–223, 233–234
p-coumaric acid, 219–222, 233
ferulic acid, 223, 228, 234
high-performance liquid chromatography analysis of products, 214, 219, 222–223, 228
mass spectrometry analysis of products, 214–215, 221–222, 228
reaction conditions, 214
spectral characterization of products, 214, 219–221, 223, 228
2-Hydroxyethylamidoethyl thionophosphate
antioxidant activity mechanism, 294–295
peroxyl radical reactivity, 299
reactivity
hydrogen peroxide decomposition, 296
lipid hydroperoxide reduction, 297–300
sodium hypochlorite, 296–297
synthesis, 295
Hydroxyl radical
electron spin resonance assay of scavenging activity
ascorbic acid contribution, 28, 32, 34
electron spin resonance measurements, 29–30
reaction conditions, 30
sample preparation, 29
standards, 30
formation in cells, 3

L

LDL, see Low-density lipoprotein
Lipid peroxidation, see also Low-density lipoprotein
amidothionophosphate reduction, 297–300

assays
gas chromatography headspace assay of volatile oxidation products
diet effects on lipid peroxidation, 198–200
instrumentation, 197
overview, 196–197
phenolic antioxidant effects, *in vitro* assays, 200
sample preparation, 197–198
wine phenolics, evaluation, 199–201
overview, 191–192
thiobarbituric acid assay, 191–193
initiation and propagation, 190–191
polyunsaturated fatty acid chemistry and antioxidant action, 190–191
retinol interactions with α-tocopherol in liposome oxidation prevention
chemicals and equipment, 422
oxidation conditions
liposomes, 422–423
retinal membrane oxidation, 423–424
regeneration, 421–422, 428, 430
retinal membrane protection, 425, 427–429
soybean phosphatidylcholine liposome protection, 424–425, 429
vitamin analysis with high-performance liquid chromatography, 424
Lipoic acid
antioxidant activity, 240
cellular reduction to dihydrolipoic acid, 246
high-performance liquid chromatography with electrochemical detection
cell culture, 243
chromatography, 241
coulometric detection, principle, 240–241
current–voltage response curve, 242
electrodes, 240
extraction, 243, 245–246
instrumentation, 241
standards, 242–243
protein modification, 239
Tetrahymena thermophila assay
basal medium for assay, 287–288
growth culture, 287
maintenance culture, 287
principle, 287

sample preparation
  blood, 292
  cerebrospinal fluid, 292
  liver, 293
  overview, 290-291
  urine, 292
  standards, 288-289
Low-density lipoprotein
  chemiluminescence, free radical measurement, 4-6
  dietary effects on fatty acid composition and oxidation, 194-195
  oxidation assays
    gas chromatography headspace assay of volatile oxidation products
      diet effects on lipid peroxidation, 198-200
      instrumentation, 197
      overview, 196-197
      phenolic antioxidant effects, *in vitro* assays, 200
      sample preparation, 197-198
      wine phenolics, evaluation, 199-201
    overview, 193-196
  oxidation in atherosclerosis, 35, 192, 362
  plasma oxidizability assays
    Alzheimer's disease patients, 46
    coronary heart disease patients, 46, 48
    hyperlipidemic patients, 46, 48
    overview, 36-37
    photometric detection of conjugated dienes
      comparison with other indices of lipid peroxidation, 41, 43-44, 47
      data acquisition, 38-39
      isolated low-density lipoprotein assay, 44-45
      kinetics, 39-40, 48
      oxidants, 37-40
      reproducibility, 40-41
      sample preparation, 38, 40, 49
      relationship to antioxidant and fatty acid composition, 45, 47-48
    Watanabe heritable hyperlipidemic rabbits, 46-47
  susceptibility to oxidation, 35-36, 191, 194
  α-tocopherol, prooxidant activity
    antioxidant versus prooxidant activity, 362-364, 373, 375
    assay
      high-performance liquid chromatography, 368-370
      reagents, 364-365
    dose response of lipid oxidizability, 372-373
    inverse deuterium kinetic isotope effect, 371-372
    lipoprotein preparation
      deuterium-labeled low-density lipoprotein, 367-368
      native and tocopherol-enriched low-density lipoprotein, 365-366
      tocopherol-depleted low-density lipoprotein, 366-367
      tocopherol-replenished low-density lipoprotein, 367
    mechanism, 362-364
    phase transfer activity effects, 370-371
  total peroxyl radical-trapping potential assay, 8-9
Lutein, *see* Carotenoids
Lycopene, *see also* Carotenoids
  photolysis, 419-420
  simultaneous determination with vitamin E, vitamin A, and xanthophylls with reversed-phase high-performance liquid chromatography
    calibration, 352
    extraction from plasma and serum, 352-354, 356
    instrumentation and chromatography conditions, 350-352
    peak identification, 357-360
    quality assurance, 353, 356-357, 361-362
    reagents and solutions, 349
    stability of samples, 360-361
    standards, 350

M

Mass spectrometry, *see* Gas chromatography-mass spectrometry; Matrix-assisted laser desorption ionization-mass spectrometry
Matrix-assisted laser desorption ionization-mass spectrometry, carotenoids

extraction, 391–392
fragment ion mass analysis, 394–395, 398–401, 406–408
green lettuce analysis, 403, 406
instrumentation, 392–393
molecular ion species and sensitivity, 393–394
overview of ionization techniques, 390–391
postsource decay, 391, 393
standards and samples, 391
tangerine juice concentrate analysis, 401, 403

N

NF-κB, *see* Nuclear factor-κb
NPM, *see* N-(1-Pyrenyl)maleimide
Nuclear factor-κB, Anethole dithiolthione effects, 300–301

O

ORAC, *see* Oxygen radical absorbance capacity assay
Oxidative stress, *see* Glutathione disulfide/glutathione ratio
Oxygen radical absorbance capacity assay
hydroxyl radical assay
automated assay on Cobas Fara II, 60–61
manual assay, 61
kinetics, 51–52
peroxyl radical assay
automated assay on Cobas Fara II
data acquisition, 54–56
reagents, 54
comparison with other assays, 58–60
manual assay, 56
reproducibility, 56–57
sample preparation
animal tissues, 53
biological fluids, 52–53
foods, 53
sensitivity of phycoerythrin, 57
principle, 51
transition metal assay, 61–62

P

Peroxynitrite
cellular sources, 210–211
hydroxycinnamate interactions
m-coumaric acid, 223, 234
o-coumaric acid, 222–223, 233–234
p-coumaric acid, 219–222, 233
ferulic acid, 223, 228, 234
high-performance liquid chromatography analysis of products, 214, 219, 222–223, 228
mass spectrometry analysis of products, 214–215, 221–222, 228
reaction conditions, 214
spectral characterization of products, 214, 219–221, 223, 228
synthesis, 212
tyrosine nitration
inhibitors
catechins, 215–216, 231–232
hydroxycinnamates, 216, 219, 232–233
scavenging mechanisms, 230–235
mechanism, 211, 230
product analysis, 212, 215, 228
scavenging assay
calibration, 213
high-performance liquid chromatography analysis, 213
incubation conditions, 212–213
recovery, 215, 228
Phenolic antioxidants, *see also specific compounds*
absorption and excretion in humans, 92–95
food additives, 152
gas chromatography–mass spectrometry of wine
costs, 151
extraction and characterization, 142–143
gas chromatography conditions, 141
instrumentation, 139, 141–142
linearity, 148
mass spectrometry conditions, 142
materials, 139
peak identification, 143–145
precision, 149–150
recovery, 148–149

resolution, 145, 147–148
sample preparation, 139
sensitivity, 149–150
standards, 139–140
high-performance liquid chromatography
  chromatography conditions, 103–105
  food sample preparation, 100, 103
  retention times, 105–106
  standards, 100
  urine sample
    dietary manipulation, 97, 100, 106
    preparation, 103
  wine analysis
    absorbance characterization of peaks, 119–120, 132–133
    anthocyanins, 119–120, 135–136
    catechins, 119
    difficulty of separations, 113–114
    flavonols, 121
    instrumentation, 124, 131
    linearity, 125, 127–129, 132
    low pH gradients, 114, 116–117, 124–125, 131
    pH shift gradient, 114–116
    polystyrene reversed-phase chromatography, 117
    precision, 129, 132
    procyanidins, 137
    recovery, 129, 132
    resolution, 125, 132
    sample preparation, 123
    sensitivity, 122–123, 130, 132
    standards, 118–119, 123–124, 130–131, 133, 135
oxidation, pH dependence, 153
quinone formation and phenol regeneration, 153
radical scavenging activity, 91–92
total phenol analysis with Folin–Ciocalteu reagent
  advantages, 177–178
  chemistry of reaction, 160–161
  cinchonine precipitations, 175–176
  comparison with other techniques, 169–171
  flavonoid–nonflavonoid separations, 170–173
  flow automation, 156–158
  Folin–Denis reagent comparison, 154–155

  incubation conditions and color development, 155–156
  interference
    ascorbic acid, 166–167
    nucleic acids, 164
    oxidation, 164
    proteins, 164–165
    sugars, 165–166
    sulfite, 167–169
    sulfur dioxide, 167–168
  molar absorptivity of specific phenols, 161–163
  phloroglucinol addition, 173
  preparation of reagent, 155
  sample preparation, 159
  standards and blanks, 158–159
  tannin separations, 174–175
  values for various wines, 176–177
  volume of reaction, 156
Photobleaching, *see* Carotenoids
Phycoerythrin, fluorescence assay of oxygen radical absorbing capacity, *see* Oxygen radical absorbance capacity assay
Piceid
  isomers, 186
  wine assays
    content in wine, 186
    gas chromatography–mass spectrometry, 186–187
    high-performance liquid chromatography, 187–188, 190
    standards, 188, 190
Plasma oxidizability, assays
  Alzheimer's disease patients, 46
  coronary heart disease patients, 46, 48
  hyperlipidemic patients, 46, 48
  overview, 36–37
  photometric detection of conjugated dienes
    comparison with other indices of lipid peroxidation, 41, 43–44, 47
    data acquisition, 38–39
    isolated low-density lipoprotein assay, 44–45
    kinetics, 39–40, 48
    oxidants, 37–40
    reproducibility, 40–41
    sample preparation, 38, 40, 49
    relationship to antioxidant and fatty acid composition, 45, 47–48

Watanabe heritable hyperlipidemic rabbits, 46–47
Premature infants, oxidative stress, 341–342, 348
Proanthocyanidin, see Tannin
Procyanidins, see Phenolic antioxidants
N-(1-Pyrenyl)maleimide, high-performance liquid chromatography of thiol derivatives
   derivatization reaction, 259, 261–264
   glutathione disulfide quantitation, 259–260, 262
   materials, 260
   metabolic modulation studies, 265–266
   sample preparation, 264
   standard curves, 261–262

R

Reactive oxygen species, see also Hydroxyl radical; Superoxide anion radical
   classification, 293
   disease association, 3
   sources in cells, 3–4
   total antioxidant power assays, see Chemiluminescence; Ferric reducing/antioxidant power assay; Oxygen radical absorbance capacity assay; Total peroxyl radical-trapping potential assay; Trolox equivalent antioxidant capacity assay
Resveratrol
   anticancer properties, 185
   antiinflammatory properties, 185
   antioxidant activity, 184–185
   food distribution, 185–186
   isomers, 186
   wine assays
      content in wine, 186
      gas chromatography–mass spectrometry, 186–187
      high-performance liquid chromatography, 187–188, 190
      standards, 188, 190
Retinoic acid
   absorbance properties, 431–432
   isomer separation
      data analysis, 437–438

      extraction, 433–435, 441
      high-performance liquid chromatography, 434–437, 441
      resolution of peaks, 438–440
      sensitivity and reproducibility, 440–441
      standards, 431, 433
   receptors, 430–431
Retinol, see also Vitamin A
   absorbance properties, 431–432
   antioxidant activity, 421
   interactions with α-tocopherol in liposome oxidation prevention
      chemicals and equipment, 422
      oxidation conditions
         liposomes, 422–423
         retinal membrane oxidation, 423–424
      regeneration, 421–422, 428, 430
      retinal membrane protection, 425, 427–429
      soybean phosphatidylcholine liposome protection, 424–425, 429
      vitamin analysis with high-performance liquid chromatography, 424
   receptors, 430–431
ROS, see Reactive oxygen species

S

Sulfarlem, see Anethole dithiolthione
Superoxide anion radical, electron spin resonance assay of scavenging activity
   ascorbic acid contribution, 28, 32, 34
   calibration, 32
   electron spin resonance measurements, 29–30
   reaction conditions, 28–31
   sample preparation, 29

T

Tannin
   extraction and purification from tissue, 180
   gel-filtration chromatography, 179
   proanthocyanidin classification and structures, 178–179

SUBJECT INDEX

procyanidin separation by high-performance liquid chromatography, 137
separations in Folin–Ciocalteu reagent total phenol analysis, 174–175
size separation by high-performance liquid chromatography
  degree of polymerization, analysis, 182, 184
  gradient, 182
  peak identification, 180–181
  sample preparation, 180–181
Tea, high-performance liquid chromatography of catechins
  chromatography conditions, 109, 111, 203–204, 206
  comparison of green teas, 112–113, 203
  extraction, 204
  instrumentation, 108–109
  precision, 204, 206
  quantitative analysis, 111–112
  sample extraction, 109
  stability of samples, 206
  standards, 109, 204
TEAC assay, see Trolox equivalent antioxidant capacity assay
Tetrahymena thermophila, see Lipoic acid
Thin-layer chromatography, carotenoid photobleaching products, 418
Thioctic acid, see Lipoic acid
TLC, see Thin-layer chromatography
α-Tocopherol, see also Vitamin E
  gas chromatography–mass spectrometry
    calibration curve preparation, 311, 317–318
    extraction and derivatization, 311–312
    instrumentation, 312
    oxidation product analysis from rat liver, 314–317
    selected ion monitoring, 313–314
  high-performance liquid chromatography with electrochemical detection, 309–310
  interactions with retinol in liposome oxidation prevention
    chemicals and equipment, 422
    oxidation conditions
      liposomes, 422–423
      retinal membrane oxidation, 423–424
    regeneration, 421–422, 428, 430
    retina membrane protection, 425, 427–429
    soybean phosphatidylcholine liposome protection, 424–425, 429
    vitamin analysis with high-performance liquid chromatography, 424
  peroxyl radical trapping and products, 309
  prooxidant activity in low-density lipoprotein
    antioxidant versus prooxidant activity, 362–364, 373, 375
    assay
      high-performance liquid chromatography, 368–370
      reagents, 364–365
    dose response of lipid oxidizability, 372–373
    inverse deuterium kinetic isotope effect, 371–372
    lipoprotein preparation
      deuterium-labeled low-density lipoprotein, 367–368
      native and tocopherol-enriched low-density lipoprotein, 365–366
      tocopherol-depleted low-density lipoprotein, 366–367
      tocopherol-replenished low-density lipoprotein, 367
    mechanism, 362–364
    phase transfer activity effects, 370–371
  synthesis of deuterium-labeled compound, 310–311
Total antioxidant power assays, see Chemiluminescence; Ferric reducing/antioxidant power assay; Oxygen radical absorbance capacity assay; Total peroxyl radical-trapping potential assay; Trolox equivalent antioxidant capacity assay
Total peroxyl radical-trapping potential assay
  chemiluminescence enhancement
    accuracy assessment, 9
    contributions of individual components, 10–14
    effects on results
      age, 11–12
      antioxidant supplementation of diet, 10
      disease states, 12, 14

fasting, 9–10
gender, 11–12
metabolic rate, 14
pH of assay buffer, 13
smoking, 10
low-density lipoprotein assay, 8–9
plasma assay, 7–8
principle, 6–7
stoichiometric peroxyl radical-scavenging factors, 10
total antioxidant reactivity, 12–13
electrode stability, 50
TRAP assay, see Total peroxyl radical-trapping potential assay
7-Trifluoromethyl-4-chloro-N-methylquinolium, colorimetric assay of thiols
absorbance properties, 277, 279–281
glutathione-specific assay
incubation conditions, 284–286
instrumentation, 283
interferences, 281–283
reaction mechanism, 280, 286
reagents, 283
reproducibility, 281
sample preparation, 283–284
sensitivity, 281, 286
reaction with thiols, 277–278
total mercaptan assay, 277–280, 284
N,N',N''-Tripropylamidothionophosphate
antioxidant activity mechanism, 294–295
peroxyl radical reactivity, 299
reactivity
hydrogen peroxide decomposition, 296
lipid hydroperoxide reduction, 297–300
sodium hypochlorite, 296–297
synthesis, 296
Trolox equivalent antioxidant capacity assay
comparison with oxygen radical absorbance capacity assay, 58–60
decolorization assay, 384, 389
extraction and analysis of tomato, 383–385, 387, 389
materials, 381
preparation of preformed 2,2'-azinobis(3-ethylbenzothiazoline-6-sulfonic acid) radical cation, 381

principle, 50–51, 380
standards, preparation, 381–383
Tyrosine nitration, see Peroxynitrite

V

Vitamin A, see also Carotenoids; Retinoic acid; Retinol
absorbance properties, 431–432
antioxidant activity, 421
receptors, 430–431
simultaneous determination with vitamin E, lycopene, and xanthophylls with reversed-phase high-performance liquid chromatography
calibration, 352
extraction from plasma and serum, 352–354, 356
instrumentation and chromatography conditions, 350–352
peak identification, 357–360
quality assurance, 353, 356–357, 361–362
reagents and solutions, 349
stability of samples, 360–361
standards, 350
Vitamin C, see Ascorbic acid
Vitamin E, see also α-Tocopherol
components and structures, 318–319, 330
high-performance liquid chromatography
diol column chromatography, 320, 323–325
fluorescence detection, 325
materials, 321
milk replacer diet effects in piglet tissues, 328–329
oils and diets, extraction, 322
separation of components, 323–324
simultaneous determination with coenzyme Q
animal maintenance, 334–335
chromatography conditions, 337–338, 340, 344, 347
electrochemical detection, 333–334, 342, 344–345
extraction, 331–332
materials, 334

SUBJECT INDEX

plasma sample preparation, 344
recovery and reproducibility, 340–341, 345
saponification, 332–333
standards, 336–337, 342–344, 347
storage of samples, 332
simultaneous determination with vitamin A, lycopene, and xanthophylls with reversed-phase high-performance liquid chromatography
  calibration, 352
  extraction from plasma and serum, 352–354, 356
  instrumentation and chromatography conditions, 350–352
  peak identification, 357–360
  quality assurance, 353, 356–357, 361–362
  reagents and solutions, 349
  stability of samples, 360–361
  standards, 350
standards, 322–323
tissue pulverization and extraction, 321–322, 329
triacylglycerol influence on quantitation, 325–329
saponification in extraction, 318–320

W

Wine
  antioxidant components different wine types, 137–138, 153–154, 176–177
  food additives, 152
  gas chromatography–mass spectrometry of phenolic antioxidants
    costs, 151
    extraction and characterization, 142–143
    gas chromatography conditions, 141
    instrumentation, 139, 141–142
    linearity, 148
    mass spectrometry conditions, 142
    materials, 139
    peak identification, 143–145
    piceid, 186–187
    precision, 149–150
    recovery, 148–149
    resolution, 145, 147–148
    resveratrol, 186–187
    sample preparation, 139
    sensitivity, 149–150
    standards, 139–140
  high-performance liquid chromatography of phenolic antioxidants
    absorbance characterization of peaks, 119–120, 132–133
    anthocyanins, 119–120, 135–136
    catechins, 119
    difficulty of separations, 113–114
    flavonols, 121
    instrumentation, 124, 131
    linearity, 125, 127–129, 132
    low pH gradients, 114, 116–117, 124–125, 131
    pH shift gradient, 114–116
    piceid, 187–188, 190
    polystyrene reversed-phase chromatography, 117
    precision, 129, 132
    procyanidins, 137
    recovery, 129, 132
    resolution, 125, 132
    resveratrol, 187–188, 190
    sample preparation, 123
    sensitivity, 122–123, 130, 132
    standards, 118–119, 123–124, 130–131, 133, 135
  phenolic antioxidant effects on lipid peroxidation, gas chromatography assay, 199–201
  total phenol analysis with Folin–Ciocalteu reagent
    advantages, 177–178
    chemistry of reaction, 160–161
    cinchonine precipitations, 175–176
    comparison with other techniques, 169–171
    flavonoid–nonflavonoid separations, 170–173
    flow automation, 156–158
    Folin–Denis reagent comparison, 154–155
    incubation conditions and color development, 155–156
    interference
      ascorbic acid, 166–167
      nucleic acids, 164

oxidation, 164
proteins, 164–165
sugars, 165–166
sulfite, 167–169
sulfur dioxide, 167–168
molar absorptivity of specific phenols, 161–163
phloroglucinol addition, 173
preparation of reagent, 155
sample preparation, 159
standards and blanks, 158–159
tannin separations, 174–175
values for various wines, 176–177
volume of reaction, 156

X

Xanthophylls, simultaneous determination with vitamin E, lycopene, and vitamin A with reversed-phase high-performance liquid chromatography
calibration, 352
extraction from plasma and serum, 352–354, 356
instrumentation and chromatography conditions, 350–352
peak identification, 357–360
quality assurance, 353, 356–357, 361–362
reagents and solutions, 349
stability of samples, 360–361
standards, 350

Z

Zeaxanthin, *see also* Carotenoids
oxidative metabolites, 460–461
photooxidation protection of retina, 457–460
stereoisomer separations
dicarbamate diastereomers, preparation, 463–466
extraction, 462
high-performance liquid chromatography, 466–467
retinal distribution, 467
separation from lutein, 462–463
serum and tissue sampling, 462
serum distribution, 467
types of isomers, 461

ISBN 0-12-182200-1

90038